A Practical Guide to Microstructural Analysis of Cementitious Materials

A Practical Guide to Microstructural Analysis of Cementitious Materials

Edited by

Karen Scrivener
École Polytechnique Fédérale de Lausanne, Switzerland

Ruben Snellings
VITO, Mol, Belgium

Barbara Lothenbach
EMPA, Dübendorf, Switzerland

CRC Press
Taylor & Francis Group
Boca Raton London New York

CRC Press is an imprint of the
Taylor & Francis Group, an **informa** business

A SPON PRESS BOOK

CRC Press
Taylor & Francis Group
6000 Broken Sound Parkway NW, Suite 300
Boca Raton, FL 33487-2742

First issued in paperback 2017

© 2016 by Taylor & Francis Group, LLC
CRC Press is an imprint of Taylor & Francis Group, an Informa business

No claim to original U.S. Government works

ISBN-13: 978-1-4987-3865-1 (hbk)
ISBN-13: 978-1-138-74723-4 (pbk)

Visit the Taylor & Francis Web site at
http://www.taylorandfrancis.com

and the CRC Press Web site at
http://www.crcpress.com

Contents

List of abbreviations

^{1}H-NMR	proton nuclear magnetic resonance
ADC	analogue-to-digital converter
AFm	Al_2O_3–Fe_2O_3 (mono)
AFt	Al_2O_3–Fe_2O_3 (tri)
AMCSD	American Mineralogist Crystal Structure Database
ASR	alkali–silica reaction
ASTM	American Society for Testing and Materials
BABA	back-to-back pulse sequence
BET theory	Brunauer–Emmett–Teller theory
BF	bright field (in transmission electron microscopy or scanning transmission electron microscopy)
BJH model	Barret–Joyner–Halenda model
BPP theory	Bloembergen–Purcell–Pound theory
BSE	backscattered electron
CAC	calcium aluminate cement
C-A-S-H	calcium–alumino–silicate–hydrate
CEN	European Committee for Standardisation (Comité Européen de Normalisation)
CH	portlandite or calcium hydroxide
COD	Crystallography Open Database
CP	cross polarisation
CPMG	Carr Purcell Meiboom–Gill
CS	chemical shrinkage
CSA	chemical shift anisotropy
C-S-H	calcium–silicate–hydrate
DAS	dynamic-angle spinning
DF	dark field (in transmission electron microscopy or scanning transmission electron microscopy)
DFT	density functional theory
DNP	dynamic nuclear polarisation
DSC	differential scanning calorimetry
DTA	differential thermal analysis
DTG	differential thermogravimetry

EDS/EDX	energy-dispersive X-ray spectroscopy
EFG	electric field gradient
EN	European Standard
ESD	equivalent spherical diameter
EZS	electrical zone sensing
FA	fly ash
FEG	field emission gun
FIB	focused ion beam
FIB-nt	focused ion beam nanotomography
FID	free induction decay
HAADF	high-angle annular dark field
Hc	hemicarbonate
HETCOR	heteronuclear correlation
HPF	high-pressure freezing
IA	image analysis
ICDD	International Centre for Diffraction Data
ILT	inverse Laplace transform
IP	inner product (calcium–silicate–hydrate)
IPA	isopropanol
IR	inversion recovery
ISO	International Organisation for Standardisation
ITZ	Interfacial Transition Zone
IUPAC	International Union of Pure and Applied Chemistry
KOSH	KOH/sugar (extraction)
LALLS	low-angle laser light scattering
LD	laser diffraction
MAC	mass (X-ray) absorption coefficient
MAS	magic-angle spinning
Mc	monocarbonate
MIP	mercury intrusion porosimetry
MK	metakaolin
MQ	multiple quantum
Ms	monosulfate
M-S-H	magnesium silicate hydrate
NAD	nitrogen adsorption–desorption
NaP	sodium metaphosphate
NMR	nuclear magnetic resonance
NS	number of scans
NT BUILD	Nordtest methods for building materials
OP	outer product (calcium–silicate–hydrate)
PC	portland cement
PDF	powder diffraction file
PE	polyethylene
PIPS	Precision Ion Polishing System
PONKCS	partial or no known crystal structure

PSD	particle size distribution
QE	quad-echo pulse sequence
RD	recycling delay
REAPDOR	rotational-echo adiabatic passage double resonance
REDOR	rotational-echo double resonance
rf	radio frequency
S/N	signal-to-noise ratio
SAM	salicylic acid methanol (extraction)
SCM	supplementary cementitious material
SDD	silicon drift detector
SE	secondary electron
SEM	scanning electron microscopy
SEM-EDS/EDX	scanning electron microscopy–energy-dispersive X-ray spectroscopy
SEM-IA	scanning electron microscopy–image analysis
SF	silica fume
SFEG	Schottky field emission gun
SR	saturation recovery pulse sequence
SSA	specific surface area
STEM	scanning transmission electron microscopy
T1	spin–lattice relaxation time
T2	spin–spin relaxation time
TEDOR	transferred-echo double resonance
TEM	transmission electron microscopy
TGA	thermogravimetric analysis
VT	variable temperature
w/b	water-to-binder ratio by mass
w/c	water-to-cement ratio by mass
w/s	water-to-solid ratio by mass
WC	white cement
WDS	wavelength-dispersive X-ray spectrometry
XRD	X-ray diffraction
XRF	X-ray fluorescence
XRS	X-ray gravitational sedimentation

CEMENT CHEMISTRY NOTATION

A	Al_2O_3
c	CO_2
C	CaO
F	Fe_2O_3
K	K_2O
M	MgO
N	Na_2O
s	SO_3
S	SiO_2

Preface

The purpose of this book is to provide practical guidance to the effective microstructural characterisation of cement-based materials.

Understanding the microstructure of a material is the key to understanding its properties and performance and how these are related to the process parameters which form the material. However, the microstructural characterisation of cementitious materials provides many challenges. First, the important length scales extend through many orders of magnitude, from the atomic to the metre scale. Second, water is an integral part of the structure, but it needs to be removed for many characterisation methods. Third, the majority of the phases making up a hardened cement paste are not well crystallined.

Over the past few decades there have been tremendous advances in methods to characterise the microstructure of cementitious materials, especially in making these techniques more quantitative. Quantification is essential as most commercial cements have broadly similar compositions. Unfortunately, today it is still not possible to characterise the microstructure of a cementitious material with the same precision which can be obtained in a test of mechanical performance. More precise quantification depends on good experimental methods and understanding of the workings of the different methods.

Most of microstructural characterisation techniques for cementitious materials belong to the discipline of materials science and are not well studied by civil engineers, who make up the majority of researchers on cementitious materials. At the same time most machine operators in materials science departments are not familiar with the foibles of cementitious materials. All too often this leads to poor-quality results. In this book we aim to provide practical information not usually discussed in scientific articles to help researchers obtain the best results from microstructural characterisation methods. We aim to highlight common pitfalls and give guidance on the best use of the different techniques and interpretation of the results.

The main motivation for the book comes from two research groups in Switzerland, which are among the leading practitioners of microstructural characterisation: the Laboratory of Construction Materials at École

Polytechnique Fédérale de Lausanne (Swiss Federal Institute of Technology in Lausanne), led by Professor Karen Scrivener and where Ruben Snellings worked from 2012 to 2014, and the Thermodynamic Modelling Group at Empa (Swiss Federal Institute of Material Research), led by Barbara Lothenbach. The content of most chapters comes from the students and researchers in these groups with practical day-to-day experience of using the techniques. These contributions are complemented by chapters from colleagues, experts in particular techniques, with whom we have been long associated through the Nanocem network (a consortium of 35 industrial and academic partners fostering fundamental research into cementitious materials). Here we try to write down the knowledge and tips acquired over the last 10 to 15 years. The input of the students and researchers working with these techniques every day is critical to the practical objective of the book. A critical nucleation step for the book was the work of the International Union of Laboratories and Experts in Construction Materials, Systems and Structures (RILEM) Technical Committee 238-SCM, 'Hydration and microstructure of concrete with supplementary cementitious materials', in which the need for more guidance about the use of common techniques, such as X-ray diffraction and thermogravimetric analysis, became clear.

Chapter 1 deals with the often-neglected topic of sample preparation, which is the essential first step in microstructural characterisation. The emphasis is on cement pastes – the binder phase of concrete – which are most suitable for microstructural characterisation. As concrete contains around 70% of aggregates, this dilutes the signal from the cement paste and greatly diminishes the precision of most methods. Chapters 2 and 3 look at two valuable methods – calorimetry and chemical shrinkage – to follow the hydration kinetics. These are not, strictly speaking, microstructural characterisation techniques, but as microstructural development during hydration is the most frequent topic of investigation, they are an essential first step to identifying the kinetics of the reaction and indicating at what times the microstructure should be investigated in more detail.

Chapters 4 and 5 look at the most widespread techniques for identifying the phases making up cementitious materials – X-ray diffraction and thermogravimetic analysis. The use of X-ray diffraction has been revolutionised in the past 20 years by the practical application of the Rietveld method to quantify phases, but it still has limitations for investigations of the hydrate phases, which are often poorly crystallined. Thermogravimetic analysis has the advantage of characterising also X-ray amorphous hydrate phases, because of their decomposition during heating, but quantification is more challenging.

Chapters 6 and 7 detail two very different applications of nuclear magnetic resonance. Solid-state nuclear magnetic resonance probes the local environment of atoms. As it probes short-range order, it is a powerful technique for studying, for example, the structure of the nanocrystalline calcium–silicate–hydrate phase, which makes up around half of the volume of a hardened cement paste. Unfortunately, the equipment for solid-state nuclear magnetic resonance is much less widely available than the techniques featured in the other chapters. Also, measurements can be very time consuming and demand expert interpretation. In contrast, the equipment for ^{1}H proton nuclear magnetic resonance is much less expensive and should become more widely available in coming years. This technique probes hydrogen nuclei and so has the huge advantage for looking at the water component of cementitious materials in situ and in a nondestructive manner. Recently, many questions about the interpretation of this method have been resolved to enable a complete description of the state of water in a cement paste to be obtained.

Chapter 8 looks at electron microscopy, which is the preeminent microstructural characterisation technique for cementitious materials. Not least, this is one of the few techniques which can be applied to mortars and concretes as well as cement pastes. It can provide a wealth of information about the nature and disposition of phases, but obtaining quantitative information is challenging and time consuming. Unfortunately, many researchers have a poor understanding of the potential and limitations of this technique, which leads to much misuse and wrong interpretation.

Chapter 9 looks at mercury intrusion porosimetry, which before the advent of ^{1}H nuclear magnetic resonance was the best available technique for studying the pore structure of cementitious materials. Although the technique has been much maligned, we believe it can give good quantitative information about the pore structure if specimens are prepared correctly (not oven dried) and the interpretation of the results takes into account how the technique works.

The penultimate chapter, Chapter 10, details techniques for the characterisation of particles and surfaces. Although this chapter does not directly relate to the microstructure of hardened cementitious materials, the characterisation of the powders that will transform into the cement is of great importance. Binder properties such as particle packing and rheology critically depend on the accurate characterisation of the powder mix. Again, the study of cementitious materials requires specifically adapted methodologies and care to overcome common problems.

In the final chapter, Chapter 11, we discuss phase diagrams and thermodynamic modelling, which are becoming increasingly important methods for understanding the phase compositions determined by microstructural analysis.

The different chapters cover the common techniques, which together can usually give a complete and detailed picture of the microstructure of cementitious materials. We have deliberately focused on these everyday methods, where the interpretation is well established. A wide range of more exotic techniques is now featured in publications. However, these are often only available at specialist centres such as synchrotrons and access is very limited. For this reason, such techniques can be only, at best, complementary to techniques found in individual research labs.

Above all, we hope this book will be a practical guide to all those starting out on microstructural analysis of cementitious materials. Of course, a book can never be a substitute for hands-on practical experience, but it is better than nothing. The authors of this book are also dedicated to more widespread dissemination of knowledge about microstructural characterisation through research collaborations and training courses around the world. We strongly believe that better microstructural characterisation is the key to understanding and improvement of these important materials, which make up over half of all the 'stuff' human beings produce and are essential to the infrastructure of modern civilisation.

Karen Scrivener
Ruben Snellings
Barbara Lothenbach

Editors

Karen Scrivener, PhD, graduated from the University of Cambridge in 1979 with a degree in materials science. She went on to do a PhD thesis, entitled 'The Microstructural Development during the Hydration of Portland Cement', at Imperial College, remaining there until 1995 as a Royal Society Research Fellow and then as a lecturer. In 1995 she joined the Central Research Laboratories of Lafarge near Lyon in France.

In March 2001 Dr Scrivener was appointed as professor and head of the Laboratory of Construction Materials, Department of Materials, École Polytechnique Fédérale de Lausanne in Switzerland. The work of this laboratory is focused on improving the sustainability of cementitious building materials.

Dr Scrivener is the founder and coordinator of Nanocem, a network of industry and academia for fundamental research on cementitious materials, and editor in chief of *Cement and Concrete Research*, the leading academic journal in the field. She is a fellow of the Royal Academy of Engineering, of the RILEM and of the Institute of Materials, Minerals and Mining. She is also president of École Polytechnique Fédérale de Lausanne – Women in Science and Humanities Foundation, which seeks to promote the careers of female scientists connected to the École Polytechnique Fédérale de Lausanne.

Ruben Snellings, PhD, is a researcher in the Sustainable Materials Management Unit at VITO, the Flemish Institute for Technological Research in Belgium. He earned an MSc and PhD in geology at KU Leuven, Belgium, and was formerly a postdoc at the Magnel Laboratory for Concrete Research at UGent in Belgium and a Marie Curie Fellow at the École Polytechnique Fédérale de Lausanne in Switzerland. He is a member of RILEM, active in Technical Committee 238-SCM 'Hydration and microstructure of concrete with supplementary cementitious materials'.

Barbara Lothenbach is a leader of the Thermodynamic Modelling Group of the Laboratory for Concrete and Construction Chemistry at EMPA, the Swiss Federal Institute for Materials Science and Technology. She was formerly project leader and consultant scientist for nuclear wastes at BMG Engineering Ltd., Schlieren, Switzerland. She is a member of the editorial board of *Cement and Concrete Research* and of the RILEM Technical Committees on multicomponent transport and chemical equilibrium in cement-based materials and on concrete in aggressive aqueous environments.

Contributors

Amélie Bazzoni
Laboratory of Construction
  Materials
Interdisciplinary Center for
  Electron Microscopy
École Polytechnique Fédérale de
  Lausanne
Lausanne, Switzerland

Elise Berodier
Laboratory of Construction
  Materials
Interdisciplinary Center for
  Electron Microscopy
École Polytechnique Fédérale de
  Lausanne
Lausanne, Switzerland

Julien Bizzozero
Laboratory of Construction
  Materials
Interdisciplinary Center for
  Electron Microscopy
École Polytechnique Fédérale de
  Lausanne
Lausanne, Switzerland

Paul Bowen
Powder Technology Laboratory
École Polytechnique Fédérale de
  Lausanne
Lausanne, Switzerland

Klaartje De Weerdt
Department of Structural
  Engineering
Norwegian University of Science and
  Technology
Trondheim, Norway

Paweł Durdziński
Laboratory of Construction
  Materials
Interdisciplinary Center for
  Electron Microscopy
École Polytechnique Fédérale de
  Lausanne
Lausanne, Switzerland

Mette Geiker
Department of Structural
  Engineering
Norwegian University of Science
  and Technology
Trondheim, Norway

Duncan Herfort
Cementir Technical Centre
Aalborg, Denmark

Hadi Kazemi-Kamyab
Laboratory of Construction
  Materials
Interdisciplinary Center for
  Electron Microscopy
École Polytechnique Fédérale de
  Lausanne
Lausanne, Switzerland

Barbara Lothenbach
Laboratory for Concrete and
  Construction Chemistry
Empa – Swiss Federal Laboratories
  for Materials Science and
  Technology
Dübendorf, Switzerland

Sara Mantellato
Institute of Building Materials
Eidgenössische Technische
  Hochschule Zürich
Zurich, Switzerland

Peter J. McDonald
Department of Physics
University of Surrey
Surrey, United Kingdom

Jonathan Mitchell
Schlumberger Gould Research
Cambridge, United Kingdom

Berta Mota
Laboratory of Construction
  Materials
Interdisciplinary Center for
  Electron Microscopy
École Polytechnique Fédérale de
  Lausanne
Lausanne, Switzerland

Arnaud C. A. Muller
Laboratory of Construction
  Materials
Interdisciplinary Center for
  Electron Microscopy
École Polytechnique Fédérale de
  Lausanne
Lausanne, Switzerland

Marta Palacios
Institute of Building Materials
Eidgenössische Technische
  Hochschule Zürich
Zurich, Switzerland

Kyle Riding
Department of Civil Engineering,
Kansas State University
Manhattan, Kansas

John E. Rossen
Laboratory of Construction
  Materials
Interdisciplinary Center for
  Electron Microscopy
École Polytechnique Fédérale de
  Lausanne
Lausanne, Switzerland

Paul Sandberg
Calmetrix
Arlington, Massachusetts

Axel Schöler
Laboratory for Concrete and
  Construction Chemistry
Empa – Swiss Federal Laboratories
  for Materials Science and
  Technology
Dübendorf, Switzerland

Karen Scrivener
Laboratory of Construction
    Materials
Interdisciplinary Center for
    Electron Microscopy
École Polytechnique Fédérale de
    Lausanne
Lausanne, Switzerland

Jørgen Skibsted
Department of Chemistry and
    Interdisciplinary Nano Science
    Center (iNANO)
Aarhus University
Aarhus C, Denmark

Ruben Snellings
Sustainable Materials Management
VITO – Flemish Institute for
    Technological Research
Mol, Belgium

Lars Wadsö
Building Materials
Lund University
Lund, Sweden

Frank Winnefeld
Laboratory for Concrete and
    Construction Chemistry
Empa – Swiss Federal Laboratories
    for Materials Science and
    Technology
Dübendorf, Switzerland

Chapter 1

Sample preparation

Frank Winnefeld, Axel Schöler
and Barbara Lothenbach

CONTENTS

1.1 INTRODUCTION

This chapter gives general information related to sampling, grinding, homogenisation, and storage of anhydrous binders and preparation and curing of pastes and mortars as well as stoppage of hydration, focusing on the experimental work in a research laboratory devoted to studying cementitious systems. The content given in this chapter will be restricted mostly to general considerations, while the individual chapters will give more detailed information related to the respective analytical methods. The main purpose of this chapter is that the reader should be aware that reliable analytics is impossible when the sample preparation is inappropriate. When consulting scientific literature, not only in the field of cement science, it is highly recommended to check how sample preparation was performed. For example, many papers related to cement hydration and microstructure describe harsh methods for hydration stoppage, such as drying at 105°C, used prior to mercury intrusion porosimetry (MIP). Conclusions from data obtained under these conditions can be strongly biased.

1.2 SAMPLING

Before using powders for, e.g. chemical–mineralogical analysis of particle properties, or before preparing samples of pastes or mortars, generally, representative samples need to be taken out of a bigger quantity of granular material. For example, when compressive strength is measured, e.g. according to EN 196-1 (2005), a sample of 450 g cement needs to be taken out of a 25 kg bag. In case of an X-ray diffraction (XRD) analysis, the sample size would only be a few grams. In both cases the sample needs to be representative. The reason why sampling is of such importance is that granular materials tend to segregate. For example, when they are put on a heap, they will segregate in a way that fine particles tend to remain in the centre of the heap and coarse particles will assemble at the rim as schematically shown in Figure 1.1. It is evident that in such a case representative sampling requires great care.

There is a lot of works on sampling available, e.g. in textbooks (Allen 1997; Hart et al. 2010; Masuda et al. 2006) and journal papers focused on the theory of sampling, e.g. Gy (1976) and Petersen et al. (2005). Sampling of granular material is also treated in standardisation, regarding general and theoretical aspects, e.g. in ISO 11648-1 (2003) and ISO 11648-2 (2003), and regarding practical issues, e.g. in EN 932-1 (1996) and

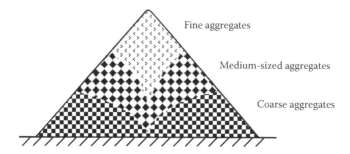

Figure 1.1 Schematic representation of a heap of granular material of different particle sizes which has dropped from a conveyor belt. (Adapted from EN 932-1, Tests for general properties of aggregates – Part 1: Methods for sampling, 1996.)

EN 932-2 (1992) related to sampling and sample splitting of mortar and concrete aggregates. The reader is referred to these documents for details and theoretical aspects, while in this section only the aspects relevant to cementitious materials are discussed.

Generally, the sample to be analysed needs to be representative of the whole batch it originates from, and to achieve this, the 'golden rules of sampling' (Allen 1997) should be applied, namely, (1) powder should be sampled while in motion and (2) several small samples should be taken at different intervals or at different positions rather than taking one larger sample. The different samples can be homogenised afterwards to obtain a single sample.

Luckily, cements and similar materials, such as slags or fly ashes, are fine powders and the very high number of particles present already in a few grams of the material (e.g. 1 g of 'monosized' cement with a particle size of 5 μm and a density of 3.15 g/cm^3 contains 2.5×10^9 particles) helps greatly to reduce sampling errors, e.g. those caused by segregation. Thus, it is generally sufficient and frequently done in the case of finely powdered cementitious materials to sample using the scoop method. The cement is mixed in the bucket by using a spatula, and then a sample is taken from the middle. Probing the surface region should be avoided as this region may be affected by segregation or alterations, such as carbonation. More accurate is the cone and quartering method (see Figure 1.2), which can be used to reduce the sample size until the quantity needed for analysis is reached after a certain number of repetitions. Moreover, many sampling devices have been developed which comply with the golden rules of sampling. Figure 1.3 shows two examples, and more examples are given in EN 932-2 (1996).

With coarser material, wider particle size distribution and variations in density, shape and chemical composition of the particles, representative samples will be larger and more difficult to achieve. For example,

(a) (b) (c)

Figure 1.2 Illustration of the cone and quartering method for reducing the sample size. (a) The granular material is poured on a flat surface so that it takes on a conical shape. The top of the cone is flattened afterwards. (b) The cone is divided into halves. (c) The cone is divided into quarters. Two opposite quarters are discarded; the other two are combined (= reduced sample). The process can be repeated until a suitable sample size is reached. See, e.g. EN 932-1 (1996) and EN 932-2 (1996) for details.

Figure 1.3 Examples of sample splitters: (a) rotating sample splitter (Courtesy of R. Snellings, Flemish Institute of Technological Research, Mol, Belgium.) and (b) static sample splitter.

sampling concrete aggregates for sieve analyses according to EN 933-1 (2012) requires a sample size of 200 g for aggregates of a maximum of 4 mm, whereas 40 kg is needed for aggregates with a maximum grain size of 63 mm. For coarse granular materials, such as aggregates, the use of sample splitters is mandatory.

1.3 GRINDING

1.3.1 Grinding of unhydrated binders

Various tools exist for crushing and grinding raw materials for cementitious binders. Typical materials that are grinded are cement clinker, blast furnace slag, fly ash and limestone. The mills used in a laboratory are usually disk mills and ball mills, which enable processing of various amounts of materials ranging from several grams to several kilograms.

Figure 1.4 shows some examples of crushers and mills used in a laboratory. Ball mills are preferable for many applications, as they generally provide a narrower particle size distribution than, e.g. disk mills, such as the one shown in Figure 1.4f, and are thus closer to the conditions in a cement plant. Disk mills, like the one shown in Figure 1.4f, provide a rapid grinding process and are suitable especially for grinding prior to chemical analysis such as X-ray fluorescence analysis.

Generally, it has to be considered that laboratory and industrial grinding are not directly comparable. This refers not only to particle size distribution but also, especially in the case of portland cement, to the composition of the calcium sulfate set regulator and its distribution within the particle

Figure 1.4 Examples of crushers and laboratory mills used for particle size reduction of raw materials for cementitious binders: (a) jaw crusher; (b) planetary ball mill; (c, d) laboratory ball mills; and (e, f) disk mills. (Courtesy of R. Snellings, Flemish Institute of Technological Research, Mol, Belgium [a, b, e], and J. Rossen, École Polytechnique Fédérale de Lausanne, Switzerland [c, d].)

blend. In portland cement technology, generally, a mix of gypsum and anhydrite is added to the portland cement clinker. In an industrial ball mill at a cement plant, the temperature at the mill outlet is usually in the order of 90 to 120°C. Under this condition the gypsum can dehydrate to hemihydrate and/or soluble anhydrite III (Hansen and Clausen 1973). In a simple, nonheatable laboratory ball mill such temperatures are usually not reached, thus preventing gypsum from dehydration. To compensate for this, either a heatable laboratory ball mill, if available (see Tang 1992), or a synthetic hemihydrate could be used in the laboratory as part of the set regulator.

When producing cements in the laboratory from clinker and calcium sulfate, intergrinding should be preferred. It has been shown that gypsum is better able to control the early hydration of C_3A when it is interground than when it is interblended with the clinker (Tang 1992) (see Figure 1.5), as the interground gypsum is more efficient in suppressing early C_3A hydration.

Depending on the type of the mill and the material to be ground, with increasing grinding time the particle fineness increases as well but approaches a threshold of a certain value as the fine particles tend to agglomerate and cannot be ground further in dry state without using grinding aids. This is illustrated in Figure 1.6, where a threshold value for the mean particle diameter is approached when a CEM I 42.5 N is ground for 5 min in a disk mill. Wet milling in a nonaqueous solvent, such as isopropanol, might be

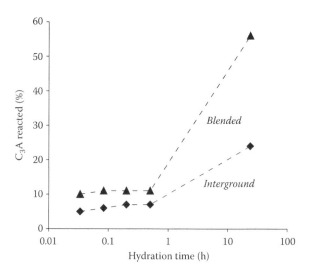

Figure 1.5 Amounts of C_3A reacted in a laboratory-made portland cement produced by blending (upper line graph) and intergrinding (lower line graph) an industrial clinker with 5.84% gypsum. Both cements have very similar fineness and were hydrated at 23°C using a water-to-cement ratio (w/c) of 0.50. C_3A content was determined by quantitative XRD analysis. (Adapted from Tang, F. J., Optimisation of sulfate form and content, *PCA Research and Development Bulletin RD105T*, Portland Cement Association, Skokie, Illinois, U.S., 1992.)

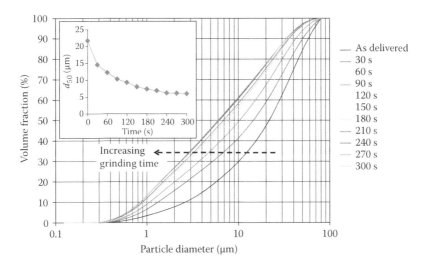

Figure 1.6 Change in the cumulative particle size distribution of a CEM I 42.5 N depending on grinding time in a disk mill (see Figure 1.4f). The inset shows the decrease in the mean particle size d_{50} with increasing grinding time, approaching a threshold value after about 5 minutes of grinding. (Courtesy of A. Zingg, Empa, Dübendorf, Switzerland.)

used to reach lower particle sizes; however, this is suitable only if the material quantities are not too high.

Special care has to be taken when grinding pure synthesised cement clinker phases. Ball mills using steel balls may contaminate the sample with traces of iron. To avoid this, grinding tools made of tungsten carbide could be used.

For obtaining a narrow particle size fraction other techniques could be used after the grinding process, such as wet sieving or sedimentation using a solvent (Costoya 2008).

For more details and general issues related to grinding technology, general textbooks on particle technology can be consulted, e.g. Masuda et al. (2006) and Rhodes (2008).

1.3.2 Grinding of hydrated pastes

Some analytical methods, such as thermal analysis or XRD (for XRD also the use of cut slices is possible), require the use of ground samples. However, cement hydrate phases are very sensitive to grinding, as the increased temperatures and the friction during intense grinding processes may cause loss of crystal water and changes in the phase assemblage. For example, gypsum loses 1.5 mol of its crystal water above 40°C, and ettringite is unstable in portland cement (PC) pastes above about 50°C (Lothenbach et al. 2007).

Thus, hydrated pastes should be ground after hydration stoppage gently by hand in an agate mortar. Grinding of pastes where the hydration was not stopped beforehand should be avoided, as in this case further hydration and carbonation of the samples may occur. Mechanical dry or wet grinding should be avoided in any case. The following two examples illustrate that gentle grinding is mandatory regarding the analysis of hydrated samples.

Figure 1.7 compares the XRD patterns of pure ettringite ground by hand and ground mechanically using a mortar grinder. The manually ground sample shows only the reflections of ettringite. In the mechanically ground sample the ettringite reflections are still present but less intense than in the case of the sample ground by hand. Additionally, reflections of mono-sulfate, gypsum and hemihydrate appeared, indicating the partial decomposition of ettringite. In addition, ettringite might have transformed into metaettringite with less crystal water, which is X-ray amorphous (Pourchez et al. 2006).

Figure 1.8 shows the XRD patterns of hydrated portland cement pastes ground manually in an agate mortar and mechanically ground using a mortar grinder. In the manually ground sample, ettringite and portlandite can be identified as crystalline hydrate phases. The ettringite content is quite low as the cement used is a CEM I with a high sulfate resistance. When ground mechanically, the reflections of ettringite slightly decrease, and monosulfate appears. The intensity of the portlandite reflection at 18° 2Θ seems to be unaffected by the grinding procedure (however, it has to be noted that this reflection has a high sensitivity to the preferred orientation).

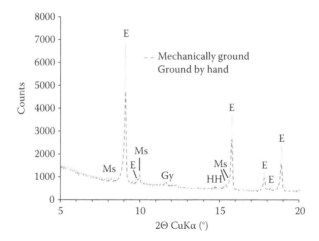

Figure 1.7 XRD patterns of ettringites ground by hand in an agate mortar and mechanically ground using a mortar grinder. E: ettringite; Gy: gypsum; HH: hemihydrate; Ms: monosulfate.

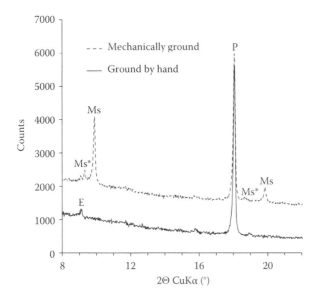

Figure 1.8 XRD patterns of portland cement pastes (CEM I 52.5 N with high sulfate resistance hydrated at w/c of 0.50 for 28 days) ground by hand in an
agate mortar and mechanically ground using a mortar grinder. E: ettringite;
Ms: monosulfate with 12 molecules of water; Ms*: monosulfate with 14 molecules of water; P: portlandite. For further details on the hydration of this
sample, see Lothenbach et al. (2007).

1.4 BLENDING OF DRY BINDERS AND MORTARS

Generally, the binders studied are blends of different powders, as even a PC
of type, e.g. CEM I is a mixture of ground clinker, calcium sulfate set regulator and very often a small percentage of a mineral addition. Producing
laboratory-made cement from a ground clinker by addition of a calcium
sulfate is a frequent task in the laboratory. Manufacturing the cement by
intergrinding (see Section 1.3.1) is preferable to manufacturing by inter
mixing, as in the first case the calcium sulfate will be better distributed
within the dry cement; thus, a better control of C_3A hydration is provided
(Tang 1992; Tang and Gartner 1987). This procedure is also closer to the
industrial process, where intergrinding of the set regulator is performed,
and the optimisation of the calcium sulfate addition is even done at the full
industrial scale (Winnefeld et al. 2005).

  While on the industrial scale at least the calcium sulfate is interground
when producing a portland cement, different powders often need to be
blended in laboratory studies in order to obtain a homogeneous mix. Such
a mix could be either a binder or a dry mix mortar. Blending is extensively
treated in textbooks related to particle technology, and the reader can be

referred to, e.g. Masuda et al. (2006) and Rhodes (2008) for more details. In the following section, some general considerations related to blending of dry binders and mortars in the laboratory are discussed.

In principle there are two different ways of blending the individual components when producing a fresh paste, mortar or concrete, as it is possible to (1) make a premix of the dry materials, which is later mixed with water and eventually aggregates to obtain the paste or the mortar, or (2) prepare directly the paste or the mortar from the individual ingredients.

The first case requires a proper homogenisation of the raw materials by using a blender suitable for producing a dry mix. Examples are shown in Figure 1.9. A cheap and simple possibility is the paint mixer shown in Figure 1.9a, whereas the other mixers shown in Figure 1.9 are more comfortable but are also more expensive.

Mixing time should be sufficient to achieve a thorough homogenisation. The homogenisation of dry binders is more difficult compared to the mixing of dry mortars, as the shear forces in the first case are lower. In order to improve the mixing of dry binders, ceramic balls or bouncy balls can be put in the dry mix. If necessary, the homogenisation procedure can be checked by adding, for example, a small percentage of a pigment, such as iron oxide, and verify by eye the proper mixing of the binder after different homogenisation times. Dry mix mortars are easier to homogenise due to the higher shear forces; however, they may segregate upon storage, which makes it difficult to take out a representative sample out of a bigger batch. This refers also to industrially premixed mortars delivered in bags. There a homogenisation of the whole bag should be performed before use, especially when coarse aggregates (2 mm and larger) are contained and only a part of the mortar is taken out, e.g. to produce prisms for strength testing. Especially when preparing dry mix mortars in the lab, the quantity to homogenise should be chosen in a way that the whole batch is used for one experiment (e.g. preparation of mortar prisms to test strength), without surplus material, in order to ensure the targeted composition of the mortar. The tendency to segregation increases with increasing particle size of the aggregate, so that it cannot be recommended to prepare dry mix mortars with maximum grain size of 2 mm and above (use a premixed binder instead and mix it with the aggregate and water).

The second case – mixing the individual components with water – requires a thorough mixing process of the fresh paste or mortar; see Section 1.6.1 for details.

Admixtures might be either added in a dry mix or dissolved in the mixing water. The first case is relevant for dry mix mortars, whereas in concrete production the admixtures generally are added in liquid form. In case the admixture is added to the dry mix, a good homogenisation needs to be ensured. In addition, it has to be kept in mind that there is a difference between adding the admixture as a solid and adding it dissolved in water, as the dry admixture needs to be dissolved first before it can interact with

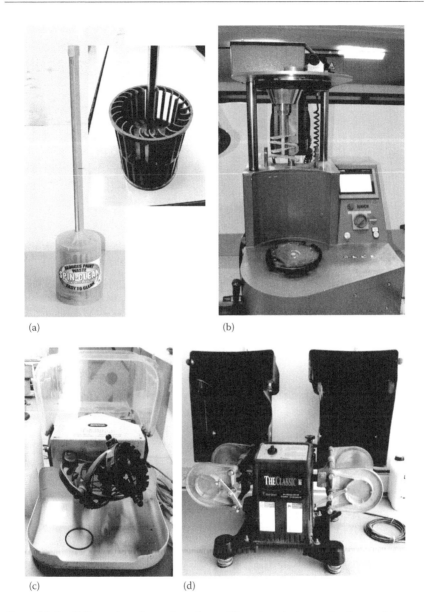

(a)

(b)

(c)

(d)

Figure 1.9 (a–d) Examples of powder mixers. The mixer in b can also be used to pre-
pare fresh pastes and mortars. (Courtesy of J. Rossen, École Polytechnique
Fédérale de Lausanne, Switzerland [a]; R. Snellings, Flemish Institute of
Technological Research, Mol, Belgium [b]; and C. Muller, Saint-Gobain Weber
AG, Winterthur [c, d].)

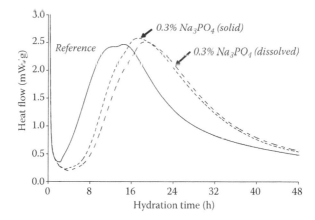

Figure 1.10 Isothermal calorimetry of CEM I 42.5 N hydrated at 20°C using w/c = 0.50 without addition (reference), with 0.3% Na₃PO₄ added as solid material (short dashed line) and with 0.3% Na₃PO₄ added dissolved (long dashed line) in the mixing water. The pastes were mixed outside the calorimeter by using a blade agitator at 500 rotations per minute (rpm) for 30 s.

the binder. Thus, results might not be the same between the two modes of addition. This is illustrated in Figure 1.10, where 0.3% Na_3PO_4 was added to a CEM I 42.5 N as solid and dissolved in the mixing water, respectively. In the case of the dissolved Na_3PO_4 the retarding effect is stronger compared to the addition as solid material.

Very fine particles such as microsilica are difficult to incorporate both in dry mixes and in fresh pastes, mortars or concretes. Further information and suggestions on how to treat such material are given in Section 1.6.1.

1.5 STORAGE AND SHELF LIFE OF UNHYDRATED BINDERS

1.5.1 Prehydration and shelf life of cementitious binders and mortars

Cements such as portland cement exhibit an ageing history between production and use as a binder in mortars or concretes. This process already starts in the cement mill, where the cement comes in contact with water in the case of internal water cooling or due to dehydration of the gypsum at high grinding temperatures (Sprung 1978; Theisen and Johansen 1975). It continues during storage in the silo, transport and until final use. Carbonation may also occur.

The changes in cement characteristics due to prehydration affect the properties of mortars and concretes. Especially when admixtures and complex binder formulations are used, a large influence of cement prehydration is found. This is reported, e.g. for premixed, rapid-hardening dry mix mortars (Schmidt et al. 2007). The changes in properties involve mainly retardation of setting and hydration (Dubina et al. 2010; Theisen and Johansen 1975; Whittaker et al. 2013; Winnefeld 2008) (see Figure 1.11); the decrease in compressive strength, especially at early ages (Winnefeld 2008); or the decrease in the heat of hydration (Schmidt et al. 2007; Winnefeld 2008). However, some more complex systems may behave differently. Prehydration may lead in certain cases to an acceleration of hydration and an increase in early strength as reported by Knöfel et al. (1997) for calcium aluminate cements.

Concerning portland cement, the free lime is the phase most sensitive to moisture uptake (Dubina et al. 2011); see Table 1.1. As the water sorption starts at such a low relative humidity, the transformation of calcium oxide to portlandite cannot really be avoided in practise. Thus, the determination of free lime in clinkers and cements by quantitative XRD should always be verified by other methods, e.g. the extraction method of Franke (1941). From the cement clinker phases the orthorhombic C_3A phase is the most sensitive to moisture. The prehydrated surfaces retard further hydration of the clinker grains (Dubina et al. 2011). Calcium silicates and calcium sulfates are involved as well in the prehydration process, as they may take up moisture already at quite low relative humidity.

Due to prehydration issues and the related changes in cement and mortar properties, the shelf lives of cementitious binders and mortars in

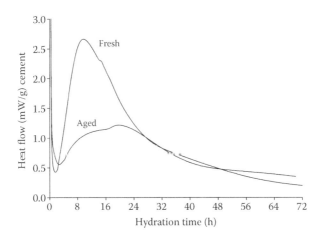

Figure 1.11 Heat flow calorimetry of 'freshly' bagged and artificially aged CEM I 42.5 N (storage at 90% relative humidity for 3 days). In case of the aged cement both the silicate and the second aluminate reaction are retarded, and the measured heat flow is lower compared to that of the fresh sample. Heat of hydration after 72 h is 248 J/g for the fresh cement, and 185 J/g for the aged cement. (Adapted from Winnefeld, F., *ZKG International*, 61 (11), 68–77, 2008.)

Table 1.1 Relative humidities at which cement constituents start to take up water
vapour, measured on a sorption balance at 20°C using the ramp mode

Cement constituent	Relative humidity at which water sorption starts (%)
CaO	< 10
$CaSO_4 \cdot 2H_2O$	24
$\beta\text{-}CaSO_4 \cdot 1/2H_2O$	34
C_3A, orthorhombic	55
$CaSO_4$	58
C_2S, monoclinic	64
C_3S, monoclinic	63
C_4AF, orthorhombic	78
Pure C_3A, cubic	80

Source: Dubina, E., L. Wadsö and J. Plank, *Cement and Concrete Research*, 41 (11), 1196–1204, 2011.

construction practise are usually restricted to 6 months to 1 year, even if
the material is properly stored.

Prehydration can be easily checked in the laboratory by thermogravimetric analysis (TGA) (Dubina et al. 2011; Theisen and Johansen 1975;
Winnefeld 2008). An example taken from Winnefeld (2008) is shown
in Figure 1.12. The fresh cement exhibits only minor weight losses,
mainly between 110 and 150°C, which can be attributed to gypsum and

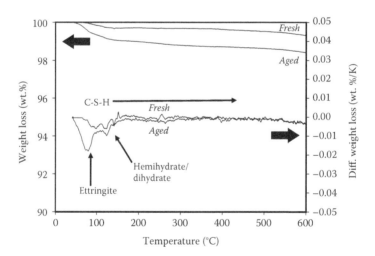

Figure 1.12 TGA of freshly bagged and artificially aged CEM I 42.5 N (storage at 90%
relative humidity for 3 days). The corrected losses on ignition (without contribution of hemihydrate, gypsum and portlandite) as defined by Theisen and
Johansen (1975) of the fresh and the aged cements are 0.18 and 1.8 wt.%,
respectively. Fresh cements typically show values between 0.15 and 0.30 wt.%.
(Adapted from Winnefeld, F., *ZKG International*, 61 (11), 68–77, 2008.)

hemihydrate. The aged sample shows clearly the formation of ettringite. It has to be kept in mind that even the fresh cement has an ageing history due to its production and storage until delivery. Cements strongly prehydrated contain hardened 'lumps' with a diameter of several millimetres, which cannot easily be crushed between the fingers. Such cement should not be used in any case.

1.5.2 Storing of unhydrated binders in the laboratory

One should start the experiments already from a sample as fresh as possible. Thus, it should be avoided to buy cement in a shop such as a building supplies store, as there it is uncertain how aged the material is and under which conditions it was stored. It is preferable to contact the cement producer directly and ask for the material.

When using a cementitious binder within a research project, usually it is not desired to change the batch of the cement, as the results might not be comparable and the characterisation of chemical–mineralogical and physical properties needs to be repeated for the new batch of cement, creating extra work. Furthermore, changing the batch will require the repetition of some of the previous experiments in order to check the repeatability of the results with the new batch. Thus, there is generally a demand to store a cement sample for several months or even for a few years, the latter being well beyond the usual shelf life of a cementitious binder or mortar. As cement bags are not water vapour tight, storage of the cement inside the bags (even if unopened) should be avoided. It is preferable to fill the cement directly after delivery in barrels or buckets, which can be closed watertight and water vapour tight, and to store them in a room with low relative humidity (if possible 35% or less; see Table 1.1). Further improvement can be made if the cement is additionally shrink-wrapped in different portions needed for certain series of experiments, and each portion is consumed within a short period. Dried silica may be applied as desiccant. The example in Figure 1.13 shows that under optimised storage conditions a portland cement may keep its setting time for a period of about 5 years.

While some materials such as calcium aluminate cements, portland cement and burnt or hydrated lime as well as dry mix mortar formulations (especially fast-setting ones) are sensitive towards alteration during storage, the sensitivity decreases generally with decreasing reactivity of the material. While slags still have a certain tendency to prehydration, fly ashes or inert fillers have a much lower or no tendency to change properties during storage. In cases of doubt, one should check prehydration phenomena by the methods used in the publications referred to previously.

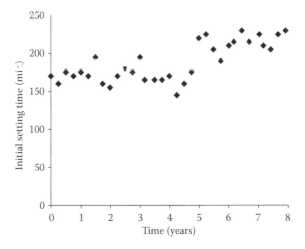

Figure 1.13 Initial setting times of a sample of a CEM I 42.5 N stored over a period of 8 years. The sample was kept in a sealed bucket at 20°C and 35% relative humidity.

1.6 MIXING, CASTING AND CURING OF PASTES AND MORTARS

1.6.1 Mixing

The goal when mixing a paste, mortar or concrete is to obtain a homogeneous material in a reproducible way. There are various mixing tools available; some examples are shown in Figure 1.14, which will be more suitable for certain materials and less suitable for other materials. This means that the appropriate mixer needs to be chosen according to the material intended to be mixed (e.g. paste or mortar) and to the type of analysis which is performed afterwards on the prepared paste or mortar. The mixing regime needs to be optimised as well, as, e.g. mixing time and intensity will be different depending on the material. For example, high-performance or self-compacting mortars or concretes require a prolonged mixing time (Chopin et al. 2004), or mixtures containing redispersible polymer powders might need a 'time of maturation' followed by a second mixing step in order to ensure the proper dissolution and redispersion of the admixtures (De Gasparo et al. 2009).

Hand mixing can be suitable for hydration studies; however, it should be avoided if, e.g. rheological measurements are done afterwards. Hand mixing or mechanical mixing with low mixing energy might not be comparable to mixing in a concrete mixer, where the aggregates present provide high shear forces, or high shear mixing in general; thus, some of the conclusions made from pastes might not directly be transferable to concrete. In

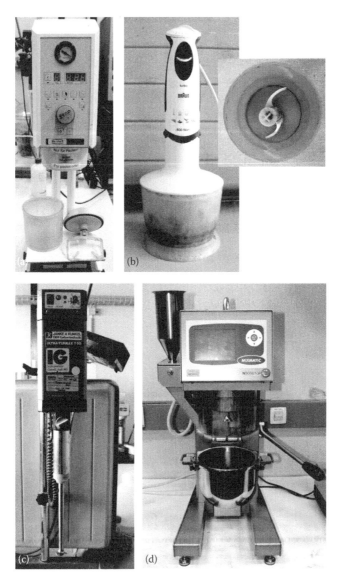

Figure 1.14 Examples of mixers used to blend cement pastes and mortars: (a) vacuum mixer; (b) kitchen blender; (c) high-performance dispersing mixer (Courtesy of R. Snellings, Flemish Institute of Technological Research, Mol, Belgium.) and (d) programmable mixer for preparing cement pastes according to EN 196-3 (2005) and cement mortars according to EN 196-1 (2005). The mixers in a through c are in particular suitable for pastes. The mixer in Figure 1.9c can be used as well to blend pastes and mortars.

such a case the use of high-shear blenders might be a suitable compromise. Hydration kinetics becomes faster when mixing energy is increased (Juilland et al. 2012); in Section 2.3 an example is given. Vacuum mixers (see Figure 1.14a) efficiently remove air voids (Dils et al. 2013) and are suitable tools for mixing pastes with low w/c (Justs et al. 2014).

Very fine particles, like silica fume, are very difficult to deagglomerate and disperse in a cement paste or mortar; densified silica fume may be especially problematic. Even in concrete, the occurrence of lumps of silica fume is reported (Diamond and Sahu 2006). It is recommended to use silica fume preferably as slurry prepared with a high-speed mixer. The mixing sequence plays a role as well (Marchuk 2002). Densified silica might be dispersed by an ultrasonic treatment (Rodriguez et al. 2012). The influence of mixing intensity (hand mixing vs. high-speed kitchen blender) on the hydration kinetics of portland cement pastes with 10% added microsilica was examined by Geiker et al. (2006). Figure 1.15 illustrates the effect of mixing on the consistency of cement paste with and without silica fume after 5 minutes of mixing. Agglomerates of microsilica can be observed in the case of hand mixing, whereas the sample appears homogeneous when high-speed mixing is performed (Figure 1.16).

(a) (b)

Figure 1.15 Cement paste (w/c = 0.35, 0.65 mass% added superplasticiser) with 10% silica fume after 5 minutes mixing (a) at low intensity (by hand) and (b) at high intensity (using a high-speed kitchen blender). In the case of (a), it was necessary to continue mixing for an additional 5 minutes to obtain a workable, visually homogeneous paste. (Reprinted from Geiker, M. R. et al., *Proceedings pro052: International RILEM Conference on Volume Changes of Hardening Concrete: Testing and Mitigation*, 20–23 August 2006, Lyngby, Denmark, RILEM Publications S.A.R.L., Bagneux, France, 303–310, 2006. With permission.)

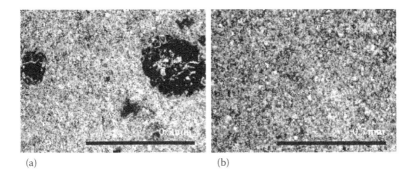

(a)   (b)

Figure 1.16 Optical microscopy of thin sections of cement pastes (w/c = 0.35, 0.65 mass% added superplasticiser) with 10% silica fume mixed (a) at low intensity (by hand) and (b) at high intensity (using a high-speed kitchen blender). Agglomerates of 60–400 μm silica fume particles are seen in the paste mixed at low intensity. (Reprinted from Geiker, M. R. et al. *Proceedings pro052: International RILEM Conference on Volume Changes of Hardening Concrete: Testing and Mitigation*, 20–23 August 2006, Lyngby, Denmark, RILEM Publications S.A.R.L., Bagneux, France, 303–310, 2006. With permission.)

1.6.2 Casting and curing

Depending on the methods used, the samples will be cast in various kinds of moulds, vessels, etc. which are often specified together with the curing conditions in standards describing the respective testing method. Thus, in the following some general aspects concerning sample casting and curing will be mentioned, without going into too much detail. For the individual methods, the reader is referred to the respective chapters.

Plastic vessels (see Figure 1.17 for an example) are suitable for many purposes, such as hydration studies. The size of the vessel should be chosen to be appropriate to the amount of the sample, as there should not be much empty space in order to minimise carbonation. A separation agent should not be used as it may influence cement hydration.

When a paste (or a mortar) is cast, special care should be taken concerning segregation or bleeding. While gentle bleeding cannot be often avoided and can be considered as acceptable, strong bleeding or segregation will cause an inhomogeneous sample composition. To overcome this, the sample mix design could be changed (which is generally not wanted) or the sample can be gently rotated while it is still in a plastic state. Vigorous rotation, however, should be avoided, as a too high centrifugal force will cause sample inhomogeneity as well.

Curing is often done under sealed conditions, as this minimises drying and carbonation. The vessel should be water vapour tight, which could be achieved using a vessel with a screw cap and sealed with Parafilm, like

Figure 1.17 (a) Hardened cement paste cylinder cast in a polyethylene (PE) vessel and after removing the vessel. The vessel has an outer diameter of 35 mm and a height of 70 mm. (b, c) From the cylinder, slices can be cut.

in Figure 1.17a. Additionally, the vessels should be stored in a humidity chamber with at least 90% relative humidity. In the long term, however, even under such condition some carbonation may occur as CO_2 can diffuse very slowly through commonly used plastic materials, such as polyethylene. Containers like the one shown in Figure 1.17a are suitable for preparing slices to be used for XRD analyses (see Chapter 4). To do so, slices of approximately 3 mm thickness are cut from the hardened cement paste cylinder by use of a disk saw. The saw blade and the sample are flushed with water to avoid heating of the sample and to dispatch the sawdust. The first slice (from the top of the sample) is usually discarded, as it might not have a representative composition. The cut slices (Figure 1.17c) might be used after hydration stoppage directly for XRD, or after breakage or grinding for SEM or XRD/TGA, respectively.

Sometimes it might be of interest to store the samples under water or in air; however, it has to be kept in mind that leaching and/or carbonation issues may occur. In order to minimise leaching during storage under water, the cylinders shown in Figure 1.17a can be demoulded after 24 hours and placed in a slightly wider vessel (approximately 1–2 mm larger in diameter). In the vessel just enough water is filled to submerge the cylinder (Figure 1.18).

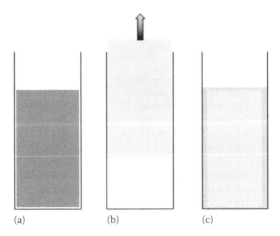

Figure 1.18 Procedure for providing water curing without excessive leaching: (a) casting of cement paste cylinder, (b) removal of the cylinder after 24 hours and (c) placement of the cylinder in a slightly wider vessel filled with water. (Courtesy of J. Rossen, École Polytechnique Fédérale de Lausanne, Switzerland.)

1.7 HYDRATION STOPPAGE

Generally, the hydration of cements is stopped before analysis. At early hydration times hydration stoppage is necessary to suppress the further progress of hydration. At hydration times of 1 month or longer, where the further progress of hydration is very slow, stoppage procedures are used to remove free pore solution, which is necessary before TGA, infrared and Raman spectroscopy and MIP or before impregnation of the samples for SEM analysis. For some techniques, such as XRD and NMR hydration stoppage is not strictly necessary but is generally done, as it enables sample storage, and the low relative humidity after stoppage helps to minimise carbonation and the characterisation of the same sample at an identical degree of hydration by different techniques. As slightly different stopping procedures are optimal for the different characterisation techniques, some general aspects of hydration stoppage are discussed in this chapter, and the effect of stoppage on a specific characterisation techniques is detailed in the respective chapters.

The primary aim of the various available methods of hydration stoppage is to remove water present in the pores without removing the water present in the hydration products to avoid alteration of the hydrates and to preserve the microstructure. The free water is generally removed either by direct drying, e.g. oven drying or freeze-drying, or by replacing the water with an organic solvent miscible with water, which is then evaporated. Isopropanol or acetone can be used to this effect. A recent review paper of Zhang and Scherer (2011) compares the effect of the different techniques

Table 1.2 Advantages and disadvantages of solvent exchange and direct drying techniques for hydration stoppage

	Solvent exchange	*Direct drying*
Advantages	• Easy to perform • Preserves chemically bound water in hydrate phases (depending on solvent) • Preserves microstructure • Removes soluble ions from pore solution	• Oven drying easy to perform, vacuum and freeze-drying need equipment • Suppresses carbonation • Short freeze-drying (1 hour) preserves water bound in hydrates
Disadvantages	• Alteration of hydration products and removal of chemically bound water by some organic solvents (esp. methanol) • Organic solvents not completely removed by drying (such as acetone)	• Generally, chemically bound water removed • Microstructure generally altered
Suggested use	• Stop samples (crushed or cut slices) for XRD, NMR, MIP and SEM analyses with isopropanol and dry afterwards a few minutes at 40°C to remove the solvent. Alternatively, when using slices, they can be placed on dry lab paper, and then compressed air can be used to quickly dry both surfaces. • For TGA remove the isopropanol with diethyl ether and then dry a few minutes at 40°C.	• Freeze-drying or other drying techniques can be used if only the portlandite content is of interest. • Gentle (short) freeze-drying for, e.g. 1 hour can be also used for TGA, XRD, NMR and SEM as mainly the free water is removed. For MIP, direct drying generally cannot be recommended (see Chapter 9).

on the preservation of the hydrate assemblage and the microstructure. In the following, the main effects of hydration stoppage by drying and by solvent exchange are summarised and illustrated (see also Table 1.2).

1.7.1 Direct drying techniques

Oven drying, vacuum drying and freeze-drying techniques remove the free water efficiently, which helps to avoid carbonation but generally structural water is also removed, leading to the decomposition of the ettringite and AFm phases (Zhang and Glasser 2000). Different freeze-drying procedures are mentioned in literature, with varying sample size (usually a few millimetres in diameter), immersion time in liquid nitrogen (typically 5 min–2 h), temperature (in the order of –40 to –80°C), duration (mostly between 1 day and 2 weeks) and pressure (in the order of 0.1–5 Pa) of freeze-drying; see, e.g. Collier et al. (2008), Gallé (2001), Gorce and Milestone (2007), Konecny and Naqvi (1993), Korpa and Trettin (2006), Zeng et al. (2013) and Zhang and Scherer (2011). Either the sample can be immersed directly in liquid nitrogen or a vessel containing the sample is immersed (Konecny and Naqvi 1993; Zhang and Scherer 2011). All the direct drying methods remove only H_2O without the ions dissolved in the pore solution, which can lead to artefacts, such as the formation of syngenite ($CaK_2(SO_4)_2 \cdot H_2O$) during drying at early hydration times (Rößler et al. 2008), when high potassium and sulfate concentrations are still present in the pore solution.

Zhang and Glasser (2000) reported oven drying to be more destructive than vacuum drying. One important problem generally associated with these drying methods is that the duration of the drying determines whether and how much of the chemically bound water is removed. Gentle drying such as storage in the moderate vacuum of an aspirator pump is efficient in preventing carbonation and preserves hydrates such as the ettringite and AFm phases well (Figure 1.19), but it is too gentle to stop the hydration immediately (as necessary for stoppage at early age) and removes only a part of the gel water associated with C-S-H (the approximately 30 mbar pressure reached by an aspirator pump will dry pores with a diameter above ≈3 Å; a higher vacuum as provided by a rotary vane pump of, e.g. 0.3 mbar would dry pores down to 1 Å). A few minutes of oven drying at 105°C or of freeze-drying may remove mainly the free water (see Figure 1.19; further examples are in Chapter 5); longer drying, however, removes all the gel water associated with C-S-H, leaving only structural and inter-layer water (L'Hopital et al. 2015). Long drying also removes water associated with the ettringite and AFm phases, which leads to decomposition and underestimation of the amount of the ettringite and AFm phases by TGA and XRD techniques (Figure 1.19), underestimation of the bound water content and an increase in porosity in MIP measurements. The ettringite, C-S-H and AFm phases are more sensitive to decomposition by dehydration than hydrogarnet, AH_3 or portlandite are.

A long vacuum drying of 7 days after a prolonged immersion of the sample in isopropanol removes some of the water associated with the C-S-H, ettringite and the AFm phases (see Figure 1.20), thus leading to an underestimation of the amount of ettringite, AFm and bound water. Similar to these observations, Khoshnazar et al. (2013a,b) detected a decrease in the water content in monosulfate and ettringite already upon prolonged storage in isopropanol. Hydration stoppage by solvent exchange using isopropanol and other organic liquids will be discussed in detail in the next section.

1.7.2 Solvent exchange

Hydration of cement paste can also be stopped by diluting and removing the water present in the pores of the cement paste by a solvent. Solvent

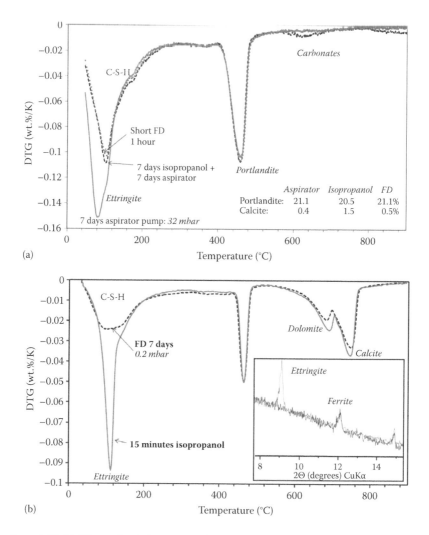

Figure 1.19 (a) Effect of a slight vacuum, freeze-drying (1 hour) or isopropanol on the water content of a portland cement observed by TGA; w/c = 0.4, hydrated for 7 days. (Courtesy of P. Durdzinski.) (b) Effect of freeze-drying during a longer period on the ettringite content observed by TGA and XRD (inset) in a portland cement hydrated for 12 hours. Further details are in Le Saoût et al. (2013). FD: freeze-drying.

exchange is gentler, changes the microstructure less and removes not only H_2O but also the pore solution including dissolved ions, thus minimising the accumulation of dissolved ions in the solid phase possible for the drying methods. Finely crushed cement paste or small discs are submerged in a relatively large amount of solvent for a certain period of time, which depends on the size of the sample (Zhang and Scherer 2011). Finely crushed

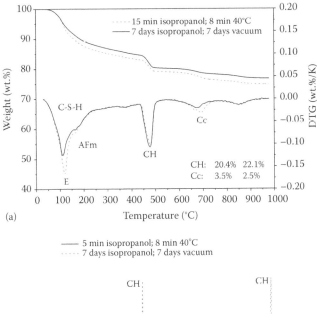

(a)

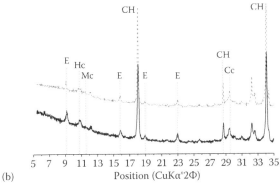

(b)

Figure 1.20 Effects of the stoppage procedure for 15 minutes in isopropanol, washing with diethyl ether and 8 minutes drying at 40°C in an aerated oven and for 7 days in isopropanol followed by 7 days in a moderate vacuum as obtained by an aspirator pump on the hydrates and bound water in a PC hydrated for 7 days: (a) thermogravimetry and derivative thermogravimetry (DTG); (b) XRD. Further details are in Schöler et al. (2015). Cc: calcite; CH: portlandite; E: ettringite; Hc: hemicarboaluminate; Mc: monocarboaluminate.

samples can be exchanged in a few minutes; cut slices as used, e.g. for XRD analysis will have to equilibrate much longer, as detailed in Chapter 4.

An ideal solvent should be miscible with water and have a relatively small molecular size such that it can enter small pores to replace the pore solution, but it should also not be too small to avoid replacement of water in the hydrates, i.e. ettringite and AFm phases. Commonly used solvents include methanol (CH_3OH), ethanol (CH_3CH_2OH), acetone ((CH_3)$_2$CO) and isopropanol ($(CH_3)_2CHOH$). Methanol, especially, replaces the pore solution but interacts too strongly with the hydrates by replacing a part of the bound water (Taylor and Turner 1987; Thomas 1989). In particular, ettringite is completely decomposed by methanol (Khoshnazar et al 2013b), and the effect on the AFm phases is less strong, although mono-sulfate dehydrates from 12 to 10 molecules of water. The effect of ethanol and isopropanol is much weaker. All of the polar solvents mentioned above have the potential to bind on the hydrates and are thus often not entirely removed, even by vacuum drying. In addition, acetone may undergo the so-called aldol condensation under alkaline conditions, which is a dimeri-sation followed by the loss of one molecule of water, forming mesityl oxide as a reaction product. Taylor and Turner (1987) identified mesityl oxide besides other condensation products in a hydrated C_3S paste stored in ace-tone for 30 days. As the condensation products have boiling points higher than that of acetone, they are much more difficult to remove by subsequent drying. Thus, stopping with acetone should be avoided in case of TGA analyses. The presence of organics in the hydrates is of little importance for XRD or Si-NMR analysis; however, in TGA the organics oxidise and the resulting carbonate reacts with portlandite or calcium oxide from C-S-H during the TGA. This results in an underestimation of portlandite content and an overestimation of calcium carbonates, as detailed in Chapter 5 and shown in Figure 1.21b. The use of acetone leads to less portlandite and an increased weight loss is observed above 600°C in the TGA measurements (Figure 1.21b), while the presence of acetone has no significant effect on the amount of ettringite, hemicarbonate or portlandite observed by XRD as visible in the comparison between the XRD signal of a slice, measured immediately after cutting, and of a powder sample (exchanged with ace-tone) in Figure 1.21a.

Figure 1.22 highlights again that any prolonged drying step, even at 40°C, decreases the amount of water loss in TGA, which is attributed to water loss from ettringite. Thus, the solvent used for stoppage should not be removed by a drying step before TGA analysis but instead by a sec-ond solvent exchange using the easily volatile diethyl ether, which can be removed by a short drying process (only several minutes at 40°C) without significantly altering the hydrate phases.

1.7.3 Sample storage

After stoppage, samples are often stored before the measurement, which requires a low relative humidity to prevent further hydration. The four main PC clinker phases do not adsorb significant amounts of water vapour at rel-ative humidity below 55%; see Dubina et al. (2011) and Table 1.1. Sample storage can lead to carbonation (see Chapter 5), especially if the sample is already ground to the high fineness needed for many of the analytical methods. Carbonation issues are especially of relevance if portlandite is

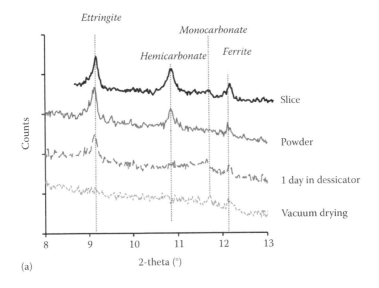

(a)

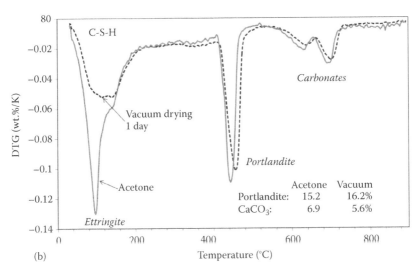

(b)

Figure 1.21 (a) XRD and (b) DTG of a portland cement containing 4 wt.% of limestone hydrated for 7 days, measured as follows: without stopping 'slice', on powder samples after solvent exchange by acetone and short drying 'powder', same as the preceding condition + 1 day storage in a desiccator and after vacuum drying for 1 day. Further details are in Lothenbach et al. (2008).

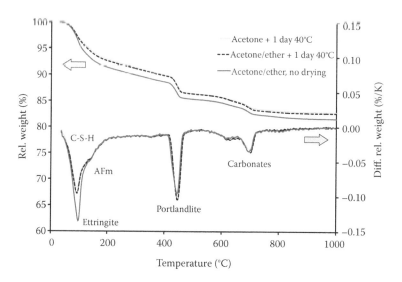

Figure 1.22 TGA of a portland cement hydrated for 3 days using w/c = 0.40. Hydration was stopped by solvent exchange using acetone and different subsequent treatments: drying at 40°C, rinsing with diethyl ether and drying at 40°C and rinsing with diethyl ether without further drying. (Courtesy of A. Gruskovnjak, Empa, Dübendorf, Switzerland.)

present and may lead also to partial decomposition of the ettringite and AFm phases. Carbonation occurs mainly at a relative humidity of between 50% and 80% (Mmusi et al. 2009); however, in a study performed by atomic force microscopy (Yang et al. 2003), carbonation products of portlandite could already be identified at a relative humidity of 30%.

The amounts of hemicarbonate and strätlingite are especially strongly influenced by storage of the sample. One day of storage in a N_2-filled desiccator results in the decomposition of hemicarbonate to poorly crystalline monocarbonate and to a reduction in the amount of ettringite (Figure 1.21b). Vacuum drying results in the decomposition of both ettringite and hemicarbonate. Hydrotalcite seems to be quite stable towards drying procedures, as shown by the example of alkali-activated slags in Figure 1.23. The C-S-H phase seems to be altered by the drying at 40°C, as a slight peak broadening and decrease in the intensity of its two main reflections can be observed. There are also indications that the strätlingite present in the alkali-activated samples may decompose during sample storage.

1.7.4 General points to consider

The different drying methods offer a fast way to remove water but they are often too harsh, leading to the removal of water in the hydrates and altering the microstructure, and it can cause microcracking; for further details see Zhang

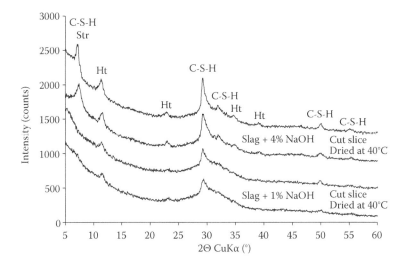

Figure 1.23 XRD of alkali-activated slag hydrated for 28 days using w/c = 0.40. The samples were either stopped by drying at 40°C or directly measured as freshly cut slices. Str: strätlingite; Ht: hydrotalcite. (Courtesy of A. Gruskovnjak, Empa, Dübendorf, Switzerland.)

and Scherer (2011). In particular, oven drying at 105°C should be avoided for all purposes. Freeze-drying or other harsh drying techniques generally suppress the carbonation of portlandite and might be the adequate technique if only the quantity of portlandite is of interest. In all other cases these techniques tend to remove chemically bound water in the ettringite and AFm phases.

Gentler techniques, such as solvent exchange, preferably by isopropanol, should be used for sample stoppage for XRD, NMR, MIP or SEM analysis. For TGA, the solvent exchange is best followed by a replacement of the isopropanol in the stopped samples with a less polar organic with a low boiling point, such as diethyl ether, to minimise the amount of organics sorbed to the hydrates. In all cases, analysis should be done as soon as possible as even storage for a few days or gentle heating results in a decrease in the amount of ettringite and in the possible destabilisation of hemicarbonate and strätlingite. If storage cannot be avoided, the sample should be stored in a desiccator under nitrogen, if possible not yet ground; the grinding should be performed directly before the measurement. In the case of XRD it is also possible to measure samples without stopping hydration and grinding procedures if cut slices are used. Figure 1.24 highlights that the results on slices and properly stopped and ground powders are quite comparable and that hemicarbonate is quite well preserved in both cases. However, it seems that ettringite quantity is lower in the case of the slices, especially when the slice is treated by isopropanol. More details on this are given in Chapter 4.

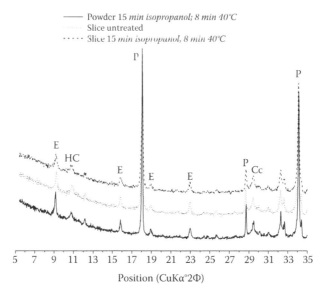

Figure 1.24 Comparison of XRD measurements of a hydrated PC paste using a powder sample where hydration was stopped by solvent exchange, a cut slice where hydration was not stopped, and a cut slice where hydration was stopped by solvent exchange. Cc: calcite; E: ettringite; HC: hemicarboaluminate; P: portlandite.

1.8 CONCLUSIONS

Besides reproducibility and repeatability of the analytical methods used to characterise cementitious materials, sample preparation and sample treatment is a mandatory prerequisite for high-quality research. Not only the accuracy of the testing methods should be assessed and verified but also sample preparation needs to be addressed thoroughly. As many experiments in the cement field are long-term studies, i.e. 28 days of hydration and longer, the appropriate way of sample preparation from raw materials until the analytics at the various testing times needs to be included in the experimental planning as early as possible. It might be helpful to perform preliminary tests in order to identify and practise the correct sample treatment. Inappropriate sample treatment will spoil the whole experimental program, waste time and create unnecessary extra work and costs. Table 1.3 summarises the important points to consider. For details the reader is referred to the preceding sections. It is also important that the conditions of sample preparation are presented in detail in publications such as scientific reports, theses or journals. Some considerations can be found in Table 1.4. Thus, the reader has the full information under which conditions the experimental data were derived.

Table 1.3 General considerations about best practise for sample preparation
          when studying cement hydration

Topic	General considerations
Sampling	• Cements and fine powders in general are quite homogeneous. Sampling with the scoop method is often sufficient. • If your sample is inhomogeneous, use appropriate sampling procedures (golden rules of sampling).
Grinding	• Use appropriate grinding tools and control your particle size distribution. • Check if your material is altered by the grinding process (e.g. dehydration of gypsum). • Intergrinding gypsum is preferable to interblending when producing laboratory cement from a clinker and a calcium sulfate set regulator. • Never apply mechanical grinding to hydrated cementitious materials. You will alter your hydrate assemblage and destroy some of your hydrate phases.
Blending	• Use appropriate tools for blending different powders to obtain a homogeneous mix. • A paint mixer is a simple and cheap tool in case nothing else is available. • Prepare just the amount of material used for a certain series of experiments.
Storage of unhydrated binders	• Store your materials in sealed buckets or shrink-wrapped bags, if possible in a room with low relative humidity (35% or below). • Storage in moist environment will prehydrate your cement. Its properties will be severely altered.
Mixing, casting and curing of pastes and mortars	• Use appropriate mixing tools and procedures. • A premixed binder is a choice better than adding the components during the mixing process. • It might be difficult to disperse fine powders, such as microsilica.
Hydration stoppage	• Use solvent exchange with isopropanol or short freeze-drying. • Do not apply harsh procedures such as drying at $105°C$; they will destroy a part of the hydrates and alter the microstructure. • Measure your samples after stoppage as soon as possible (ideally on the same day); otherwise, a part of your hydrates will be destroyed.

Table 1.4 General considerations about reporting of sample preparation
in publications

Topic	What should be reported
Raw materials	• What is the chemical composition (XRF, XRD/Rietveld, ...)? • What are their physical characteristics (specific surface, density, particle size distribution, ...)? • How were they treated before use (grinding, blending, heating)? • Do not report brand names.
Mix design	• Give the exact mix composition. • Specify the w/c. • Specify where you refer the dosage of admixtures to (cement, binder, mortar, ...). • In case of liquid admixtures: Does the dosage refer to the solid content or to the admixture as delivered? Is the water content of the admixture included in the w/c?
Mixing	• Which type of mixer was used? • Which operation conditions were chosen (rpm, mixing duration, mixing sequence, temperature, quantity of material, ...)?
Moulding	• Which kind of mould, vessel, etc. was used? • Were separation agents applied?
Curing	• Report temperature, relative humidity and curing time. • Were the samples sealed or in contact with the environment (air, water)? • How was the demoulding done?
Hydration stoppage	• Which procedure was used? • Was the same procedure used for all methods applied? • How and how long were the samples stored after hydration stoppage?

ACKNOWLEDGEMENTS

Special thanks to Paul Bowen, Paweł Durdziński, Astrid Gruskovnjak, Christian Müller, John Rossen, Ruben Snellings, Paul Sandberg and Anatol Zingg for providing measurement data and photos. Paul Bowen, John Rossen and Karen Scrivener are acknowledged for helpful discussions.

REFERENCES

Allen, T. (1997). *Particle Size Measurement*. Fifth Edition, Chapman and Hall, New York.

Chopin, D., F. de Larrard and B. Cazacliu (2004). 'Why do HPC and SCC require a longer mixing time'? *Cement and Concrete Research* **34**(12): 2237–2243.

Collier, N. C., J. H. Sharp, N. B. Milestone, J. Hill and I. H. Godfrey (2008). 'The influence of water removal techniques on the composition and micro-structure of hardened cement pastes'. *Cement and Concrete Research* **38**(6): 737–744.

Costoya Fernandez, M. M. (2008). Effect of particle size on the hydration kinetics and microstructural development of tricalcium Silicate, PhD thesis No. 4102, École Polytechnique Fédérale de Lausanne, Switzerland.

De Gasparo, A., M. Herwegh, R. Zurbriggen and K. Scrivener (2009). 'Quantitative distribution patterns of additives in self-leveling flooring compounds (under-layments) as function of application, formulation and climatic conditions'. *Cement and Concrete Research* 39(4): 313–323.

Diamond, S., and S. Sahu (2006). 'Densified silica fume: Particle sizes and dispersion in concrete'. *Materials and Structures* 39(9): 849–859.

Dils, J., V. Boel and G. De Schutter (2013). 'Influence of cement type and mix-ing pressure on air content, rheology and mechanical properties of UHPC'. *Construction and Building Materials* 41: 455–463.

Dubina, E., L. Black, R. Sieber and J. Plank (2010). 'Interaction of water vapour with anhydrous cement minerals'. *Advances in Applied Ceramics* 109(5): 260–268.

Dubina, E., L. Wadsö and J. Plank (2011). 'A sorption balance study of water vapour sorption on anhydrous cement minerals and cement constituents'. *Cement and Concrete Research* 41(11): 1196–1204.

EN 196-1 (2005). Method of testing cement – Part 1: Determination of strength.

EN 196-3 (2005). Method of testing cement – Part 3: Determination of setting times and soundness.

EN 932-1 (1996). Tests for general properties of aggregates – Part 1: Methods for sampling.

EN 932-2 (1996). Tests for general properties of aggregates – Part 2: Methods for reducing laboratory samples.

EN 933-1 (2012). Tests for geometrical properties of aggregates; Determination of particle size distribution: Sieving method.

Franke, B. (1941). 'Determination of calcium oxide and calcium hydroxide also on water-free and watery calcium silicate'. *Zeitschrift für Anorganische und Allgemeine Chemie* 247(1/2): 180–184.

Gallé, C. (2001). 'Effect of drying on cement-based materials pore structure as identified by mercury intrusion porosimetry – A comparative study between oven-, vacuum- and freeze-drying'. *Cement and Concrete Research* 31(10): 1467–1477.

Geiker, M. R., A. Bøhm and A. M. Kjeldsen (2006). 'On the effect of mixing on property development of cement pastes'. *Proceedings pro052: International RILEM Conference, Volume Changes of Hardening Concrete: Testing and Mitigation*, 20–23 August 2006, Lyngby, Denmark, RILEM Publications S.A.R.L., Bagneux, France: 303–310.

Gorce, J. P., and N. B. Milestone (2007). 'Probing the microstructure and water phases in composite cement blends'. *Cement and Concrete Research* 37(3): 310–318.

Gy, P. M. (1976). 'The sampling of particulate materials – A general theory'. *International Journal of Mineral Processing* 3(4): 289–312.

Hansen, F. E., and H. J. Clausen (1973). 'Entwässerung von Gips'. *Zement-Kalk-Gips* 26: 223–226.

Hart, J. R., Y. Zhu and E. Pirard (2010). 'Advances in the characterisation of indus-trial minerals – Particle size and shape characterisation: Current technology and practise'. *EMU Notes in Mineralogy* – Volume 9.

ISO 11648-1 (2003). Statistical aspects of sampling from bulk materials – Part 1: General principles.

ISO 11648-2 (2003). Statistical aspects of sampling from bulk materials – Part 2: Sampling of particulate materials.

Juilland, P., A. Kumar, E. Gallucci, R. J. Flatt and K. L. Scrivener (2012). 'Effect of mixing on the early hydration of alite and OPC systems'. *Cement and Concrete Research* **42**(9): 1175–1188.

Justs, J., M. Wyrzykowski, F. Winnefeld, D. Bajare and P. Lura (2014). 'Influence of superabsorbent polymers on hydration of cement pastes with low water-to-binder ratio'. *Journal of Thermal Analysis and Calorimetry* **115**(1): 425–432.

Khoshnazar, R., L. Raki, J. Beaudoin and R. Alizadeh (2013a). 'Solvent exchange in sulphoaluminate phases; Part I: Ettringite'. *Advances in Cement Research* **25**(6): 314–321.

Khoshnazar, R., L. Raki, J. Beaudoin and R. Alizadeh (2013b). 'Solvent exchange in sulphoaluminate phases; Part II: Monosulfate'. *Advances in Cement Research* **25**(6): 322–331.

Knöfel, D., S. Duckwitz and T. Bier (1997). 'Optimizing the early strength of high alumina cement'. *ZKG International* **50**(8): 454–462.

Konecny, L., and S. J. Naqvi (1993). 'The effect of different drying techniques on the pore-size distribution of blended cement mortars'. *Cement and Concrete Research* **23**(5): 1223–1228.

Korpa, A., and R. Trettin (2006). 'The influence of different drying methods on cement paste microstructures as reflected by gas adsorption: Comparison between freeze-drying (F-drying), D-drying, P-drying and oven-drying methods'. *Cement and Concrete Research* **36**(4): 634–649.

L'Hopital, E., B. Lothenbach, D. Kulik and K. Scrivener (2015). 'Influence of the Ca/Si on the aluminium uptake in C-S-H'. *Cement and Concrete Research*: submitted.

Le Saoût, G., B. Lothenbach, A. Hori, T. Higuchi and F. Winnefeld (2013). 'Hydration of Portland cement with additions of calcium sulfoaluminates'. *Cement and Concrete Research* **43**: 81–94.

Lothenbach, B., G. Le Saout, E. Gallucci and K. Scrivener (2008). 'Influence of limestone on the hydration of Portland cements'. *Cement and Concrete Research* **38**(6): 848–860.

Lothenbach, B., F. Winnefeld, C. Alder, E. Wieland and P. Lunk (2007). 'Effect of temperature on the pore solution, microstructure and hydration products of Portland cement pastes'. *Cement and Concrete Research* **37**(4): 483–491.

Marchuk, V. (2002). 'Dispersibility of silica fume slurry in cement paste and mortar'. *Beton* **29**(7/8): 393–398.

Masuda, H., K. Higashitani and H. Yoshida (2006). *Powder Technology Handbook*, CRC Press, Boca Raton.

Mmusi, M. O., M. G. Alexander and H. D. Beushausen (2009). 'Determination of critical moisture content for carbonation of concrete'. *2nd International Conference on Concrete Repair, Rehabilitation and Retrofitting, ICCRRR-2*, 24–26 November 2008, Cape Town, South Africa: 359–364.

Petersen, L., P. Minkkinen and K. H. Esbensen (2005). 'Representative sampling for reliable data analysis: Theory of sampling'. *Chemometrics and Intelligent Laboratory Systems* **77**(1–2): 261–277.

Pourchez, J., F. Valdivieso, P. Grosseau, R. Guyonnet and B. Guilhot (2006). 'Kinetic modelling of the thermal decomposition of ettringite into metaettringite'. *Cement and Concrete Research* **36**(11): 2054–2060.

Rhodes, M. J. (2008). *Introduction to Particle Technology*, Second Edition. John Wiley & Sons Ltd., Chichester, U.K.

Rodriguez, E. D., L. Soriano, J. Paya, M. V. Borrachero and J. M. Monzo (2012). 'Increase of the reactivity of densified silica fume by sonication treatment'. *Ultrasonics Sonochemistry* 19(5): 1099–1107.

Rößler, C., A. Eberhardt, H. Kučerová and B. Möser (2008). 'Influence of hydration on the fluidity of normal Portland cement pastes'. *Cement and Concrete Research* 38(7): 897–906.

Schmidt, G., T. A. Bier, K. Wutz and M. Maier (2007). 'Characterisation of the ageing behaviour of premixed dry mortars and its effect on their workability properties'. *ZKG International* 60(6): 94–103.

Schöler, A., B. Lothenbach, F. Winnefeld and M. Zajac (2015). 'Hydration of quaternary Portland cement blends containing blast-furnace slag, siliceous fly ash and limestone powder'. *Cement and Concrete Composites* 55: 374–382.

Sprung, S. (1978). 'Effect of the storage conditions on properties of cement'. *Zement-Kalk-Gips* 31(6): 305–309.

Tang, F. J. (1992). 'Optimisation of sulfate form and content'. *PCA Research and Development Bulletin RD105T*, Portland Cement Association, Skokie, Illinois, U.S.

Tang, F. J., and E. M. Gartner (1987). 'Influence of sulfate source on Portland cement hydration'. *PCE R&D Serial No. 1876*, Construction Technology Laboratories, Inc., Skokie, Illinois, U.S.

Taylor, H. F. W., and A. B. Turner (1987). 'Reactions of tricalcium silicate paste with organic liquids'. *Cement and Concrete Research* 17(4): 613–623.

Theisen, K., and V. Johansen (1975). 'Prehydration and strength development of Portland cement'. *American Ceramic Society Bulletin* 54(9): 787–791.

Thomas, M. D. A. (1989). 'The suitability of solvent exchange techniques for studying the pore structure of hardened cement paste'. *Advances in Cement Research* 2(1): 29–34.

Whittaker, M., E. Dubina, F. Al-Mutawa, L. Arkless, J. Plank and L. Black (2013). 'The effect of prehydration on the engineering properties of CEM I Portland cement'. *Advances in Cement Research* 25(1): 12–20.

Winnefeld, F. (2008). 'Influence of cement ageing and addition time on the performance of superplasticisers'. *ZKG International* 61(11): 68–77.

Winnefeld, F., B. Lothenbach, R. Figi, G. Rytz and M. Plötze (2005). 'The influence of different calcium sulfates on the hydration of Portland cement – A practical study'. *ZKG International* 58(3): 62–70.

Yang, T., B. Keller, E. Magyari, K. Hametner and D. Gunther (2003). 'Direct observation of the carbonation process on the surface of calcium hydroxide crystals in hardened cement paste using an atomic force microscope'. *Journal of Materials Science* 38(9): 1909–1916.

Zeng, Q., K. F. Li, T. Fen-Chong and P. Dangla (2013). 'Water removal by freeze-drying of hardened cement paste'. *Drying Technology* 31(1): 67–71.

Zhang, J., and G. W. Scherer (2011). 'Comparison of methods for arresting hydration of cement'. *Cement and Concrete Research* 41(10): 1024–1036.

Zhang, L., and F. P. Glasser (2000). 'Critical examination of drying damage to cement pastes'. *Advances in Cement Research* 12(2): 79–88.

Chapter 2

Calorimetry

*Lars Wadsö, Frank Winnefeld,
Kyle Riding and Paul Sandberg*

CONTENTS

2.1 INTRODUCTION

Calorimetry is the measurement of heat and heat production rate. It is a generic way of studying processes, as all processes (physical, chemical and biological) are generally related to enthalpy changes. One of the oldest and most common applications of calorimetry is to study the hydration of cement, and this chapter is an applied overview of calorimetry in the cement field.

There are many different types of calorimeters that can be used in the cement field, and some of them have many uses. The most common technique is isothermal (heat conduction) calorimetry in which the heat production rate (thermal power) from small samples of paste or mortar is directly measured. This technique is the focus of this chapter as it is the most versatile calorimetric technique in the cement field. Typical commercial instruments of this type used in the cement field are TAM Air (TA Instruments, U.S.), I-Cal (Calmetrix, U.S.), MC CAL (C3 Prozess- und Analysentechnik, Germany), ToniCAL III and ToniCAL TRIO (Toni Technik, Germany) and C80 (Setaram, France).

In semiadiabatic calorimeters, samples of concrete or mortars are insulated so that their hydration can be followed by measuring the temperature change of the sample. This is a relatively simple technique for following the hydration of large specimens (concrete), but it is also used to determine the heat of hydration of mortars, for example, in the standardised Langavant calorimeter according to EN 196-9 (2010).

For studies of mass concrete – where the heat losses are negligible – perfectly adiabatic calorimeters can be used. Instead of being insulated, like semiadiabatic calorimeters, fully adiabatic calorimeters are surrounded by an adiabatic shield that prevents any flow of heat from the sample; see Gibbon et al. (1997) for an example of such a calorimeter.

Solution calorimeters are used for determination of heat of hydration typically after 7 days as specified, e.g. in European Standard EN 196-8 (2010) or in American Standard ASTM C186 (2013). Heat of hydration is calculated from the measured heats of dissolution of a cement paste sample that has hydrated for 7 days and of the dry cement powder. Solution calorimetry has significantly decreased in use in the last decades as the acids needed to dissolve the samples pose severe occupational hazards.

Differential scanning calorimetry (DSC) is used to study the heat effects of changing the temperature. This is a very common technique in many fields; in the cement field typical applications are, e.g. quantifying gypsum and hemihydrate in anhydrous cement (Dunn et al. 1987) or determining the free (freezable) water via low-temperature DSC, assessing freezing processes (Kaufmann 2004) or hydration kinetics (Ridi et al. 2003).

As isothermal and semiadiabatic calorimetries are by far the most frequent calorimetric methods for studying cement hydration and equipment for these is present in almost every laboratory related to inorganic construction materials, this chapter will focus on these two methods. Some applications of DSC in the field of cementitious materials are mentioned in Chapter 5.

Isothermal and semiadiabatic calorimeters both quantify cement hydration kinetics, but they do so in different ways. This is illustrated in Figure 2.1. In isothermal (heat conduction) calorimetry the heat production rate (P) in a small sample (S) is measured by a heat flow sensor as heat is conducted to a heat sink that is placed in a thermostated environment. It is also necessary to have a reference sample (R) with the same properties (especially the same heat capacity) as the sample but without any heat production. This arrangement significantly reduces the noise in the measurements. The output from the calorimeter is the difference between the sample signal and the reference signal.

In a semiadiabatic calorimeter the sample (S) is insulated and the temperature (T) increase during hydration is assessed as a measure of the rate of the hydration. Corrections must be made for the heat losses to the surroundings. The heat loss rate can be found by calculations using a measured heat loss coefficient or it can be directly measured with heat flow sensors placed in the insulation.

One main difference between isothermal calorimetry and semiadiabatic calorimetry is the sample size. Isothermal calorimetry uses small (1–100 g) samples of paste or mortar and, in some cases, small aggregate concrete. In semiadiabatic calorimetry, the use of concrete with aggregates up to approximately 16 mm is possible if one is interested to assess just retardation or acceleration phenomena. For a better signal-to-noise ratio, it is better to remove the coarse aggregates >4 mm by using a sieve. In semiadiabatic calorimeters larger samples (around 500–1000 g cement mortar to 10 kg of concrete) are used; thus, isothermal calorimetry tends to be used in cement research and development, while semiadiabatic calorimetry is also used by mortar and concrete manufacturers and in the field.

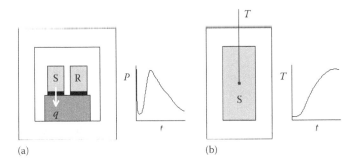

Figure 2.1 Schematic illustrations of (a) isothermal (heat conduction) calorimetry and (b) semiadiabatic calorimetry. *P*: thermal power; R: reference; S: sample; *T*: temperature; *t*: time.

Isothermal calorimeters measure thermal power (heat production rate), while (semi)adiabatic calorimeters measure temperature (change). It is possible to calculate one of these from the other, but to do so we need to take the derivative (to go from semiadiabatic to isothermal) or integrate (to go from isothermal to semiadiabatic); and in both cases we need the heat capacity of the sample. The thermal power signal from an isothermal calorimeter shows more details than the temperature signal from an adiabatic calorimeter as the former directly assesses the rate of the process, while an adiabatic calorimeter measures the integral of the rate. Isothermal calorimetry is thus a more generally useful analytical method than semiadiabatic calorimetry.

For all types of calorimeters in the cement field except DSC, the purpose of calorimetry is to study the kinetics and extent of the cement hydration reaction. This is shown in Figure 2.2 for isothermal calorimetry. In Figure 2.2a the typical heat production rate (thermal power) as a function of time is shown for a rapidly reacting and a slowly reacting cement. From such calorimetric curves we can divide the cement hydration into different phases, for example, early reactions, induction period, accelerating phase and decelerating phase (Bensted 1987; Jansen et al. 2012a,b; Taylor 1997), as seen in Figure 2.3. Note that the separation between the early phases is rather arbitrary in most cases.

Figure 2.2b shows the same data as in Figure 2.2a but shows, instead of thermal power, heat, the integral of thermal power. It is seen here that the rapid cement produces more heat during the first days of hydration; the slag has a slower reaction, but this reaction will continue for a much longer time. The heat per gram of cement produced after 1, 3 or 7 days of hydration is a standardised measure of the reactivity of cements according to ASTM C1702 (2014). The data in Figure 2.2b start at zero heat at time zero and there is an initial rapid increase in heat during the first minutes after water and cement have made contact. The corresponding very high thermal power peak at time zero is not shown in Figure 2.2a (see Section 2.3).

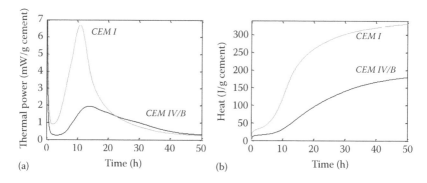

Figure 2.2 Typical results from 2 days of measurements of (a) thermal power and (b) heat of hydration with an isothermal calorimeter for a rapid-hardening cement (CEM I) and a slow-hardening cement (CEM IV/B).

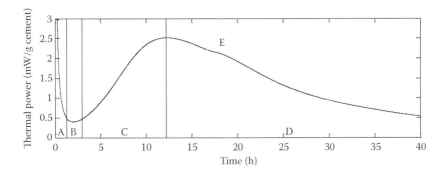

Figure 2.3 Different phases of cement hydration as seen in an isothermal calorimetric measurement. Four phases are shown: initial period (A), induction period (B), accelerating period (C) and decelerating period (D). We have also marked the 'sulfate depletion peak' as E. This is a feature often found in measurements of portland cement systems and related to the start of secondary aluminate hydration.

Under the assumption that the rate of hydration is a function only of the degree of hydration and the temperature, it is possible to calculate the semiadiabatic result from the isothermal one, and vice versa (heat capacity and activation energy need to be known). Examples of such calculations are given by Wadsö (2003).

To summarise, isothermal, semiadiabatic and adiabatic calorimeters are all used to study the cement hydration process, but they follow different time–temperature trajectories. This is illustrated in Figure 2.4.

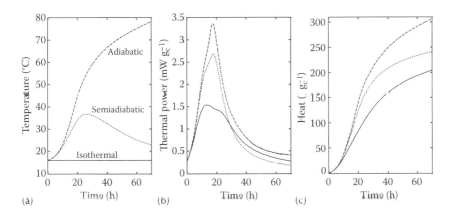

Figure 2.4 An example of how (a) temperature, (b) thermal power and (c) heat develop over time in an isothermal, a semiadiabatic and an adiabatic calorimeter. Note that paste or mortar is generally used in isothermal calorimetry, while mortar or concrete is used in (semi)adiabatic calorimetry.

2.2 OPERATIONAL ISSUES

Calorimetric measurements are often quite simple to perform, but as heat is such a ubiquitous phenomenon, there are many sources of error, and measurements should be made following proven operational procedures. We discuss here some of the most important operational issues with isothermal and semiadiabatic calorimetries and give some short notes on the heat-of-solution method as well.

2.2.1 Isothermal conduction calorimetry

2.2.1.1 Temperature stability and accuracy

A calorimeter measures heat and it is therefore important that the calorimeter is placed in a constant-temperature environment, usually a high-precision thermostat. It is important to check that the temperature of the thermostat is accurate. For this, one needs a calibrated thermometer that can be inserted into the calorimeter thermostat.

A stable laboratory temperature can improve the precision of calorimetric measurements, but it is not needed for qualitative and many quantitative applications since the calorimeter has its own temperature control, which for most calorimeters is far more precise than normal laboratory air-conditioning. However, several standards place demands on laboratory temperature; for example, ASTM C1679 (2014) demands the requirements of ASTM C511 (2013) for cement mixing rooms (23.0 ± 4.0°C and a relative humidity of not less than 50%) to reduce the disturbance when charging samples.

2.2.1.2 Calibration

To use a calorimeter for quantitative measurements, it has to be calibrated; i.e. parameters that make it possible to evaluate the results in terms of standard units need to be determined. This involves measuring calibration coefficients, baselines and (sometimes) the time constants. The calibration coefficient is the parameter that transforms the voltage from the heat flow sensor in the calorimeter to thermal power. It is often called ε (unit: watts/volt) and is usually measured by electrical calibrations. The baseline is the signal from the calorimeter when no heat is produced in the sample. The baseline U_0 (unit: volts) is measured without any heat production in the sample ampoule. For applications where an accurate value of the baseline is important – such as determinations of 7 days' heat of hydration – a baseline measurement should be made for 1–2 days with inert material in the sample vial.

When voltage U (the output from the heat flow sensors) has been measured, thermal power P is calculated by the following equation:

$$P = \varepsilon(U - U_0), \tag{2.1}$$

where P is thermal power (in watts), ε is calibration coefficient (in watts/volt), U is voltage (in volts) and U_0 is baseline voltage (in volts).

It is also common to give the specific thermal power:

$$P = \frac{\varepsilon(U - U_0)}{m}, \qquad (2.2)$$

where m (in grams) usually is the mass of cement (or binder) in the sample.

The third calibration parameter, which is needed when rapidly changing processes are studied, is the time constant. This is a measure of the thermal inertia of the sample that 'blurs' details in rapid events. The time constant is used in the Tian equation – named after a pioneer in isothermal calorimetry – to correct for this. Typical time constants in isothermal calorimeters are 100–1000 s. As the main hydration has timescales much longer than this, the Tian equation is not needed in cement calorimetry when the main hydration is studied, but it is needed when early reactions are studied. Further information on the Tian equation is given by Wadsö (2005), and other similar methods are discussed by Evju (2003).

It is common that commercial calorimeters have internal, automatic calibration. Although this makes a calorimeter user friendly, it is problematic if the user does not know whether the calibrations are accurate. One way to check whether the instrument is working properly and whether the user is performing the measurement in a correct way is to run a validation procedure, i.e. an experiment with a known outcome (proficiency test). A number of such 'chemical calibration' systems are described by Wadsö and Goldberg (2001); however, none is similar to cement hydration measurements. It is therefore of interest to establish reference cements – or other similar systems, for example, based on calcium hemihydrate – that can be used to validate the quality of calorimetric cement measurements in a laboratory.

Isothermal calorimeters are often very stable instruments in which the calibration parameters remain rather constant for many years (Wadsö 2010). One may be tempted therefore to skip calibrations, but this is a bad strategy as, sooner or later, all instruments will malfunction. In the case of calorimeters this can come suddenly and be easy to detect, or gradually, so that a user will continue to do measurements without noticing that the calorimeter is not working properly. Calibration, therefore, should be seen not only as a way of getting correct coefficients for the evaluation but also as a way of validating that the instrument is working well. If the value of a calibration coefficient starts to change, this is an indication that something is not well. Regular calibrations at intervals of approximately 3 months are generally a suitable choice (Wadsö 2010). A new baseline calibration should be done if, e.g. the temperature or the type of vial is changed or a totally different type or mass of sample is used.

Isothermal calorimeters have two vial positions: one for the sample vial and one for a reference vial. The latter should be charged with a substance with a similar heat capacity as the sample but with no heat production. Convenient materials are water and quartz sand and a method for selecting a proper reference is discussed by Wadsö (2010). Do not use old hydrated cement paste samples as such samples may still produce heat or produce heat when the temperature changes.

The measured signal is the difference between the sample signal and the signal from the (inert) reference. Because of this, the reference removes a large part of the noise with external origin (including cross talk in multi-channel instruments) as this influences both the sample and the reference in a similar way (if the heat capacities are matched). Failure to use a proper reference will give noisy results.

There is a second important reason for using a reference: a proper reference reduces drift in the measurements. This has to do with the fact that the temperature of the heat sink is not exactly constant because of the heat produced in the sample, but as both sample and reference are connected to the same heat sink, this small temperature change does not affect a measurement.

2.2.1.3 Precision

It is not trivial to state how good a certain type of calorimeter or a certain type of calorimetric measurement is in terms of precision (or accuracy), as many factors (room temperature stability, reference balance, etc.) influence this. A problem is also that different errors influence different types of measurements in different ways. For example, an error in the calibration coefficient contributes the same error to 1 and 7 days' measured heats; but an error in the baseline is much worse for the 7 days' heat as the thermal powers are low in the end of measurement. For a short (30 min) measurement of initial reactions, the baseline error is normally of no importance as very high thermal powers are measured; more important is the time constant and how well the Tian equation can deconvolute the rapidly changing signal.

The precision of isothermal calorimetry has mainly been assessed in round-robin studies of heat of hydration. As an example, a German study with international participation (Lipus and Baetzner 2008) in 2006/2007 gave a standard deviation for repeatability of 5–7 J g^{-1} (within one lab) and a standard deviation for reproducibility (between the labs) of 13.6 J g^{-1}, the latter when regarding only the device used by most of the participants. This value is slightly better than those for the other calorimetric techniques.

2.2.1.4 Practical notes

It is important to always use proper references, i.e. references with a similar heat capacity as the sample, but with no heat production. If the same sample sizes and the same water-to-cement ratio (w/c) are used repeatedly, the references do not need to be changed but can rest in the calorimeter. If different sample sizes are used, the references need to be changed accordingly. It can be convenient to prepare reference vials with different masses of water corresponding to the different cement paste/mortar sample sizes measured on; see Wadsö (2010). Note that the matching does not have to be perfect; in most cases a heat capacity within 5%–10% is good enough.

One should always use vapour-tight vials as the vaporisation of water will produce high thermal powers. If in doubt, first measure the baseline with empty vials in both sample and reference positions and then add a few drops of water to the sample vial. If water leaks from the vial, the baseline with water will be lowered compared to that of the empty vial (the actual water loss rate can be calculated from the enthalpy of vaporisation of about 2400 J g^{-1}).

Baselines should be measured over a weekend to be sure that you have the correct value. If the baseline takes a long time to stabilise (more than half a day), you may have a bad reference.

When a sample is charged into a calorimeter, there will always be a sharp peak (the instrument may even go out of range) as the sample temperature is different from the calorimeter temperature. If the sample is warmer than the calorimeter (for example, if the lab temperature is 25°C and the calorimeter is at 20°C), the peak will be positive (like for an exothermal process); if the sample is at a temperature lower than that of the calorimeter (for example, if you are conducting measurements at 40°C), the peak will be negative. Note that these peaks often do not have much to do with heat of hydration and should then not be interpreted as such. However, the initial peak shows the true heat production (including heat of vaporisation and mixing) if mixing is made inside the vial in the calorimeter; see Wadsö (2005). A method for significantly reducing the initial disturbance is to thermostat the preweighed materials (especially water, which has a high heat capacity) in the calorimeter, take them up and quickly mix them and charge the sample. This method works well for larger samples.

In cases where the initial reactivity may be high, and the later reactivity is low, it is often useful to prepare two separate samples of each mixture: one smaller sample for capturing the initial reactivity and one larger sample for capturing later age reactivity. Examples of such materials include calcium aluminate or sulfoaluminate-based binders and high-calcium fly ash.

The heat flow curves can be integrated in order to obtain the cumulative heat of hydration. In case of samples mixed externally the integration should not start too early after charging the sample, typically 30 min later (see Wadsö 2005); otherwise, the heat introduced due to the external mixing will influence the values of heat of hydration. In the case of portland cement, the first 30 min of hydration contribute less than 10 J/g to the heat of hydration (Justs et al. 2014) and can generally be neglected.

2.2.2 Semiadiabatic calorimetry

We discuss here operational issues of semiadiabatic calorimetry by taking the standardised Langavant calorimeter according to EN 196-9 (2010) as an example. The temperature increase of a mortar inside the calorimeter is measured over time and compared to that of an inert sample with the same heat capacity (e.g. a completely hydrated cement mortar or preferably an aluminium cylinder) inside a reference calorimeter placed nearby (distance about 120 mm). Maximum temperature increases are on the order of 25–35 K, depending on the type of cement. As the calorimeter was calibrated before, the heat of hydration can be calculated. The mortar is composed according to EN 196-1 (2005) but using slightly less material (360 g cement, 180 g water, 1080 g standard sand 0–2 mm, mixed in a Hobart mixer by using a shortened mixing procedure compared to EN 196-1 [2005]). The heat of hydration after 41 h is used to assess whether a cement can be classified as a cement with low or very low heat of hydration.

Simpler and less expensive devices perform only temperature measurements without calibration. They are used to explore the hydration kinetics of mortars of various compositions and quantities, often in industrial applications such as product development and quality control.

2.2.2.1 Temperature stability and accuracy

Temperature stability is crucial also for semiadiabatic calorimetry, especially when heat of hydration should be quantified. In the case of EN 196-9 (2010) the calorimeters are placed in a climate room operated at 20 ± 1°C with an accuracy of ±0.5°C. In order to have constant conditions concerning heat loss, the velocity of the air next to the calorimeter should be below 0.5 m/s.

2.2.2.2 Calibration

The calibration of a Langavant calorimeter is quite complicated and time consuming, as it takes several days. It uses a calibration cylinder which is placed in the calorimeter. The calibration cylinder is electrically heated, and the electrical energy for maintaining a constantly increasing temperature of the calibration cylinder (increase by 10–40 K, five different values) is measured. The heat loss coefficient and the heat capacity of the calorimeter are determined by measuring the temperature decrease inside the calorimeter after the electrical heating is turned off. For more details on the calibration procedure see EN 196-9 (2010) and Balbas et al. (1991).

2.2.2.3 Calculation of heat of hydration

From the measured temperature over time the heat of hydration Q (in joules/gram) can be calculated using the following formula:

$$Q = \frac{c}{m_c} \cdot \Delta T + \frac{1}{m_c} \cdot \int_0^t \alpha \cdot \Delta T \cdot dt, \tag{2.3}$$

where m_c is the mass of cement (in grams), t is the hydration time (in hours), c is the heat capacity of the calorimeter including the sample (in joules/kelvin), α is the heat loss coefficient of the calorimeter (in joules/kelvin-hours) and ΔT is the temperature difference between the calorimeter containing the sample and the reference calorimeter (in kelvins).

The heat capacity of the sample is calculated assuming heat capacities of 0.8 J K^{-1} g^{-1} for cement and sand, 3.8 J K^{-1} g^{-1} for water (this value is slightly lower than that of free water as hydrated water has a decreased heat capacity) and 0.50 J K^{-1} g^{-1} for the mortar container (steel). For more details the reader is referred to EN 196-9 (2010) and Gascon and Varade (1993).

2.2.2.4 Precision

EN 196-9 (2010) gives a standard deviation for repeatability of 5 J/g. Thus, two measurements done in the same laboratory on the same cement should not deviate more than 14 J/g. Accuracy in interlaboratory tests is moderate: For standard deviations in interlaboratory tests a value of 15 J/g is given, and measurements in two laboratories should not deviate more than 42 J/g.

2.2.3 Heat-of-solution method

The heat-of-solution method is standardised in ASTM C186 (2013) and EN 196-8 (2010). As very hazardous chemicals (hydrofluoric and nitric acid) are used, this method tends to be replaced by the methods mentioned previously. A cement paste hydrated for usually 7 days using a w/c of 0.40 at 20°C (EN method) or 23°C (ASTM method) is dissolved in a solution calorimeter containing a mixture of hydrofluoric and nitric acid. The determined heat is compared to the heat generated by dissolving anhydrous cement according to the same procedure. The heat of hydration after 7 days is used to assess the heat of hydration of a portland cement. Accuracy is similar as that of the Langavant method. For more details the reader is referred to the respective standards.

2.3 SAMPLE PREPARATION – MIXING
AND CHARGING

This section is mainly relevant for isothermal calorimetry, however, many aspects are also valid for semiadiabatic measurements.

There are several possibilities of mixing cement pastes and mortars for isothermal calorimetry:

- Make a sample from a larger mix; a drawback with this may be that it is difficult to take out small representative samples, both concerning the water content and the sand content (for mortars), and it might be difficult to know how much cement there is in the sample, especially if the sample shows segregation. Quantitative measurements are thus difficult in such cases, but qualitative ones (for example, to determine retardation – see in the following) are still possible. Nevertheless, if it is possible to take a homogeneous sample out of the mix, it is possible to perform quantitative measurements.

- Weigh constituents in the calorimeter vial and mix inside the vial with a plastic spatula (that is left inside vial) or on a test tube shaker. This is a rapid way of starting measurements that is useful for tests that do not require concrete-like high-shear mixing, for example, when testing binders without admixture for sulfate optimisation or heat of hydration. Especially for larger samples, it is advantageous to equilibrate the water inside the calorimeter thermostat. This water can then be added to a known mass of cement in another vial, mixed and quickly charged into the calorimeter. As the major part of the heat capacity in cement paste and mortar comes from the water, this method produces limited initial disturbances if made quickly.

- Internal mixing can be made inside the vial when it is in the calorimeter (Wadsö 2005). This is the best way of studying the initial processes (0–10 min), but the mixing intensity is very low as the calorimeter will be disturbed or damaged if normal mixing methods are used inside the calorimeter. Because of this, the hydration rate will be lower than if standard methods of mixing are used. When cement is mixed with water, a number of rapid processes take place, for example, sulfate dissolution and hydration of aluminate phases that are of importance for both the fresh concrete rheology and the later main strength-giving reactions. To study these early processes (<10 min), it is most convenient to use isothermal calorimetry and mix inside the vial in the calorimeter. To do this, cement and water must be kept separate until the time of mixing, and the vessel must be equipped with a mixer. See Wadsö (2005) for an example of this. This method is not suitable for 'sticky' mixtures, e.g. with low w/c, as proper mixing cannot be guaranteed then. If early processes are not of interest, external mixing is always the preferable choice.

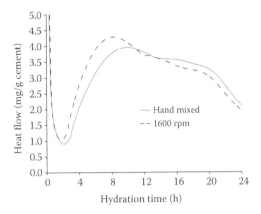

Figure 2.5 Influence of mixing on rate of hydration. Twelve grams of cement, 20 g of quartz powder and 16 g of water (water–binder ratio = 0.40; w/c = 1.33) were either mixed by hand or mixed using a high-speed mixer at 1600 revolutions per minute (rpm). (Courtesy of Elise Berodier, École Polytechnique Fédérale de Lausanne, Switzerland.)

Note that mixing intensity ('mixing energy') influences the hydration rate (an example is given in Figure 2.5) and it is not possible to attain the mixing intensity of an industrial concrete mixer with a standard lab mixer (not to mention when mixing is made inside a calorimetric vessel). Figure 2.5 shows that mixing at higher intensity shifts both the main hydration peak, which can be attributed to alite reaction, and the sulfate depletion peak to earlier times; see also Juilland et al. (2012).

2.4 PRACTICAL APPLICATIONS AND EXAMPLES

Both isothermal and semiadiabatic calorimetries can be used for many different purposes in the cement field, some of which are highlighted in the following. The heat-of-solution method is almost exclusively used to determine heat of hydration.

2.4.1 Determination of heat of hydration

Determination of the heat of hydration after 1, 3 or 7 days is made to classify cements into different types, based on their reactivity, and this type of measurement is most important for low-heat cements used in mass concreting. Solution calorimetry according to ASTM C186 (2013) or EN 196-8 (2010) and semiadiabatic calorimetry according to EN 196-9 (2010) have been the preferred techniques for this, but isothermal calorimetry is becoming more and more used. As discussed previously, the different types of calorimeters have different uses, but for the determination of 1, 3 or 7 days' heat of hydration all three types of calorimeters can be applied. Solution calorimetry and semiadiabatic calorimetry are standardised and have been in use for a long time, although solution calorimetry is used less and less frequently as it requires the use of hazardous chemicals.

Isothermal calorimetry is standardised by ASTM C1702 (2014) (heat of hydration) and ASTM C1679 (2014) (general methodology), and there is an active task group within the European Committee for Standardization (CEN) working for a European standard. On a national basis, isothermal calorimetry is described in a Nordtest procedure in the Nordic countries NT BUILD 505 (2003) and in the Swiss national appendices of European Standards EN 196-8 (2010) and EN 196-9 (2010). As heat of hydration is an important issue in the cement industry, it is of interest to discuss it here, not the least because it is a difficult issue to compare heat of hydration measurements with semiadiabatic, isothermal calorimetry and heat-of-solution measurements. In the EN standards heat of hydration can be measured by both semiadiabatic and heat-of-solution measurements. Because of the temperature increase in the semiadiabatic measurements, the hydration proceeds faster than in the case of the isothermal curing in the heat-of-solution method. Thus, heat of hydration after 41 h is used to assess low-heat cements in the case of semiadiabatic calorimetry, and 7 days (168 h) are used in the case of the heat-of-solution method. Several interlaboratory tests, e.g. those by Mengede et al. (2006), which are summarised by Lipus and Baetzner (2008) (see Figure 2.6), have shown that the correlation between 41 h semiadiabatic calorimetry and 7-day heat-of-solution method is often moderate. In Figure 2.6a (CEM I) the correlation between the two methods is quite poor, whereas in Figure 2.6b (CEM III) the correlation is much better.

As the apparent activation energy (see Section 2.4.2) influences the semiadiabatic measurements but not the isothermal ones, it may be an impossible task to prepare a general procedure by which these two methods can be compared. From a theoretical point of view, solution calorimetry and isothermal calorimetry should give the same result as shown by Killoh (1988) as both are isothermal methods. However, this is not the case for the data presented in Figure 2.6, which might be related to the lower w/c of 0.40 used in the case of the heat-of-solution method compared to isothermal calorimetry (0.50).

Heat of hydration can be converted approximately into degree of hydration if the hydration enthalpies of the different clinker phases are known. Values for the clinker phases can be taken from Taylor (1997); see Table 2.1. These values can be used to calculate the enthalpy of complete hydration if the phase composition of a portland cement is known, e.g. by (extended) Bogue calculation (Taylor 1997) or by quantitative X-ray diffraction (XRD) analysis (see Chapter 5). This value can be used to roughly estimate the hydration degree from the cumulated heat development. For

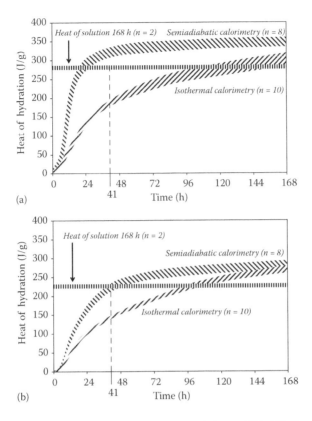

Figure 2.6 Comparison between isothermal, semiadiabatic (EN 196-9) and heat-of-solution calorimetry (EN 196-8 [2010]) of (a) a CEM I and (b) a CEM III/B type of cement. The hatched areas envelop the measured data (*n* = number of individual measurements carried out by the participating laboratories). (Adapted from Mengede, M. et al., 16. *Internationale Baustofftagung (ibausil)*, Weimar, Germany, I, 527–534, 2006.)

Table 2.1 Enthalpy of complete hydration

Phase	Enthalpy of complete hydration (J/g)
C_3S	-517 ± 13
$\beta\text{-}C_2S$	-262
C_3A	-1144^a to -1672^b
C_4AF	-418^c

Source: Taylor, H. F. W., *Cement Chemistry*, Thomas Telford Publishing, London, U.K., 1997.

[a] Reaction with gypsum to give monosulfate.
[b] Reaction with gypsum to give ettringite.
[c] Reaction in presence of excess portlandite to give hydrogarnet.

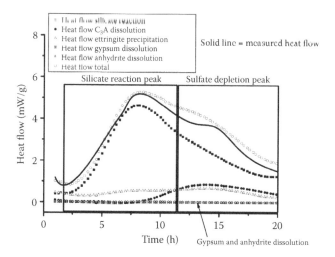

Figure 2.7 Comparison between measured and calculated heat flows of a portland cement hydrated at 23°C using a w/c of 0.50. Heat flow was calculated from quantitative XRD analyses using the dissolution enthalpies of the clinker phases and the precipitation enthalpies of the hydrate phases. (From Jansen, D. et al., *Cement and Concrete Research*, 42(1), 134–138, 2012a. With permission.)

a more accurate calculation the different reaction kinetics of the clinker phases need to be taken into account; see, e.g. Copeland et al. (1960). Recent work showed that isothermal heat flow curves can be calculated using the dissolution kinetics of the clinker phases derived from quantitative XRD analyses (Bizzozero 2014; Hesse et al. 2011; Jansen et al. 2012a,b), taking into account the dissolution enthalpies of the clinker phases and the precipitation enthalpies of the hydrates. Figures 2.7 and 2.8 show that the calculated developments of heat flow and heat of hydration are able to represent the general trends in the experimental curves.

2.4.2 Influence of temperature on hydration kinetics

The total amount of heat released from a cement at a given point in time is a function of the amount of cement that has reacted up to that point in time and independent of the sample temperature history up to that point in time. With increasing temperature, the dissolution of the anhydrous cement and the precipitation of hydrates proceed faster. This acceleration can be easily seen by calorimetry; see Figure 2.9. Both the main hydration peak and the sulfate depletion peak are shifted to earlier hydration times. The increase in hydration rate with temperature can be accounted for by using a transformed scale that combines the effects of time and temperature called maturity. According to this methodology, the cement degrees of

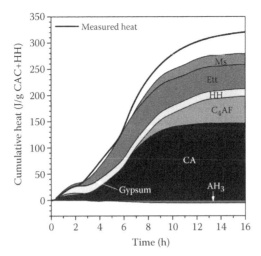

Figure 2.8 Comparison between measured and calculated heats of hydration of a calcium aluminate cement blended with hemihydrate (with a ratio of calcium aluminate to hemihydrate of 80:20 by mass). The sample was hydrated at 20°C using a water–solid ratio of 0.40. Heat flow was calculated from quantitative XRD analyses by using the dissolution enthalpies of the anhydrous phases and the precipitation enthalpies of the hydrate phases. CA: calcium aluminate; Ett: ettringite; HH: hemihydrate; Ms: monosulfate. (From Bizzozero, J., 'Hydration and dimensional stability of calcium aluminate cement based systems', PhD Thesis no. 6336, École Polytechnique Fédérale de Lausanne, Switzerland, 2014.)

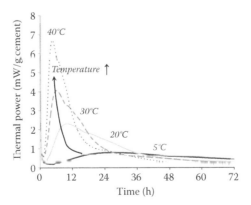

Figure 2.9 Influence of temperature on the heat flow of a CEM I 42.5 N (w/c = 0.40). (Adapted from Lothenbach, B. et al., *Beton*, 55(12), 604–609, 2005.)

reaction and the consequent cumulative heats released up to a given point for two samples of the same cementitious system at the same maturity but with different temperature histories are assumed to be equivalent.

The equivalent age maturity method was developed to account for the nonlinear effects of temperature and time on the rate of concrete property development (Carino and Lew 2001). In this method, the concrete equivalent age t_e (in hours) describes the amount of time that a concrete mixture would need to be cured at a constant reference temperature in order to achieve the same degree of hydration as the same concrete mixture cured at a different temperature. The equivalent age method, based on the Arrhenius equation that describes the temperature dependence of a chemical reaction, is given as follows (ASTM C1074 2011):

$$t_e = \sum_0^t e^{\frac{-E_a}{R}\left(\frac{1}{T_c} - \frac{1}{T_r}\right)}\Delta t, \qquad (2.4)$$

where E_a is the concrete apparent activation energy (in joules/mole), R is the universal gas constant (8.314 J mol^{-1} K^{-1}), T_c is the concrete average absolute temperature during the time interval Δt (in kelvins) and T_r is the absolute reference temperature (in kelvins).

The apparent activation energy is an empirical coefficient that describes the concrete hydration or property development rate dependence on temperature. The term *activation energy* according to the Arrhenius equation is used to describe the temperature dependence of a single chemical reaction. Because cement hydration is a heterogeneous system that involves multiple concurrent chemical reactions, the term *apparent activation energy* is used to describe the system reaction rate temperature dependence. E_a is typically calculated from concrete property development measurements taken with time at several (usually three to five) different isothermal temperatures; however, E_a values calculated from different property measurements such as strength, time of 'set', and heat of hydration measurements will not be the same (Siddiqui and Riding 2012). Isothermal calorimetry is often used to quantify the apparent activation energy for use in concrete hydration or temperature simulations. The methodology used to calculate the activation energy from isothermal calorimetry data is described by Poole et al. (2007).

Calorimetric data can be used to calculate the temperature development in heavy castings. As the heat produced from hydration heats up a cast concrete, calorimetric measurements are needed as input data to calculations of temperature rise, strength development, risk of cracking, etc. Traditionally, semiadiabatic and adiabatic calorimetries have been used for this as these techniques can be used on concrete samples (Kada-Benameur et al. 2000; Poole et al. 2007; Riding et al. 2011). Furthermore, since concrete typically is made up of large proportions of aggregates and sand, one needs accurate information of the thermal properties of all constituents in the concrete mixture. As a result, many users use a combination of isothermal calorimetry for the effect of temperature on the heat of hydration rate and semiadiabatic or fully adiabatic calorimetry for the resulting concrete temperature development; see e.g. Riding et al. (2012).

2.4.3 Influence of water–cement ratio on hydration kinetics

Figure 2.10a shows the influence of w/c on the heat flow. A decreasing w/c increases early hydration kinetics, which is probably because of the higher concentration of alkali ions in the pore solution, promoting the dissolution of the anhydrous phases (Danielson 1962). However, the long-term degree of hydration is decreased (see Figure 2.10b), as less water is available (below a w/c of about 0.4, full hydration cannot be reached in the case of portland cement and self-desiccation may occur) and less space for hydrates to grow is available (Lothenbach et al. 2011).

Another similar application of isothermal calorimetry is the assessment of the thermal power at different curing regimes; see Figure 2.11. Sealed conditions, which are the usual case in isothermal calorimetry, are compared to pastes with extra water added on top of the sample. This extra water causes an increase in cumulative heat and thus an increased hydration degree. The effect becomes more pronounced at lower w/c, where the samples tend to undergo self-desiccation. The same effect can be seen also with other types of cement, such as calcium sulfoaluminate cements. As this type of cement needs a higher w/c for complete hydration compared to portland cement (around 0.5–0.6 instead of 0.4; see Winnefeld and Lothenbach 2010), the effect of the extra water is already very evident at a w/c of 0.7; see Figure 2.12.

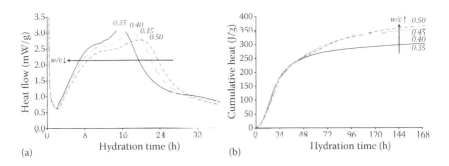

Figure 2.10 Influence of the w/c on (a) hydration heat flow and (b) cumulative heat of a CEM I 52.5 N. (Adapted from Lura, P. et al., *Journal of Thermal Analysis and Calorimetry*, in preparation, 2015.)

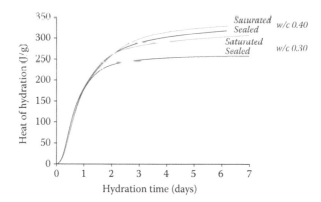

Figure 2.11 Influence of extra water on portland cement hydration. Sealed: cured in vials; saturated: cured with added water on top of sample in vial. Normal calorimetry conditions are equivalent to sealed samples: this can have an important effect at low w/c. (Adapted from Chen, H. et al., *Cement and Concrete Research*, 49, 38–47, 2013.)

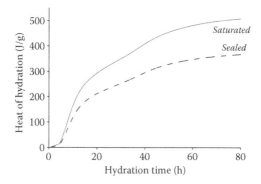

Figure 2.12 Influence of extra water on hydration of a calcium sulfoaluminate cement at a w/c of 0.70. Sealed: cured in vials; saturated: cured with added water on top of sample in vial. (Adapted from Lura, P. et al., *Journal of Thermal Analysis and Calorimetry*, 101(3), 925–932, 2010).

2.4.4 Correlation with setting and strength development

There is no direct correlation between cement hydration and set, since the relative proportions of binder, water and aggregates determines how far the cement hydration needs to proceed to create enough rigidity of a mixture to reach set. However, for a given mixture design it is clear that initial and final sets both occur at a specific point on the accelerating phase (see Figure 2.3). Typically, depending on the type and amount of binder and water content, the initial set for most conventional concrete mixtures at w/c

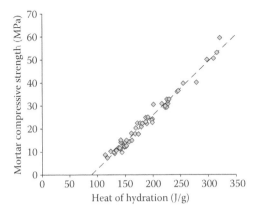

Figure 2.13 Correlation between compressive strength of EN 196 standard mortars and standard heat of hydration according to ASTM C1702 (2014), excluding any heat produced up to 90 minutes after mixing. (Courtesy of Aalborg Portland, Aalborg, Denmark.)

of 0.35–0.50 occurs soon after the main reaction has started and the final set occurs closer to the maximum thermal power.

As with setting, there is no general relation between heat production and strength of concrete (as aggregate, water and porosity have a very strong effect on strength). As with setting, it is possible to correlate strength with heat for a given mixture design, such as standard mortars, simply by correlating the measured heat of hydration with the measured strength at different hydration times. Since heat of hydration by isothermal calorimetry is much more repeatable and a less labour-intensive method compared to, for example, compressive strength at 24 h, this is of interest for cement producers. Figure 2.13 shows the correlation between compressive strength of EN 196 mortars and heat of hydration (ASTM C1702 2014) measured from 1 to 7 days curing at 20°C during cement production. As with traditional concrete maturity, the initial heat produced before the acceleration phase (see Figure 2.3) is not considered useful for strength development and therefore not included.

2.4.5 Sulfate optimisation for portland cement

Sulfate optimisation of hydraulic binders is one of the main tasks in cement technology; see, e.g. Lerch (1946), Tang (1992) and Tang and Gartner (1987). The optimisation using heat of hydration is an alternative to the strength-based method used by most cement producers. A second method based on the secondary aluminate hydration at sulfate depletion (see Figure 2.3) described in ASTM C1679 (2014) is also used mainly for portland cements. Figure 2.14 shows an example for a fly ash–blended cement with

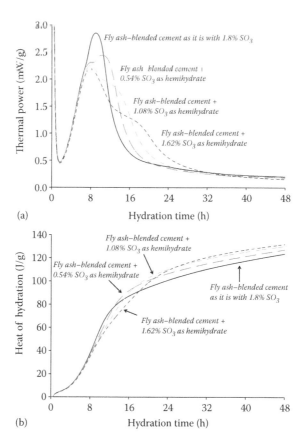

Figure 2.14 Sulfate optimisation of a fly ash–blended cement type CEM II/A-W using isothermal calorimetry. Sulfate has been added in increments to determine the optimal level. (a) Thermal power; (b) heat of hydration (excluding the initial 30 minutes).

a high cement aluminate content that displays an easily identifiable sulfate depletion peak (Lerch 1946). Calcium sulfate hemihydrate was added in increments of 0.5% SO_3 by mass of cement and the response was measured at 20°C using both isothermal calorimetry and compressive strength determination. Figure 2.14a gives thermal power curves, showing the delay of sulfate depletion as hemihydrate is added to the cement. As a rule of thumb the sulfate depletion peak for a well-balanced cement should appear several hours after the maximum of the main hydration peak (see Figure 2.3). In Figure 2.14b the corresponding heat of hydration as a function of total SO_3 content is shown. Figure 2.15 shows the optimum SO_3 curves at 24 hours using heat of hydration and compressive strength. It is seen that calorimetry and strength measurements give very similar results and that sulfate optimisation by calorimetry should work as well as that by strength measurements.

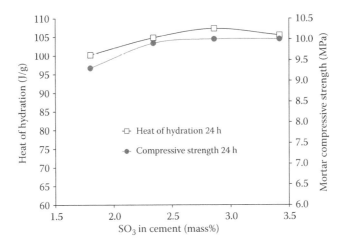

Figure 2.15 Sulfate optimum curves for a fly ash–blended cement type CEM II/A-W using heat of hydration and compressive strength. The strength measurements give an optimal SO₃ content of 2.9%; the heat measurements give 3.0%.

2.4.6 Reactivity of supplementary cementitious materials

Isothermal calorimetry on cements with different concentrations of supplementary cementitious materials (SCMs) can be used to assess the reactivity of SCMs. Measurements are typically made with different fractions of SCM so that one can see how the SCM addition changes the rate of hydration and the produced heat. Measurements can be made either with the same mass of cement and different amounts of the SCM or with a constant mass of cement plus SCM. An inert reference material, such as quartz powder (ideally the quartz powder has a particle size distribution very similar to that of the examined SCM) can be used to distinguish the so-called filler effect from the contribution of a reactive mineral addition such as slag or fly ash on the heat of hydration. The term *filler effect* describes, as summarised by Berodier and Scrivener (2014), Deschner et al. (2012), Lothenbach et al. (2011) and Scrivener et al. (2015), the effect of adding a material which (initially) does not react itself but (1) provides additional nucleation sites for hydrate phases and (2) increases the w/c, which increases the long-term hydration degree by providing more space for the hydrates to precipitate (see Figure 2.10b). The partial replacement of portland cement by quartz in Figure 2.16 causes an increase in cumulative heat due to the filler effect. Fly ash and slag show, besides the filler effect, an additional contribution to the heat of hydration which is related to their participation in the hydration reactions. There are several attempts to extract the reaction degree of SCM from calorimetric measurements described in literature; see, e.g.

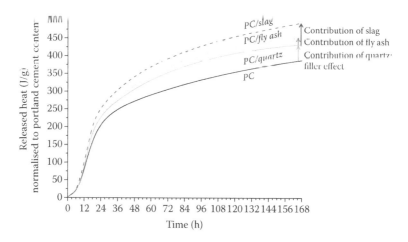

Figure 2.16 The effect of additions of different SCMs (substitution level 50%) on the hydration heat of a CEM I 52.5 R. The binders were hydrated at 20°C using a water–solid ratio of 0.50. The released heat is normalised to the amount of portland cement. (Unpublished calorimetric data of the cements used in Schöler, A. et al., *Cement and Concrete Composites*, 55, 374–382, 2015. With permission.)

Kocaba (2009), Kocaba et al. (2012), Lothenbach et al. (2011), Pane and Hansen (2005), Poppe and De Schutter (2005) and Scrivener et al. (2015). For slags, Kocaba (2009) extracted the heat attributed to their participation in the hydration reactions as described previously; see Figure 2.16. To convert heat to reaction degree, the enthalpy of slag reaction would be needed; however, reliable data are lacking in literature. Instead, the reaction degree of the slag was determined by image analysis using scanning electron microscopy (SEM)-BSE images of polished sections (see Chapter 8) and used for calibration, as shown in Figure 2.17. A drawback of the calorimetric methods for determining the degree of hydration of SCM is that long-term effects (due to the inaccuracy of calorimetric measurements lasting longer than approximately 14 days) and low-reactivity SCM such as many fly ashes (see, e.g. Figure 2.16) cannot be assessed. Regarding long-term effects, chemical shrinkage (see Chapter 3) may be used instead of calorimetry (Kocaba 2009; Kocaba et al. 2012).

Figure 2.18 shows an example of the use of calorimetry to measure the activity of a biomass fly ash in water without portland cement at 20°C. This approach is sometimes useful to isolate the comparably low fly ash heat of hydration. Figure 2.19 shows a similar example where three different samples of high-calcium fly ash are hydrating in a 'simulated portland cement environment' including 50% calcium hydroxide with 0.5 M NaOH at w/c of 1.0, thus using heat of fly ash hydration as a test method for fly ash activity in a well-defined chemical environment, excluding interference from portland cement clinker.

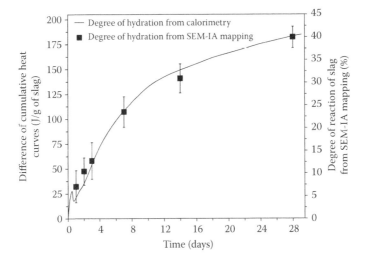

Figure 2.17 Calorimetric curves of portland cement–slag blends (60:40 by mass) cali-brated with SEM–IA. The left *y* axis shows the contribution of the slag to the heat of hydration; the right *y* axis, the hydration degree of the slag as determined by SEM-IA. The values after 28 days were used for calibra-tion. (Adapted from Kocaba, V., 'Development and evaluation of methods to follow microstructural development of cementitious systems including slags', PhD Thesis no. 4523, École Polytechnique Fédérale de Lausanne, Switzerland, 2009.)

2.4.7 Cement–admixture interactions

One of the main purposes of organic or inorganic admixtures is to con-trol the hydration kinetics of cementitious systems; see, e.g. Cheung et al. (2011) and Ramachandran (1996). Isothermal calorimetry is ideally suited to quantifying how additives and admixtures change the rate of hydration (Wadsö 2005). One case of special importance is the possibility of detecting incompatible combinations of binder and admixtures. Most admixtures adsorb on the surface of cement hydrates and therefore alter the rate and sometime also the path of cement hydration. Parameters such as admixture type, dose, amount, addition time relative to water, mixing intensity and temperature are important and can be systematically assessed using iso-thermal calorimetry. Most undesired interactions occur as a result of the mixture of binder and admixture becoming undersulfated with respect to the reactivity of the aluminate phases at any stage of the early cement hydra-tion. Isothermal calorimetry is an excellent technique for studying this as the early reactions – and hence the early heat production rate – are quite different when not enough sulfate is accessible to control the aluminate hydration (Sandberg and Roberts 2005). This is illustrated in Figure 2.20a. Without the admixture, the cement shows normal hydration behaviour. The admixture accelerates C_3A reaction, and an uncontrolled aluminate

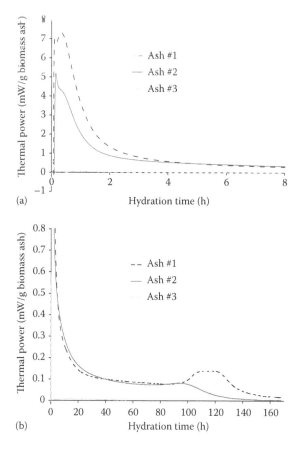

Figure 2.18 Biomass fly ash hydration power in water, comparing two reactive samples with one nonreactive sample (Ash #3): (a) 0–8 h, capturing the initial hydration; (b) 0–170 h capturing later reactivity. Note the different scales of the y axes to highlight the difference in hydration between the two reactive biomass ashes starting at approximately 100 h.

reaction occurs, which is also due to the presence of a low amount of soluble sulfate. In the case of the higher admixture dosage, the system even shows a flash set, and no significant strength-giving alite hydration occurs during the first 20 hours. When 0.5 mass% additional soluble sulfate is added, the early aluminate reaction is reduced, and less retardation of the alite hydration occurs (Figure 2.20b).

Figure 2.21 shows the heat flow development of cement pastes without and with 0.3% of superplasticiser in simultaneous and in delayed addition mode (Winnefeld 2008). Products based on polycarboxylate ether (PCE) and sulfonated naphthalene-formaldehyde polycondensate (SNF) were applied. Compared to the plain paste, the pastes with admixtures show a retardation of the setting and a delay of the main hydration peak. In

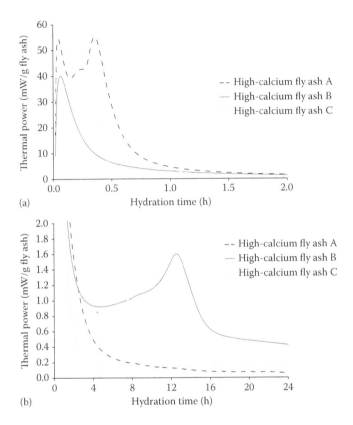

Figure 2.19 Three different samples of high-calcium fly ash hydrating in a simulated portland cement pore solution environment, with excess water, calcium hydroxide and 0.1 M NaOH. (a) Initial hydration 0–2 h after mixing. (b) Residual
hydration 0–24 h after mixing. Note that the scales of the y axes are different in a and b.

simultaneous addition mode, the SNF causes less retardation compared to
the PCE at the same dosage. In the case of the PCE, the delayed addition
mode causes almost no alteration to the shape of the heat flow curve when
compared to the direct addition mode. In the case of the SNF, however,
the delayed addition causes a much stronger retardation compared to the
simultaneous addition. The admixture adsorption was lower in the case of
the delayed addition compared to the direct addition; see also Uchikawa et
al. (1995). The stronger retardation in the delayed addition mode can be
explained by the fact that less admixtures is incorporated in early hydration products than in the case of direct addition; thus, more superplasticiser
is available to retard cement hydration (Sandberg and Roberts 2005).

Further examples of cement–admixture interaction studies are given, e.g.
by Cheung et al. (2011), Ramachandran (1996) and Sandberg and Roberts
(2005) and in ASTM C1679 (2014).

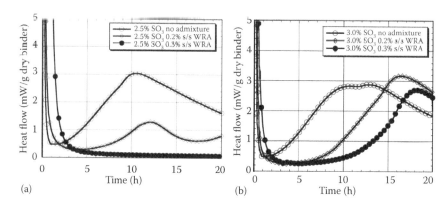

Figure 2.20 Heat flow calorimetry of a portland cement without and with a water reducing admixture (WRA) added that accelerates the aluminate reaction. The cement used meets the specifications of ASTM II and V portland cement. (a) The cement was used as it is (2.5 mass% SO_3); (b) 0.5 mass% soluble sulfate was added to the cement (total sulfate content 3.0 mass%). (Reprinted from Sandberg, P. J., and L. R. Roberts, *Journal of ASTM International*, 2(6), 1–14, 2005. With permission.)

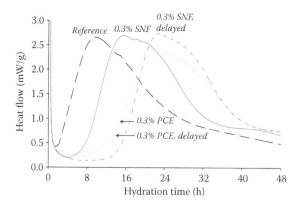

Figure 2.21 Heat flow of a CEM I 42.5 hydrated at 20°C using a w/c of 0.35. Superplasticisers based on PCE and SNF were added at a dosage of 0.3 mass% (dry mass) referred to the cement. The admixtures were either dissolved in the mixing water or added dissolved in 10% of the mixing water 30 s after starting the mixing process (delayed addition). (Adapted from Winnefeld, F., *ZKG International*, 61(11), 68–77, 2008.)

Note that tests with dispersing or retarding admixtures using cement paste are generally not recommended to solve issues with concrete unless special care is taken to properly shear the cement grains in a concrete-like manner. This is because the adsorption of the admixture on cement hydrates and the resulting retardation are highly dependent on the mixing intensity.

2.4.8 Dry mix mortars

Dry mix mortars often exhibit a quite complex mix composition, especially if they are rapid setting and/or rapid hardening. In the latter case, they generally contain binary or ternary binders based on calcium aluminate or calcium sulfoaluminate cements in blends with calcium sulfate without and with portland cement. Isothermal calorimetry is an efficient method to use for optimising mix designs of such mortars with respect to the hydration kinetics. As only small cement mortar or paste samples are used, the influence of the binder composition as well as of different combinations of accelerators, retarders, water reducers, plasticisers, etc. can quickly be tested. Two examples of how the amount of calcium sulfate addition is able to influence hydration kinetics are shown for blends of calcium aluminate cement with hemihydrate (Figure 2.22) and ternary binders based on portland cement, calcium sulfoaluminate cement and anhydrite (Figure 2.23).

Figure 2.24 shows that some dry mix mortars, in this case a self-levelling compound based on a ternary binder composed of portland cement, calcium aluminate cement and calcium sulfate, may show very complex heat flow patterns, which cannot be interpreted without further analysis, e.g. by XRD or TGA. Figure 2.24 highlights as well the retarding influence of casein, which is often used as plasticiser in self-levelling compounds, on hydration. More examples of this are given, e.g. by Anderberg and Wadsö (2009).

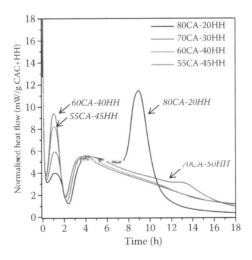

Figure 2.22 Hydration kinetics of various blends between calcium aluminate cement and hemihydrate. The numbers given refer to the ratio between calcium aluminate (CA) and hemihydrate (HH). The samples were hydrated at 20°C using a water–solid ratio of 0.40. (From Bizzozero, J., 'Hydration and dimensional stability of calcium aluminate cement based systems', PhD Thesis no. 6336, École Polytechnique Fédérale de Lausanne, Switzerland, 2014.)

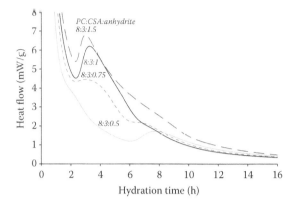

Figure 2.23 Influence of the amount of anhydrite added on the hydration kinetics of ternary binders based on portland cement (PC), calcium sulfoaluminate cement (CSA) and anhydrite hydrated at 20°C and at a water–binder ratio of 0.50. Citric acid (0.27 mass% referred to binder) was added as set retarder. (Adapted from Pelletier, L. et al., *Cement and Concrete Composites*, 32(7), 497–507, 2010.)

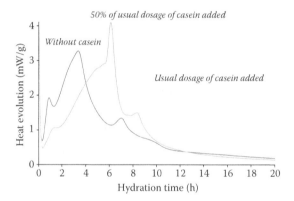

Figure 2.24 Hydration heat flow of three different self-levelling compounds based on ternary binders portland cement–calcium aluminate cement–calcium sulfate (portland cement dominated) which differ only in the amount of the flow agent casein. A formulation without casein is compared to two formulations with half the usual and the usual dosage of casein. (Adapted from Kighelman, J., 'Hydration and structure development of ternary binder system as used in self-levelling compounds', PhD Thesis no. 37777, École Polytechnique Fédérale de Lausanne, Switzerland, 2007.)

2.5 CONCLUSIONS AND OUTLOOK

Calorimetric techniques have been used for 100 years in the cement field. Today the heat-of-solution method is being phased out, while isothermal and (semi)adiabatic calorimetry will continue to be used both for heat of hydration measurements and as a general monitoring of concrete hydration rate.

Table 2.2 summarises the most important issues related to the different methods. Isothermal calorimetry has seen a recent increase in use as an analytical instrument as more user-friendly instruments have come to the market.

Table 2.2 Comparison of the different calorimetric methods

Method	Advantages	Drawbacks
Isothermal calorimetry	Very repeatable at early age due to testing at near-constant temperature Easy to test at different temperatures Easy to measure the effect of SCM and admixtures Popular but relatively new standards in ASTM; CEN effort in progress Well established for providing binder and admixture data for prediction of temperature and risk of thermal cracking in field structures Multicell instruments very suitable for large matrix testing, for example, binder and admixture selection prior to concrete testing Can be combined with other measurement techniques	Low signal after several days requires costly instrumentation Caution required if testing samples that may generate erroneous data due to temperature mismatch between sample and calorimeter; typically an issue during the first 1–2 h when testing concrete or even paste with a binder with a high heat of hydration (especially when the heat is released rapidly) Small samples extracted from larger sample preparation batches may result in nonuniform binder content, in which case the results are only semiquantitative, unless the binder content is directly measured and accounted for Small paste samples may give erroneous kinetic effects of admixtures when compared to concrete
Semiadiabatic calorimetry	Historically well-established standards cement (CEN Langavant method) Simple and cheap equipment for testing in industrial lab (usually not calibrated) Well established for providing concrete data for prediction of temperature and risk of thermal cracking in field structures	Indirectly calculated heat from temperature rise and estimated heat loss Extensive calibration procedure when heat of hydration should be determined Rather poor repeatability Temperature rise alters hydration kinetics Temperature-induced effects may not be relevant for field applications
Heat-of-solution method	Historically well established in both ASTM and CEN standards Does not tie up instrument for the time of hydration	Requires use of hazardous chemical (hydrofluoric acid, nitric acid) Rather poor repeatability Discontinuous measurement

Table 2.3 General considerations about best practice for (isothermal) calorimetry

Topic	General considerations
Instrument	• Use an appropriate instrument (with respect to sensitivity, sample quantity, temperature range, stability, etc.). • Check the temperature of the thermostat by an independent temperature measurement. • Measure and, if needed, adjust for the error caused by water vapour leakage from sample vials if performing tests with very low signal, such as when testing normal cementitious mixes beyond 48 hour. • Stable climate is needed for precise long-term measurements. • If you measure at low temperature (e.g. 5°C–10°C), place your instrument in a climate room of similar temperature. Otherwise condensation will occur inside your machine, spoiling not only your measurements but also the instrument.
Calibration	• Calibration coefficients and baselines have very stable values when the same setup is used. • A calibration interval of, e.g. 3 months is appropriate. • Recalibrate when you change the experimental conditions (type of vials, different temperature, highly different sample quantity, etc.). • Baselines should be measured for at least 10 hour; over the weekend is better.
Mixing and charging	• Ensure proper mixing. The mixing process influences kinetics. • Take care that your sample is representative when you take material from a larger mix. • Stiff mixes are not suitable for internal mixing. • Charge your samples as rapidly as possible. • Charging introduces extra heat (temperature difference between calorimeter and climate room, mixing energy, early hydration reactions). This can be reduced by thermostating the raw materials in the calorimeter. • One smaller sample may be charged for capturing the initial reactivity, and one larger sample for capturing later age reactivity. • Do not use too large samples – your conditions might not be isothermal anymore and an acceleration of hydration kinetics might occur (could be checked by temperature measurements).
Measurement	• Use an appropriate reference with the same (±10%) heat capacity as your sample. Use, e.g. quartz sand or water. The use of an old cement paste is not recommended, as it still releases heat. • For long-term measurements (7–14 days) a stable baseline, an appropriate reference, a suitable sample quantity and a stable climate room are needed.
Evaluation	• Refer your measured thermal power to the mass of cement, paste, mortar, etc. • For determination of cumulative heat in the case of external mixing, start the integration at a reasonable point in time (e.g. 30 min) after charging. • For very rapid processes apply the Tian correction. The time constant of the calorimeter in the experimental setup used (type of vial, heat capacity of the sample) needs to be determined.

The calorimetric curve upon cement hydration, which is generally continuously monitored, provides valuable kinetic information. Thus, the hydration times can be identified when further discontinuous hydration experiments, e.g. using XRD or TGA, can be carried out. There have also been interesting combinations of isothermal calorimetry and other measurement techniques published (mainly in fields other than cement science), such as the measurement of pressure, ion concentration, pH, relative humidity, rheology and chemical shrinkage (Champenois et al. 2013; Johansson and Wadsö 1999; Lura et al. 2010; Minard et al. 2007; Wadsö and Anderberg 2002). Recent work focused also on the calculation of heat flow curves from in situ hydration experiments done by quantitative XRD analysis (Bizzozero 2014; Hesse et al. 2011; Jansen et al. 2012a,b).

Table 2.3 gives the important points to consider when carrying out calorimetric measurements. The information given refers to isothermal calorimetry, but the considerations are also mainly valid for semiadiabatic calorimetry. For reporting experimental conditions in journal papers and other kinds of reports, Table 2.4 can be consulted.

Table 2.4 General considerations about reporting the experimental conditions of (isothermal) calorimetry in publications

Topic	What should be reported
Mix design	• Give the exact mix composition (see Table 1.4). • Did you use paste or mortar?
Mixing	• Was external or internal mixing applied? • Which type of mixer was used and what were the operation conditions?
Measurement conditions	• Which instrument did you use? If it is not a commercial one, describe it as accurately as possible (design, reference samples, calibration procedure, etc.). • The reader will assume that calibration was done according to the state of the art and that appropriate reference samples were chosen (see Table 2.3). It is recommended to briefly describe how calibration was made. • What was the temperature? • At what time after mixing were the samples charged? • Which sample quantities were used?
Evaluation	• Specify where you refer heat flow and heat of hydration to – is it referred to cement, binder, total solid, total paste or mortar? • When was integration of the heat flow curves started to determine heat of hydration? • Mention if you applied Tian correction. The time constant of the calorimeter also needs to be reported in this case.

ACKNOWLEDGEMENTS

Special thanks to Elise Berodier, Julian Bizzozero, Axel Schöler and Karen Scrivener for providing calorimetric data. Marta Palacios, Karen Scrivener and Ruben Snellings are acknowledged for helpful discussions.

REFERENCES

Anderberg, A., and L. Wadsö (2009). 'Using a standard mix design to study properties of a flooring compound'. *Nordic Concrete Research* **40**: 21–32.
ASTM C186 (2013). Standard test method for heat of hydration of hydraulic cement.
ASTM C511 (2013). Standard specification for mixing rooms, moist cabinets, moist
   rooms, and water storage tanks used in the testing of hydraulic cements and
   concretes.
ASTM C1074 (2011). Standard practice for estimating concrete strength by the
   maturity method.
ASTM C1679 (2014). Standard practice for measuring hydration kinetics of hydraulic cementitious mixtures using isothermal calorimetry.
ASTM C1702 (2014). Standard test method for measurement of heat of hydration
   of hydraulic cementitious materials using isothermal conduction calorimetry.
Balbas, M., J. I. Diaz, F. Gascon and A. Varade (1991). 'The systematic uncertainty in
   the calibration of the Langavant calorimeter'. *Review of Scientific Instruments*
   62(11): 2795–2800.
Bensted, J. (1987). 'Some applications of conduction calorimetry to cement hydration'. *Advances in Cement Research* **1**(1): 35–44.
Berodier, E., and K. Scrivener (2014). 'Understanding the filler effect on the nucleation and
   growth of C-S-H'. *Journal of the American Ceramic Society* **97**(12): 3764–3773.
Bizzozero, J. (2014). 'Hydration and dimensional stability of calcium aluminate
   cement based systems'. PhD Thesis no. 6336, École Polytechnique Fédérale de
   Lausanne, Switzerland.
Carino, N. J., and H. S. Lew (2001). 'The maturity method: From theory to application'. *Proceedings of the 2001 Structures Congress & Exposition*, 21–23 May
   2001, Washington, DC, U.S.: 1–19.
Champenois, J. B., C. Cau dit Coumes, A. Poulesquen, P. Le Bescop and D. Damidot
   (2013). 'Beneficial use of a cell coupling rheometry, conductimetry, and calorimetry to investigate the early age hydration of calcium sulfoaluminate
   cement'. *Rheologica Acta* **52**(2): 177–187.
Chen, H., M. Wyrzykowski, K. Scrivener and P. Lura (2013). 'Prediction of self-
   desiccation in low water-to-cement ratio pastes based on pore structure evolution'. *Cement and Concrete Research* **49**: 38–47.
Cheung, J., A. Jeknavorian, L. Roberts and D. Silva (2011). 'Impact of admixtures
   on the hydration kinetics of Portland cement'. *Cement and Concrete Research*
   41(12): 1289–1309.
Copeland, L. E., D. L. Kantro and G. Verbeck (1960). 'Chemistry of hydration of
   Portland cement'. *Proceedings of the 4th International Symposium on the
   Chemistry of Cement*, Washington, DC, U.S.: Monograph 43, Vol. I, Session
   N, Paper IV-43: 429–465.

Danielson, U. H. (1962). 'Heat of hydration of cement as affected by water–cement ratio'. *Proceedings of the 4th International Symposium on the Chemistry of Cement*, Washington DC, U.S., Paper IV-S7: 519–526.

Deschner, F., F. Winnefeld, B. Lothenbach, S. Seufert, P. Schwesig, S. Dittrich, F. Goetz-Neunhoeffer and J. Neubauer (2012). 'Hydration of Portland cement with high replacement by siliceous fly ash'. *Cement and Concrete Research* 42(10): 1389–1400.

Dunn, J., K. Oliver, G. Nguyen and I. Sills (1987). 'The quantitative determination of hydrated calcium sulfates in cement by DSC'. *Thermochimica Acta* 121: 181–191.

EN 196-1 (2005). Method of testing cement – Part 1: Determination of strength.

EN 196-8 (2010). Method of testing cement – Part 8: Heat of hydration – Solution method.

EN 196-9 (2010). Method of testing cement – Part 8: Heat of hydration – Semi-adiabatic method.

Evju, C. (2003). 'Initial hydration of cementitious systems using a simple isothermal calorimeter and dynamic correction'. *Journal of Thermal Analysis and Calorimetry* 71(3): 829–840.

Gascon, F., and A. Varade (1993). 'The systematic uncertainty of the heat of hydration of cement'. *Review of Scientific Instruments* 64(8): 2353–2360.

Gibbon, G. J., Y. Ballim and G. R. H. Grieve (1997). 'A low-cost, computer-controlled adiabatic calorimeter for determining the heat of hydration of concrete'. *Journal of Testing and Evaluation* 25(2): 261–266.

Hesse, C., F. Goetz-Neunhoeffer and J. Neubauer (2011). 'A new approach in quantitative in-situ XRD of cement pastes: Correlation of heat flow curves with early hydration reactions'. *Cement and Concrete Research* 41(1): 123–128.

Jansen, D., F. Goetz-Neunhoeffer, B. Lothenbach and J. Neubauer (2012a). 'The early hydration of ordinary Portland cement (OPC): An approach comparing measured heat flow with calculated heat flow from QXRD'. *Cement and Concrete Research* 42(1): 134–138.

Jansen, D., J. Neubauer, F. Goetz-Neunhoeffer, R. Haerzschel and W. D. Hergeth (2012b). 'Change in reaction kinetics of a Portland cement caused by a super-plasticizer – Calculation of heat flow curves from XRD data'. *Cement and Concrete Research* 42(2): 327–332.

Johansson, P., and I. Wadsö (1999). 'An isothermal microcalorimetric titration/perfusion vessel equipped with electrodes and spectrophotometer'. *Thermochimica Acta* 342(1–2): 19–29.

Juilland, P., A. Kumar, E. Gallucci, R. J. Flatt and K. L. Scrivener (2012). 'Effect of mixing on the early hydration of alite and OPC systems'. *Cement and Concrete Research* 42(9): 1175–1188.

Justs, J., M. Wyrzykowski, F. Winnefeld, D. Bajare and P. Lura (2014). 'Influence of superabsorbent polymers on hydration of cement pastes with low water-to-binder ratio'. *Journal of Thermal Analysis and Calorimetry* 115(1): 425–432.

Kada-Benameur, H., E. Wirquin and B. Duthoit (2000). 'Determination of apparent activation energy of concrete by isothermal calorimetry'. *Cement and Concrete Research* 30(2): 301–305.

Kaufmann, J. P. (2004). 'Experimental identification of ice formation in small concrete pores'. *Cement and Concrete Research* 34(8): 1421–1427.

Kighelman, J. (2007). 'Hydration and structure development of ternary binder system as used in self-levelling compounds'. PhD Thesis no. 37777, École Polytechnique Fédérale de Lausanne, Switzerland.

Killoh, D. C. (1988). 'A comparison of conduction calorimeter and heat of solution methods for measurement of the heat of hydration of cement'. *Advances in Cement Research* **1**(3): 180–186.

Kocaba, V. (2009). 'Development and evaluation of methods to follow microstructural development of cementitious systems including slags'. PhD Thesis no. 4523, École Polytechnique Fédérale de Lausanne, Switzerland.

Kocaba, V., E. Gallucci and K. L. Scrivener (2012). 'Methods for determination of degree of reaction of slag in blended cement pastes'. *Cement and Concrete Research* **42**(3): 511–525.

Lerch, W. (1946). 'The influence of gypsum on the hydration and properties of Portland cement pastes'. *Proceedings of the American Society for Testing Materials* **46**: 1252.

Lipus, K., and S. Baetzner (2008). 'Determination of the heat of hydration of cement by isothermal conduction calorimetry'. *Cement International* **6**(4): 92–103.

Lothenbach, B., C. Alder, F. Winnefeld and P. Lunk (2005). 'Einfluss der Temperatur und Lagerungsbedingungen auf die Festigkeitsentwicklung von Mörteln und Betonen (Influence of temperature and curing conditions on strength development of mortars and concretes)'. *Beton* **55**(12): 604–609.

Lothenbach, B., K. Scrivener and R. D. Hooton (2011). 'Supplementary cementitious materials'. *Cement and Concrete Research* **41**(12): 1244–1256.

Lura, P., F. Winnefeld and X. Fang (2015). 'A simple method for determining the total amount of physically and chemically bound water of different cements'. *Journal of Thermal Analysis and Calorimetry*: In preparation.

Lura, P., F. Winnefeld and S. Klemm (2010). 'Simultaneous measurements of heat of hydration and chemical shrinkage on hardening cement pastes'. *Journal of Thermal Analysis and Calorimetry* **101**(3): 925–932.

Mengede, M., F. Moro and F. Winnefeld (2006). 'Prüfverfahren zur Bestimmung der Hydratationswärme – Erfahrungen in der Schweiz' (Testing methods for determination of heat of hydration – Experiences in Switzerland). *16. Internationale Baustofftagung (ibausil)*, Weimar, Germany. **1**: 527–534.

Minard, H., S. Garrault, L. Regnaud and A. Nonat (2007). 'Mechanisms and parameters controlling the tricalcium aluminate reactivity in the presence of gypsum'. *Cement and Concrete Research* **37**(10): 1418–1426.

NT BUILD 505 (2003). Measurement of heat of hydration of cement with heat conduction calorimetry.

Pane, I., and W. Hansen (2005). 'Investigation of blended cement hydration by isothermal calorimetry and thermal analysis'. *Cement and Concrete Research* **35**(6): 1155–1164.

Pelletier, L., F. Winnefeld and B. Lothenbach (2010). 'The ternary system Portland cement–calcium sulphoaluminate clinker–anhydrite: Hydration mechanism and mortar properties'. *Cement and Concrete Composites* **32**(7): 497–507.

Poole, J. L., K. A. Riding, K. J. Folliard, M. G. G. Juenger and A. K. Schindler (2007). 'Methods for calculating activation energy for portland cement'. *ACI Materials Journal* **104**(1): 86–94.

Poppe, A.-M., and G. De Schutter (2005). 'Cement hydration in the presence of high filler contents'. *Cement and Concrete Research* 35(12): 2290–2299.

Ramachandran, V. S. (1996). *Concrete Admixtures Handbook*. Noyes Publication, Park Ridge, New Jersey, U.S.

Ridi, F., L. Dei, E. Fratini, S. H. Chen and P. Baglioni (2003). 'Hydration kinetics of tri-calcium silicate in the presence of superplasticizers'. *Journal of Physical Chemistry B* 107(4): 1056–1061.

Riding, K. A., J. L. Poole, K. J. Folliard, M. C. G. Juenger and A. K. Schindler (2011). 'New model for estimating apparent activation energy of cementitious systems'. *ACI Materials Journal* 108(5): 550–557.

Riding, K. A., J. L. Poole, K. J. Folliard, M. C. G. Juenger and A. K. Schindler (2012). 'Modeling hydration of cementitious systems'. *ACI Materials Journal* 109(2): 225–234.

Sandberg, P. J., and L. R. Roberts (2005). 'Cement–admixture interactions related to aluminate control'. *Journal of ASTM International* 2(6): 1–14.

Schöler, A., B. Lothenbach, F. Winnefeld and M. Zajac (2015). 'Hydration of quaternary Portland cement blends containing blast-furnace slag, siliceous fly ash and limestone powder'. *Cement and Concrete Composites* 55: 374–382.

Scrivener, K. L., B. Lothenbach, N. De Belie, E. Gruyaert, J. Skibsted, R. Snellings and A. Vollpracht (2015). 'TC 238-SCM: Hydration and microstructure of concrete with SCMs – State of the art on methods to determine degree of reaction of SCMs'. *Materials and Structures* 48(4): 835–862.

Siddiqui, S. and K. A. Riding (2012). 'Effect of calculation methods on cement paste and mortar apparent activation energy'. *Advances in Civil Engineering Materials* 1(1): 1–19.

Tang, F. J. (1992). 'Optimization of sulfate form and content'. PCA Research and Development Bulletin RD105T, Portland Cement Association, Skokie, Illinois, U.S.

Tang, F. J., and E. M. Gartner (1987). 'Influence of sulfate source on Portland cement hydration'. PCE R&D Serial No. 1876, Construction Technology Laboratories, Inc., Skokie, Illinois, U.S.

Taylor, H. F. W. (1997). *Cement Chemistry*. Thomas Telford Publishing, London, U.K.

Uchikawa, H., D. Sawaki and S. Hanehara (1995). 'Influence of kind and added timing of organic admixture on the composition, structure and property of fresh cement paste'. *Cement and Concrete Research* 25(2): 353–364.

Wadsö, L. (2003). 'An experimental comparison between Isothermal calorimetry, semi-adiabatic calorimetry and solution calorimetry for the study of cement hydration'. Nordtest Report TR 522, available at http://www.nordtest.info/index .php/technical-reports/category/building/2.html.

Wadsö, L. (2005). 'Applications of an eight-channel isothermal conduction calorimeter for cement hydration studies'. *Cement International* 5(3): 94–101.

Wadsö, L. (2010). 'Operational issues in isothermal calorimetry'. *Cement and Concrete Research* 40(7): 1129–1137.

Wadsö, L., and A. Anderberg (2002). 'A method for simultaneous measurements of heat of hydration and relative humidity'. *Self-Desiccation and Its Importance in Concrete Technology, Proceedings of the Third International Research Seminar in Lund*, 14–15 June 2002: 103–112.

Wadsö, I., and R. N. Goldberg (2001). 'Standards in isothermal microcalorimetry (IUPAC technical report)'. *Pure and Applied Chemistry* **73**(10): 1625–1639.

Winnefeld, F. (2008). 'Influence of cement ageing and addition time on the performance of superplasticizers'. *ZKG International* **61**(11): 68–77.

Winnefeld, F., and B. Lothenbach (2010). 'Hydration of calcium sulfoaluminate cements – Experimental findings and thermodynamic modelling'. *Cement and Concrete Research* **40**(8): 1239–1247.

Chapter 3

Characterisation of development of cement hydration using chemical shrinkage

Mette Geiker

CONTENTS

3.1 INTRODUCTION

Characterisation of cement hydration is essential for both researchers and practitioners. Measurement of chemical shrinkage provides information on the overall development of hydration, assuming that the products are independent of the degree of hydration. Determination of the degree of hydration requires knowledge of the ultimate value of chemical shrinkage, which depends on both binder composition and temperature. Measurement of chemical shrinkage is undertaken on a saturated sample.

Cementitious materials exhibit what is known as chemical shrinkage during hydration because the products take up less space compared to the reactants. Chemical shrinkage is illustrated in Figure 3.1. Based on the reaction scheme, the volume change associated with hydration of C_3S is calculated from the molar weights and the densities of the reactants and products. A figure often used to illustrate the development of chemical shrinkage during hydration of cement pastes is given in Figure 3.2. The figure is based on work by Powers and coworkers, e.g. Powers (1935), and shows the volumetric proportions of the constituents – cement, capillary water, gel water, solid hydration products – and chemical shrinkage, as a function of the degree of hydration. Before hydration starts, the volume of capillary water is equal to the amount of initial mixing water.

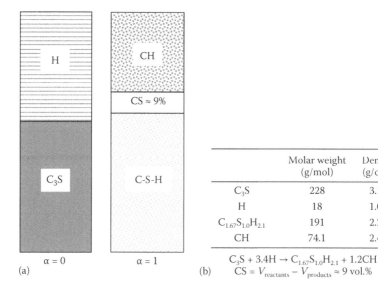

	Molar weight (g/mol)	Density (g/cm³)
C_3S	228	3.12
H	18	1.00
$C_{1.67}S_{1.0}H_{2.1}$	191	2.24
CH	74.1	2.44

$$C_3S + 3.4H \rightarrow C_{1.67}S_{1.0}H_{2.1} + 1.2CH$$
$$CS = V_{reactants} - V_{products} \approx 9 \text{ vol.}\%$$

(a) $\alpha = 0$ (b) $\alpha = 1$

Figure 3.1 (a) Change in volume (chemical shrinkage, CS) due to hydration of C_3S and (b) molar weights and densities from CEMDATA07 database (Lothenbach et al. 2008; Matschei et al. 2007). The dimensions are calculated assuming a C-S-H composition of $C_{1.67}S_{1.0}H_{2.1}$. α = degree of hydration. Cement chemistry notation used. (Courtesy of Axel Schöler.)

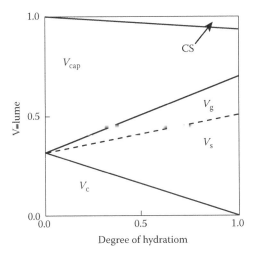

Figure 3.2 Principle of the development of the volumetric proportions of the constituents of the cement paste during hydration. CS: chemical shrinkage. Volumes: V_c is for cement; V_{cap}, capillary water; V_g, gel water; V_s, solid hydration products. (Based on Powers, T. C., *Industrial and Engineering Chemistry*, 27(7), 790–794, 1935.)

The overall change in molar volumes can be measured to follow the progress of hydration. A principle of measuring is illustrated in Figure 3.3; the amount of water sucked into a sample of cement paste is monitored. Before setting, when the sample is in the plastic state, chemical shrinkage might be measured as the volume decrease of the bulk sample (assuming no constraints). However, as soon as setting has taken place, chemical shrinkage causes some

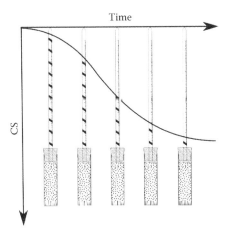

Figure 3.3 Principle of measurement of chemical shrinkage (CS) as volume decrease of a saturated cement paste to follow the overall development of hydration as a function of time.

of the pores in the solid matrix to become either filled with water from the exterior or void. Thus, after setting, chemical shrinkage can be measured only while water is being sucked into the sample, refilling the emptied pores. Therefore, measurement of chemical shrinkage is undertaken on a saturated sample with a limited sample size to avoid the emptying of water filled pores.

Measurement of chemical shrinkage is – like measurement of the heat of hydration and the nonevaporable water content – an indirect method of obtaining information on the degree of hydration of portland cement and other cementitious systems. All three methods provide data which are reasonable measures of hydration; see, e.g. Parrott et al. (1990). The advantages of chemical shrinkage measurements are that measurements may be undertaken continuously and over relatively long periods; the equipment is relatively inexpensive and making the measurements requires only limited training. The disadvantages are that only saturated samples can be measured; there are limitations on minimum water-to-cement ratio (w/c) and maximum sample size; the initial reaction is not measured; there is a risk of alkali silica reaction in glassware and the correlation between chemical shrinkage and degree of hydration is not fully known.

A sealed sample will undergo self-desiccation. Depending on the amount of chemical shrinkage, the pore structure and the pore liquid composition self-desiccation will cause change and autogenous (bulk) deformation. In the plastic state (before setting) the chemical shrinkage and the autogenous shrinkage of a sealed and unrestrained paste sample are equal. After setting, only part of the chemical shrinkage is reflected in autogenous shrinkage; the remaining part of the chemical shrinkage in a nonsaturated sample results in empty pores. Drying of a sample will result in drying shrinkage. Both autogenous shrinkage and drying shrinkage cause deformation of the bulk sample; these phenomena should not be confused with chemical shrinkage.

Comparing the data obtained by different methods, it should be kept in mind that the moisture state of the hydrating material (saturated or self-desiccated) affects the development of hydration. Figure 3.22 illustrates possible differences between the developments of hydration of sealed and saturated (open) samples.

Other names have been used for chemical shrinkage in English: *water absorption* (Fulton 1962; Powers 1935; Swayze 1942); *contraction* (Hemeon 1935); *volume contraction* (Mills 1962); *autogenous volume change* (Swayze 1942) and *pycnometry* (Gartner and Pham 2005).

3.2 DETERMINATION OF TOTAL CHEMICAL SHRINKAGE AND DEGREE OF HYDRATION

In order to determine hydration degree over time, the chemical shrinkage at full hydration must be determined. The chemical shrinkage of a fully hydrated cement can be estimated based on assumptions of type of reaction products and information on the molar volumes and the densities of the reactants and products; see, e.g. Bentz et al. (2005); Hansen (1995); Justnes et al. (1998) and Mounanga et al. (2004). Apparently, sufficient data are available only for room-temperature curing of portland cement; and only limited data exist for supplementary cementitious materials (Bentz 2007; Lura et al. 2003). Chemical shrinkage of partially hydrated cement paste can be calculated from the degree of hydration of the phases. Combined modelling of reaction kinetics and phase assemblage provides a means of estimating the chemical shrinkage of a given system at a given time (Lothenbach 2010; Winnefeld and Lothenbach 2010); see Figure 3.4.

The chemical shrinkage of a fully hydrated cement can also be estimated based on assumptions of hydration models. Considering the combined effect of hydration kinetics and the broad particle size distribution of portland cement, Knudsen (1980; 1984) proposed the dispersion model to be used for the quantification of the kinetics of cement hydration and the ultimate value of the property investigated. The dispersion model is given by Knudsen (1984) as follows:

$$\frac{CS(t)}{CS_{max}} = \frac{(t-t_0)^n}{t_1^n + (t-t_0)^n}, \qquad (3.1)$$

where
$CS(t)$ = chemical shrinkage at time t (in millilitres/gram cement)
CS_{max} = ultimate chemical shrinkage (in millilitres/gram cement)
t_0 = duration of the induction period (in hours)
t_1 = reaction time $(t - t_0)$ to reach $CS/CS_{max} = 0.5$ (in hours)
n = constant describing the kinetics; $n = 1$ for linear kinetics and $n = \frac{1}{2}$
    for parabolic kinetics

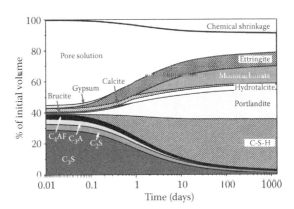

Figure 3.4 Thermodynamic modelling of phase development in a portland cement paste hydrated at 20°C. Cement composition and modelling are detailed in (Lothenbach et al. 2008). (Courtesy of Barbara Lothenbach.)

Because of the difference between the shapes of hydration curves and of the dispersion model, the first part of the hydration curve cannot be fitted by the model; this is illustrated in Figure 3.5. For portland cement at 20°C, the part which cannot be fitted corresponds to the degree of hydration below 10%–15%. In using the dispersion model, two different types of kinetics were observed, linear and parabolic (Geiker 1983; Knudsen 1984). To allow for differentiation, data for up to a minimum of 80% degree of hydration need to be available. For some systems also, combinations of linear and parabolic kinetics were observed; for more information see Knudsen (1984).

Fu et al. (2012) suggested the use of a hyperbolic-like function proposed by Xiao et al. (2009) and measurement for a minimum of 14 days using 3 mm thick samples to predict the long-term chemical shrinkage:

$$CS(t) = \frac{CS_{max} x t^{a}}{t^{a} + b} \quad \text{or} \quad \frac{CS(t)}{CS_{max}} = \frac{t^{a}}{t^{a} + b}, \tag{3.2}$$

where

    $CS(t)$ = chemical shrinkage at time t (in millilitres/gram cement)
    CS_{max} = long-term chemical shrinkage (in millilitres/gram cement)
    a and b = hydration constants related to the cementitious materials'
        properties

It can be observed that the hyperbolic-like function proposed by Xiao et al. (2009) to a large extent resamples the dispersion model proposed by Knudsen (1980, 1984).

Studying the impact of temperature on early-stage hydration of oil-well cements and assuming no temperature impact on total heat development, Pang et al. (2013) found that the total chemical shrinkage decreases linearly with increasing temperature at a rate of 0.600% ± 0.046% per

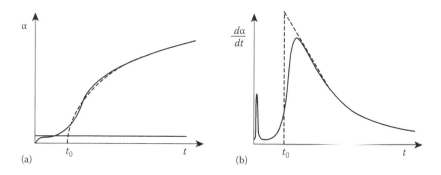

Figure 3.5 Hydration curve compared with the dispersion model (Equation 3.1): (a) integral curve; (b) first derivative. α = degree of hydration. Solid line: measured chemical shrinkage; dashed line: modelled chemical shrinkage using Equation 3.1.

degree Celsius. This is in the same range as earlier observed by Geiker and Knudsen (1982) for grey and white portland cements.

3.3 SETUPS FOR MEASURING CHEMICAL SHRINKAGE

Swayze 1942 described three principles for measuring chemical shrinkage, which are called dilatometry, gravimetry and pycnometry:

- Dilatometry: measurement of the change in water (or oil) level in an initially filled tube; no exchange of material with the environment takes place (see Figure 3.6)
- Gravimetry: measurement of the change in buoyancy of a sample of cement paste plus extra water; no exchange of material with the environment takes place (see Figures 3.7 and 3.8)
- Pycnometry: measurement of the change in weight of a sample of cement paste plus extra water (or oil); additional liquid is added to a constant volume before measurement

As far as the author knows, Le Chatelier (1900) was the first to report measurements of chemical shrinkage in cement paste; he used dilatometry. This measuring principle is used in the ASTM standard for testing early-age hydration, ASTM C1608 (2012), 'Test method for chemical shrinkage

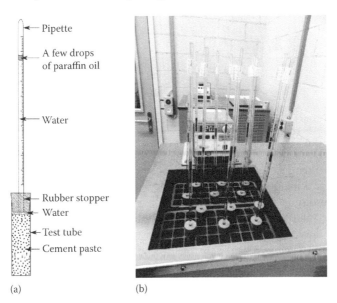

Figure 3.6 Devices for measurement of chemical shrinkage using dilatometry: (a) by Geiker (1983) and (b) Courtesy of Axel Schöler.

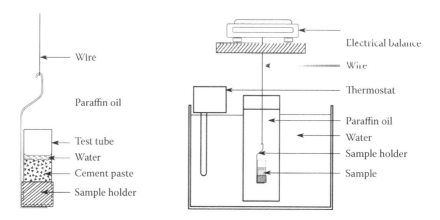

Figure 3.7 Devices for measurement of chemical shrinkage using gravimetry (Geiker 1983).

Figure 3.8 Setup for automatic measurement of chemical shrinkage using gravimetry.

of hydraulic cement paste'. Dilatometry and gravimetry give comparable results (Geiker 1983).

Knudsen and Geiker (1985) proposed a setup for automatic measurement of multiple samples using gravimetry. The 'gravimeter' was developed as a pilot test but was never put into production. Instead Knudsen and Lolk developed an automatic apparatus, the 'konometer', which measures the displacement of the pistons in special test syringes; see Figure 3.9. Increased focus was given to the possible use of chemical shrinkage for the quality control of cementitious materials and aggregates (Knudsen 1985), and the so-called konometer is in Denmark used for testing the alkali-aggregate reactivity in only 20 hours by means of the chemical shrinkage method (Danish Technological Institute 1989). Comparing test methods for

Figure 3.9 (Well-used) Konometer for automatic measurement of chemical shrinkage. Here it is used for testing the alkali reactivity of sand. (Courtesy of Søren Lolk.)

early-age behaviour of cementitious materials Bullard et al. (2006) applied automatic gravimetry and rather large samples of approximately 200 g paste but with limited height to facilitate correct measurements (see Section 3.4).

An automatic setup for measuring chemical shrinkage using dilatometry and a webcam has been developed at the École Polytechnique Fédérale de Lausanne (EPFL) (Costoya 2008) based on the protocol by Geiker (1983); see Figure 3.10. A modification of the EPFL setup is used at the Oregon State University (Fu et al. 2012).

Selected protocols for measuring chemical shrinkage using dilatometry are compared in Table 3.1. In Section 3.9 guidelines for analysis using manual dilatometry and automatic gravimetry are given based on protocols of Geiker (1983).

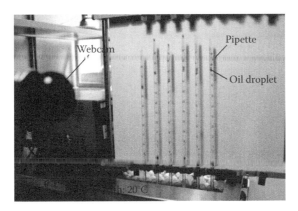

Figure 3.10 Setup for automatic measurement of chemical shrinkage using dilatometry (Berodier 2014b).

Table 3.1 Comparison of selected protocols for measuring chemical shrinkage of cementitious materials by dilatometry

	ASTM C1608 (2012)	Empa (Schöler et al. 2014)	EPFL (Berodier 2014b; after Costoya 2008)
Temperature control	Water bath, 23.0 ± 0.5°C	Water bath, 20°C	Water bath, 20°C
	Cover water with floating balls or lid	Refill water bath when needed	
	Temperature-controlled room (23 ± 2°C)		
Test vials	Glass, diameter 22 mm, height 55 mm	Glass, diameter 20 mm, height 60 mm	Plastic, diameter 20 mm, height 50 mm
Graduated capillary tube/pipette	Typical capacity 1.0 mL, graduation ≤ 0.01 mL	Capacity 1 mL, grading 0.1 mL	Capacity 1 mL
Stoppers	Rubber; pipette is fixed in stopper by using suitable adhesive	Rubber; pipette is inserted in the stopper	
Water	Deaerated (boiled and cooled sealed) clean water	Decarbonated (boiled) deionised water	
Paste		Mix in vacuum mixer, 2 min	High-speed mixer, 1600 rpm for 2 min
	5–10 mm (≤ 3 mm if w/c ≤ 0.40)	10 mm, ~3 g	<10 mm high, ~5 g
	Consolidate by tapping, vibrating or similar action	Consolidated by manually tapping	Tap manually and place in ultrasonic bath to remove air bubbles

(Continued)

Table 3.1 (Continued) Comparison of selected protocols for measuring chemical shrinkage of cementitious materials by dilatometry

	ASTM C1608 (2012)	Empa (Schöler et al. 2014)	EPFL (Berodier 2014b; after Costoya 2008)
Assembly	Carefully add deaerated water	Carefully add decarbonated water • Fill vial fully • 1 mL water; fill test tube with paraffin oil (not used by Schöler, Lothenbach and Winnefeld [2014])	Add water drop by drop to minimise disturbance of paste
	Place the stopper with the inserted pipette tightly in the glass vial; avoid entrapped air bubbles	Seal glass vial with rubber stopper with pipette mounted in	Seal filled plastic vial with rubber stopper with pipette mounted in
	Place a drop of paraffin oil in the top of the pipette to limit evaporation	Place some coloured paraffin oil on top of water in pipette to limit evaporation and facilitate readings	Place a few drops of coloured oil on top of the water by using a syringe
Number of replicates	Minimum of two	Three	Three, preferably in different water baths
Control of measurements	$\leq$0.02 g weight change of device during test period		
Recording	Manual recording	Manual recording	Automatic recording by webcam; optimise contrast and brightness
	Every 30–60 min for first 8 h (zero point 1 h after mixing), including 1 h after mixing, thereafter every 8 h, including 24 h after mixing	Every 30–60 min first day (zero point 1 h after mixing), three readings on second day, reduced sequence later	Every 15 min; up to 6 months

Note: Empa, Swiss Federal Laboratories for Materials Science and Technology.

3.3.1 Calculations

Chemical shrinkage is computed as the measured millilitres of sorbed water, i.e. the volume decrease per gram of cement in the paste sample. The mass of cement powder in the vial is given by (assuming density of water at 1000 kg/m³) (ASTM 2012) as

$$M_{cement} = (M_{vial+paste} - M_{vial})/(1 + w/c), \tag{3.3}$$

where

M_{cement} = mass of cement in the vial (in grams)
$M_{vial+paste}$ = mass of vial with paste (in grams)
M_{vial} = mass of empty vial (in grams)
w/c = water–cement ratio by mass of paste

The chemical shrinkage per unit mass of cement at time t is (ASTM C1608 2012)

$$CS(t) = \frac{h(t) - h(t_{ref})}{M_{cement}} \tag{3.4}$$

where

$CS(t)$ = chemical shrinkage at time t (in millilitres/gram cement)
$h(t)$ = water level in pipette at time t (in millilitres)
$h(t_{ref})$ = water level in pipette at the reference time t_{ref} (in millilitres)

The first reading can be done when the samples is in temperature equilibrium with the surroundings. In ASTM C1608 (2012) a reference time of 1 hour is used. For portland-based cements hydrated at 20°C the degree of reaction developed during the first hour is relatively limited; see, e.g. Justs et al. (2014). However, for hydration at higher temperatures or more reactive systems, attempts should be made to obtain temperature equilibrium earlier (e.g. temperature conditioning the constituent materials before mixing and using [a series of] smaller samples).

3.4 SAMPLE DIMENSIONS

The sample dimensions are critical when measuring chemical shrinkage for longer periods and on dense pastes. The size should be sufficiently large to provide accurate measurements and sufficiently small to reduce sources of error as possible temperature effects and self-desiccation due to depercolation of the capillary porosity.

Examples of the impact of samples height and paste denseness (combination of w/c and fineness) are given in Figure 3.11. Initially, the measured chemical shrinkage is independent of sample height, but after a certain time

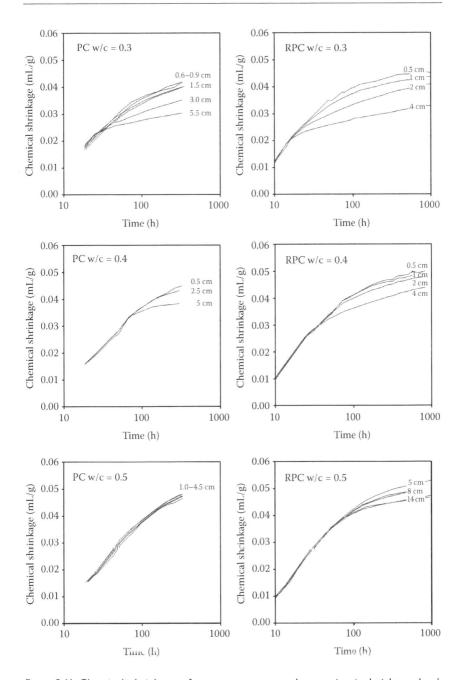

Figure 3.11 Chemical shrinkage of cement paste samples varying in height and w/c (20°C). Left graphs: ordinary portland cement (PC); right graphs: rapid-hardening portland cement (RPC) (Geiker 1983).

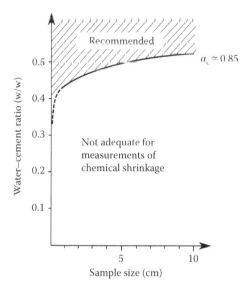

Figure 3.12 Correlated values of water–cement ratio and sample size not causing an apparent change in kinetics for degree of hydration $\alpha < 0.85$ of rapid-hardening portland cement at 20°C. The solid line indicates the recommended combination of minimum w/c and maximum sample height to be used (Geiker 1983).

chemical shrinkage development of the high samples reaches a plateau and the actual chemical shrinkage is no longer measured. The higher the sample and the denser the paste, the earlier this plateau is reached. To ensure that the developed chemical shrinkage can be measured, the thickness of the sample must be sufficiently small to allow the exterior water to refill all pores being emptied due to chemical shrinkage. For a given cement paste, the maximum sample height should be determined depending on the w/c of the paste and the required period of measuring/degree of hydration; see Figure 3.12. Similar observations have been reported by, e.g. Chen et al. (2013) and Sant et al. (2006).

3.5 SOURCES OF ERROR

Potential sources of error when measuring chemical shrinkage are summarised in Table 3.2.

The importance of including blank samples (without paste) in each measuring series and correcting for possible volume changes due to causes other than chemical shrinkage is illustrated in Figure 3.13. This figure shows (1) the measured development of chemical shrinkage of anhydrite due to the reaction to gypsum ($CaSO_4 + 2H_2O \rightarrow CaSO_4 \cdot 2H_2O$) measured according

Table 3.2 Sources of error when measuring chemical shrinkage

Related to	Topic	Possible impact on measured chemical shrinkage	Suggested solution
General	Sample height	Reduction; see Figure 3.11	Limit sample height.
	Air bubbles in paste	Reduction	Limit possible air bubbles by vacuum mixing or using freshly boiled water and carefully cast and compact samples.
	Fluid composition and amount	Increase; see Figure 3.15	Limit amount of excess fluid.
	Lack of temperature equilibrium at early age		Use small samples; for gravimetry, several samples can be placed in one sample holder. Consider using cooled water to target curing temperature after mixing.
Gravimetry	Fluid composition and amount	Increase; see Figure 3.15	Limit amount of excess fluid to calculated maximum chemical shrinkage plus a few drops for wetting of test vial surfaces.
Automatic gravimetry	Drift of scale due to temperature fluctuations in room	Varies	Place scale in temperature controlled room and limit draft.
Dilatometry	Fluid composition and amount	Increase; see Figure 3.15	Cut glasses to fit to sample height. Oil might be used as excess fluid, but stoppers have to be stable (see Figures 3.13 and 3.14).
	Disturbance of surface when adding excess fluid causing particles to end up in pipette	Increase	Add fluid along cylinder side, leave to sediment for a few minutes and remove possible particles at the fluid surface.
	Temperature fluctuations	Varies	Place setup in temperature-stable environment. Cover water bath with floating balls or lid to limit evaporation. Mount extra water supply to ensure constant water level in the water bath. Include blank samples in each test series. Limit excess water in device.

(Continued)

Table 3.2 (Continued) Sources of error when measuring chemical shrinkage

Related to	Topic	Possible impact on measured chemical shrinkage	Suggested solution
Dilatometry	Swelling of stoppers or otherwise unstable stoppers; see Figure 3.14	Typically increase; see Figure 3.13 and, e.g. Justnes et al. (1994)	Use nonswelling and stable stoppers. Consider using stoppers with screw cap (avoid entrapped air bubbles) or sealing them with wax. Use similar stoppers (type and age) for all samples and include blank samples in each test series.
	Deformation of plastic vials (Justnes et al. 1994)	Increase	Use nondeformable plastic vials or glass vials.
	Evaporation	Increase (middle to long term)	Place a few drops of oil at the top of the pipette
	Reaction of glass vial and pipettes with alkaline solution	Increase (long term)	Use alkali-resistant glass.
	Reuse of pipettes	Increase if pipettes contaminated; decreases if pipettes partly dissolved	Use clean and undamaged pipettes.
Automatic measurements using webcams.	Limited contrast of position indicator (coloured oil on water in pipette)	Noisy curves	Optimise contrast and brightness (Costoya 2008). Use only transparent suspension; avoid agglomerates of pigment in oil. Change the resolution of the software to at least 800 × 600. Use white light. Avoid reflection of lamp on pipettes (Myers 2013).
	Movement of camera	Varies	Fix position of camera thoroughly (Myers 2013).
	Pipettes not perpendicular to axis of camera; distance between samples changes	Deviations between curves	Align sample holder (Myers 2013).
	Focus varies (typically after zooming)	Varies	Adjust the distance of the camera instead of zooming. Clean the lens and add a clear reference point on top of the pipette (Myers 2013).
	Camera 'freezes'	No data	Change camera. A polynomial approximation might be used for restoring parts of hydration curves (Myers 2013).

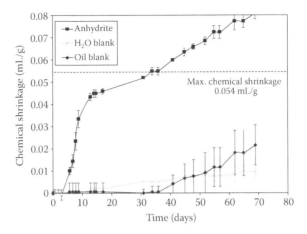

Figure 3.13 Chemical shrinkage of anhydrite pastes (water-to-solid ratio [w/s] = 2) hydrated at 20°C and volume change (per mass of anhydrite) of blank samples. Dilatometry was done using the Empa procedure (see Table 3.1) on three replicate samples; the error bars indicate the standard deviation. The calculated maximal chemical shrinkage for full hydration of the anhydrite (0.054 mL/g) is indicated with a dotted line. (Courtesy of Barbara Lothenbach.)

to the Empa procedure but with little water and mainly oil (see Table 3.1) before correcting for the volume change of the blank samples and (2) the measured volume change (per mass of anhydrite) of the blank samples with either oil or water plus a drop of oil at the top (standard Empa procedure; see Table 3.1). Figure 3.14 illustrates the swelling of stoppers after storage in oil for several months. The stoppers were found stable for short-term (a few days') exposure to oil.

Figure 3.14 Stoppers before and after exposure to oil for several months.

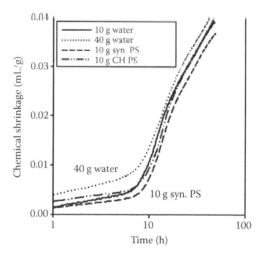

Figure 3.15 Effect of amount and composition of fluid on top of cement paste on measured chemical shrinkage. w/c = 0.30; 25 g paste; water: deionised water, syn. PS: synthetic pore solution, CH PS: calcium hydroxide saturated pore solution. (After Sant, G., P. Lura and J. Weiss, *Journal of the Transportation Research Record*, 1979, 21–29, 2006.)

The composition and the amount of fluid applied on top of the cement paste affect the measured chemical shrinkage. The larger the amount and the more diluted the fluid, the higher the initial rate of chemical shrinkage. (Sant et al. 2006); see Figure 3.15. To reduce the possible impact of the fluid, the excess fluid should be limited; see Table 3.2 and the guidelines in Section 3.9.

3.6 GENERAL REPRODUCIBILITY

Based on an interlaboratory study including seven laboratories, the repeatability and the reproducibility (standard deviation) of chemical shrinkage measurements 24 h after mixing using ASTM C1608 (2012) (dilatometry) were 0.00193 and 0.00240 mL/g, respectively. Assuming 50% hydration and a total chemical shrinkage of 0.06 mL/g, this corresponds to approximately 7%. Bullard et al. (2006), in undertaking a multilaboratory study, found a 5% reproducibility of automatic gravimetry measurements based on three tests between 1 and 24 h.

An example of the accuracy of automatic dilatometry using a webcam and the EPFL procedure (see Table 3.1) on replicate samples of portland cement paste is shown in Figure 3.16. The accuracy of the Empa procedure (see Table 3.1) is illustrated in Figure 3.17.

A comparison of the volume changes in the PC-Qz and H_2O-blank samples (i.e. in a water-filled vessel) illustrates also that for the applied

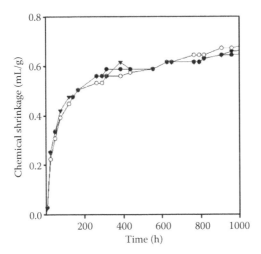

Figure 3.16 Automatic dilatometry using webcam and the EPFL procedure (see Table 3.1) on replicate samples of portland cement paste (CEM I 52.5R, w/c = 0.4, mixed 2 min at 1600 rpm). (After Berodier, E., Impact of the supplementary cementitious materials on the kinetics and microstructure development of cement hydration, PhD thesis, École Polytechnique Fédérale de Lausanne, Switzerland, 2014.)

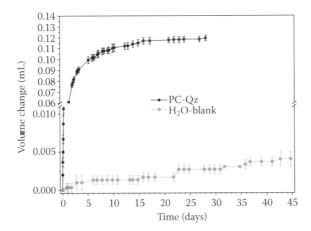

Figure 3.17 Chemical shrinkage of portland cement–quartz (PC-Qz) pastes (water-to-binder ratio [w/b] = 0.5, CEM I 52.5 R) hydrated at 20°C. Chemical shrinkage is given as volume change to compare with measured volume change of blank samples. Dilatometry was done using the Empa procedure (see Table 3.1) on three replicate samples; the error bars indicate the standard deviation. (Courtesy of Axel Schöler.)

setup and investigated system, measurements of chemical shrinkage should be limited to approximately a month as at later times when little further reaction occurs, the general drift might become comparable to the measured changes.

3.7 CASE STUDIES

3.7.1 Effect of mineral additions on chemical shrinkage development

Studying the reaction in blended systems, Schöler et al. (2015), among others, used chemical shrinkage measurements. They observed that both inert and reactive additives caused increased chemical shrinkage development of Portland cement pastes (see Figure 3.18) and concluded that supplementary cementitious materials both have a physical and a chemical effect on blended Portland cement-based systems. The replacement of a fraction of the cement by quartz results in a higher effective water–cement ratio, inducing a higher degree of cement reaction as mirrored in the higher chemical shrinkage for the quartz-blended cement. The difference between the chemical shrinkages of the slag-blended and quartz-blended cements indicates the reaction of the slag. Similar observations were reported by Berodier (2014a).

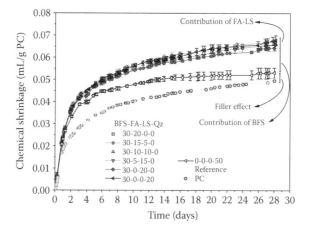

Figure 3.18 Chemical shrinkage of portland cement–slag–fly ash–limestone–quartz (PC-BFS-FA-LS-Qz) systems. Pastes were hydrated at 20°C (w/c = 0.5, CEM I 52.5 R, 50% cement by mass in blended systems; details are in Schöler et al. [2015]). Dilatometry was done using the Empa procedure (see Table 3.1) on three replicate samples; the error bars indicate the standard deviation. (Courtesy of Axel Schöler.)

3.7.2 Effect of temperature on chemical shrinkage development

Studying, among others, the impact of temperature on the rate of cement hydration, Geiker (1983) used the measurement of chemical shrinkage in combination with modelling using the dispersion model developed by Knudsen (1984) (Equation 3.1). Figure 3.19 illustrates typical data obtained for pastes of rapid-hardening portland cement (w/c = 0.5) using gravimetry and a maximum sample height of 20 mm. The measured data illustrate the impact of temperature on compressive strength, while the relative chemical shrinkage data illustrate the impact of temperature on the rate of reaction.

Based on the data, an apparent activation energy of 50–60 kJ/mol was found for the rapid-hardening portland cement tested, assuming that the rate of reaction can be described by the reaction time t_1 obtained by the dispersion model (Equation 3.1) and using the Arrhenius equation (Castellan 1970)

$$k = A \exp(-E_a/RT) \Leftrightarrow \ln(k) = \ln(A) - E_a/RT, \tag{3.5}$$

where

k = the rate of reaction
A = constant
E_a = apparent activation energy (in kilojoules/mole)
R = gas constant (8.31 J/K mol)
T = temperature (in kelvins)

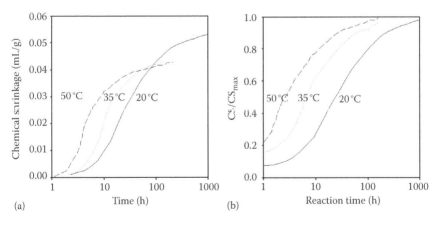

(a)   (b)

Figure 3.19 Chemical shrinkage development of rapid-hardening portland cement cured at 20, 35 and 50°C (w/c = 0.5, high-speed mixing, gravimetry, maximum height 20 mm): (a) as measured; (b) relative chemical shrinkage (CS/CS$_{max}$) versus reaction time according to the dispersion model (Equation 3.1) (Geiker 1983).

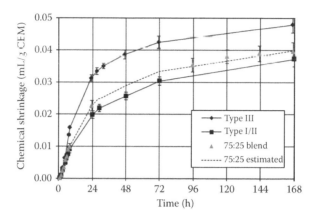

Figure 3.20 Comparison of measured chemical shrinkages of cement pastes of ASTM C150 types I/II and III and a 75:25 blend (w/c = 0.4) and predicted chemical shrinkages of the 75:25 blend by applying the law of mixtures. Error bars indicate ±1 standard deviation for three replicate specimens for each cement paste (Bentz 2010). (Courtesy of Dale Bentz.)

Similar values for the apparent activation energy of portland cement were determined using isothermal calorimetry; see, e.g. Kada-Benameur et al. (2000) and Schindler (2004).

3.7.3 Predicting chemical shrinkage of various cement blends

Bentz (2010) suggested the engineering of the performance of concrete by blending cements differing in fineness. In his study he showed that the heat development, the chemical shrinkage and even the compressive strength of the blended systems could be accurately predicted using a law of mixtures. Figure 3.20 illustrates the applicability of the law of mixtures for a 75:25 blend of a coarse type I/II and a fine type III ASTM C150 cement.

3.8 COMPARISON OF CHEMICAL SHRINKAGE WITH OTHER METHODS OF HYDRATION CHARACTERISATION

If the hydration reactions were not changing during the development of hydration, different characterisation methods should provide similar information on the degree of hydration. However, the hydrates do not react simultaneously and the reaction products vary. Nevertheless, reasonable correlation between data obtained with different characterisation methods is observed.

Selected experimental methods of monitoring portland cement hydration were compared by Parrott et al. (1990). Near-linear relations were found between chemical shrinkage and heat of hydration and between chemical shrinkage and quantitative X-ray diffraction (XRD) data. The cements tested were an ordinary portland cement, a high-belite cement and a Danish white portland cement.

Combined isothermal calorimetry and chemical shrinkage is increasingly used for kinetic studies, e.g. for investigations of the effect of particle size on the hydration kinetics and microstructural development of alite pastes (Costoya 2008). For measuring the degree of reaction of slag in blended pastes, isothermal calorimetry and automatic chemical shrinkage using dilatometry and webcam in combination with scanning electron microscopy (SEM) mapping using backscattered electrons with image analysis (IA) for calibration were found promising (Kocaba 2009; Kocaba et al. 2012); see Figure 3.21. To determine the contribution of the slag, data for cement–filler systems were subtracted from data for cement–slag systems.

Studying the impact of temperature on early-stage hydration of oil-well cements, Pang et al. (2013) found that the ratio between total chemical shrinkage and total heat release at complete hydration (CS_{max}/H_{max}) varies slightly with cement composition (mainly C_3A content) and decreases with increasing curing temperature. See Section 3.3 for further information on the temperature dependency.

Simultaneous measurements of chemical shrinkage and heat development illustrate the importance of comparable curing conditions (Lura et al. 2010). Sealed cured samples showed lower heat development compared to saturated (open) samples; see Figure 3.22.

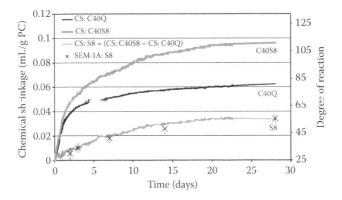

Figure 3.21 Chemical shrinkages (CS) of blends of cement and slag (C40S8) and cement and quartz (C40Q) and calculated chemical shrinkage originating from slag reaction versus degree of slag reaction determined by SEM-IA mapping. (After Kocaba, V., Development and evaluation of methods to follow microstructural development of cementitious systems including slags, PhD thesis, École Polytechnique Fédérale de Lausanne, Switzerland, 2009.)

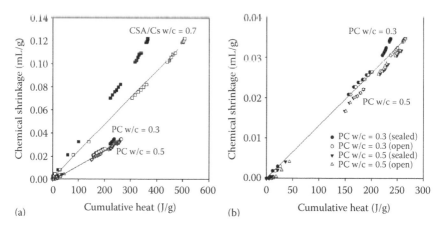

Figure 3.22 Chemical shrinkage as a function of cumulative heat of hydration (a) for calcium sulfoaluminate cement (CSA) and portland cement (PC) pastes and (b) for two portland cement pastes with different w/c. The open symbols show the results of saturated (open) samples where chemical shrinkage and heat flow were measured simultaneously, while the full symbols represent the heat of hydration of samples with no water on top (sealed) (Lura et al. 2010).

3.9 GUIDELINES

Before using chemical shrinkage for investigating the development of hydration, the required accuracy needs to be considered. For example, in comparing different binder systems, the expected difference should be measurable, i.e. larger than the accuracy of the method; see Section 3.6.

3.9.1 Guidelines to analysis using gravimetry

3.9.1.1 Purpose

The purpose of the method is to measure the chemical shrinkage development of cement paste.

3.9.1.2 Principle

Chemical shrinkage is the decrease in total volume taking place during hydration. During hydration, possible water from the exterior may be sucked into the cement paste, refilling pores being emptied by chemical shrinkage. For portland cement the ultimate chemical shrinkage at 20°C is approximately 0.6×10^{-4} m³/kg (0.06 mL/g).

According to the law of Archimedes, the weight of a body submerged in liquid, compared to its weight in air, will decrease by the weight of the amount of liquid the body displaces. Hence, a system of hydrating cement paste with a surplus of water will gain weight during hydration. Measuring chemical shrinkage by measuring this weight gain of a saturated sample immersed in an inert liquid is called gravimetry.

3.9.1.3 Materials and equipment for one test

- Balance (accuracy ±0.001 g) with automatic weight change registration
- Temperature-controlled oil bath
- Sample holder
- Cylindrical test vial: h = 35–50 mm, inner ø = 17–22 mm
- Distilled, boiled and cooled water
- Materials to be tested

3.9.1.4 Procedure

- Before testing, check the setup with a test vial with a small amount of water (similar to paste samples; see below).
- Before testing, select the height of the sample based on a combined evaluation of the w/c and the sample height; see, e.g. Section 3.4 for guidance.
- Mark the test vial (with the name and the height of paste) and weigh it.
- Mix paste; see Chapter 1 for guidance.
- Add cement paste to the test vial until the selected height is reached. Make sure that no paste is spilled on the sides above the paste. Consolidate by tapping, vibrating or similar action. Weigh the test vial + cement paste.
- Carefully place a small amount* of distilled, boiled and cooled water on the top of the sample with a pipette (slide the water down the side of the glass). Weigh the test vial + cement paste + water.
- Carefully fill the top with paraffin oil before placing the sample in a paraffin oil bath.
- Carefully submerge the sample in the oil bath by using a large pair of tweezers, keeping the test vial at an angle.
- Place the sample in the sample holder and follow the on-screen instructions to collect data by a connected personal computer.
- See Section 3.3.1 for guidelines on calculations.

---

* With 1 cm sample height in a test tube of ø 22 mm, the maximum amount of cement will be approximately 2.5 g, corresponding to approximately 0.16 mL of ultimate shrinkage, corresponding to 0.43 mm in height. With the given glass radius of 1.1 cm, the smallest height of water to be placed on top of the sample is thus 1 mm, as a small surplus needed.

3.9.2 Guidelines to analysis using dilatometry

3.9.2.1 Purpose

The purpose of the method is to measure the chemical shrinkage development of cement paste.

3.9.2.2 Principle

Chemical shrinkage is the decrease in total volume taking place during hydration. During hydration, possible water from the exterior may be sucked into the cement paste, refilling pores being emptied by chemical shrinkage. For portland cement the ultimate chemical shrinkage at 20°C is approximately 0.6×10^{-4} m³/kg (0.06 mL/g).

Measuring chemical shrinkage by measuring the amount of water being sucked into a saturated sample is called dilatometry.

For short-term measurements ASTM C1608 (2012) might be used. For long-term measurements the following procedure is recommended. A webcam might be added to allow for automatic measurement; see the EPFL procedure in Table 3.1. Measurements should preferably be performed in a temperature-controlled room.

3.9.2.3 Materials and equipment for one test

- Graduated capillary tube/pipette: capacity and grading depending on amount of paste (typical capacity 1.000 ± 0.5 mL); alkali-resistant glass
- Temperature-controlled bath: cover water bath with floating balls or lid to limit evaporation; mount extra water supply to ensure constant water level in the water bath
- Sample holder
- Cylindrical test vial: $h = 35$–50 mm; inner $\varnothing = 17$–22 mm; height to be adjusted to fit sample height (see in the following)
- Stoppers with central hole: use rubber or silicone and test stability before use
- Distilled, boiled and cooled water
- Materials to be tested

3.9.2.4 Procedure

- Use three replicate samples.
- Include a set of blank samples in all measurement series. Use similar stoppers and pipette, but adjust the size of the test vial to allow for only a few millimetres of water below the stopper.
- Before testing, select the height of the sample based on a combined evaluation of the w/c and the sample height; see, e.g. Section 3.4 for guidance.

- Cut the cylindrical test vial to height to allow for only a few millimetres of water below the stopper.
- The nonpointed end of the pipette is inserted in the stopper, ensuring no recess where possible air bubbles might get caught.
- Mark the test vial (with the name and the height of paste) and weigh it.
- Mix paste; see Chapter 1 for guidance.
- Add cement paste to the test vial until the selected height is reached. Make sure that no paste is spilled on the sides above the sample. Consolidate by tapping, vibrating or similar action. Weigh the test vial + cement paste.
- Carefully add deaerated water by letting the water flow along the side of the vial. Completely fill the vial. Allow possible cement grains to settle for a few minutes before wiping off possible cement grains stuck to the surface.
- Place the stopper with the inserted pipette tightly in the test vial. Avoid entrapped air bubbles. When placing the stopper, water will be pushed through the pipette. Ensure that the pipette is almost filled at the start of measurements.
- Place a drop of paraffin oil in the top of the pipette to limit evaporation.
- Manual recording: Place the samples in the sample holder to allow for possible lifting without touching the pipettes.
- Make ≥3 readings per day (including readings around 10%–15% of maximum chemical shrinkage; 0.5–1 h after mixing, depending on temperature equilibrium and use of data); reduce sequence later.
- See Section 3.3.1 for guidelines on calculations.

3.9.3 Guidelines to reporting

The following information should be included in the reporting:

- Principle and procedure of measuring (including accuracy of balance (gravimetry) or pipettes (dilatometry)
- Method of mixing
- Composition and identification (ID) of paste
- ID of companion blank sample
- Time of mixing
- Amount of paste and sample dimensions; amount of additional water
- Readings (time, measurement and temperature)
- Person undertaking the work

3.10 CONCLUSIONS AND OUTLOOK

Cementitious materials exhibit so-called chemical shrinkage during hydration because the products take up less space than the reactants. Approximately

100 years after Le Chatelier (1900) reported measurement of chemical shrinkage using dilatometry, the measuring principle is used as the ASTM standard C1608 for testing early-age hydration, ASTM C1608 (2005), now ASTM C1608 (2012). As chemical shrinkage measurements have a good reproducibility (around 5% after 24 h), can be undertaken continuously and with equipment of reasonable price (dilatometry with webcam) their application is increasing.

Measurement of chemical shrinkage of cementitious materials is used for quality control, kinetic studies and prediction of self-desiccation and proportioning of internal curing water.

ACKNOWLEDGEMENTS

Assistance from Axel Schöler, Barbara Lothenbach, Dale Bentz, Elise Berodier, Frank Winnefeld, Gaurav Sant, Jason Weiss, John Rossen and Pietro Lura in identifying literature and providing protocols, photos and data is greatly appreciated. Also, the review comments by Barbara Lothenbach and Frank Winnefeld are highly valued.

REFERENCES

ASTM C1608 (2005). Test method for chemical shrinkage of hydraulic cement paste.
ASTM C1608 (2012). Test method for chemical shrinkage of hydraulic cement paste.
Bentz, D. P. (2007). 'Internal curing of high-performance blended cement mortars'. *ACI Materials Journal* 104(4):408–414.
Bentz, D. P. (2010). 'Blending different fineness cements to engineer the properties of cement-based materials'. *Magazine of Concrete Research* 62(5):327–338.
Bentz, D. P., P. Lura and J. W. Roberts (2005). 'Mixture proportioning and internal curing'. *Concrete International* 27(2):35–40.
Berodier, E. (2014a). Impact of the supplementary cementitious materials on the kinetics and microstructure development of cement hydration. PhD thesis, École Polytechnique Fédérale de Lausanne, Switzerland.
Berodier, E. (2014b). Personal communication.
Bullard, J. W., M. D'Ambrosia, Z. Grasley, W. Hansen, N. Kidner, D. Lange, P. Lura et al. (2006). A comparison of test methods for early-age behavior of cementitious materials. In *Second International Symposium on Advances in Concrete through Science and Engineering*, Quebec City, Canada: RILEM.
Castellan, G. W. (1970). *Physical chemistry*. Fourth ed. Addison-Wesley Publishing Company, Boston.
Chen, H., M. Wyrzykowski, K. Scrivener and P. Lura (2013). 'Prediction of self-desiccation in low water-to-cement ratio pastes based on pore structure evolution'. *Cement and Concrete Research* 49:38–47.
Costoya, M. (2008). Effect of particle size on the hydration kinetics and microstructural development of tricalcium silicate. PhD thesis, École Polytechnique Fédérale de Lausanne, Switzerland.

Danish Technological Institute (1989). TK 84 (89) Prøvningsmetode Kemisk svind. Tåstrup, Denmark: Danish Technological Institute.

Fu, T. F., T. Deboodt and J. H. Ideker (2012). 'Simple procedure for determining long-term chemical shrinkage for cementitious systems using improved standard chemical shrinkage test'. *Journal of Materials in Civil Engineering* 24(8):989–995.

Fulton, F. S. (1962). 'The rate of hydration of Portland cement'. In *Laboratory Report*. Portland Cement Institute, Johannesburg, South Africa.

Gartner, E., and G. Pham (2005). Comparison of enthalpy and volume changes in early-age hydration, Lafarge Centre de Recherche, Paris, France.

Geiker, M. (1983). Studies of portland cement hydration: Measurements of chemical shrinkage and a systematic evaluation of hydration curves by means of the dispersion model. PhD thesis, The Institute of Mineral Industry, The Technical University of Denmark, Kongens Lyngby, Denmark.

Geiker, M., and T. Knudsen (1982). 'Chemical shrinkage of portland cement pastes'. *Cement and Concrete Research* 12(5):603–610.

Hansen, P. F. (1995). Materialefysik for bygningsingeniører. Beregningsgrundlag, SBI-anvisning 183. Statens Byggeforskningsinstitut, Hørnsholm, Denmark.

Hemeon, W. C. L. (1935). 'Setting of Portland cement'. *Industrial and Engineering Chemistry*:694–699.

Justnes, H., B. Reyniers, D. Van Loo and E. J. Sellevold (1994). 'An evaluation of methods for measuring chemical shrinkage of cementitious paste'. *Nordic Concrete Research*:45–61.

Justnes, H., E. J. Sellevold, B. Reyniers, D. Van Loo, A. Van Gemert, F. Verboven and D. Van Gemert (1998). The influence of cement characteristics on chemical shrinkage. In *International Workshop on Autogenous Shrinkage of Concrete*. Hiroshima, Japan.

Justs, J., M. Wyrzykowski, F. Winnefeld, D. Bajare and P. Lura (2014). 'Influence of superabsorbent polymers on hydration of cement pastes with low water-to-binder ratio'. *Journal of Thermal Analysis and Calorimetry* 115(1):425–432.

Kada-Benameur, H., E. Wirquin and B. Duthoit (2000). 'Determination of apparent activation energy of concrete by isothermal calorimetry'. *Cement and Concrete Research* 30(2):301–305.

Knudsen, T. (1980). On the size distribution in cement hydration. In *7th International Conference on Chemistry of Cement*. Paris, France.

Knudsen, T. (1984). 'The dispersion model for hydration of portland-cement: 1. General concepts'. *Cement and Concrete Research* 14(5):622–630.

Knudsen, T. (1985). 'On the possibility of following the hydration of fly-ash microsilica and fine aggregates by means of chemical shrinkage'. *Cement and Concrete Research* 15(4):720–722.

Knudsen, T., and M. Geiker (1985). 'Obtaining hydration data by measurement of chemical shrinkage with an archimeter'. *Cement and Concrete Research* 15(2):381–382.

Kocaba, V. (2009). Development and evaluation of methods to follow microstructural development of cementitious systems including slags, PhD thesis, École Polytechnique Fédérale de Lausanne, Switzerland.

Kocaba, V., E. Gallucci and K. L. Scrivener (2012). 'Methods for determination of degree of reaction of slag in blended cement pastes'. *Cement and Concrete Research* 42(3):511–525.

Le Chatelier, M. H. (1900). 'Sur les chargements de volume qui accompagnement le durcissement des ciment'. *Société d'Encouragement pour l'Industrie National Bulletin, Arts Chimiques*.

Lothenbach, B. (2010). 'Thermodynamic equilibrium calculations in cementitious systems'. *Materials and Structures* 43(10):1413–1433.

Lothenbach, B., G. Le Saout, E. Gallucci and K. Scrivener. (2008) 'Influence of limestone on the hydration of Portland cements'. *Cement and Concrete Research* 38(6):848–860.

Lothenbach, B., T. Matschei, G. Möschner and F. P. Glasser (2008). 'Thermodynamic modelling of the effect of temperature on the hydration and porosity of Portland cement'. *Cement and Concrete Research* 38(1):1–18.

Lura, P., O. M. Jensen and K. van Breugel (2003). 'Autogenous shrinkage in high-performance cement paste: An evaluation of basic mechanisms'. *Cement and Concrete Research* 33(2):223–232.

Lura, P., F. Winnefeld and S. Klemm (2010). 'Simultaneous measurements of heat of hydration and chemical shrinkage on hardening cement pastes'. *Journal of Thermal Analysis and Calorimetry* 101(3):925–932.

Matschei, T., B. Lothenbach and F. P. Glasser (2007). 'Thermodynamic properties of Portland cement hydrates in the system $CaO–Al_2O_3–SiO_2–CaSO_4–CaCO_3–H_2O$'. *Cement and Concrete Research* 37(10):1379–1410.

Mills, R. H. (1962). 'The relationship between the reduction in specific volume of the products of hydration and strength of pastes, mortars, and concretes made with Portland cement, Portland-blastfurnace cement and mixtures of Portland cement and blastfurnace slag'. *Die Sievile Ingeniur in South-Africa*: 125–132.

Mounanga, P., A. Khelidj, A. Loukili and V. Baroghel-Bouny (2004). 'Predicting $Ca(OH)_2$ content and chemical shrinkage of hydrating cement pastes using analytical approach'. *Cement and Concrete Research* 34(2):255–265.

Myers, N. (2013). Hydration kinetics study: Methods and reproducibility for blended cements. MSc thesis, École Polytechnique Fédérale de Lausanne, Switzerland.

Pang, X. Y., D. P. Bentz, C. Meyer, G. P. Funkhouser and R. Darbe (2013). 'A comparison study of Portland cement hydration kinetics as measured by chemical shrinkage and isothermal calorimetry'. *Cement and Concrete Composites* 39:23–32.

Parrott, L. J., M. Geiker, W. A. Gutteridge and D. Killoh (1990). 'Monitoring Portland-cement hydration – Comparison of methods'. *Cement and Concrete Research* 20(6):919–926.

Powers, T. C. (1935). 'Absorption of water by Portland cement paste during the hardening process'. *Industrial and Engineering Chemistry* 27(7):790–794.

Sant, G., P. Lura and J. Weiss (2006). 'Measurement of volume changes in cementitious materials at early ages: Review of testing protocols and interpretation of results'. *Journal of the Transportation Research Record* 1979:21–29.

Schindler, A. K. (2004). 'Effect of temperature on hydration of cementitious materials'. *Materials Journal* 101(1):72–81.

Schöler, A., B. Lothenbach and F. Winnefeld (2014). Hydration of multi-component cements containing cement clinker, slag, type-V fly ash and Limestone: CEM X – M (S, V, LL, x). Progress Report 1 (2013–2014), Empa.

Schöler, A., B. Lothenbach, F. Winnefeld and M. Zajac (2015). 'Hydration of quaternary Portland cement blends containing blast-furnace slag, siliceous fly ash and limestone powder'. *Cement and Concrete Composites* **55**:374–382.

Swayze, M. A. (1942). 'Early concrete volume changes and their control'. *Journal of the American Concrete Institute* (13):425–440.

Winnefeld, F., and B. Lothenbach (2010). 'Hydration of calcium sulfoaluminate cements – Experimental findings and thermodynamic modelling'. *Cement and Concrete Research* **40**(8):1239–1247.

Xiao, K. T., H. Q. Yang and Y. Dong (2009). 'Study on the influence of admixture on chemical shrinkage of cement based materials'. *Ultra-High-Pumpability and High Performance Concrete Technology* **405–406**:226–233.

Chapter 4

X-ray powder diffraction applied to cement

Ruben Snellings

CONTENTS

4.1 INTRODUCTION

X-ray diffraction (XRD) is one of the most prominent analytical techniques in the characterisation of crystalline, fine grained materials, such as cements. The power of XRD is in the rapid and, if carried out appropriately, reliable delivery of quantitative data on crystal structural properties and abundances of individual phases contained in cements. In cements the technique is mostly used for qualitative, i.e. phase identification, and quantitative phase analysis (QPA). The relatively recent extension to the quantitative study of hydrated cements opens up a wide range of opportunities for groundbreaking research in cementitious materials. The diffraction of X-rays by a crystalline material produces an XRD pattern consisting of peaks of varying intensities at characteristic diffraction angles. The diffraction angle or position of the peaks is determined by the symmetry and the size of the unit cell through Bragg's law, while the intensities of the peaks relate to the nature and disposition of the atoms within the unit cell of the crystalline material. This way XRD produces patterns of peak positions and relative intensities that characterise different crystal structures and enable identifying their presence in unknown samples. In mixtures of phases a straightforward relationship exists between the overall peak intensities of a phase and its weight fraction in the mixture. This constitutes the basis of QPA.

Recent years have seen a surge in cement hydration studies focusing on the characterisation and quantification of the development of the hydrate phase assemblage (Antoni et al. 2012; De Weerdt et al. 2011; Jansen et al. 2012; Korpa et al. 2009; Lothenbach, Le Saout et al. 2008; Martín-Sedeño et al. 2010; Scrivener et al. 2004; Soin et al. 2013). Among the most prolific laboratory techniques is XRD. XRD can be used to quantify the degree of hydration of the anhydrous cement and can provide information on the formation of individual hydrate phases. While excellent textbooks and reviews are available that describe the physical principles (Dinnebier and Billinge 2008; Klug and Alexander 1974) and/or general data analysis methods such as the Rietveld method (McCusker et al. 1999; Young 1993) in detail, few texts have treated the specific problems encountered in sample preparation, measurement and data analysis when using XRD to study (hydrated) cements.

This chapter, therefore, describes and compares practical approaches in sample preparation, data collection and data analysis used for quantification and characterisation of (hydrated) cements by XRD. Specific attention is paid to the effects of hydration stoppage and practical strategies towards data collection are described. The aim of the chapter is not only to present best practise but also to inform about compromises inherent in the adopted approaches and to draw attention to potential pitfalls. As an aid to data analysis, a database is given that provides crystal structures and powder diffraction identification files of common cement anhydrous and hydrate phases. Finally, the application of XRD to the study of cements is illustrated by two typical case studies of QPA on (1) an anhydrous portland cement and (2) a hydrated portland cement.

It should be highlighted that a number of review papers concerning XRD QPA of anhydrous cements are available (Le Saoût et al. 2011; Stutzman 2011; Taylor et al. 2002) and that a review of the application of Rietveld QPA to clinkers, cements and hydrated pastes was published recently (Aranda et al. 2012).

4.2 QUALITATIVE PHASE ANALYSIS

Qualitative phase analysis is based on a comparison of the peaks in a measured XRD pattern to a database containing peak patterns of known phases. Usually a general-purpose database, such as the powder diffraction file (PDF) published by the International Centre for Diffraction Data is used, often in combination with chemical or categorical filters, to restrict the number of candidate patterns. As a general aid to phase identification, a list of phases encountered in cements is supplied with the corresponding PDF reference numbers in Tables A4.1 through A4.7. It should be noted that alternative open-source databases exist, such as the Crystallography Open Database (COD; www.crystallography.net) and the American Mineralogist Crystal Structure Database (AMCSD; rruff.geo.arizona.edu/AMS/amcsd.php); these databases have their separate referencing systems. To track the structures in alternative databases, Tables A4.1 through A4.7 also refer to the original paper in which the structure was first reported. Tables A4.1 through A4.7 can be used to build a custom database for cement phases to be used in combination with any of the available search/match software. Alternatively, a classical and, in some cases, more rapid manual identification routine can benefit from a comparison of the peak positions in the recorded XRD pattern with the d values of the most intense XRD reflections of common cement phases in the search list in Table A4.8. The reflection lines in Table A4.8 are ordered according to d value. Three peak list categories are distinguished: (1) the major clinker phases (calcium silicate, aluminate and ferrite phases), (2) the minor clinker and cement phases (sulfates, oxides and aluminosilicates) and (3) the hydrate phases. In addition, an indication of the relative intensities of the reflections is given, with 10 corresponding to the most intense reflection and linearly scaling down for less intense reflections. As regular search/match programs perform well in identifying major phases but have difficulties in identifying minor or trace phases partially hidden by peak overlap with major phases, the search list provided in Table A4.8 is particularly useful in finding suitable candidates for minor unidentified peaks or features in the XRD patterns.

Cements, both anhydrous and hydrated, are complex mixtures of phases. In powder diffraction scans the reflection lines of the cement phases often show considerable overlap and are difficult to observe separately. For illustration, Figure 4.1 shows the diffraction scans of a range of classical cements used to date in construction practise. All cements consist of at least three major phases, supplemented with a variable number of minor phases. A multitude

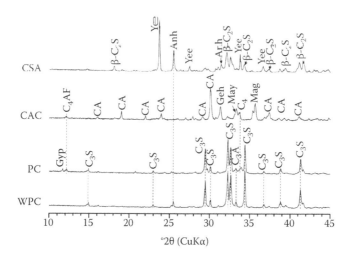

Figure 4.1 XRD scans (10°–45° 2θ range) of typical industrially produced cements: white portland cement (WPC), plain portland cement (PC), calcium aluminate cement (CAC) and calcium sulfoaluminate cement (CSA). The diffraction peaks of the main phases are indicated: alite (C_3S M3), belite (β-C_2S), aluminate (C_3A), ferrite (C_4AF), calcium aluminate (CA), ye'elimite (Yee), anhydrite (Anh), gypsum (Gyp), gehlenite (Geh), mayenite (May) and magnetite (Mag).

of diffraction peaks can be observed; most peaks are composite and result from the overlap of diffraction peaks of several phases. Peak assignment for all but the dominant phases is therefore often not straightforward.

The extent of peak overlap is illustrated by the pattern decomposition of an XRD scan of a plain portland cement in Figure 4.2. By using Rietveld analysis (cf. Section 4.4), the XRD scan in Figure 4.2 is decomposed into the calculated contributions of the phases that build up the cement. In practise, phase identification, therefore, often remains a nontrivial task, in particular for phases present in minor amounts. Often the search/match algorithm for phase identification will supply only a list of suggestions for minor phases and it is up to the analyst to decide, preferably based on supplemental information, such as additional diffraction experiments on selectively enriched fractions, sample chemistry, microscopic observations or general experience with the material. In the case of anhydrous cements, controlled phase enrichment by selective dissolution treatments is particularly useful to constrain the identification of minor or trace alkali sulfates or to verify the presence of polymorphism of major phases, such as C_2S or C_3A. Suitable selective dissolution procedures are the KOH/sugar (glucose) (KOSH) treatment that concentrates the silicate phases in the anhydrous cement and the salicylic acid/methanol (SAM) treatment that results in a residue containing the aluminate, ferrite, sulfate and other minor/trace phases. The example in Figure 4.3 illustrates the added resolution gained

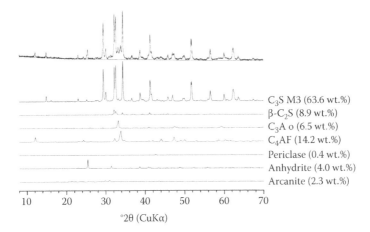

Figure 4.2 XRD scan of a portland cement (CEM I 52.5 N) (top) and its pattern decomposition calculated by Rietveld QPA (downwards from XRD scan). At right the QPA results are presented for indicative purposes. Arcanite was identified by means of a salicylic acid/methanol selective dissolution treatment.

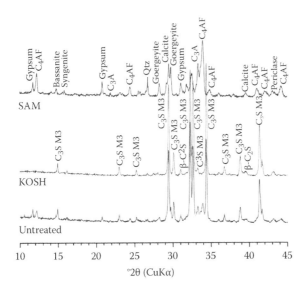

Figure 4.3 XRD scans showing the effect of selective dissolution treatments for a portland cement. Top: residues of SAM; middle: residues of KOSH treatment; bottom: original untreated portland cement. The KOSH treatment dissolves the aluminate, ferrite, sulfate and most minor phases, leaving only the calcium silicate phases. In contrast, the SAM treatment dissolves the calcium silicate phases and thus concentrates the aluminate, ferrite and minor phases in the residue. In this particular example, the SAM treatment led to the clear identification of goergeyite ($K_2Ca_5(SO_4)_6 \cdot H_2O$) as a minor phase in the cement.

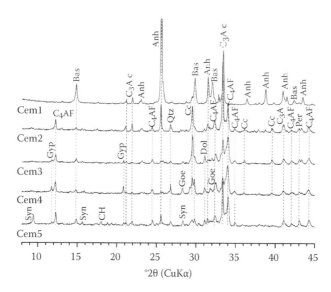

Figure 4.4 XRD scans of SAM selective dissolution treatment residues of a series of five different portland cements. The concentration of minor phases in the residues exposes significant differences between cements. The differences in minor phase composition result in variations in the early-age hydration of the cements (setting, early-age strength development). The main diffraction peaks of the phases are assigned as follows: bassanite (Bas), aluminate (C₃A), anhydrite (Anh), ferrite (C₄AF), quartz (Qtz), calcite (Cc), periclase (Per), gypsum (Gyp), dolomite (Dol), goergeyite (Goe), syngenite (Syn) and portlandite (CH).

by the selective dissolution treatments. Figure 4.4 compares the residues of the SAM treatment for a series of portland cements and illustrates the significant differences in the minor phase content. The selective dissolution procedures are described in detail by Gutteridge (1979).

Finally, the XRD pattern-fitting procedures in QPA often indicate the presence, in the difference curve, of minor or trace phases previously hidden by extensive peak overlap with major phases. An iterative procedure, going back and forth between phase identification and quantification, is therefore very common and instrumental in obtaining a complete analysis.

4.3 QUANTITATIVE PHASE ANALYSIS: THE RISE OF RIETVELD

QPA by XRD has become a prominent characterisation technique in cement science and more recently also in cement production. Properties such as strength development or durability are closely related to the cement phase composition. Early analyses were mostly based on the Bogue calculation

(Bogue 1929) or its later modifications (Taylor 1989) that estimate the phase composition from the chemical bulk composition. Relatively large discrepancies between Bogue estimates and phase abundances from direct measurements, such as microscopy and quantitative XRD, are common and derive from deviations from the assumptions underlying the Bogue calculations. The largest discrepancies result from the fact that the calculation does not take into account solid solution of minor elements into the major clinker phases and generally disregards the presence of minor phases, which may lead to significant biases in, for instance, the alite-to-belite ratio (Barry and Glasser 2000; Crumbie et al. 2006). Current developments in clinker production that implement a high degree of flexibility and variability in the use of alternative fuels and raw materials will inevitably change trace and minor element abundances significantly and will further undermine the validity of the Bogue assumptions. As a countermeasure, the production process is controlled more closely by implementing QPA online in production and quality control (Chatterjee 2011). Direct quantitative phase abundances are usually obtained through XRD, even though a good approximation of the phase abundances can be obtained by the modified Bogue method when the actual composition of the clinker phases is measured by energy-dispersive X-ray spectroscopy (EDX) micro-analyses (Le Saoût et al. 2011). However, the latter renders the method comparatively time consuming. For similar reasons, direct determination of phase abundances by optical or electron microscopy is impractical as large amounts of images need to be processed by point counting or image analysis methods before a statistically representative estimate is produced (cf. Chapter 8). Moreover, microscopic QPA encounters problems in resolving fine intergrowths of phases and minor phases or in analysing finely ground cements. Because of the limitations of both Bogue calculations and microscopic analyses, XRD methods are nowadays the most widely used for characterisation and quantification of the phase composition of cements and hydrated cements.

Traditional XRD QPA methods that use single-peak heights or integrated peak heights, compared to an internal standard peak, have not been very successful for cements (Aldridge 1982; Strubble 1991). The high degree of peak overlap between the main peaks of the major phases renders this approach difficult and prone to significant error (Kristmann 1977) (e.g. Figure 4.2). Because of the important overlap of the main peaks, smaller 'freestanding' peaks need to be used in this method. The much reduced signal-to-background ratio of these smaller peaks significantly reduces the accuracy and sensitivity of this approach (Gutteridge 1984). Additional problems include the variability in composition and crystal structure that can result in significant variations in peak positions and relative intensity. In the case of portland cements, solid solution of minor elements (alkalis, magnesium, sulfate, etc.) in the major phases (alite, belite and aluminate) is common and leads to the stabilisation of different polymorphic forms. The differences in crystal structure are reflected in the diffraction patterns as exemplified for alite (C$_3$S) polymorphs in Figure 4.5. An incorrect identification of the polymorph obviously leads to an additional bias in the quantification results. Finally, single-peak methods are very sensitive to preferred orientation effects (cf. Section 4.6.2), which is problematic particularly in the case of quantification of the calcium sulfate additions in cement.

The use of whole-powder-pattern-fitting methods enabled to largely overcome peak overlap problems and to mitigate the effects of preferred orientation. In whole-powder-pattern-fitting, all reflections of all phases are considered and an algorithm is used to scale the reflection intensities of the phases to match the measured pattern. The calculated pattern can be produced from a combination of measured standard diffraction patterns (Smith et al. 1987) or it can be calculated from crystal structure data as in the Rietveld method. The former approach has not found widespread application in cement as it demands the measurement of standard patterns of pure phases that are as similar as possible to the phases in the unknown mix. As the main phases in cements and clinkers show important variability in composition, the construction of a database of standard patterns is time consuming. Moreover, the synthesis or separation of the required pure phases is experimentally challenging. In contrast, whole-powder-pattern fitting using the Rietveld method has found widespread applications on cementitious materials. Early applications involving QPA of portland clinkers and cements (Mansoutre and Lequeux 1996; Meyer et al. 1998;

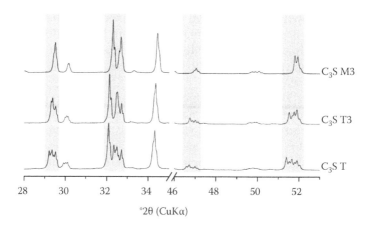

Figure 4.5 XRD scans of alite (C$_3$S) polymorphs showing clear differences in characteristic peak profiles with changing crystal structure. The comparison illustrates a clear increase in the number of diffraction peaks with decreasing crystal symmetry from monoclinic to triclinic (M3 > T3 > T). In industrial portland cements, solid solution of magnesium and sulfate into the C$_3$S structure normally leads to the stabilisation of monoclinic (M1 and M3) alite polymorphs.

Scarlett et al. 2001; Taylor and Aldridge 1993; Walenta and Füllmann 2004) were later extended towards studying the hydration of cementitious systems using both synchrotron and laboratory X-ray sources (Hesse et al. 2011; Merlini et al. 2007; Mitchell et al. 2006; Scrivener et al. 2004). QPA based on the Rietveld method has become a popular technique in studying the hydration of cementitious materials as clearly evidenced by an increasing number of publications and applications of the technique. In addition to its increasing use in research, Rietveld QPA by XRD is now being used as an online tool for quality and process control in the cement industry (Scarlett et al. 2001). This success is partially rooted in recent developments in detector technologies that have enabled a strong reduction of data collection times (Gualtieri and Brignoli 2004), while simultaneous advances in hardware computing power and calculation speed and the stability of the available Rietveld programs have rendered data analysis more efficient and accessible.

4.4 THE RIETVELD METHOD IN QUANTITATIVE PHASE ANALYSIS

The Rietveld method was originally devised for the refinement of crystal structures using neutron powder diffraction (Rietveld 1969) and was later extended to XRD (Young et al. 1977) and eventually to QPA (Bish and Howard 1988; Hill and Howard 1987). For an in-depth description of the various aspects and applications of the Rietveld method, the reader is referred to Dinnebier and Billinge (2008) and Young (1993). The following notes will focus on the application of the Rietveld method to QPA.

The Rietveld method is different from integrated peak intensity methods in that it minimises the difference between measured and calculated patterns at each data point i in the diffraction pattern using a least-squares approach. Here the sum of the statistically weighted (w_i) squared differences between observed $y_i(\text{obs})$ and calculated $y_i(\text{calc})$ intensities is minimised over the whole pattern:

$$S_y = \sum_i w_i [y_i(\text{obs}) - y_i(\text{calc})]^2. \tag{4.1}$$

The calculated intensity at a given data point is obtained as the sum of the contributions of the background bkg_i and all neighbouring reflections p of all phases j as in

$$y_i(\text{calc}) = \sum_{j=1}^{N\,\text{phases}} S_j \sum_{p=1}^{N\,\text{peaks}} m_{p,j} Lp_p \left| F_{\mathbf{p,i}} \right|^2 G_{p,j}(2\theta_i - 2\theta_{p,j}) P_{p,j} + bkg_i, \tag{4.2}$$

where S_j is the scale factor, $m_{p,i}$ is the reflection multiplicity factor, Lp_n is the Lorentz polarisation factor, $F_{p,j}$ is the structure factor, $G_{p,j}(2\theta_i - 2\theta_{p,j})$ is the peak shape function and $P_{p,j}$ is the preferred orientation correction. The calculation of the structure factors requires the crystal structures of all phases of interest to be known. The square of the structure factor F_p relates directly to the phase-specific distribution of intensities over the reflections p with reflection indices hkl. The structure factor F_p is calculated as a summation over all atoms n in the unit cell:

$$F_p = \sum_n f_n o_n \exp\left(2\pi i \left\{ hx_n + ky_n + lz_n \right\}\right) \exp(-B_n), \tag{4.3}$$

where f_n is the atomic X-ray form factor (depends on the X-ray wavelength); o_n represents the fractional site occupancy; x_n, y_n and z_n are the fractional coordinates of atom n in the unit cell and B_n represents the atomic displacement parameter.

The angular positioning θ of the hkl reflection peaks is determined by Bragg's law:

$$\lambda = 2d_{hkl} \sin\theta, \tag{4.4}$$

where λ is the wavelength of the incident X-ray radiation and the d_{hkl}-spacing values are determined by the lattice symmetry group and the unit cell parameters. Table 4.1 summarises the crystal structure parameters that impact peak intensities, positions and profiles and indicates whether they are refined in a regular Rietveld QPA.

The parameters that are usually varied in QPA are the so-called global parameters, such as the background and the sample displacement correction,

Table 4.1 Relationship between reflection peak properties and crystal structure parameters and their refinement in a standard Rietveld QPA of cements

Reflection peak properties	Crystal structure parameters	Refined in Rietveld QPA
Position	Lattice symmetry	Not applicable, fixed
	Unit cell parameters	Yes
Intensity	Atom type	No
	Atomic fractional coordinates	No
	Site occupancies	No
	Atomic displacement parameters	No
Profile shape, width	Crystallite size	Yes, usually part of analytical peak shape function
	Lattice strain	No

and phase-specific parameters, such as the scale factors that fit to the observed peak intensities, the unit cell parameters that relate to the peak positions and the shape of the peak profile function. The implementation of intensity corrections such as a preferred orientation function may be necessary depending on the nature of the phases present in the sample. Strongly one- or two-dimensionally shaped particles may tend to assume a nonisotropic orientation. The refinement of other crystal structure parameters, such as atomic positions, site occupancies or atomic displacement parameters, is not recommendable for complex mixtures of phases such as cements as this may lead to erroneous results because of strong correlations in parameter variation during fitting as the ratio between variables and independent data strongly increases and the optimisation problem becomes underconstrained.

The scale factors obtained in the Rietveld refinement scale linearly with the phase weight fraction W_j as in (Hill and Howard 1987)

$$ S_j = K \left(\frac{W_j}{\rho_j V_j^2 \mu_m} \right), \tag{4.5} $$

where K is a constant determined by the experimental diffractometer conditions, ρ_j corresponds to the unit cell density of phase j, V_j designates the unit cell volume and μ_m is the sample mass absorption coefficient (MAC). As K and μ_m cannot be obtained from the diffraction experiment, the phase weight fractions cannot be calculated directly from the scale factors. A number of practical solutions to this problem exist. The most common approach is to cancel out the unknown terms by constraining the sum of the crystalline phases considered in the refinement to 100%. This assumes that the sample contains only the considered crystalline phases. In this case the phase weight fraction is calculated by

$$ W_j = \frac{S_j \rho_j V_j^2}{\sum_j^{N \, phases} S_j \rho_j V_j^2}. \tag{4.6} $$

It should be stressed that only relative proportions of weight fractions can be obtained if the sample contains amorphous phases or if any additional crystalline phases were not included into the Rietveld refinement. This approximation is often used in QPA of portland cements or clinker when only negligible levels of amorphous phases are expected to be present (León-Reina et al. 2009; Le Saoût et al. 2011; Peterson et al. 2006; Taylor et al. 2002). As these assumptions are no longer valid for cements blended with mostly amorphous supplementary cementitious materials (SCMs) or for hydrated cements, different methods were developed to take into account the presence of nanocrystalline or amorphous phases. Amorphous phases can be visually recognised in diffraction patterns when they constitute more than 10–20 wt.% of the sample. This is the case for most SCMs. Figure 4.6 shows a series of diffraction patterns for common SCMs. The broad peak relates to the diffuse scattering maximum that indicates the presence of an amorphous phase.

4.4.1 Internal standard method

The most common approach to XRD QPA of samples containing amorphous phases is the internal standard method (Madsen et al. 2011; Scarlett et al. 2002). This method requires the sample to be 'spiked' with a known weight fraction of a crystalline standard W_s. This procedure cancels out both the K and μ_m parameters in Equation 4.5. If the standard is homogeneously mixed into the sample, the following expression can be used to calculate the absolute phase weight fractions:

$$W_j = W_s \frac{S_j \rho_j V_j^2}{S_s \rho_s V_s^2},$$ (4.7)

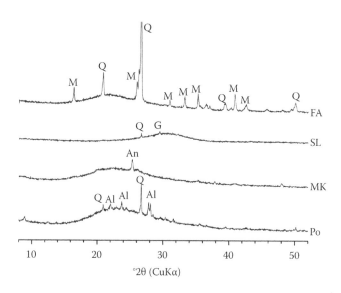

Figure 4.6 XRD scans of common SCMs: siliceous fly ash from coal combustion (FA), blast furnace slag from iron smelting (SL), thermally activated metakaolin (MK) and a volcanic tuff natural pozzolan (Po). The broad hump beneath the diffraction peaks of the crystalline phases indicates the presence of one (or more) amorphous phases. The main crystalline phases identified are mullite (M), quartz (Q), gehlenite (G), anatase (An) and albite (Al).

where S_s, ρ_s and V_s correspond to the scale factor, the unit cell density and the unit cell volume of the internal standard, respectively. Alternatively, the apparent weight fractions can be calculated using Equation 4.6; the apparent weight fraction of the internal standard can then be

1. Higher than the weighted amount, indicating the presence of amorphous or unidentified crystalline material;
2. Equal to the weighted amount, indicating that there are no amorphous or unidentified materials present; or
3. Lower than the weighted amount, indicating a problem in the experimental or data analysis procedure.

If the comparison indicates the presence of unknown or amorphous phases, then the apparent weight fractions can be corrected using the following relationship:

$$W_j = W_{j,app} \frac{W_s}{W_{s,app}} . \tag{4.8}$$

The total weight fraction of all amorphous and/or unidentified components W_{au} can then be calculated as the difference between the sum of all crystalline components and 1:

$$W_{au} = 1 - \sum_{j}^{N \text{ phases}} W_j. \tag{4.9}$$

The internal standard method requires appropriate sample preparation to homogeneously mix the standard into the sample. If the sample and standard are ground together, care should be taken that no overgrinding of the sample takes place by cogrinding with a harder standard material. Overgrinding may lead to peak broadening and partial amorphisation of softer phases in the sample. Hydrated phases in cement pastes are especially vulnerable to decomposition because of dehydration and carbonation during sample manipulation and special care should be taken to avoid sample alteration. The internal standard should be of known crystallinity (i.e. the amorphous content should be known) and should, preferably, not be present in the sample. Recently the crystallinity of the National Institute of Standards and Technology (NIST) standard reference material 676a for QPA, α-Al_2O_3, has been recertified to be 99.02% ± 1.11% (Cline et al. 2011), which enables to benchmark the crystallinity of other standard materials. Preferably a standard material should present high symmetry and have peaks that do not overlap with the main reflection peaks of the phases present in the material. In addition, microabsorption problems may occur in coarse powders if a large X-ray absorption contrast exists between the phases present in the sample. The X-ray absorption of a phase depends on the elemental composition and is a function of the X-ray wavelength, in the following X-ray absorption coefficients are given for CuKα radiation. In general X-ray absorption contrast between phases leads to an underestimation of the high-absorbing phases. This effect can be countered by decreasing the particle size of the mixture and by selecting a standard with a MAC similar to that of the material under investigation. For anhydrous cements and clinkers (μ_m = 90–100 cm^2/g), rutile (μ_m = 124.6 cm^2/g) may be a good choice; for hydrated cement pastes (μ_m = 50–80 cm^2/g), corundum (μ_m = 31.7 cm^2/g) may be the better option in terms of absorption contrast. As most standard materials are usually very fine grained (2–5 μm), an appropriate choice of standard material and adequate sample preparation allow disregarding of tedious microabsorption corrections.

The obtainable accuracy of the quantification of the amorphous and unknown phases strongly depends on the amount of internal standard added. A mathematical development of the equations by Westphal et al. (2009) demonstrated that small quantities of amorphous phases (small percentages) are very difficult to quantify precisely using the internal standard approach. Higher levels of amorphous phases (above 10 wt.%) can be quantified with greater accuracy. Therefore, it has been suggested that in order to carry out a high-accuracy quantification of the amorphous phase in an unknown sample, it is advisable to carry out a two-step analysis procedure: a first step to estimate the amorphous level, using a relatively large proportion of internal standard (50 wt.%), and a second step in which the amount of the internal standard is adapted to obtain the optimal accuracy. Obviously, one of the drawbacks of using high additions of internal standard is that the sample will be diluted in the analysis and that the accuracy of the results on the crystalline materials in the sample will be lowered. For hydrating (blended) cements, which may contain between 30% and 90% of amorphous, nanocrystalline or unknown phases depending on hydration age and formulation, a good compromise would be to use additions of internal standard of around 20 wt.%. As it has been demonstrated that the presence of a significant amount of internal standard in a hydrating paste changes the kinetics of hydration through the filler effect (Gutteridge and Dalziel 1990), an internal standard should best be added after stopping the hydration. In mixing the standard into the hydrated sample, care should be taken not to alter the hydration products. Mixing operations should be thorough but gentle and sample exposure to air should be limited to avoid both hydrate decomposition and carbonation. As the latter can be tedious to achieve in practise, the less labour-intensive external standard method may be preferable for routine analyses.

4.4.2 External standard method

In the external standard approach, phase quantification is carried out through the comparison of the phase scale factors of a sample to the scale factor of a well-characterised standard material measured under identical diffractometer conditions. In this procedure only the constant parameter K is cancelled out and the different MACs of the sample, μ_m, and the standard, μ_s, need to be taken into account. The following expression can be developed from Equation 4.5 (Man Suherman et al. 2002; O'Connor and Raven 1988):

$$W_j = \frac{S_j \rho_j V_j^2}{S_s \rho_s V_s^2} w_s \frac{\mu_m}{\mu_s}, \tag{4.10}$$

where w_s is the weight fraction of the standard phase in the external standard material, which also takes into account the crystallinity of the standard material. The crystallinity of the standard material can be assessed by comparison to the available NIST-certified standard reference material.

Although the obvious advantage of this method is that the sample does not need to be intermixed with an internal standard and thus avoids additional sample preparation and homogenisation problems, this approach does demand that the MAC of the sample be known. In the most practical way it is obtained by calculating the MAC for the used X-ray wavelength from the chemical composition, for cements usually available from X-ray fluorescence (XRF) spectrometry (Jansen et al. 2011). X-ray absorption coefficients were tabulated by Creagh and Hubbell (2006). This approach was applied to anhydrous cements (Jansen et al. 2011) and to early in situ hydration experiments recently (Bergold et al. 2013; Jansen et al. 2012). In studying hydrated cements, the calculation of the sample MAC should also take into account the water content of the sample. For closed, undried samples the initial water content of the mix may be adopted; for samples from which water is removed to stop the hydration reactions an additional measurement of the remaining bound water content, for instance, by thermogravimetry, is needed (Chapter 5).

4.5 DATA COLLECTION

Before data are collected, it should be ascertained that the diffractometer system is well aligned and that a suitable measurement configuration is chosen. In case of a flat-plate reflection geometry, a correct combination of optic elements (slits and monochromator system) and sample size should ensure that neither sample transparency nor beam overflow at low angles occurs. The optimal configuration depends on the diffractometer and the purpose of the analysis. A thorough treatment of diffractometer alignment and measurement configurations is provided in reference textbooks on XRD (Dinnebier and Billinge 2008; Jenkins and Snyder 1996; Klug and Alexander 1974) and is not repeated here.

Data collection strategies should aim at collecting a maximum of useful information from the sample. The data collection range for portland cements is usually chosen to be within the range 7°–70° 2θ for CuKα radiation but can be wider or more restricted depending on the problem. The 7°–70° 2θ (CuKα) range provides important information on hydrate phase reflections at low angles and covers the most intense reflections of both anhydrous and hydrates phases. For most in-house diffractometers a step size of about 0.02° 2θ is used and the counting time per step is usually a compromise between the optimal signal-to-noise ratio and practical considerations, such as available measurement time or acceptable sample exposure time. The development of high-performance X-ray detector systems has allowed to drastically reduce measurement times, thus minimising the exposure time of hydrated samples to the low-humidity environment inside the diffractometer. It is advisable to spin the sample on the horizontal plane to improve the particle statistics of the measurement.

The quality of the final characterisation and quantification results depends on the quality of the collected data. The peak-to-background intensity ratio and the peak resolution are the most important parameters for assessing the quality of a diffraction scan. Increasing peak-to-background ratios can be done by increasing measurement times, increasing X-ray source intensity or decreasing background scattering intensity. Conditioning of the incoming and reflected X-rays by diffractometer optics (slits, mirrors, etc.) is in this respect of great importance and should be optimised in view of the application. Increasing the peak resolution often involves cutting the part of the incoming and reflected intensity that leads to peak broadening and thus reduces overall peak intensities. Thus, in practise a compromise needs to be found among different data quality measures, also taking into account practical considerations, such as available measurement time and acceptable sample exposure time. An example of the dependence of data quality on equipment is presented in Figure 4.7. Here, a comparison is made between the diffraction patterns of two different anhydrous portland cements, one measured by a conventional laboratory diffractometer and the other by a dedicated synchrotron XRD beamline (European Synchrotron Radiation Facility [ESRF], ID31 beamline). The main advantage of using synchrotron facilities is the much higher monochromaticity and brilliance (intensity) of the X-ray radiation. This greatly improves peak resolution and enables much shorter data collection times (few minutes), which is useful for early-age cement hydration experiments (Cuberos et al. 2009; Merlini et al. 2007; Snellings, Mertens et al. 2010).

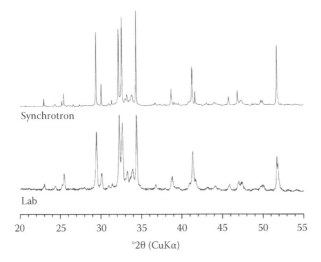

Synchrotron

Lab

°2θ (CuKα)

Figure 4.7 Comparison of XRD scan qualities obtainable at a conventional laboratory diffractometer and a dedicated synchrotron facility (ESRF, ID31). The synchrotron data are replotted on a degree 2θ (CuKα) basis for comparison. Note that the synchrotron X-ray source is monochromatic, while the laboratory equipment collects both CuKα$_1$ and CuKα$_2$ radiations.

4.6 SAMPLE PREPARATION

Appropriate sample preparation is critical in X-ray powder diffraction analysis. A prerequisite for accurate and precise analysis is that the intensities and peak positions in the primary data are unbiased by sample preparation or sample-inherent problems. As quantitative powder diffraction analysis relies on the fitting of representative reflection intensities, two basic conditions need to be satisfied, i.e. (1) that the number of crystallites (coherently diffracting domains) able to diffract at a certain angle is infinite or very large and (2) that the crystallites are randomly oriented (Bish and Reynolds 1989). The former is commonly referred to as particle statistics and the latter as preferred orientation and it is the purpose of sample preparation and data collection strategies that ideal conditions are approached as closely as possible for a specific material. Typically, this requires a powder to be finely ground, ideally to particle sizes of less than 5 μm. In addition, for XRD experiments in flat-plate reflection geometry (Bragg–Brentano configuration), the preferred orientation should be minimised while mounting the sample and the sample surface should be flat and smooth. The sample surface area and thickness should be large enough to avoid beam overflow and sample transparency aberrations. For cement samples, a sample thickness of at least several hundred micrometres is generally sufficient when using CuKα radiation on laboratory instruments. General guidelines regarding sample preparation for XRD analysis have been described elsewhere (Bish and Reynolds 1989; Buhrke et al. 1998). However, due to the sensitive nature of many phases in hydrated cements, some additional issues in the sample preparation procedure need to be considered. The following sections will focus therefore on sample preparation strategies for (hydrated) cements.

4.6.1 Particle statistics

The repeatability of the intensities in a XRD measurement depends on the number of randomly oriented crystallites. The higher the amount of crystallites in the sample, the lower is the scatter in intensities and the better is the repeatability of the analysis. Elton and Salt (1996) estimated that the particle statistics error in measured intensities would be around 0.6% in pure quartz samples for a mean particle size of 5 µm, increasing to 13.4% for 40 µm particle sizes. In the case of anhydrous portland cement, for an alite phase present at 60 wt.%, the higher MAC would lead to higher particle statistics errors of 1.3% and 28.8% for the respective particle sizes of 5 and 40 µm. Fortunately, whole-pattern methods average scale factors over a large number of peaks and thus partially mitigate the effect of particle statistics errors on the quantification results for coarser samples. In addition, sample spinning during data collection enables more particles to diffract at a given angle and reduces the error in the measured intensities by a factor of 5 (Elton and Salt 1996). Sample spinning during data collection is strongly encouraged therefore (Aranda et al. 2012). Klug and Alexander (1974) reported that repeatable intensities were obtainable for particle size below 15 µm for quartz samples (μ_m = 36.0 cm^2/g) using CuKα radiation. Acceptable particle sizes for more highly absorbing materials such as anhydrous cements ($\mu_m \approx$ 100 cm^2/g) should be expected to be even lower (below 10 µm).

Insufficient sample fineness is one of the most common sources of error in QPA of portland cements. As a considerable fraction of commercial cements is coarser than 10 µm and alite crystallites especially can be much larger, particle statistics errors can be significant and result in spiky patterns in which some peaks may be unexpectedly sharp, while other peaks may be too low. Differences in peak shapes may lead to problems in profile fitting and the artificial detection of an amorphous phase (Snellings et al. 2014a). Figure 4.8 reports the diffraction scans of synthetic alite ground for 30 s (d_{50} of 36 µm) and 20 min (d_{50} of 3.5 µm) in a McCrone micronizing mill® using isopropanol as the grinding agent. Adequate sample preparation leads to a more regular peak shape and improves the peak profile fit and thus the quantification results. Sample comminution by further grinding is therefore advisable to improve the precision and accuracy of the QPA by XRD. Also, microabsorption (Brindley 1945) and extinction (Sabine 1993) effects are best mitigated by reducing the particle size of the powder,

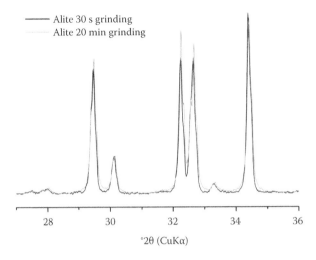

——— Alite 30 s grinding
········ Alite 20 min grinding

°2θ (CuKα)

Figure 4.8 Selected angular range of the XRD scans of synthetic alite (C_3S M3) ground for 30 s and 20 min. The reduction in particle size leads to a more representative and more regular peak profile that is better fitted by analytical peak functions.

as applying any postmeasurement corrections is experimentally unfeasible for complex multicomponent systems (Le Saoût et al. 2011).

Grinding should occur without sample fractionation (Gutteridge 1984), phase transformation, decomposition or amorphisation (e.g. transformation of gypsum to bassanite or anhydrite during dry grinding [Füllmann and Walenta 2003]). These overgrinding issues can be overcome by using a liquid grinding aid that dissipates the heat produced on grinding impact and largely prevents structural damage to the sample (Petrovich 1981). For portland clinkers and cements, isopropanol, ethanol or acetone can be used. Chemical reactions between the cement components and the grinding aid must be avoided. After grinding, the ground powder is usually recovered by evaporation of the grinding aid. Unfortunately, fine wet grinding cannot be applied to hydrated cements without problems caused by the dehydration and the decomposition of hydrate phases during grinding in the solvent and by the sample sensitivity to carbonation during drying or liquid separation. To avoid changing the hydrate assemblage during grinding, currently reported sample preparation procedures do not involve fine grinding of hydrated samples. Instead, either relatively coarsely ground powders (Lothenbach, Matschei et al. 2008; Winnefeld and Lothenbach 2010) or slices cut from a hardened paste sample (Scrivener et al. 2004) have been used. As both powders and slices represent aggregates of many smaller crystallites, particle fineness problems will usually not be exacerbated compared to the original anhydrous cement.

4.6.2 Sample mounting

In mounting the obtained powder/slice onto the sample holder for quantitative XRD measurement, the following points should be addressed: In flat-plate reflection geometry, care should be taken that the sample surface area and thickness are large enough to avoid beam overflow at low angles or sample transparency problems. Too small sample surface areas will lead to an underestimation of reflection intensities at low angles and may also result in spurious diffraction peaks or an increased background originating from the sample holder. In case smaller sample holders are used, the diffractometer optics should be adjusted by choosing smaller divergence slits. Especially for phases that show prominent peaks in low-angle regions such as the ettringite or AFm phases, beam overflow will lead to an underestimation of their content in QPA. In case of fine, well-packed samples, sample thickness problems are less important as most commercially supplied sample holders are at least 1 or 2 mm deep and the X-ray penetration depth will not exceed 500 μm in general. Sample transparency may form a problem for coarse and very loosely packed samples or shallow sample holders.

Other requirements are that the mount surface should be flat, without roughness and not tilted, in order to avoid peak displacement or broadening associated with sample height differences. For hydrated cements, cut slices can be polished using sandpaper to remove surface inhomogeneities. The slice surface should be carefully positioned at the aligned sample height for measurement. Rough powder surfaces prepared by front-loading can be smoothened using a spatula or frosted glass surface. Care should be taken not to induce the preferred orientation of sample particles during sample loading. Preferred orientation of platy or needlelike particles is easily produced by front-loading techniques by pressing the sample into the holder. In anhydrous cements, particularly gypsum (0 2 0), bassanite (0 0 1), anhydrite (2 0 0) and calcite (1 0 4) particles can be affected by preferred orientation (Le Saoût et al. 2011). The effect of sample loading on the preferred orientation of anhydrite and alite is illustrated in Figure 4.9, where XRD scans of front-pressed, front-loaded and back-loaded sample mounts are compared.

In hydrated cements, platy crystallites of portlandite (0 0 2) and AFm phases (0 0 *l*) and needlelike ettringite (1 0 0) crystals may undergo preferred orientation. Preferred orientation effects on peak intensities can be corrected to a certain extent using a March–Dollase function (Dollase 1986) or a spherical harmonics correction (Ahtee et al. 1989). As phase scale factors and QPA results are strongly correlated with the preferred orientation functions, their unconstrained use should be discouraged to prevent parameter drift and false convergence. The use of preferred orientation corrections should be limited as much as possible by appropriate sample mounting techniques. As a general rule, it is better to prepare the sample appropriately than to try and correct preferred orientation problems

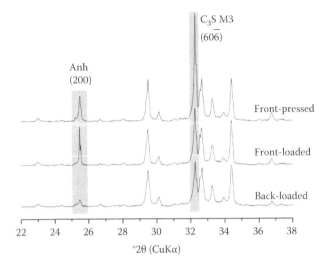

Figure 4.9 Effect of sample loading on preferred orientation of anhydrite and alite.

in the analysis. Compared to the standard front-loading technique, back or side loading has shown to strongly reduce preferred orientation effects for cement powders (Le Saôut et al. 2011). Compared to packed powders, using cut slices of hydrated cements has the advantage of having preserved the more or less random particle orientation of the cement paste after mixing. A potential disadvantage of using slices is that some unground systems may show the presence of large crystals of hydration products. This has been shown to be the case for portlandite in monophase C_3S/alite systems (Gallucci and Scrivener 2007), resulting in a significant bias (underestimation) in portlandite content in an early age in situ hydration study of alite pastes by XRD (Bergold et al. 2013).

4.6.3 Hydration stoppage

Most characterisation techniques of hydrated cements need the hydration of the cement to be stopped or the sample to be dried. Even if hydration stoppage is not strictly necessary for XRD-based studies, this is common practise mostly for practical reasons. Arresting the hydration enables sample storage and the characterisation of the same sample at an identical degree of hydration by a range of complementary techniques (e.g. thermogravimetric analysis, electron microscopy and nuclear magnetic resonance [NMR] spectroscopy). The primary aim of the various available methods of hydration stoppage is that the properties of interest of the hydrating cement, the phase composition in case of XRD, are preserved as much as possible. To stop the hydration of cement, unbound water present in

the capillary pores needs to be removed. Water present in the smaller gel
pores or chemically bound into the hydration products should preferably be
retained as its removal may cause hydrate alteration or decomposition and
microstructural changes. Two main ways of removing water exist, either by
direct drying (oven drying, D-drying, vacuum drying or freeze drying) or
by solvent exchange in which water is first replaced by an organic solvent
(e.g. isopropanol, acetone, diethyl ether or pentane), which is then evaporated. A recent review paper provides an extensive comparison of the effects
of the different approaches on the preservation of the hydrate assemblage
and the microstructure (Zhang and Scherer 2011). Below, only the effects
of different hydration stoppage methods/sample preparation procedures on
the hydrate phase assemblage as measured by XRD will be considered.
Hydration stoppage procedures and their more general effects are treated
in detail in Chapter 1.

Zhang and Glasser (2000) considered the effects of direct drying methods on the stability of ettringite/AFm mixes as model systems for cements.
They observed that oven drying and vacuum drying resulted in the removal
of structural water and decomposition of the ettringite and AFm phases.
Oven drying was observed to be more destructive than vacuum drying. It
appears that all direct drying methods result in the (partial) dehydration
of cement hydrates. In the case of crystalline products, this may lead to
decomposition and an underestimation of the phase content on hydration-
stopped samples. Ettringite and AFm are much more sensitive to decomposition by dehydration than portlandite.

Solvent exchange techniques represent a more gentle way of removing
water that better preserves the microstructure (Zhang and Scherer 2011).
The rate of solvent exchange is governed by diffusion and depends on the
sample dimensions. Solvent exchange on cut slices may take 1 week or longer to be completed. This is too slow to monitor early hydration processes;
in that case faster solvent exchange can be obtained by crushing the sample
followed by a shorter period of exchange. Samples crushed to pass a 63 μm
sieve can thus be exchanged in less than 1 hour (following the approach of
Zhang and Scherer [2011] to calculate the required time for exchange); a
50 μm cement particle can be expected to be isopropanol exchanged in less
than 3 min. The effect of hydration stoppage by solvent exchange on the
hydrate assemblage is illustrated in Figures 4.10 and 4.11 by XRD patterns
and quantification results for a 28 days hydrated blended cement containing fly ash and limestone. A comparison is made between cut slices and
crushed powders that were measured either fresh immediately after cutting
or crushing or after hydration stoppage by isopropanol exchange during
1 week. The cut slices were measured with and without surface polishing.
The measurement duration was 30 min. It was observed that isopropanol
exchange of 1 week (slice thickness is 2 mm) resulted in a reduction in
the ettringite and hemicarbonate quantification results of about 50%. The
monocarbonate and portlandite phases appear more resistant to solvent

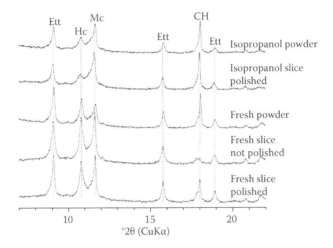

Figure 4.10 Effect of the sample preparation procedure on the hydrate assemblage as measured by XRD. The following hydrate phases were monitored: ettringite (Ett), hemicarbonate (Hc), monocarbonate (Mc) and portlandite (CH). (Courtesy of P. Durdziński.)

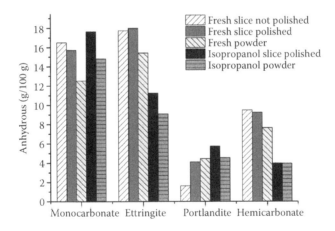

Figure 4.11 Effects of sample preparation on the quantification of the hydrate assemblage. Rietveld analysis and the external standard method were used for QPA. (Courtesy of P. Durdziński.)

exchange. Sample crushing led to a decrease in monocarbonate peak intensities. Unpolished fresh slices were deficient in portlandite and contained 1.5% more calcite (not shown), indicating rapid surface carbonation and/or portlandite dissolution during cutting. It should be noted that procedures that involve a shorter solvent exchange as described in Chapter 5 result in less ettringite/AFm decomposition and are suitable for XRD studies as well.

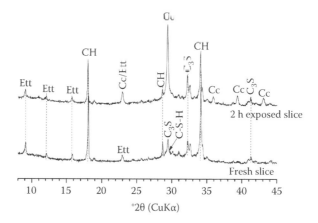

Figure 4.12 Effect of atmospheric exposure on the hydrate assemblage of a blast furnace slag-blended cement hydrated for 3 days. Clearly, portlandite (CH) is
affected by atmospheric carbonation and calcite (Cc) is formed. Ettringite
(Ett) is more resistant to atmospheric exposure.

Storage conditions should be closely monitored to minimise the rate of
sample degradation. In general, samples are stored in low-humidity, low-
vacuum desiccators. Exposure to ambient air and humidity should be limited; the hydrate assemblage of samples will be preserved over longer times
when the samples are stored in block form and ground freshly before measurement rather than when they are kept in a powdered form. The duration
of storage before analysis should be limited. Exposure of non-hydration-
stopped samples to the atmosphere leads to a quick degradation of the
hydrate phases as shown in Figure 4.12. The same slice sample remeasured
after 2 h exposure to ambient air shows the conversion of portlandite and
other hydrates to calcite. As XRD in regular flat-plate reflection geometry
mainly measures the upper micrometres of the sample, effects such as carbonation are readily detected. This is why XRD measurements not only on
fresh but also on hydration-stopped samples should be limited in time, for
instance, to 15–30 min.

4.6.4 Guidelines for reporting X-ray diffraction data collection

Enabling the reproduction of results by others is an essential part of the
scientific method. Therefore, when reporting results in a report, thesis or
publication, it is of great importance to carefully report how the results
were obtained. In this respect, the data acquisition equipment and procedure should be concisely but completely described in a methods section.
Table 4.2 shows a list of settings and parameters that should be detailed
when reporting of the measurement procedure.

Table 4.2 Reporting guidelines for XRD data collection

Data collection properties and settings		Examples
Equipment	Manufacturer	PANalytical, Bruker, Rigaku
	Model	X'Pert³ Powder, D8 Advance
Diffractometer geometry	Measurement setup	Bragg–Brentano, transmission; θ–θ, θ–2θ
	Goniometer radius	240 mm
X-ray source	X-ray radiation	$CuK\alpha_{1,2}$ (λ = 1.5408 Å)
	Generator operation	45 kV, 40 mA
Diffractometer optics (devices depend on equipment)	Incident divergence slit	0.5° (fixed, programmable)
	Incident antiscatter slit	0.5
	Incident Soller slits	0.04 rad
	Specific beam-conditioning devices	Johansson monochromator, Göbbel mirror
	Receiving antiscatter slit	0.5° (fixed, programmable)
	Receiving Soller slits	0.04 rad
Detector	Type	Point scintillation counter, linear position sensitive (one dimensional [1D]), two-dimensional position sensitive
	Model	X'Celerator, Lynxeye
	Scanning mode	Step, continuous
	Detector length	2.122° 2θ (1D detector)
Sample	Dimensions	26 mm diameter, 25 × 15 mm²
	Spinning speed	8 rpm
	Sample type	Powder, slice
	Sample pretreatment	Grinding, hydration stoppage (cf. Chapter 1)
	Sample loading	Back loading, side loading
Scan parameters	Angular range	5°–70° 2θ
	Step size	0.02° 2θ
	Time per step	30 s
	Total measurement time	30 min

4.7 GUIDELINES FOR QUANTITATIVE ANALYSIS (RIETVELD METHOD)

The QPA of hydrated cements by XRD is not straightforward. It is important to stress that the XRD data of complex systems should not be analysed by 'black box' approaches. Round-robins on QPA have revealed that the accuracy of the results, even for simple systems, is highly dependent on the adopted analysis strategy and the skill and experience of the analyst (Madsen et al. 2001; Scarlett et al. 2002). The experimenter should have a basic understanding of crystallography and cement chemistry to be able to guide the Rietveld refinement and to judge the quality of the results. Practical experience can be gained by practising on simple model mixtures and/or training by experienced researchers/groups. Many of the guidelines described below have been developed within one group; however, they generally comply with earlier published accounts (Aranda et al. 2012; Le Saoût et al. 2011; Stutzman 2011) (not specifically applied to hydrated cements), and can be regarded as having a general value. It should be noted, though, that specific details in the refinement strategies will depend on the system under investigation and the functionalities/capabilities of the Rietveld analysis programs used. A Rietveld QPA can be broadly subdivided into a phase identification and crystal structure selection part and a quantitative analysis part in which selected parameters are varied in a data-fitting procedure. In the case of hydrating cements the amount of solids varies because of the bonding of water into hydration products. An additional step should then be to rescale the results to a common basis (e.g. the initial clinker/cement content) to enable easy comparison between different stages of reaction. Practical guidelines for the different stages in Rietveld QPA are given in the following sections.

As a general comment, it is highly recommended to consistently use the same data analysis strategy for similar systems and to compare the quantification outcome with the results of other independent techniques, such as thermogravimetry, microscopy, calorimetry and NMR spectroscopy. For hydrated cements, the analysis results can also be checked by mass balancing the hydration reactions if sufficient information on the reactants and products is available.

4.7.1 Crystal structure models

Once a phase has been identified to be present, a suitable crystal structure description needs to be chosen to enable the calculation of the structure factors in the Rietveld master equation (Equation 4.2). The crystal structure models form the basis of the Rietveld QPA. A large number of published crystal structures have been collected or deposited in commercial and open-source databases (ICSD, AMCSD and COD). Available crystal structures for a specific phase can be searched and compared. Changes in the crystal structural description can have an important impact on the quantification results. In the case of cements, some crystal structure descriptions are more complete than others or represent better the phases encountered. For a specific phase, the crystal structure models may differ significantly in terms of the atomic displacement parameters, in the reporting of the position of light elements such as hydrogen and in the site occupancies for solid solution series such as ferrite (Al-Fe substitution) or hydrogarnet (Si-H substitution). In the case of complex structures, such as alite, different (superstructure) symmetries can be reported. Selecting the most appropriate crystal structures is therefore of great importance in obtaining good quantification results. Important selection criteria are the correspondence between the calculated crystallographic density and the measured density, the reporting of chemically reasonable atomic displacement parameters and, obviously, the attainment of a good fit to the available experimental data. The use of chemically reasonable atomic displacement parameters is of particular importance as they are directly correlated with the calculated scale factors. Too high values will result in an overestimation of the phase content; too low values, in an underestimation (Snellings et al. 2010). As a general aid to crystal structure selection, Tables A4.1 through A4.7 present a database of phases encountered in anhydrous cements, in SCMs such as slags or fly ashes and in hydrated cements. Some hydrate phases are found only in hydrated cements deteriorated by chemical attack. The provided list is not exhaustive. Other phases can be present in special cements, and phases encountered in natural pozzolans or raw materials for activated clays have not been included.

Finally, the selected crystal structure models are imported or entered in a so-called control file that acts as an input initial model for the Rietveld QPA. In importing the crystal structure files, care should be taken that all parameters are entered/transferred correctly. Some software does not automatically import the atomic displacement parameters or may not insert them in the correct format.

4.7.2 Refinement strategies

Having compiled a control file containing the structural information of the identified phases, an XRD pattern can be calculated using the Rietveld method to compare to the measured pattern. Through a least-squares fitting algorithm, selected parameters are optimised to obtain the best possible fit between the calculated pattern and the measured pattern. In QPA of cementitious systems the phase-dependent parameters that are varied should be limited to (1) the phase scale factors, (2) the unit cell parameters, (3) the peak shape parameters or microtextural parameters and, if needed, (4) the preferred orientation parameters. Next to phase-dependent parameters, global parameters also need to be refined: (1) a specimen displacement correction and (2) coefficients of the background function. In complex systems such as cements, the refinement of crystal structural parameters such as the atomic positions and displacement factors should best be avoided. Crystal structure refinement by the Rietveld method should best be reserved to pure systems.

The order in which the parameters are varied generally depends on the distance to their final values, the so-called parameter shift. Parameters that will experience a large shift need to be refined first. Depending on the performance of the optimisation algorithm, damping factors may need to be placed to limit the parameter variation between the iterative optimisation cycles.

Usually, phase scale factors and global parameters are refined first, while phase-dependent lattice parameters and peak shape parameters are refined in later stages. Intensity corrections, such as preferred orientation functions, are optimised last. Depending on the program, this parameter variation sequence can be preprogrammed or needs to be carried out manually.

Parameter correlation during the optimisation routine may result in false results. For instance, lattice parameters may diverge from their original values to such an extent that chemical bond lengths and coordination within the structure are beyond what is chemically acceptable. Therefore, to control such parameter drift during refinement, some parameters can be constrained to remain within a certain value interval considered as being chemically reasonable (e.g. a relative interval of 1% around the initial value). Phase-dependent lattice and peak shape parameters may be constrained as such. If during refinement some parameters converge to the set limits, this may indicate a problem such as the presence of an additional phase or the use of an incorrect structure model.

Not all phase-dependent parameters can be optimised for minor phases, for instance, because of peak overlap problems with major phases. This may be solved by resolving the crystal structural parameters for minor phases in concentrates such as selective dissolution residues (Le Saoût et al. 2011). The optimised crystal structural parameters may then be fixed in the refinement of the original bulk sample and will allow obtaining better constrained quantification results for the concerned minor phases. This strategy is particularly valuable for resolving alkali sulfates in anhydrous cements.

A similar strategy may be adopted for hydrated cements. When analysing a hydration time series of a specific system, such as in Figure 4.13 for a zeolite-blended cement, all parameters of the anhydrous phases may be fixed except for the scale factors. The phase-dependent parameters of the hydration products should be allowed to vary as their stoichiometry and microtexture may change during hydration. This approach strongly limits parameter drift and is very instrumental in obtaining consistent trends in hydration studies.

One of the most important problems in quantitative analysis of hydrated cements by XRD is the presence of nanocrystalline phases such as C-S-H, the main portland cement hydration product. Additional amorphous phases may be present, such as unreacted blast furnace slag or the glassy phases of fly ash in blended cements. Diffuse scattering from these phases results in the appearance of broad peaks that cannot be calculated by the conventional Rietveld method that assumes three-dimensional long-range order. The major difficulty in QPA is that the broad peaks overlap with peaks from other crystalline phases, for instance, C-S-H and belite, and that the cumulated intensities should be distributed correctly over the overlapping phases. This may also occur for ternary blended cements consisting of cement and two different amorphous phases. Figure 4.14 presents the decomposition of the diffraction pattern of a ternary metakaolin–blast

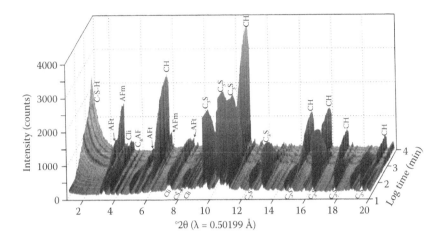

Figure 4.13 Hydration stack over time of a zeolite-blended portland cement as measured by synchrotron radiation; indicated are ettringite (AFt), monosulfate (AFm), clinoptilolite (zeolite; Cli), portlandite (CH) and alite (C_3S). (Reprinted from *Cement and Concrete Research*, 40, Snellings, R., G. Mertens, O. Cizer and J. Elsen (2010), Early age hydration and pozzolanic reaction in natural zeolite blended cements: Reaction kinetics and products by in situ synchrotron x-ray powder diffraction, 1704–1713, Copyright 2010, with permission from Elsevier.)

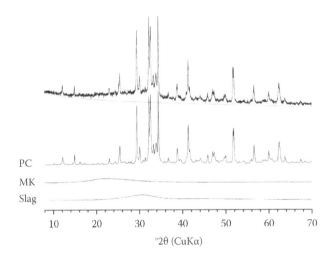

Figure 4.14 XRD pattern of a ternary blended cement and its decomposition into the component diffraction patterns of portland cement, metakaolin (MK) and blast furnace slag. The diffraction patterns of the component phases were measured separately and recombined in the decomposition procedure.

furnace slag–portland cement into its component phases. The diffraction patterns of the separate component phases were available as input for the decomposition.

If intensities are wrongly distributed, this results in a significant bias in the quantification results. This problem has received relatively little attention so far but appears to be treated either (1) by manual interpolation of the background (Kocaba 2009) or (2) by using a combination of an analytical background function and a set of broad peaks to cover the contribution of the C-S-H and potential residual amorphous phases (Snellings et al. 2010). A visual interpolation of the background has the advantage of being flexible; however, it depends strongly on operator-based decisions and is very difficult for patterns that show strong overlap. Background levels may be taken too high if many weaker peaks overlap and will consequently result in an underestimation of the crystalline peak intensities and phase levels. On the other hand, the use of an analytical background function combined with separate peak contributions may cope better with peak overlap but may be sensitive to parameter correlation. A promising development here is the blending of a profile summation method with the Rietveld method using the so-called partial or no known crystal structure (PONKCS) approach (Scarlett and Madsen 2006). This method takes into account the contribution of a phase that has no or no fully known crystal structure by calibration of the 'phase constant', the $\rho_j V_j^2$ in Equation 4.6, in a mixture in which the content of the phase is known. This phase constant can then be used in combination with the refined scale factor to calculate the phase weight fraction in unknown mixes. This method was recently applied to the quantification of the degree of reaction of metakaolin in alkali activation (Williams et al. 2011) and in the quantification by XRD of C-S-H in the early hydration of alite (Bergold et al. 2013). The precision and accuracy of the PONKCS approach to the quantification of amorphous SCM (blast furnace slag and metakaolin) levels in blended cements was assessed recently in synthetic model mixes (Snellings et al. 2014b). For mixes in which the SCMs were the sole unknown/amorphous components next to a series of crystalline phases, excellent precision (around 1 wt.%) and accuracy (2–3 wt.%) of the SCM quantification results were found. This analytical performance is similar to the expected errors for quantitative XRD on the crystalline phases in anhydrous cements (León-Reina et al. 2009; Le Saoût et al. 2011).

In more complex mixes the errors and detection limits are expected to be larger. A particular difficulty in decomposing hydrated cements is acquiring a suitable model for the C-S-H phase. Figure 4.15 shows a white portland cement hydrated for 7 years. C-S-H is the main amorphous phase in this sample and a C-S-H peak model, thus, can be fitted and used to decompose similar hydrating cements. The situation becomes more complex for C-S-H formed in blended cements, which may show a slightly different peak profile and therefore quantification errors may increase.

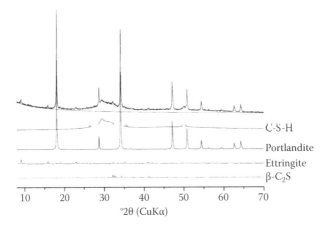

Figure 4.15 Pattern decomposition of a 7-year-hydrated white Portland cement. C-S-H is the main component of the cement and fitted using a 'peak model'. The other phases are crystalline and do not affect the C-S-H peak profile.

4.7.3 Back calculation of Rietveld quantitative phase analysis results

The total mass of solids does not stay constant during cement hydration; free water is bound into hydration products (Figure 4.16) and the initial porosity is filled. Rietveld results normalised on a total solids base for dried samples will need to be corrected. Therefore, the final step in the data analysis is to recalculate the QPA to a common basis. This base of comparison may be the initial paste (free water + anhydrous binder) content, the initial binder (SCM + portland cement) or the initial portland cement or clinker content. The recalculation depends on whether free water is removed to stop the hydration reactions. If the sample is assumed to be undried and

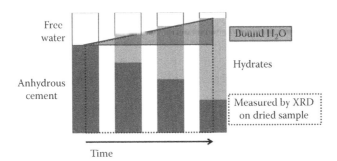

Figure 4.16 Diagram illustrating the need for back calculation of the Rietveld quantification results to a common basis (for instance, 100 g of anhydrous cement or 100 g of cement paste) to enable comparison of time series data.

cut as a fresh slice, the correction factor contains the water-to-cement ratio (w/c) of the paste. If the sample is dried, a separate measurement of the bound water content (H_2O_{bound}) by thermogravimetry is needed to rescale the results. The following expressions were developed for the case of an undried, fresh hydrated sample and for a dried hydrated sample, the bound water content should be expressed on an ignited basis:

- Fresh specimen:

Per 100 g paste: $W_{j,rescaled} = W_{j,Rietveld}$ (4.11)

Per 100 g anhydrous: $W_{j,rescaled} = W_{j,Rietveld} (1 + w/c)$ (4.12)

- Dried specimen:

Per 100 g paste: $W_{j,rescaled} = W_{j,Rietveld}/[(1 - H_2O_{bound})(1 + w/c)]$ (4.13)

Per 100 g anhydrous: $W_{j,rescaled} = W_{j,Rietveld} (1 - H_2O_{bound})$ (4.14)

4.7.4 Guidelines for reporting X-ray diffraction analysis results

When reporting the results of qualitative and quantitative phase analyses, a minimal amount of information should be presented regarding the data analysis procedure to enable verification, reproduction and comparison of the results. As a guideline, Table 4.3 supplies a list of properties and settings that characterise a QPA routine; some parameters are specific to Rietveld analysis, but most are applicable to other XRD analysis methods as well.

Finally, the presented practical guidelines are illustrated by two typical applications of quantitative phases analysis in the study of cements: (1) analysis of a CEM I 52.5 N anhydrous cement and (2) analysis of the same cement hydrated for 3 days. The chemical composition of the cement is given in Table 4.4, together with the calculation of the cement's MAC. The latter is calculated in a straightforward manner using a linear combination of the element or oxide weight fraction X_i and the respective oxide or element MAC μ_i as in

$$MAC_{sample} = \sum_{i=1}^{n} \mu_i X_i. \tag{4.15}$$

The calculation in Table 4.4 departs from the oxide composition as measured by XRF. The loss of ignition (LOI) (2 h at 1050°C) was assigned to

Table 4.3 Guidelines for XRD data analysis

Data analysis properties and settings		Examples
Analysis software	Qualitative analysis	Diffrac.Suite Eva, HighScore
	Quantitative analysis	Topas (Academic), HighScore (Plus), GSAS, FullProf
Phase information	Pattern references	PDF number, COD number
	Crystal structure references	ICSD number, publication reference
Refinement strategy	Refined global parameters and sequence of refinement	Specimen displacement, background type and number of coefficients[a]
	Refined phase-specific parameters and sequence of refinement	Scale factors, lattice parameters, peak shape type and parameters, site occupancies,[a] preferred orientation correction[a]
	Parameter constraints	Limits on lattice parameter shift, peak shape parameters
	Fixed parameters	Atomic positions, atomic displacement parameters
Figures	Measured diffraction patterns, calculated patterns and difference plot	At least one figure enabling a visual assessment of the data collection and analysis quality
Numerical criteria of fitting	Agreement indices	R_{wp}[b], R_p[c]
Result presentation	Back-calculation approach	g/100 g anhydrous g/100 g paste

[a] If applicable.

[b] $R_{wp} = \left\{ \sum w_i \left[y_i(\text{obs}) - y_i(\text{calc}) \right]^2 \Big/ \sum w_i \left[y_i(\text{obs}) \right]^2 \right\}^{1/2}$.

[c] $R_p = \sum \left| y_i(\text{obs}) - y_i(\text{calc}) \right| \Big/ \sum y_i(\text{calc})$

Table 4.4 Anhydrous cement (CEM I 52.5 N) chemical composition and MAC

Oxide	wt.%		μ (cm²/g)	MAC (CuKα) (cm²/g)
CaO	63.2		124.04	78.4
SiO$_2$	19.3		36.03	7.0
Al$_2$O$_3$	3.7		31.69	1.2
Fe$_2$O$_3$	4.1		214.9	8.8
MgO	2.1		28.6	0.6
Na$_2$O	0.28		24.97	0.1
K$_2$O	0.88		122.3	1.1
SO$_3$	4.0		44.46	1.8
TiO$_2$	0.29		124.6	0.4
P$_2$O$_5$	0.17		39.66	0.1
LOI	1.9	(H$_2$O)	10.07	0.2
Total	99.9			99.5

H_2O, assigning the LOI to CO_2 ($\mu = 9.42$ cm^2/g) instead would not significantly affect the total MAC.

In the following case studies, the sample preparation, data collection and data analysis procedures are described in a stepwise, in-depth fashion. The presented protocols were adapted to the equipment and experience available to the author and may not necessarily represent the best available practise for other laboratories. QPA of cements, hydrated cements in particular, cannot be used, as yet, as an operator independent black box characterisation technique. Considerable time and effort should be spent on know-how buildup of operators and analysts. Cross-checking with other characterisation techniques is paramount in building confidence in the analysis procedure and further optimising measurement and analysis protocols.

4.7.5 Example 1: Anhydrous cement

4.7.5.1 Sample preparation

The commercial portland cement was received as finely ground powder. The particle size distribution measured by laser diffractometry indicated a d_{50} of 12.1 μm, which is generally sufficiently fine for a regular QPA. To increase the accuracy of the QPA even further, an additional grinding step can be considered. Note that not all portland cements are of sufficient fineness; depending on the application and strength class, the d_{50} value may vary from less than 10 to over 30 μm. In the case of insufficient fineness an additional grinding step should be considered for optimal results. In the presented example, the cement was further wet ground to obtain a final d_{50} of 3.2 μm. To this purpose, 3 g of cement was wet ground for 10 min in a McCrone micronizing mill using 4 mL isopropanol as grinding medium. The isopropanol was evaporated under a fume hood. As a precautionary measure towards potential sample reaction/interaction during grinding, XRD scans of the as-received and ground samples were compared to check for the appearance, disappearance or alteration of certain phases. No significant differences were noted.

As an aid in minor phase identification, a SAM extraction was performed on the as-received cement. To this purpose, 15 g of cement was mixed with a solution of 900 mL methanol and 75 g of salicylic acid. The suspension was stirred and left to react for 30 min; then it was filtered and rinsed twice with methanol. The filter residue was dried at 60°C and measured by XRD.

4.7.5.2 Data collection

The powder samples were back loaded in the sample holders to mitigate preferred orientation effects for XRD data collection. A PANalytical X'Pert Pro diffractometer in conventional Bragg–Brentano θ-2θ geometry with a

240 mm goniometer radius was used for data collection. CuKα X-rays were generated using 40 mA and 45 kV tube operating conditions. Incident beam Soller slits of 0.04 rad were used and the incident divergence and antiscatter slits were fixed at 0.5°. Air scattering was reduced using a beam knife. The receiving antiscatter slit was fixed at 1° and receiving Soller slits limiting the axial divergence to 0.04 rad were positioned in the diffracted beam path. An X'Celerator linear position-sensitive X-ray detector with a length of 2.122° 2θ was used for data acquisition. Data were collected over an angular range of 7° to 70° 2θ with 0.017° 2θ step size and an accumulated time per step of 29.8 s, resulting in a total measurement time of about 15 min. During measurement the samples were spun around the vertical goniometer axis to improve particle statistics.

4.7.5.3 Data analysis

Phase identification and Rietveld QPA were carried out using the X'Pert High Score Plus v3.0e software package by PANalytical; alternatively the Topas Academic v4.1 software was used to cross-check the quantification results. When similar refinement strategies were followed, the results of both software packages were found to be very similar.

A typical refinement starts with a phase identification step, followed by the quantitative analysis. Unmatched peaks of minor phases not identified in the first round can be checked and added later:

1. In the first step, C_3S (M3) and C_4AF were identified as major phases. Both phases were included in a first refinement cycle. Phase-specific parameters that were varied were the phase scale factors, lattice parameters and a Lorentzian peak-broadening parameter. To constrain parameter variation and prevent parameter drift, hard limits were placed on the variation intervals: e.g. for lattice parameters a 1% variation from the literature values was allowed. The refined global variables were a specimen displacement error and a background polynomial of two coefficients and a $1/X$ term. The results of this first refinement round are shown in Figure 4.17a. The graph shows that the majority of diffraction peaks has been matched; however, the difference curve clearly shows that many smaller peaks are not or not fully accounted for.

2. In the second round, the identification routine was focused on remaining minor peaks. C_3A was identified as a plausible phase and added to the quantification refinement. Belite is typically not recognised by the identification algorithms because of peak overlap with alite but was added through experience with regular cement compositions. The refinement results including C_3A and belite are shown in Figure 4.17b. The fit between the observed and calculated patterns improved visibly, which is also reflected in a decrease in the R_{wp} agreement index.

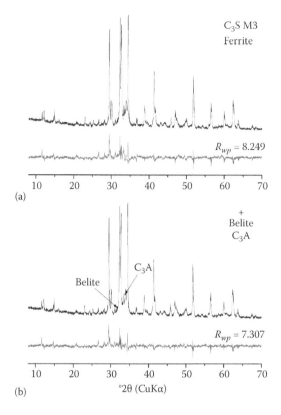

Figure 4.17 Rietveld refinement rounds on anhydrous portland cement: (a) first round; (b) second round. The R_{wp} agreement index and the difference curve (blue, below XRD scan) are given to indicate the progress in fitting. In black the observed XRD pattern is displayed, in red the calculated pattern.

(Continued)

Still, close inspection of the observed and fitted patterns reveals many minor peaks not being covered. A straightforward assignment of these minor peaks can become difficult at this point. Looking into the XRD scan of the SAM residue can be therefore of great value as an aid in minor phase identification.

3. A refinement of the SAM residue was used to identify minor phases such as anhydrite, periclase, calcite, quartz, gypsum, bassanite, goergeyite and syngenite. In addition, the SAM extraction can be used to verify if one or more C_3A polymorphs are present. The ferrite Fe-to-Al ratio can also be refined more reliably in the SAM residue. The refinement results of the SAM residue are given in Figure 4.17c. A good fit was obtained (flat difference curve, low R_{wp}) upon inclusion of all

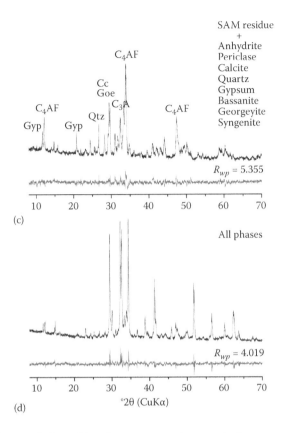

Figure 4.17 (Continued) Rietveld refinement rounds on anhydrous portland cement: (c) third round; (d) fourth round. The third round uses the XRD pattern of the SAM selective dissolution residue. The R_{wp} agreement index and the difference curve (blue, below XRD scan) are given to indicate the progress in fitting. In black the observed XRD pattern is displayed, in red the calculated pattern.

minor phases. Many of the minor phases were not directly observable in the untreated cement. Including an analysis of the SAM residue therefore leads to considerably lower phase detection limits.

4. In the final refinement round the results of the SAM residue analysis are integrated into the XRD analysis of the untreated cement. Phase-specific parameters, such as lattice and peak-broadening parameters, were kept fixed for minor phases refined in the SAM residue analysis. This measure is implemented to reduce the number of varied parameters in the final refinement and mitigates parameter drift risks. The comparison between the observed and final calculated patterns is given in Figure 4.17d. Clearly, over the subsequent analysis rounds, the fit to the observed pattern was much improved. At this stage, more

variables such as additional peak profile parameters or preferred orientation corrections may be included to further improve the fit; however, their impact on the quantification results will be rather limited in general. March–Dollase–preferred orientation corrections can be added for alite M3 (606), gypsum (0 2 0), bassanite (0 0 1), anhydrite (2 0 0) and calcite (1 0 4). Overly using these additional parameters will increase risks on parameter drift and may cause inconsistencies between refinements of similar samples. Therefore, a general guideline is to use these parameters sparingly and wisely, for instance by setting limits.

The final quantification results are given in Table 4.5. The results are normalised to 100%; no amorphous content was considered. This is in line with a previous work by the author that indicated that amorphous contents of a series of regular portland cements were below detection limits (Snellings et al. 2014a). The results are compared to a standard Bogue calculation (Bogue 1929) of the clinker phase composition highlighting the poor performance of the Bogue approach. Note that the estimation of the clinker content can be improved if instead of theoretical end-member compositions for the clinker phases, microchemical data (obtained by SEM-EDX or electron microprobe analysis) of the clinker phases are used in an extended, but labour-intensive, Bogue calculation (Le Saôut et al. 2011). Very useful is to cross-check the XRD results with the XRF data by calculating the oxide composition departing from the XRD data. Table 4.5

Table 4.5 QPA results of the portland cement in Example 1

Phase	XRD Rietveld (wt.%)	Bogue calculation (wt.%)	Oxide	Calculated from XRD (wt.%)	Difference (XRD − XRF) (wt.%)
Alite	69.7	79.8	CaO	66.0	2.8
Belite	5.1	−4.9	SiO_2	20.5	1.2
Aluminate	4.3	2.9	Al_2O_3	3.9	0.2
Ferrite	10.8	12.5	Fe_2O_3	3.6	−0.5
Calcite	2		MgO	0.2	−1.9
Dolomite	0.8		Na_2O	0.0	−0.3
Quartz	0.4		K_2O	0.4	−0.5
Gypsum	1.8		SO_3	3.6	−0.4
Bassanite	1.8		TiO_2	0.0	−0.3
Anhydrite	0.5		P_2O_5	0.0	−0.2
Syngenite	0.7		LOI	1.8	−0.1
Goergeyite	2.1		Total	100.0	

Note: The XRD results are compared to a standard Bogue calculation of the phase content. As an additional check, a back calculation of the oxide chemical composition from the XRD results is carried out and compared to the XRF results.

shows a relatively close match between the two; the largest differences, i.e. CaO and MgO, are the result of the well-known solid solution of MgO into the alite structure (Taylor 1997).

4.7.6 Example 2: Hydrated cement

In this second example, the portland cement analysed in Example 1 is studied after 3 days of hydration. The additional difficulty here is that next to the original anhydrous cement phases a range of hydrate phases appear. The example starts from the anhydrous cement refinement and adds the hydrate phases in steps while highlighting potential pitfalls. The presented sample preparation/data collection/data analysis strategy is founded on the external standard method aiming at minimal interference with the hydrate phase assemblage by XRD sample preparation and measurement procedures. Other analysis strategies have been proposed in the literature; the strategy presented here has the advantage of being practical and relatively fast and safe in terms of sample preparation. The disadvantage is that the amorphous content cannot be calculated directly as both residual water and solid amorphous phases make up the residual noncrystalline fraction of the hydrated sample. The latter can be calculated when using the internal standard approach in combination with a hydration-stopped sample. The latter procedure has the disadvantage of being more time consuming and prone to mixing problems and alterations of the hydrate assemblage.

4.7.6.1 Sample preparation

The anhydrous portland cement was mixed with water at a 0.4 w/c. The resulting paste was mixed for 2 min at 1600 rpm. The paste was cast into a plastic container, sealed with plastic paraffin film and cured at 20°C. The inner diameter of the plastic container matched the inner diameter of the XRD sample holder, enabling direct mounting and subsequent measurement of the sliced disc. The sample studied in this example was cut after 3 days of hydration; distilled water was used during cutting of the sample disc of 2–3 mm thickness. Immediately after the cutting, a fresh sample surface was uncovered by a brief wet polishing step on extrafine sandpaper. Directly after, the sample was mounted on the sample holder and measured by XRD. This sample preparation procedure was found to minimise sample desiccation and carbonation over the duration of a typical XRD scan (cf. Figures 4.10 and 4.11).

4.7.6.2 Data collection

The sample was measured using the same diffractometer operating conditions and optics as in Example 1. Data were measured from 7° to 70° 2θ with 0.017° 2θ step size and an accumulated time per step of 29.8 s, resulting in a total measurement time of about 15 min per scan. Sample spinning was applied to improve particle statistics. A rutile external standard was measured separately under identical measurement conditions. Note that this should occur shortly before/after measurement of the sample because of the decreasing X-ray intensity over time (weeks to months) due to the ageing of the X-ray tube filament.

4.7.6.3 Data analysis

The data analysis of the XRD scan of the hydrated cement was carried out using Topas Academic v4.1 software. The data analysis departed from the results of the anhydrous cement analysis and moves through a series of steps as reflected in the following:

1. In the first cycle the crystalline clinker phases found and refined in the analysis of the anhydrous cement are used. Minor, soluble phases that are consumed in the early stages of cement hydration, such as sulfates, are left out. During the fitting procedure, only global scan parameters and phase scale factors are refined. (Micro)Structural phase-specific parameters, such as lattice and peak shape parameters, are kept fixed at the values refined in the analysis of the anhydrous cement. As in Example 1, reducing the number of refined parameters minimises spurious parameter correlation and drift and maximises consistency between different analyses on the same starting materials (e.g. a time series). Also, in case of the clinker phases, one can reasonably assume that their crystal structure does not change during hydration. The fitting results of this first cycle are graphically presented in Figure 4.18a. Only a minority of peaks is well fitted; clearly the major crystalline hydrate phases need to be added to the refinement.

2. Going back to a phase identification routine, the unassigned peaks were matched to the common cement hydrates portlandite, ettringite and monosulfate. The identified phases were added to the Rietveld refinement. In the case of the hydrate phases, lattice and peak shape parameters were varied within imposed ranges (see Example 1). In general, it is advised to enable the fitting of crystal structural parameters for hydrates as they may undergo changes (crystallisation, solid solution, etc.) with hydration age. The resulting calculated pattern is compared to the observed scan in Figure 4.18b. Now, all main peaks are fitted. However, there are still clear differences around 27°–32° 2θ and 50° 2θ related to the presence of the nanocrystalline C-S-H phase. Not accounting for the C-S-H phase will result in an overestimation of the alite and belite weight fractions as the fitting routine will raise their scale factors to partially compensate for the lacking C-S-H contribution. Several approaches can be used to overcome this problem. A manual or automatically fitted background may be used. Manual

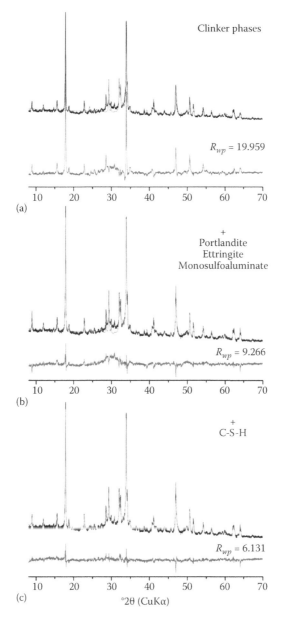

Figure 4.18 Progressive refinement cycles in the data analysis of an XRD scan of a hydrated cement: (a) fitting result when adding only the clinker phases refined in the original anhydrous cement; (b) addition of the crystalline cement hydrate phase to the refinement; (c) fitting result including a C-S-H peaks phase extracted from a 7-year-hydrated white cement. (From Snellings, R., A. Salze and K. Scrivener. *Cement and Concrete Research* 64: 89–98, 2014b.)

fitting has the disadvantage of being operator dependent and difficult to reproduce. Automatic fitting may lead to parameter correlation or imperfect fits that do not completely account for the C-S-H contribution. Here, a different approach is taken by including a separate C-S-H pattern extracted from a nearly completely hydrated white cement.

3. The addition of the C-S-H phase to the refinement led to a considerable improvement of the fit, as represented in Figure 4.18c. The C-S-H phase was added as a so-called peaks phase. The contribution of the peaks phase is not calculated from an underlying crystal structure but is merely an ensemble of peaks with fixed intensities and peak widths. The overall scale factor of the peaks phase was refined to come to a best fit with the observed XRD scan. In a sense, the combination of peaks phases and Rietveld analysis engenders a hybrid quantification approach that joins elements from pattern decomposition and Rietveld analysis.

At this point the XRD analysis was considered to be finished in terms of pattern fitting. The next step is now to extract the data and recalculate them using the external standard approach and a back-calculation procedure to meaningful, comparable quantification results.

4. The background of the external standard method is described in Section 4.4.2; the recalculation formula is presented as Equation 4.10. In order to use this equation, the calculated density, volume and scale factor need to be extracted for all phases included in the refinement. The phase-specific data obtained in the refinement of the hydrated cement example are summarised in Table 4.6. The phase scale factors are recalculated to phase weight fractions by comparison to the scale factor of the external rutile standard (Kronos 2300 TiO_2) and taking into account the ratio of the MACs of the sample μ_m to the standard μ_s in Equation 4.10. It should be noted that the MAC of the hydrated cement is lower than that of the anhydrous cement due to the bound water content. The water content of the sample thus needs to be known for each hydration age to calculate the corresponding MACs. In the case of hydration-stopped samples, where water is removed from the sample by (freeze-)drying or solvent exchange, the water content needs to be measured separately, for instance, by thermogravimetry. If, as in this example, fresh slices are measured, the initial w/c can be used to calculate the MAC if drying of the slice is minimised, e.g. by wet cutting and polishing and limiting the measurement duration to 15–30 min. The crystallinity of the rutile standard was calibrated to 96.3 wt.% using the external standard approach and the certified NIST Standard Reference Material 676a α-Al_2O_3. All input data for the calculation are given in Table 4.6; the sample MAC was calculated from the cement MAC and a w/c of 0.4. It is assumed that water loss by evaporation is minimal over the duration of the XRD measurement. Carrying out the calculation leads to quantification results on

Table 4.6 Demonstration of the external standard method recalculation procedure to obtain XRD QPA results

Refined phase-specific data					Sample MAC (cm²/g) 73.95		
Phase	Density (g/cm³)	Volume (Å³)	Scale factor		Phase	g/100 g paste	g/100 g anhydrous
C_3S M3	3.163	4314.91	9.8146×10^{-7}		C_3S M3	13.0	18.2
β-C_2S	3.316	344.99	3.9037×10^{-5}		β-C_2S	3.5	4.9
C_3A	3.041	3540.87	1.5607×10^{-7}		C_3A	1.3	1.9
C_4AF	3.704	430.21	4.2555×10^{-5}		C_4AF	6.6	9.2
Calcite	2.719	366.71	1.7284×10^{-5}		Calcite	1.4	2.0
Portlandite	2.241	54.9	1.0390×10^{-2}		Portlandite	15.8	22.1
Ettringite	1.786	2333.55	4.4815×10^{-6}		Ettringite	9.8	13.7
Monocarbonate	2.176	435.4	3.6482×10^{-6}		Monocarbonate	0.3	0.5

External standard			
	Density (g/cm³)	Volume (Å³)	Scale factor
Rutile	4.247	62.46	0.01531
	Crystallinity	μ_s (cm²/g)	
Rutile	0.963	124.6	

Degree of hydration	
	%
Cement (3 days)	64

Note: The quantification results are rescaled on an anhydrous base to enable the calculation of the cement degree of hydration.

a g/100 g cement paste base. A back calculation to an anhydrous cement base (Equation 4.12) enables to compare the quantification results to the anhydrous cement composition in Table 4.6.

Here we observe that, in line with expectations, C_3S and C_3A have reacted to an important degree at 3 days, while β-C_2S and C_4AF have reacted only marginally. Finally, a degree of hydration can be calculated by comparing the sum of the remaining anhydrous cement phases to the initial anhydrous cement composition. In the case of the studied CEM I 52.5 N, we obtain a hydration degree of 64% at 3 days of hydration (Table 4.6).

Cross-checking the obtained quantification results against other techniques is strongly encouraged. For instance, portlandite contents obtained by XRD can be compared to quantification results by thermogravimetry.

Clinker phase reaction degrees may be verified by SEM-IA or solid-state NMR spectroscopy. Mass balance calculations can be used as a first-line check for internal consistency. For example, portlandite and ettringite quantification results can be compared to theoretically calculated portland-ite and ettringite contents calculated from the reacted amounts of C_3S/C_2S and C_3A/C_4AF, respectively.

4.8 CONCLUSIONS AND OUTLOOK

XRD is quickly becoming a standard tool in cement materials science. If appropriately used, a wealth of information on cement phase composition can be retrieved, both qualitatively and quantitatively. However, cement and hydrated cements in particular are complex multicomponent materials. The application of a common material characterisation technique such as XRD to the quantitative analysis of cement is therefore fraught with inherent challenges and practical problems. For the large majority of research problems, standardised black box operation and data analysis procedures cannot deliver meaningful results and deeper insights and practical experience with the technique are a prerequisite for good analytical practise.

The challenges posed by the cement as a complex material are manifold. Anhydrous and hydrated cements consist of a multitude of phases. These phases are often part of solid solution series and may change their polymorphic form in response to changes in composition. The XRD profiles of typical cement phases show a large degree of overlap. In effect, most of the peaks discerned in the cement XRD pattern are composite, and decomposition of the pattern into its component phase profiles demands powerful XRD analysis software and some degree of analyst experience. To a similar extent, a detailed phase analysis that identifies and quantifies the minor phases also demands a solid blend of expertise in sample preparation, software operation and knowledge of the basics of cement chemistry. A final inherent challenge specific to SCM blended and hydrated cements is the additional presence of amorphous or nanocrystalline phases. The challenges outlined here need to be accounted for in the QPA routine to obtain reasonably accurate and meaningful results.

Building on a concise theoretical base, this chapter discloses practical insights into state-of-the-art XRD analysis methods applied to cement that usually never make it into the experimental descriptions in the related scientific literature. The presented examples and recommendations aim to help the reader in addressing the challenges inherent to anhydrous and hydrated cements. In doing so, the steps that are taken in a typical analysis are described, documented and exemplified in detail. Guidelines to both qualitative and quantitative phase analyses are given, paying attention to different approaches and methods so as to provide the reader with the arsenal to flexibly and creatively find the optimal analytical procedure adapted to the particular problem faced. Reference databases for both identification (PDF numbers and common phases identification table) and quantification (cf. References) are provided as an aid in setting up data analysis routines. An emphasis here is placed on good practise in XRD data analysis by the Rietveld method.

Next to challenges inherent to the material, the application of XRD analysis to cements also poses practical problems in terms of sample preparation and measurement. The problems faced are most critical with regard to the analysis of hydrated cements. As interest is often directed towards resolving the reaction kinetics of hydration, the intact preservation of the reaction system, or at least of the parameter under investigation, is of great importance. The potential adverse effects of sample preparation (grinding, hydration stoppage, sample mounting, etc.) have been demonstrated and various solutions with their inherent pros and cons were outlined.

Finally, this chapter aims to contribute to the bright future of XRD analysis applied to cements by detailing and exemplifying refinement strategies that have proven to deliver robust and reliable results for a great number, but not necessarily all, cementitious systems.

4.8.1 Outlook

Definitely, XRD analysis will find an even broader use and acceptance in the cement science community of the future as more researchers will secure access to the equipment and the supporting data analysis software will become ever more powerful and user friendly. This will not mean, however, that the cement-inherent challenges and practical problems will disappear. On the contrary, even more attention will need to be paid to the training and guidance of novices in the field to safeguard the quality of the results and to realise the full potential of XRD.

Major changes in XRD analysis on cements are difficult to anticipate. However, the predictable future path of XRD data analysis in cement science will likely involve improvements of current methods and databases to facilitate a broader applicability, more automation and improved reliability, accuracy and precision. New and further improved crystal structures will be introduced that enable to further enhance the accuracy of the results and widen the scope of samples and problems that can be addressed. It is also to be expected that cross-checks of the XRD results with other independent techniques will become implemented and automatised in the analysis software packages. The more widespread introduction of programmed statistics may further help to guide the analyst by proposing most likely solutions, e.g. by combining data of the chemical composition and XRD peak lists in the identification of minor phases in the cement. On a more methodological level, hybrid quantification methods such as PONKCS, where Rietveld and pattern summation methods are combined, will be developed further and adopted to confront the most challenging systems that remain today out of reach of QPA by XRD. It is clear that the XRD analysis of cements still has not reached its full potential and that future developments and breakthroughs are lying ahead.

ACKNOWLEDGEMENTS

The author gratefully acknowledges the provision of challenges, materials, data and support by Paweł Durdziński, Julien Bizzozero, Arnaud Muller, Amélie Bazzoni, Suhua Ma, Berta Mota, Elise Berodier, Mathieu Antoni and Hadi Kazemi-Kamyab. The painstaking laboratory preparation work of Anthony Salze, Nicholas Myers, Tiphaine Paulhiac and Vincent Fays contributed in many ways to this work. Finally, the support, expert advice and comments of Karen Scrivener, Barbara Lothenbach, Frank Winnefeld and Xuerun Li were greatly appreciated. This work was initiated through discussions in the International Union of Laboratories and Experts in Construction Materials, Systems and Structures (RILEM) TC238-SCM committee. The author acknowledges the support of the European Research Agency under the Seventh Framework Programme (FP7)–Marie Curie Intra-European Fellowship grant 298337.

REFERENCES

Ahtee, M., M. Nurmela, P. Suortti and M. Järvinen (1989). 'Correction for Preferred Orientation in Rietveld Refinement'. *Journal of Applied Crystallography* **22**: 261–268.

Aldridge, L. P. (1982). 'Accuracy and Precision of Phase Analysis in Portland Cement by Bogue, Microscopic and X-Ray Diffraction Methods'. *Cement and Concrete Research* **12**: 381–398.

Allmann, R. (1977). 'Refinement of the Hybrid Layer Structure $(Ca_2Al(OH)_6)^+$ $(0.5SO_4 \cdot 3H_2O)^-$'. *Neues Jahrbuch für Mineralogie Monatshefte* **4**: 136–144.

Angel, R. J. (1988). 'High-Pressure Study of Anorthite'. *American Mineralogist* **73**: 1114–1119.

Antoni, M., J. Rossen, F. Martirena and K. Scrivener (2012). 'Cement Substitution by a Combination of Metakaolin and Limestone'. *Cement and Concrete Research* **42** (12): 1579–89.

Aranda, M. A. G., Á. G. De la Torre and L. León-Reina (2012). 'Rietveld Quantitative Phase Analysis of OPC Clinkers, Cements and Hydration Products'. *Reviews in Mineralogy and Geochemistry* **74**: 169–209.

Ballirano, P., G. Belardi and A. Maras (2005). 'Refinement of the Structure of Synthetic Syngenite $K_2Ca(SO_4)2(H_2O)$ from X-Ray Powder Diffraction Data'. *Neues Jahrbuch Fuer Mineralogie, Abhandlungen* **182**: 15–21.

Ballirano, P., and R. Caminiti (2001). 'Rietveld Refinements on Laboratory Energy Dispersive X-Ray Diffraction (EDXD) Data'. *Journal of Applied Crystallography* **34**: 757–762.

Barry, T. I., and F. P. Glasser (2000). 'Calculations of Portland Cement Clinkering Reactions'. *Advances in Cement Research* **12** (1): 19–28.

Bellotto, M., B. Rebours, O. Clause, J. Lynch, D. Bazin and E. Elkaïm (1996). 'A Reexamination of Hydrotalcite Crystal Chemistry'. *Journal of Physical Chemistry* **100**: 8527–34.

Bergold, S. T., F. Goetz-Neunhoeffer and J. Neubauer (2013). 'Quantitative Analysis of C–S–H in Hydrating Alite Pastes by in-Situ XRD'. *Cement and Concrete Research* **53**: 119–126.

Bezou, C., A. Nonat, J.-C. Mutin, A. N. Christensen and M. S. Lehmann (1995). 'Investigation of the Crystal Structure of γ-CaSO$_4$, CaSO$_4 \cdot 0.5$ H$_2$O, and CaSO$_4 \cdot 0.6$H$_2$O by Powder Diffraction Methods'. *Journal of Solid State Chemistry* **117**: 165–176.

Bish, D. L., and S. A. Howard (1988). 'Quantitative Phase Analysis Using the Rietveld Method'. *Journal of Applied Crystallography* **21** (2): 86–91.

Bish, D. L., and R. C. Reynolds (1989). 'Sample Preparation for X-Ray Diffraction'. *Reviews in Mineralogy* **20**: 73–99.

Boeyens, J. C. A., and V. V. H. Ichharam (2002). 'Redetermination of the Crystal Structure of Calcium Sulphate Dihydrate, CaSO$_4$-2H$_2$O'. *Zeitschrift für Kristallographie – New Crystal Structures* **217**: 9–10.

Bogue, R. H. (1929). 'Calculation of the Compounds in Portland Cement'. *Industry & Engineering Chemical Analytical Edition* **1** (4): 192–197.

Bonaccorsi, E., S. Merlino and A. R. Kampf (2005). 'The Crystal Structure of Tobermorite 14 A (Plombierite), a C-S-H Phase'. *Journal of the American Ceramic Society* **88** (3): 505–512.

Bonaccorsi, E., S. Merlino and H. F. W. Taylor (2004). 'The Crystal Structure of Jennite, Ca$_9$Si$_6$O$_{18}$(OH)$_6 \cdot 8$H$_2$O'. *Cement and Concrete Research* **34** (9): 1481–1488.

Brindley, G. W. (1945). 'The Effect of Grain or Particle Size on X-Ray Reflections from Mixed Powders and Alloys Considered in Relation to the Quantitative Determination of Crystalline Substances by X-Ray Methods'. *Philosophical Magazine* **36**: 347–369.

Brotherton, P. D., J. M. Epstein, M. W. Pryce and A. H. White (1974). 'Crystal Structure of Calcium Sulphosilicate, Ca$_5$(SiO$_4$)$_2$(SO$_4$)'. *Australian Journal of Chemistry* **27**: 657–660.

Buhrke, V. E., R. Jenkins and D. K. Smith (1998). *A Practical Guide for the Preparation of Specimens for X-Ray Fluorescence and X-Ray Diffraction Analysis*. New York: Wiley-VCH.

Calos, N. J., C. H. L. Kennard, A. K. Whittaker and R. L. Davis (1995). 'Structure of Calcium Aluminate Sulfate Ca$_4$Al$_6$O$_{16}$S'. *Journal of Solid State Chemistry* **119**: 1–7.

Carbonin, S., F. Martignano, G. Menegazzo and A. dal Negro (2002). 'X-Ray Single-Crystal Study of Spinels: In Situ Heating'. *Physics and Chemistry of Minerals* **29**: 503–514.

Caspi, E. N., B. Pokroy, P. L. Lee, J. P. Quintana and E. Zolotoyabko (2005). 'On the Structure of Aragonite'. *Acta Crystallographica Section B: Structural Crystallography and Crystal Chemistry* **61**: 129–132.

Catti, M., G. Ferraris, S. Hull and A. Pavese (1992). 'Static Compression and H Disorder in Brucite, Mg(OH)$_2$, to 11 GPa: A Powder Neutron Diffraction Study'. *American Mineralogist* **77**: 1129–1132.

Chaix-Pluchery, O., I. Pannetier, J. Bouillot and J.-C. Niepce (1987). 'Structural Prereactional Transformations in Ca(OH)$_2$'. *Journal of Solid State Chemistry* 67: 225 234.

Chatterjee, A. K. (2011). 'Chemistry and Engineering of the Clinkerisation Process – Incremental Advances and Lack of Breakthroughs'. *Cement and Concrete Research* 41 (7): 624–641.

Cheng, G. C. H., and J. Zussman (1963). 'The Crystal Structure of Anhydrite (CaSO$_4$)'. *Acta Crystallographica* 16: 767–769.

Cherginets, V. L., V. N. Baumer, S. S. Galkin, L. V. Glushkova, T. P. Rebrova and Z. V. Shtitelman (2006). 'Solubility of Al$_2$O$_3$ in Some Chloride-Fluoride Melts'. *Inorganic Chemistry* 45: 7367–7371.

Cline, J. P, R. B. Von Dreele, R. Winburn, P. W. Stephens and J. J. Filliben (2011). 'Addressing the Amorphous Content Issue in Quantitative Phase Analysis: The Certification of NIST Standard Reference Material 676a'. *Acta Crystallographica Section A: Foundations of Crystallography* 67: 357–367.

Creagh, D. C., and J. H. Hubbell (2006). 'X-Ray Absorption (or Attenuation) Coefficients'. In *International Tables for Crystallography Volume C: Mathematical, Physical and Chemical Tables*, edited by Prince, E., Sixth ed., 220–229. International Union of Crystallography.

Crumbie, A., G. Walenta and T. Füllmann (2006). 'Where Is the Iron? Clinker Microanalysis with XRD Rietveld, Optical Microscopy/Point Counting, Bogue and SEM-EDS Techniques'. *Cement and Concrete Research* 36 (8): 1542–1547.

Cuberos, A. J. M., Á. G. De la Torre, M. C. Martín-Sedeño, L. Moreno-Real, M. Merlini, L. M. Ordónez and M. A. G. Aranda (2009). 'Phase Development in Conventional and Active Belite Cement Pastes by Rietveld Analysis and Chemical Constraints'. *Cement and Concrete Research* 39 (10): 833–842.

De la Torre, Á. G., S. Bruque, J. Campo and M. A. G. Aranda (2002). 'The Superstructure of C$_3$S from Synchrotron and Neutron Powder Diffraction and Its Role in Quantitative Phase Analysis'. *Cement and Concrete Research* 32: 1347–1356.

De la Torre, Á. G., R. N. De Vera, A. J. M. Cuberos and M. A. G. Aranda (2008). 'Crystal Structure of Low Magnesium-Content Alite: Application to Rietveld Quantitative Phase Analysis'. *Cement and Concrete Research* 38 (11): 1261–1269.

De Noirfontaine, M.-N., M. Courtial, F. Dunstetter, G. Gasecki and M. Signes-Frehel (2012). 'Tricalcium Silicate Ca$_3$SiO$_5$ Superstructure Analysis: A Route towards the Structure of the M1 Polymorph'. *Zeitschrift für Kristallographie* 227: 102–112.

De Weerdt, K., M. Ben Haha, G. Le Saout, K. O. Kjellsen, H. Justnes and B. Lothenbach (2011). 'Hydration Mechanisms of Ternary Portland Cements Containing Limestone Powder and Fly Ash'. *Cement and Concrete Research* 41 (3): 279–291.

Dilnesa, B. Z. (2011). 'Fe-Containing Hydrates and Their Fate during Cement Hydration: Thermodynamic Data and Experimental Study'. PhD thesis, Ecole Polytechnique Fédérale de Lausanne, Switzerland.

Dilnesa, B. Z., B. Lothenbach, G. Le Saout, G. Renaudin, A. Mesbah, Y. Filinchuk, A. Wichser and E. Wieland (2011). 'Iron in Carbonate Containing AFm Phases'. *Cement and Concrete Research* 41 (3): 311–323.

Dilnesa, B. Z., B. Lothenbach, G. Renaudin, A. Wichser and E. Wieland (2012). 'Stability of Monosulfate in the Presence of Iron'. *Journal of the American Ceramic Society* **95** (10): 3305–3316.

Dinnebier, R. E., and S. J. L. Billinge (2008). *Powder Diffraction Theory and Practice*. Cambridge: Royal Society of Chemistry Publishing.

Dollase, W. A. (1986). 'Correction of Intensities for Preferred Orientation in Powder Diffractometry: Application of the March Model'. *Journal of Applied Crystallography* **19**: 267–272.

Elton, N. J., and P. D. Salt (1996). 'Particle Statistics in Quantitative X-Ray Diffractometry'. *Powder Diffraction* **11**: 218–229.

Ferro, O., E. Galli, G. Papp, S. Quartieri, S. Szakall and G. Vezzalini (2003). 'A New Occurrence of Katoite and Re-Examination of the Hydrogrossular Group'. *European Journal of Mineralogy* **15**: 419–426.

Finger, L. W., and R. M. Hazen (1980). 'Crystal Structure and Isothermal Compression of Fe_2O_3, Cr_2O_3, and V_2O_3 to 50 kbars'. *Journal of Applied Physics* **51**: 5362–5367.

François, M., G. Renaudin and O. Evrard (1998). 'A Cementitious Compound with Composition $3CaO \cdot Al_2O_3 \, CaCO_3 \cdot 11H_2O$'. *Acta Crystallographica Section C: Crystal Structure Communications* **54** (9): 1214–1217.

Füllmann, T., and G. Walenta (2003). 'The Quantitative Rietveld Phase Analysis in Industrial Applications'. *ZKG International* **56**: 45–53.

Gallucci, E., and K. Scrivener (2007). 'Crystallisation of Calcium Hydroxide in Early Age Model and Ordinary Cementitious Systems'. *Cement and Concrete Research* **37** (4): 492–501.

Gemmi, M., M. Merlini, G. Cruciani and G. Artioli (2007). 'Non-Ideality and Defectivity of the Akermanite-Gehlenite Solid Solution: An X-Ray Diffraction and TEM Study'. *American Mineralogist* **92**: 1685–1694.

Goetz-Neunhoeffer, F., and J. Neubauer (2006). 'Refined Ettringite $(Ca_6Al_2(SO_4)_3(OH)_{12} \cdot 26H_2O)$ Structure for Quantitative X-Ray Diffraction Analysis'. *Powder Diffraction* **21** (1): 4–11.

Golovastikov, N. I., R. G. Matveeva and N. V. Belov (1975). 'Crystal Structure of the Tricalcium Silicate $(CaO)_3SiO_2 = C_3S$'. *Kristallografiya* **20**: 721–729.

Gualtieri, A. F., and G. Brignoli (2004). 'Rapid and Accurate Quantitative Phase Analysis Using a Fast Detector'. *Journal of Applied Crystallography* **37** (1): 8–13.

Guirado, F., and S. Galí (2006). 'Quantitative Rietveld Analysis of CAC Clinker Phases Using Synchrotron Radiation'. *Cement and Concrete Research* **36** (11): 2021–2032.

Gutteridge, W. A. (1979). 'On the Dissolution of the Interstitial Phases in Portland Cement'. *Cement and Concrete Research* **9**: 319–324.

Gutteridge, W. A. (1984). 'Quantitative X-Ray Powder Diffraction in the Study of Some Cementive Materials'. In *The Chemistry and Chemically Related Properties of Cement*, edited by Glasser, F. P., 11–23. British Ceramic Society.

Gutteridge, W. A., and J. A. Dalziel (1990). 'Filler Cement: The Effect of the Secondary Component on the Hydration of Portland Cement; Part I. A Fine Non-Hydraulic Filler'. *Cement and Concrete Research* **20**: 778–782.

Hanic, F., M. Handlovic and I. Kapralik (1980). 'The Structure of a Quaternary Phase $Ca_{20}Al_{3-2v}Mg_vSi_vO_{68}$'. *Acta Crystallographica Section B: Structural Crystallography and Crystal Chemistry* **36**: 2863–2869.

Hawthorne, F. C., and R. B. Ferguson (1975). 'Anhydrous Sulphates: I. Refinement of the Crystal Structure of Celestite with an Appendix on the Structure of Thenardite'. *Canadian Mineralogist* **13**: 289–292.

Hesse, C., F. Goetz Neunhoeffer and J. Neubauer (2011). 'A New Approach in Quantitative in-Situ XRD of Cement Pastes: Correlation of Heat Flow Curves with Early Hydration Reactions'. *Cement and Concrete Research* **41** (1): 123–128.

Hill, R. J., and C. J. Howard (1987). 'Quantitative Phase Analysis from Neutron Powder Diffraction Data Using the Rietveld Method'. *Journal of Applied Crystallography* **20**: 467–474.

Hoerkner, W., and H. Mueller Buschbaum (1976). 'Zur Kristallstuktur von $CaAl_2O_4$'. *Journal of Inorganic and Nuclear Chemistry* **38**: 983–984.

Huang, Q., O. Chmaissem, J. J. Caponi, C. Chaillout, M. Marezio, J. L. Tholence and A. Santoro (1994). 'Neutron Powder Diffraction Study of the Crystal Structure of $HgBa_2Ca_4Cu_5O_{12+d}$ at Room Temperature and at 10K'. *Physica C* **227**: 1–9.

Il'inets, A. M., Y. A. Malinovskii and N. N. Nevskii (1985). 'The Crystal Structure of the Rhombohedral Modification of Tricalcium Silicate'. *Doklady Akademii Nauk SSSR* **281**: 332–336.

Jacobsen, S. D., J. R. Smyth and R. J. Swope (2003). 'Thermal Expansion of Hydrated Six-Coordinate Silicon in Thaumasite'. *Physics and Chemistry of Minerals* **30**: 321–329.

Jansen, D., F. Goetz-Neunhoeffer, B. Lothenbach and J. Neubauer (2012). 'The Early Hydration of Ordinary Portland Cement (OPC): An Approach Comparing Measured Heat Flow with Calculated Heat Flow from QXRD'. *Cement and Concrete Research* **42**: 134–138.

Jansen, D., F. Goetz-Neunhoeffer, C. Stabler and J. Neubauer (2011). 'A Remastered External Standard Method Applied to the Quantification of Early OPC Hydration'. *Cement and Concrete Research* **41** (6): 602–608.

Jansen, D., Ch. Stabler, F. Goetz-Neunhoeffer, S. Dittrich and J. Neubauer (2011). 'Does Ordinary Portland Cement Contain Amorphous Phase? A Quantitative Study Using an External Standard Method'. *Powder Diffraction* **26** (1): 31–38.

Jegou Saint-Jean, S., and S. Hansen (2005). 'Nonstoichiometry in Chlorellestadite'. *Solid State Sciences* **7**: 97–102.

Jenkins, R., and R. L. Snyder (1996). *Introduction to Powder Diffractometry*. New York: John Wiley and Sons.

Kahmi, S.R. (1963). 'On the Structure of Vaterite, $CaCO_3$'. *Acta Crystallographica* **16**: 770–772.

Klug, H. P., and L. E. Alexander (1974). *X-Ray Diffraction Procedures for Polycrystalline and Amorphous Materials*, Second edition. Weinheim: Wiley-VCH.

Kocaba, V. (2009). 'Development and Evaluation of Methods to Follow Micro-structural Development of Cementitious Systems Including Slags'. Ecole Polytechnique Fédérale de Lausanne, Switzerland.

Korpa, A., T. Kowald and R. Trettin (2009). 'Phase Development in Normal and Ultra High Performance Cementitious Systems by Quantitative X-Ray Analysis and Thermoanalytical Methods'. *Cement and Concrete Research* **39** (2): 69–76.

Kristmann, M. (1977). 'Portland Cement Clinker: Mineralogical and Chemical Investigations; Part I: Microscopy, X-Ray Fluorescence and X-Ray Diffraction'. *Cement and Concrete Research* **7**: 649–658.

Kusachi, I. (1975). 'The Structure of Rankinite'. *Mineralogical Journal* **8** (1): 38–47.

Kusaka, K., K. Hagiya, M. Ohsama, Y. Okano, M. Mukai, K. Iishi and N. Haga (2001). 'Determination of Structures of $Ca_2CoSi_2O_7$, $Ca_2MgSi_2O_7$, and $Ca_2(Mg_{0.55}Fe_{0.45})Si_2O_7$ in Incommensurate and Normal Phases and Observation of Diffuse Streaks at High Temperature'. *Physics and Chemistry of Minerals* **28**: 150–166.

Lager, G. A., R. T. Downs, M. Origlieri and R. Garoutte (2002). 'High-Pressure Single Crystal X-Ray Diffraction Study of Katoite Hydrogarnet: Evidence for a Phase Transition from Ia3-D – I4-3d Symmetry at 5 GPa'. *American Mineralogist* **87**: 642–647.

Le Page, Y., and G. Donnay (1976). 'Refinement of the Crystal Structure of Low-Quartz'. *Acta Crystallographica Section B: Structural Crystallography and Crystal Chemistry* **32**: 2456–2459.

Le Saoût, G., V. Kocaba and K. Scrivener (2011). 'Application of the Rietveld Method to the Analysis of Anhydrous Cement'. *Cement and Concrete Research* **41** (2): 133–148.

León-Reina, L., A. G. De la Torre, J. M. Porras-Vázquez, M. Cruz, L. M. Ordonez, X. Alcobé, F. Gispert-Guirado, A. Larrañaga-Varga, M. Paul, T. Fuellmann, R. Schmidt and M. A. G. Aranda (2009). 'Round Robin on Rietveld Quantitative Phase Analysis of Portland Cements'. *Journal of Applied Crystallography* **42** (5): 906–916.

Lothenbach, B., G. Le Saout, E. Gallucci and K. Scrivener (2008). 'Influence of Limestone on the Hydration of Portland Cements'. *Cement and Concrete Research* **38** (6): 848–860.

Lothenbach, B., T. Matschei, G. Möschner and F. P. Glasser (2008). 'Thermodynamic Modelling of the Effect of Temperature on the Hydration and Porosity of Portland Cement'. *Cement and Concrete Research* **38** (1): 1–18.

Madsen, I. C., N. V. Y. Scarlett and M. D. Cranswick (2001). 'Outcomes of the International Union of Crystallography Commission on Powder Diffraction Round Robin on Quantitative Phase Analysis: Samples 1a to 1h'. *Journal of Applied Crystallography* **34**: 409–426.

Madsen, I. C., N. V. Y. Scarlett and A. Kern (2011). 'Description and Survey of Methodologies for the Determination of Amorphous Content via X-Ray Powder Diffraction'. *Zeitschrift für Kristallographie* **226** (12): 944–955.

Man Suherman, P., A. van Riessen, B. O'Connor, D. Li, D. Bolton and H. Fairhurst (2002). 'Determination of Amorphous Phase Levels in Portland Cement Clinker'. *Powder Diffraction* **17** (3): 178–185.

Mansoutre, S., and N. Lequeux (1996). 'Quantitative Phase Analysis of Portland Cements from Reactive Powder Concretes by X-Ray Powder Diffraction'. *Advances in Cement Research* **8**: 175–182.

Martín-Sedeño, M. C., A. J. M. Cuberos, Á. G. De la Torre, G. Álvarez-Pinazo, L. M. Ordónez, M. Gateshki and M. A. G. Aranda (2010). 'Aluminum-Rich Belite Sulfoaluminate Cements: Clinkering and Early Age Hydration'. *Cement and Concrete Research* **40** (3): 359–369.

Maslen, E. N., V. A. Strel'tsov and N. R. Strel'tsova (1993). 'X-Ray Study of the Electron Density in Calcite, $CaCO_3$'. *Acta Crystallographica Section B: Structural Crystallography and Crystal Chemistry* **49**: 636–641.

McCusker, L. B., R. B. Von Dreele, D. E. Cox, D. Louër and P. Scardi (1999). 'Rietveld Refinement Guidelines'. *Journal of Applied Crystallography* **32**: 36–50.

Merlini, M., G. Artioli, C. Meneghini, T. Cerulli, A. Bravo and F. Cella (2007). 'The Early Hydration and the Set of Portland Cements: In Situ X-Ray Powder Diffraction Studies'. *Powder Diffraction* 22 (3): 201–208.

Merlino, S., E. Bonaccorsi and T. Armbruster (2001). 'The Real Structure of Tobermorite 11Å: Normal and Anomalous Forms, OD Character and Polytypic Modifications'. *European Journal of Mineralogy* 13 (3): 577–590.

Mesbah, A., M. François, C. Cau-dit-Coumes, F. Frizon, Y. Filinchuk, F. Leroux, J. Ravaux and G. Renaudin (2011). 'Crystal Structure of Kuzel's Salt $3CaO \cdot Al_2O_3 \cdot 1/2CaSO_4 \cdot 1/2CaCl_2 \cdot 11H_2O$ Determined by Synchrotron Powder Diffraction'. *Cement and Concrete Research* 41 (5): 504–509.

Meyer, H. W., J. Neubauer and S. Malovrh (1998). 'New Quality Control with Standardless Clinker Phase Determination Using the Rietveld Refinement'. *ZKG International* 51: 152–162.

Mitchell, L. D., J. C. Margeson and P. S. Whitfield (2006). 'Quantitative Rietveld Analysis of Hydrated Cementitious Systems'. *Powder Diffraction* 21 (2): 111–113.

Mondal, P., and J. W. Jeffery (1975). 'The Crystal Structure of Tricalcium Aluminate, $Ca_3Al_2O_6$'. *Acta Crystallographica Section B: Structural Crystallography and Crystal Chemistry* 31: 689–697.

Moore, P. B., and T. Araki (1972). 'Atomic Arrangement of Merwinite, $Ca_3Mg(SiO_4)_2$, an Unusual Dense-Packed Structure of Geophysical Interest'. *American Mineralogist* 57: 1355–1374.

Moore, P. B., and T. Araki (1976). 'The Crystal Structure of Bredigite and the Genealogy of Some Alkaline Earth Orthosilicates'. *American Mineralogist* 61: 74–87.

Möschner, G., B. Lothenbach, F. Winnefeld, A. Ulrich, R. Figi and R. Kretzschmar (2009). 'Solid Solution between Al-Ettringite and Fe-Ettringite $(Ca_6[Al_{1-x}Fe_x(OH)_6]_2(SO_4)_3 \cdot 26H_2O)$'. *Cement and Concrete Research* 39 (6): 482–489.

Mumme, W. G. (1995). 'Crystal Structure of Tricalcium Silicate from a Portland Cement Clinker and Its Application to Quantitative XRD Analysis'. *Neues Jahrbuch Fuer Mineralogie – Abhandlungen* 4: 145–160.

Mumme, W. G., L. Cranswick and B. Chakoumakos (1996). 'Rietveld Crystal Structure Refinement from High Temperature Neutron Powder Diffraction Data for the Polymorphs of Dicalcium Silicate'. *Neues Jahrbuch Fuer Mineralogie – Abhandlungen* 170 (2): 171–188.

Mumme, W. G., R. J. Hill, G. W. Bushnell and E. R. Segnite (1995). 'Rietveld Crystal Structure Refinements, Crystal Chemistry and Calculated Powder Diffraction Data for the Polymorphs of Dicalcium Silicate and Related Phases'. *Neue Jahrbuch Fuer Mineralogie – Abhandlungen* 169: 35–68.

Nishi, F., and Y. Takeuchi (1975). 'The Al_6O_{18} Rings of Tetrahedra in the Structure of $Ca_{8.5}NaAl_6O_{18}$'. *Acta Crystallographica Section B: Structural Crystallography and Crystal Chemistry* 31: 1169–1173.

O'Connor, B. H., and M. D. Raven (1988). 'Application of the Rietveld Refinement Procedure in Assaying Powdered Mixtures'. *Powder Diffraction* 3: 2–6.

O'Neill, H. S. C., H. Annersten and D. Virgo (1992). 'The Temperature Dependence of the Cation Distribution in Magnesioferrite $(MgFe_2O_4)$ from Powder XRD Structural Refinements and Moessbauer Spectroscopy'. *American Mineralogist* 77: 725–740.

Ojima, K., Y. Nishihata and A. Sawada (1995). 'Structure of Potassium Sulfate at Temperatures from 296 K down to 15 K'. *Acta Crystallographica Section B: Structural Crystallography and Crystal Chemistry* 51: 287–293.

Okada, K., and J. Ossaka (1980). 'Structures of Potassium Sodium Sulphate and Tripotassium Sodium Disulphate'. *Acta Crystallographica Section B: Structural Crystallography and Crystal Chemistry* 36: 919–921.

Pajares, I., Á. G. De la Torre, S. Martinez-Ramirez, F. Puertas, M. T. Blanco-Varela and M. A. G. Aranda (2002). 'Quantitative Analysis of Mineralised White Portland Clinkers: The Structure of Fluorellestadite'. *Powder Diffraction* 17: 281–289.

Peterson, V. K., B. A. Hunter and A. Ray (2004). 'Tricalcium Silicate T1 and T2 Polymorphic Investigations: Rietveld Refinement at Various Temperatures Using Synchrotron Powder Diffraction'. *Journal of the American Ceramic Society* 87 (9): 1625–1634.

Peterson, V. K., A. S. Ray and B. A. Hunter (2006). 'A Comparative Study of Rietveld Phase Analysis of Cement Clinker Using Neutron, Laboratory X-Ray, and Synchrotron Data'. *Powder Diffraction* 21: 12–18.

Petrovich, R. (1981). 'Kinetics of Dissolution of Mechanically Comminuted Rock-Forming Oxides and Silicates – I: Deformation and Dissolution of Quartz under Laboratory Conditions'. *Geochimica et Cosmochimica Acta* 45: 1665–1674.

Ponomarev, V. I., D. M. Kheiker and N. V. Belov (1970). 'Crystal Structure of Calcium Dialuminate, CA2'. *Kristallografiya* 15: 1140–1143.

Redhammer, G. J., G. Tippelt, G. Roth and G. Amthauer (2004). 'Structural Variations in the Brownmillerite Series $Ca_2(Fe_{2-x}Al_x)O_5$: Single Crystal X-Ray Diffraction at 25°C and High-Temperature X-Ray Powder Diffraction (25°C < T < 1000°C)'. *American Mineralogist* 89: 405–420.

Renaudin, G., M. Francois and O. Evrard (1999). 'Order and Disorder in the Lamellar Hydrated Tetracalcium Monocarboaluminate Compound'. *Cement and Concrete Research* 29 (1): 63–69.

Riello, P., P. Canton and G. Fagherazzi (1997). 'Calibration of the Monochromator Band Pass Function for X-Ray Rietveld Analysis'. *Powder Diffraction* 12: 160–168.

Rietveld, H. M. (1969). 'A Profile Refinement Method for Nuclear and Magnetic Structures'. *Journal of Applied Crystallography* 2: 65–71.

Rinaldi, R., M. Sacerdoti and E. Passaglia (1990). 'Straetlingite: Crystal Structure, Chemistry, and a Reexamination of Its Polytype Vertumnite'. *European Journal of Mineralogy* 2: 841–849.

Ross, N. L., and R. J. Reeder (1992). 'High-Pressure Structural Study of Dolomite and Ankerite'. *American Mineralogist* 77: 412–421.

Runčevski, T., R. E. Dinnebier, O. V. Magdysyuk and H. Pöllmann (2012). 'Crystal Structures of Calcium Hemicarboaluminate and Carbonated Calcium Hemicarboaluminate from Synchrotron Powder Diffraction Data'. *Acta Crystallographica Section B: Structural Science* 68: 493–500.

Saalfeld, H., and W. Depmeier (1972). 'Silicon-Free Compounds with Sodalite Structure'. *Kristall Und Technik* 7: 229–233.

Saalfeld, H., and M. Wedde (1974). 'Refinement of the Crystal Structure of Gibbsite, $Al(OH)_3$'. *Zeitschrift Fuer Kristallographie, Kristallgeometrie, Kristallphysik, Kristallchemie* 139: 129–135.

Sabine, T. M. (1993). 'The Flow of Radiation in a Polycrystalline Material'. In *The Rietveld Method*, edited by Young, R. A., International Union of Crystallography.

Saburi, S., A. Kawahara, C. Henmi, I. Kusachi and K. Kihara (1977). 'The Refinement of the Crystal Structure of Cuspidine'. *Mineralogical Journal* 8 (5): 286–298.

Sacerdoti, M., and E. Passaglia (1988). 'Hydrocalumite from Latium, Italy: Its Crystal Structure and Relationship with Related Synthetic Phases'. *Neues Jahrbuch Fuer Mineralogie – Abhandlungen* 10: 462–475.

Sakakura, T., K. Tanaka, Y. Takenaka, S. Matsuishi, H. Hosono and S. Kishimoto (2011). 'The Determination of the Local Structure of a Cage with an Oxygen Ion in $Ca_{12}Al_{14}O_{33}$ Crystal'. *Acta Crystallographica Section B: Structural Crystallography and Crystal Chemistry* 67 (3): 193–204.

Sasaki, S., K. Fujino and Y. Takeuchi (1979). 'X-Ray Determination of Electron-Density Distribution in Oxides, MgO, MnO, CoO, and NiO, and Atomic Scattering Factors of Their Constituent Atoms'. *Proceedings of the Japan Academy* 55: 43–48.

Scarlett, N. V. Y., and I. C. Madsen (2006). 'Quantification of Phases with Partial or No Known Crystal Structures'. *Powder Diffraction* 21 (4): 278–284.

Scarlett, N. V. Y., I. C. Madsen, L. M. D. Cranswick, T. Lwin, E. Groleau, G. Stephenson, M. Aylmore and N. Agron-Olshina (2002). 'Outcomes of the International Union of Crystallography Commission on Powder Diffraction Round Robin in Quantitative Phase Analysis: Samples 2, 3, 4, Synthetic Bauxite, Natural Granodiorite and Pharmaceuticals'. *Journal of Applied Crystallography* 35: 383–400.

Scarlett, N. V. Y., I. C. Madsen, C. Manias and D. Retallack (2001). 'On-Line X-Ray Diffraction for Quantitative Phase Analysis: Application in the Portland Cement Industry'. *Powder Diffraction* 16 (2): 71–80.

Scrivener, K. L., T. Füllmann, E. Gallucci, G. Walenta and E. Bermejo (2004). 'Quantitative Study of Portland Cement Hydration by X-Ray Diffraction/ Rietveld Analysis and Independent Methods'. *Cement and Concrete Research* 34 (9): 1541–1547.

Smith, D. K., G. G. Johnson, A. Scheible, A. M. Wims and J. L. Johnson (1987). 'Quantitative Powder Diffraction Method Using the Full Diffraction Pattern'. *Powder Diffraction* 2: 73–77.

Smith, G. W., and R. Walls (1980). 'The Crystal Structure of Goergeyite $K_2SO_4.5CaSO_4 \cdot H_2O$'. *Zeitschrift für Kristallographie* 151: 49–60.

Snellings, R., A. Bazzoni and K. Scrivener (2014a). 'The Existence of Amorphous Phase in Portland Cements: Physical Factors Affecting Rietveld Quantitative Phase Analysis'. *Cement and Concrete Research* 59: 139–146.

Snellings, R., L. Machiels, G. Mertens and J. Elsen (2010). 'Rietveld Refinement Strategy for Quantitative Phase Analysis of Partially Amorphous Zeolitised Tuffaceous Rocks'. *Geologica Belgica*: 183–196.

Snellings, R., G. Mertens, Ö. Cizer and J. Elsen (2010). 'Early Age Hydration and Pozzolanic Reaction in Natural Zeolite Blended Cements: Reaction Kinetics and Products by in Situ Synchrotron X-Ray Powder Diffraction'. *Cement and Concrete Research* 40 (12): 1704–1713.

Snellings, R., A. Salze and K. Scrivener (2014b). 'A New Method to Quantify SCM Content and Degree of Hydration in Anhydrous and Hydrated Blended Cements'. *Cement and Concrete Research* 64: 89–98.

Soin, A. V., L. J. J. Catalan and S. D. Kinrade (2013). 'A Combined QXRD/TG Method to Quantify the Phase Composition of Hydrated Portland Cements'. *Cement and Concrete Research* 48: 17–24.

Speer, D., and E. Salje (1986). 'Phase Transitions in Langbeinites I: Crystal Chemistry and Structures of K-Double Sulfates of the Langbeinite Type $M_2^{(2+)}K_2(SO_4)_3$, $M^{(2+)}$ = Mg, Ni, Co, Zn, Ca'. *Physics and Chemistry of Minerals* 13: 17–24.

Strel'tsov, V. A., V. G. Tsirel'son, R. P. Ozerov and O. A. Golovanov (1988). 'Electronic and Thermal Parameters of Ions in CaF_2: Regularised Least Squares Treatment'. *Kristallografiya* 33. 90 97.

Strubble, L. J. (1991). 'Quantitative Phase Analysis of Clinker Using X-Ray Diffraction'. *Cement and Concrete Composites* 13: 97–102.

Stutzman, P. E. (2011). 'Direct Determination of Phases in Portland Cement by Quantitative X-Ray Powder Diffraction'. *NIST Technical Note* 1692: 59.

Subbotin, K. A., L. D. Iskhakova, E. V. Zharikov and S. V. Lavrishchev (2008). 'Investigation of the Crystallisation Features, Atomic Structure, and Microstructure of Chromium-Doped Monticellite'. *Crystallography Reports* 53 (7): 1107–1111.

Takeuchi, Y., F. Nishi and I. Maki (1980). 'Crystal Chemical Characterisation of the $(CaO)_3{}^*(Al_2O_3)$–(Na_2O) Solid-Solution Series'. *Zeitschrift für Kristallographie* 152: 259–307.

Taylor, H. F. W. (1997). *Cement Chemistry*. New York: Academic Press.

Taylor, H. F. W. (1989). 'Modification of the Bogue Calculation'. *Advances in Cement Research* 2 (6): 73–77.

Taylor, J. C., and L. P. Aldridge (1993). 'Full Profile Rietveld Quantitative X-Ray Powder Diffraction Analysis of Portland Cement: Standard XRD Profiles for the Major Phase Tricalcium Silicate'. *Powder Diffraction* 8: 138–144.

Taylor, J. C., L. P. Aldridge, C. E. Matulis and I. Hinczak (2002). 'X-Ray Powder Diffraction Analysis of Cements'. In *Structure and Performance of Cements*, edited by Bensted, J., and P. Barnes, Second edition, 420–441. London: Spon Press.

Terzis, A., S. Filippakis, H.-J. Kuzel and H. Burzlaff (1987). 'The Crystal Structure of $Ca_2Al(OH)_6{}^*2H_2O$'. *Zeitschrift für Kristallographie* 181: 29–34.

Voll, D., C. Lengauer, A. Beran and H. Schneider (2001). 'Infrared Band Assignment and Structural Refinement of Al-Si, Al-Ge, and Ga-Ge Mullites'. *European Journal of Mineralogy* 13: 591–604.

Walenta, G., and T. Füllmann (2004). 'Advances in Quantitative XRD Analysis for Clinker, Cements, and Cementitious Additions'. *Powder Diffraction* 19 (1): 40–44.

Wechsler, B. A., D. H. Lindsley and C. T. Prewitt (1984). 'Crystal Structure and Cation Distribution in Titanomagnetites $(Fe_{(3-x)}Ti_{(x)}O_4)$'. *American Mineralogist* 69: 754–770.

Westphal, T., T. Füllmann and H. Pöllmann (2009). 'Rietveld Quantification of Amorphous Portions with an Internal Standard – Mathematical Consequences of the Experimental Approach'. *Powder Diffraction* 24 (3): 239–243.

Williams, R. P., R. D. Hart and A. van Riessen (2011). 'Quantification of the Extent of Reaction of Metakaolin-Based Geopolymers Using X-Ray Diffraction, Scanning Electron Microscopy, and Energy-Dispersive Spectroscopy'. *Journal of the American Ceramic Society* 94 (8): 2663–2670.

Winnefeld, F., and B. Lothenbach (2010). 'Hydration of Calcium Sulfoaluminate Cements – Experimental Findings and Thermodynamic Modelling'. *Cement and Concrete Research* 40 (8): 1239–1247.

Yamamoto, A. (1982). 'Modulated Structure of Wustite (Fe$_{1-x}$O) (Three-Dimensional Modulation)'. *Acta Crystallographica Section B: Structural Crystallography and Crystal Chemistry* **38**: 1451–1456.

Yamanaka, T., N. Hirai and Y. Komatsu (2002). 'Structure Change of Ca$_{1-x}$Sr$_x$TiO$_3$ Perovskite with Composition and Pressure'. *American Mineralogist* **87**: 1183–1189.

Young, R. A. (1993). *The Rietveld Method*. Oxford: Oxford University Press.

Young, R. A., P. E. Mackie and R. B. Von Dreele (1977). 'Application of the Pattern-Fitting Structure Refinement Method for X-Ray Powder Diffractometer Patterns'. *Journal of Applied Crystallography* **10**: 262–269.

Zhang, L., and F. P. Glasser (2000). 'Critical Examination of Drying Damage to Cement Pastes'. *Advances in Cement Research* **12**: 79–88.

Zhang, J., and G. W. Scherer (2011). 'Comparison of Methods for Arresting Hydration of Cement'. *Cement and Concrete Research* **41** (10): 1024–1036.

APPENDIX

Table A4.1 Calcium silicate phases

Phase name	Chemical formula	Space group	Lattice parameters						ICSD no.	PDF no.	Reference
			a	b	c	α	β	γ			
Alite R	Ca_3SiO_5	R3m	7.057	–	24.974	–	–	–	22501	16-406	Il'inets et al. 1985
Alite M3	Ca_3SiO_5	Cm	33.108	7.036	18.521	–	94.137	–	94742	85-1378	De la Torre et al. 2002
Alite M1	Ca_3SiO_5	Pc	27.874	7.059	12.258	–	116.03	–	–	–	de Noirfontaine et al. 2012
Alite M	Ca_3SiO_5	Cm	12.235	7.073	9.298	–	116.31	–	81100	42-551	Mumme 1995
Alite T3	Ca_3SiO_5	P-1	11.629	14.172	13.643	104.98	94.62	90.11	162744	–	De la Torre et al. 2008
Alite T2	Ca_3SiO_5	P-1	11.742	14.279	13.773	105.13	94.41	89.89	–	–	Peterson et al. 2004
Alite T	Ca_3SiO_5	P-1	11.67	14.24	13.72	105.5	94.33	90	4331	01-070-1846	Golovastikov et al. 1975
Belite α	Ca_2SiO_4	P6$_3$/mmc	5.42	–	7.027	–	–	–	81099	23-1042	Mumme et al. 1995
Belite α'_H	Ca_2SiO_4	Pnma	6.767	5.519	9.303	–	–	–	81097	20-237	Mumme et al. 1995

(Continued)

Table A4.1 (Continued) Calcium silicate phases

Phase name	Chemical formula	Space group	Lattice parameters						ICSD no.	PDF no.	Reference
			a	b	c	α	β	γ			
Belite α_L''	Ca_2SiO_4	$Pna2_1$	20.527	9.496	5.590	—	—	—	82996	36-642	Mumme et al. 1996
Belite β	Ca_2SiO_4	$P2_1/n$	5.512	6.758	9.314	—	94.58	—	81096	33-302	Mumme et al. 1995
Belite γ	Ca_2SiO_4	$Pbnm$	5.0821	11.224	6.764	—	—	—	81095	24-34	Mumme et al. 1995
Rankinite	$Ca_3Si_2O_7$	$P2_1/a$	10.6	8.92	7.89	—	119.6	—	34338	22-539	Kusachi 1975
Cuspidine	$Ca_4(Si_2O_7)F_{1.5}(OH)_{0.5}$	$P2_1/c$	7.51	10.52	10.91	—	70.7	—	34339	41-1474	Saburi et al. 1977
Bredigite	$Ca_7Mg(SiO_4)_4$	$P2nn$	10.909	18.34	6.739	—	—	—	9828	35-260	Moore and Araki 1976
Merwinite	$Ca_3Mg(SiO_4)_2$	$P2_1/a$	13.254	5.293	9.328	—	91.9	—	26002	35-591	Moore and Araki 1972
Akermanite	$Ca_2MgSi_2O_7$	$P\text{-}42_1m$	7.835	—	5.009	—	—	—	94140	—	Kusaka et al. 2001
Monticellite	$CaMgSiO_4$	$P2/c$	6.357	4.816	11.039	—	90.3	—	173714	—	Subbotin et al. 2008

Note: ICSD: Inorganic Crystal Structure Database.

Table A4.2 Calcium aluminates/ferrites

Phase name	Chemical formula	Space group	Lattice parameters						ICSD no.	PDF no.	Reference
			a	b	c	α	β	γ			
C_3A cubic	$Ca_3Al_2O_6$	Pa-3	15.26	–	–	–	–	–	1841	38-1429	Mondal and Jeffery 1975
C_3A orthorhombic	$Ca_{8.5}Na_1Al_6O_{18}$	Pbca	10.875	10.859	15.105	–	–	–	1880	01-070-0859	Nishi and Takeuchi 1975
C_3A monoclinic	$Ca_{8.25}Na_{1.5}Al_6O_{18}$	$P2_1/a$	10.877	10.854	15.135	–	–	–	100221	01-083-1360	Takeuchi et al. 1980
Monoaluminate	$CaAl_2O_4$	$P2_1/n$	8.7	8.092	15.191	–	90.17	–	260	23-1036	Hoerkner and Mueller Buschbaum 1976
Grossite	$CaAl_4O_7$	C2/c	12.866	8.879	5.44	–	106.75	–	16191	46-1475	Ponomarev et al. 1970
Mayenite	$Ca_{12}Al_{14}O_{33}$	I4-3d	11.989	–	–	–	–	–	261586	–	Sakakura et al. 2011
Ferrite/Brownmillerite	$Ca_2Al_{1.35}Fe_{0.651}O_5$	Ibm2	5.510	14.443	5.299	–	–	–	98839	01-074-3675	Redhammer et al. 2004
Ferrite	$Ca_2Al_{0.99}Fe_{1.01}O_5$	Ibm2	5.555	14.495	5.337	–	–	–	98836	01-074-3672	Redhammer et al. 2004
Ferrite	$Ca_2Al_{0.55}Fe_{1.45}O_5$	Ibm2	5.591	14.613	5.381	–	–	–	98830	01-074-3666	Redhammer et al. 2004
C_2F	$Ca_2Fe_2O_5$	Pamn	5.597	14.763	5.427	–	–	–	98822	01-074-3658	Redhammer et al. 2004

Table A4.3 Aluminosilicates

Phase name	Chemical formula	Space group	Lattice parameters						ICSD no.	PDF no.	Reference
			a	b	c	α	β	γ			
Gehlenite	$Ca_2Al(AlSi)O_7$	$P\text{-}42_1m$	7.69	–	5.064	–	–	–	158171	–	Gemmi et al. 2007
Mullite	$Al_{2.34}Si_{0.66}O_{4.83}$	Pbam	7.565	7.688	2.885	–	–	–	158098	–	Völl et al. 2001
Anorthite	$Ca_2Al_2Si_2O_8$	I-1	8.175	12.873	14.17	93.11	115.89	91.28	202710	01-086-1705	Angel 1988
Pleochroite Q phase	$Ca_{20}Al_{26}Mg_3Si_3O_{68}$	Pmmn	27.638	10.799	5.123	–	–	–	26353	01-074-0695	Hanic et al. 1980

Table A4.4 Oxides

Phase name	Chemical formula	Space group	Lattice parameters						ICSD no.	PDF no.	Reference
			a	b	c	α	β	γ			
Lime	CaO	Fm-3m	4.797	–	–	–	–	–	75786	43-1001	Huang et al. 1994
Periclase	MgO	Fm-3m	4.217	–	–	–	–	–	9863	43-1022	Sasaki et al. 1979
α-Quartz	SiO_2	P3_2 2	4.913	4.913	5.405	–	–	–	174	46-1045	Le Page and Donnay 1976
Rutile	TiO_2	P42/mnm	4.593	–	2.961	–	–	–	93097	21-1276	Ballirano and Caminiti 2001
Corundum	Al_2O_3	R-3c	4.760	–	12.992	–	–	–	77810	10-173	Riello et al. 1997
Hematite	Fe_2O_3	R-3c	5.035	–	13.747	–	–	–	201096	33-664	Finger and Hazen 1980
Wüstite	$Fe_{0.902}O$	Fm-3m	4.3	–	–	–	–	–	27237	46-1312	Yamamoto 1982
Magnetite	Fe_3O_4	Fd-3m	8.366	–	–	–	–	–	30860	19-629	Wechsler et al. 1984
Spinel	$MgAl_2O_4$	Fd-3m	8.084	–	–	–	–	–	97180	01-072-7280	Carbonin et al. 2002
Magnesioferrite	$MgFe_2O_4$	Fd-3m	8.381	–	–	–	–	–	40672	36-398	O'Neill et al. 1992
Perovskite	$CaTiO_3$	Pbnm	5.404	5.422	7.651	–	–	–	94568	22-153	Yamanaka et al. 2002

Table A4.5 Alkali phases

Phase name	Chemical formula	Space group	Lattice parameters						ICSD no.	PDF no.	Reference
			a	b	c	α	β	γ			
Arcanite	K_2SO_4	Pmcn	5.770	10.071	7.478	–	–	–	79777	5-613	Ojima et al. 1995
Thenardite	Na_2SO_4	Fddd	9.829	12.302	5.868	–	–	–	28056	37-1465	Hawthorne and Ferguson 1975
Ca langbeinite	$Ca_2K_2(SO_4)_3$	$P2_12_12_1$	10.33	10.5	10.18	–	–	–	40989	20-867	Speer and Salje 1986
Halite	NaCl	Fm-3m	5.641	–	–	–	–	–	41411	5-628	Strel'tsov et al. 1988
Sylvine	KCl	Fm-3m	6.285	–	–	–	–	–	165593	–	Cherginets et al. 2006
Syngenite	$K_2Ca(SO_4)_2(H_2O)$	$P2_1/m$	9.771	7.145	6.247	–	103.99	–	157072	–	Ballirano et al. 2005
Goergeyite	$K_2Ca_5(SO_4)_6(H_2O)$	C2/c	17.51	6.822	18.21	–	113.3	–	30790	18-997	Smith and Walls 1980
Aphthitalite	$KNa(SO_4)$	P3m	5.607	–	7.177	–	–	–	26014	20-928	Okada and Ossaka 1980
Aphthitalite	$K_3Na(SO_4)_2$	P-3m	5.680	–	7.309	–	–	–	26018	20-928	Okada and Ossaka 1980

Table A4.6 Calcium sulfates/sulfoaluminates/sulfosilicates/carbonates

Phase name	Chemical formula	Space group	Lattice parameters						ICSD no.	PDF no.	Reference
			a	b	c	α	β	γ			
Anhydrite	$CaSO_4$	Amma	6.991	6.996	6.238	–	–	–	15876	37-1496	Cheng and Zussman 1963
Bassanite/ Hemihydrate	$CaSO_4 \cdot 0.5(H_2O)$	I2	12.032	6.927	12.671	–	90.27	–	79529	41-224	Bezou et al. 1995
Gypsum	$CaSO_4 \cdot 2(H_2O)$	C2/c	6.284	15.2	6.523	–	127.41	–	409581	33-311	Boeyens and Ichharam 2002
Ye'elimite cubic	$Ca_4Al_6O_{12}(SO_4)$	I4-3m	9.205	–	–	–	–	–	9560	33-256	Saalfeld and Depmeier 1972
Ye'elimite orthorhombic	$Ca_4Al_6O_{12}(SO_4)$	Pcc2	13.028	13.037	9.161	–	–	–	80361	85-2210	Calos et al. 1995
Sulfate spurrite	$Ca_5(SiO_4)_2(SO_4)$	Pcmn	10.182	15.398	6.85	–	–	–	4332	40-393	Brotherton et al. 1974
Ellestadite–Cl	$Ca_{10}(SiO_4)_3(SO_4)_3Cl_2$	$P6_3/m$	9.677	–	6.859	–	–	–	154205	25-167	Jegou Saint-Jean and Hansen 2005
Ellestadite–F	$Ca_{10}(SiO_4)_3(SO_4)_3F$	$P6_3/m$	9.442	9.442	6.940	–	–	–	97203	01-072-7301	Pajares et al. 2002
Calcite	$CaCO_3$	R-3c	4.991	–	17.062	–	–	–	73446	5-586	Maslen et al. 1993
Aragonite	$CaCO_3$	Pmcn	4.962	7.969	5.743	–	–	–	170225	01-073-3251	Caspi et al. 2005
Vaterite	$CaCO_3$	$P6_3/mmc$	4.13	4.13	8.49	–	–	–	15879	33-628	Kahmi 1963
Dolomite	$CaMg(CO_3)_2$	R-3	4.806	–	16.006	–	–	–	66333	36-426	Ross and Reeder 1992

Table A4.7 Hydrates

Phase name	Chemical formula	Space group	Lattice parameters						ICSD no.	PDF no.	Reference
			a	b	c	α	β	γ			
Portlandite	$Ca(OH)_2$	P-3m	3.592	—	4.906	—	—	—	202220	4-733	Chaix-Pluchery et al.1987
Brucite	$Mg(OH)_2$	P-3m	3.150	—	4.770	—	—	—	79031	7-239	Catti et al.1992
Gibbsite	$Al(OH)_3$	P2$_1$/n	8.684	5.078	9.736	—	94.54	—	6162	01-070-2038	Saalfeld and Wedde 1974
Tobermorite 11A	$Ca_4(Si_6O_{15})(OH)_2(H_2O)_5$	Bm	6.735	7.385	22.487	—	—	123.25	87690	86-2275	Merlino et al. 200
Tobermorite 14A	$Ca_5(Si_6O_{16})(OH)_2(H_2O)_7$	Bb	6.735	7.425	27.987	—	—	123.25	152489	—	Bonaccorsi et al. 2005
Jennite	$Ca_9(Si_6O_{18})(OH)_6(H_2O)_8$	P-1	10.576	7.265	10.931	101.3	96.98	109.65	151413		Bonaccorsi et al. 2004
Ettringite	$Ca_6Al_2(SO_4)_3(OH)_{12}(H_2O)_{26}$	P3c	11.229	—	21.478	—	—	—	155395	41-1451	Goetz-Neunhoeffer and Neubauer 2006
Thaumasite	$Ca_3Si(OH)_6(CO_3)(SO_4)(H_2O)_{12}$	P6$_3$	11.054	—	10.411	—	—	—	98394	01-074-3266	Jacobsen et al. 2003
Kuzelite/Monosulfate	$Ca_4Al_2(OH)_{12}(SO_4)(H_2O)_6$	R-3	5.759	—	26.795	—	—	—	100138	01-083-1289	Allmann 1977
Kuzel's salt	$Ca_2Al(OH)_6(SO_4)_{0.25}Cl_{0.5}(H_2O)_{2.5}$	R-3	5.751	—	50.419	—	—	—	—	19-0203	Mesbah et al. 201
Hemicarbonate	$Ca_4Al_2(OH)_{12}(CO_3)_{0.5}(H_2O)_5$	R-3c	5.776	—	48.812	—	—	—	—	41-0221	Runčevski et al. 2012
Monocarbonate – ordered	$Ca_4Al_2(OH)_{12}(CO_3)(H_2O)_5$	P1	5.775	8.469	9.923	64.77	82.75	81.43	59327	01-087-0493	François et al. 1998

(Continued)

Table A4.7 (Continued) Hydrates

Phase name	Chemical formula	Space group	a	b	c	α	β	γ	ICSD no.	PDF no.	Reference
Monocarbonate-disordered	$Ca_4Al_2(OH)_{12}(CO_3)(H_2O)_5$	P-1	5.742	5.744	15.091	92.29	87.45	119.54	–	–	Renaudin et al. 1999
Hydrocalumite	$Ca_8Al_4(OH)_{24}(CO_3)Cl_2(H_2O)_{9.6}$	P2/c	10.02	11.501	16.286	–	104.22	–	63251	42-558	Sacerdoti and Passaglia 1988
Friedel's salt	$Ca_2Al(OH)_6Cl(H_2O)_2$	A2/n	9.979	5.751	16.32	–	104.53	–	62363	01-078-1219	Terzis et al. 1987
Strätlingite	$Ca_2Al((AlSi)_{1.11}O_{2.1})(OH)_{12}(H_2O)_{2.25}$	R-3m	5.745	–	37.77	–	–	–	69413	29-285	Rinaldi et al. 1990
Hydrotalcite	$Mg_2Al(OH)_6(CO_3)_{0.5}(H_2O)_{1.5}$	R-3m	3.046	–	22.772	–	–	–	81963	01-089-0460	Bellotto et al. 1996
CAH8	$CaAl_2(OH)_8(H_2C)_{1.84}$	P6₃/m	16.387	–	8.279	–	–	–	407150	01-088-1410	Guirado and Galí 2006
Katoite	$Ca_3Al_2(OH)_{12}$	Ia-3d	12.573	–	–	–	–	–	94630	01-074-3032	Lager et al. 2002
Katoite-Si	$Ca_{2.92}Al_2Si_{1.11}O_{12}F_{7.56}$	Ia-3d	12.27	–	–	–	–	–	172077	–	Ferro et al. 2003
Fe monocarboferrate	$Ca_4Fe_2(OH)_{12}(CO_3)(H_2O)_{6.18}$	R-3c	5.920	–	47.790	–	–	–			Dilnesa et al. 2011
Fe monosulfate	$Ca_4Fe_2(OH)_{12}(SO_4)(H_2O)_6$	R-3c	5.887	–	26.598	–	–	–			Dilnesa et al. 2012
Fe-hydrogarnet	$Ca_3Fe_2(SiO_4)_{2.16}(OH)_{3.36}$	Ia-3d	12.542	–	–	–	–	–			Dilnesa 2011
Fe ettringite	$Ca_6Fe_2(SO_4)_3(OH)_{12}(H_2O)_{26}$	P6₃/mmc	11.19	–	22.0	–	–	–		38-1480	Möschner et al. 2009

Table A4.8 Search table

d	Major cement phases	d	Minor cement phases	d	Cement hydrates
				12.57	Strätlingite (10)
				9.72	Ettringite (10)
				9.55	Thaumasite (10)
				8.9	Monosulfate (10)
				8.6–8.3	AFm solid solution (SO_4-CO_3-OH)
				8.2	Hemicarbonate (10)
				7.92	C_4AH_{13} (10)
				7.58	Hydrotalcite (10)
		7.56	Gypsum (9)	7.57	Monocarbonate (10)
7.3	C_4AF (3)				
				6.26	Strätlingite (4)
		5.98	Bassanite (10)		
		5.71	Syngenite (6)	5.6	Ettringite (6)
		5.37	Mullite (10)	5.51	Thaumasite (4)
				5.13	Katoite (10)
4.88	Mayenite (10)			4.89	Portlandite (7–10)
4.66	CA (2)	4.65	Thenardite (6)	4.77	Brucite (8)
4.45	Rankinite (4)			4.45	Monosulfate (3)
4.43	CA_2 (4)				
4.35	C_2S γ (3)	4.34	Tridymite (10)		
		4.27	Gypsum (10)		
		4.25	Quartz (5)		
		4.21	Ca-langbeinite (3)		
		4.16	Arcanite (4)	4.18	Strätlingite (5)
4.07	C_3A cubic (2)	4.09	Tridymite (8)	4.1	Hemicarbonate (9)
		4.08	Aphthitalite (3)		
		4.04	Cristobalite (10)		
				3.99	Monosulfate (3)

(Continued)

Table A4.8 (Continued) Search table

d	Major cement phases	d	Minor cement phases	d	Cement hydrates
				3.99	C_4AH_{13} (8)
				3.88	Ettringite (5)
3.83	C_2S γ (3)	3.83	Tridymite (3)		
3.83	Rankinite (6)			3.79	Hydrotalcite (5)
				3.79	Thaumasite (4)
		3.74	Ye'elimite (10)	3.78	Monocarbonate (9)
		3.73	Arcanite (3)		
		3.7	Gehlenite (2)		
3.49	CA_2 (10)	3.49	Anhydrite (10)		
		3.48	Corundum (8)		
		3.4	Mullite (9)		
		3.34	Quartz (10)		
		3.31	Ca-langbeinite (9)		
		3.26	Ca-langbeinite (6)		
		3.26	Cuspidine (3)		
		3.248	Rutile (10)		
		3.22	Ca-langbeinite (10)		
		3.19	Sulfate spurrite (8)		
3.18–3.16	Rankinite (8)	3.18	Thenardite (5)		
		3.15	Syngenite (8)		
		3.14	Sylvine (10)		
3.07	CA_2 (2)	3.05	Gypsum (7)	3.06	C-S-H (10)
3.04	C_3S M (9)	3.05	Cuspidine (10)	3.03	Calcite (10)
3.04	C_2S γ (7)				
		2.99	Arcanite (9)		
2.97	CA (10)	2.99	Bassanite (7)		
		2.965	Magnesioferrite (4)		
2.95	CF_2 (5)	2.96	Magnetite–maghemite (3)		
		2.93	Aphthitalite (6)		
2.91	C_2S α (7)	2.90–2.86	Spinels (4)		

(Continued)

Table A4.8 (Continued) Search table

d	Major cement phases	d	Minor cement phases	d	Cement hydrates
2.89	Rankinite (8)	2.89	Arcanite (10)	2.88	Monosulfate (3)
2.88	C_2S β (3)	2.87	Gypsum (5)	2.88	Hemicarbonate (5)
		2.87	Ellestadite (7)	2.87	C_4AH_{13} (6)
		2.87	Cuspidine (5)	2.87	Strätlingite (5)
		2.85	Syngenite (10)		
		2.84	Aphthitalite (10)		
		2.84	Anhydrite (3)		
		2.84	Sulfate spurrite (10)		
		2.84	Gehlenite (10)		
		2.81	Halite (10)	2.81	Katoite (2)
		2.8	Bassanite (6)	2.8	C-S-H (3)
2.79	C_2S α'_L (10)	2.79	Ellestadite (10)		
2.79	C_4AF (3)	2.78	Thenardite (10)		
2.79–2.74	C_2S β (10)	2.78	Lime (3)		
2.78–2.74	C_3S M (10)	2.73	Syngenite (5)		
2.77	C_2S γ (6)				
2.76	C_2S α (10)				
2.75	C_2S α'_L (9)				
2.75	CA_2 (3)				
2.74	C_2S γ (10)				
2.73	Bredigite (10)				
2.72	C_3A O/M (5)				
2.71	CA_2 (3)				
2.7	Rankinite (10)				
2.7	C_3A cubic (10)				
2.69	C_3A O/M (10)	2.69	Mullite (7)		
2.69	C_4AF (4)	2.69	Hematite (10)		
2.68	Mayenite (6)				
2.67	Bredigite (9)				
2.66	C_4AF (10)	2.65	Ye'elimite (3)		

(Continued)

Table A4.8 (Continued) Search table

d	Major cement phases	d	Minor cement phases	d	Cement hydrates
		2.64	Thenardite (4)	2.62	Portlandite (10)
2.61	C_3S M (8)				
2.61	C_2S β (6)				
2.59	CA_2 (4)				
2.58	Rankinite (6)			2.57	Hydrotalcite (3)
2.58	CF_2 (10)	2.57	Sulfate spurrite (6)	2.56	Ettringite (3)
		2.55	Corundum (9)		
		2.54	Mullite (6)		
2.53–2.51	CA (3)	2.53	Magnesioferrite (10)		
		2.52	Magnetite– maghemite (10)		
2.51	Rankinite (6)	2.51	Hematite (7)		
		2.49	Wüstite (6)		
		2.487	Rutile (4)		
2.45	Mayenite (4)	2.47–2.44	Spinels (10)		
		2.42	Arcanite (3)		
2.4	CA (2)	2.405	Lime (10)		
		2.38	Corundum (4)	2.37	Brucite (10)
2.32	C_2S α'_L (2)	2.33	Anhydrite (2)		
				2.295	Kat-oite (2)
				2.28	Calcite (2)
				2.28	Hydrotalcite (3)
2.23	Bredigite (3)	2.22	Sylvine (6)		
2.22	C_2S α'_L (3)	2.2	Mullite (6)		
2.19	C_2S Beta (6)	2.2	Hematite (3)		
2.18	C_3S M (3)	2.16	Ye'elimite (2)		
		2.16	Wüstite (10)		
		2.106	Periclase (10)		
		2.1	Magnetite– maghemite (2)		
		2.05–2.02	Spinel (5)		

(Continued)

Table A4.8 (Continued) Search table

d	Major cement phases	d	Minor cement phases	d	Cement hydrates
		2.04	Aphthitalite (5)		
2	$C_2S \, \alpha$ (3)	1.99	Halite (6)		
1.98	$C_2S \, \beta$ (3)				
1.97	$C_2S \, \alpha'_L$ (3)				
1.94	C_4AF (4)				
1.924	Bredigite (3)			1.93	Hydrotalcite (4)
1.92	$C_2S \, \gamma$ (5)			1.92	Portlandite (4)
1.91	Rankinite (5)			1.91	Calcite (2)
1.91	C_3A cubic (4)				
		1.86	Thenardite (4)	1.87	Calcite (2)
		1.838	Hematite (3)	1.83	C-S-H (3)
1.81	Rankinite (5)	1.82	Quartz (3)	1.79	Brucite (5)
1.77	C_3S M (6)	1.75	Gehlenite (2)	1.79	Portlandite (3)
1.72	CF_2 (2)	1.74	Corundum (5)		
		1.7	Lime (5)		
		1.692	Hematite (5)		
		1.69	Rutile (5)	1.66	C-S-H (1)
1.65	$C_2S \, \gamma$ (2)			1.66	C_4AH_{13} (5)
		1.615	Magnetite–maghemite (3)		
		1.61	Magnesioferrite (3)		
		1.601	Corundum (8)		
1.56	Bredigite (2)	1.576–1.555	Spinels (4)	1.57	Brucite (3)
1.56	C_3A cubic (4)				
		1.52	Mullite (5)		
		1.52	Wüstite (5)		
		1.489	Periclase (5)		
		1.484	Hematite (2)		
		1.48	Magnetite–maghemite (4)		
		1.48	Magnesioferrite (4)		
		1.448–1.429	Spinels (6)		

Chapter 5

Thermogravimetric analysis

Barbara Lothenbach, Paweł Durdziński
and Klaartje De Weerdt

CONTENTS

5.1 INTRODUCTION

Thermogravimetric analysis (TGA) is a widely applied technique in the field of cement science. Measurements of bound water and portlandite content by TGA are often used to follow the reaction of portland cement or to evaluate the reactivity of supplementary cementitious materials (SCMs), such as fly ash and blast furnace slags. TGA is able to identify X-ray amorphous hydrates, such as C-S-H or AH_3, and can be used complementarily to other techniques such as X-ray diffraction (XRD).

The present chapter outlines the main factors influencing TGA. The TGA signals of the most common minerals observed in cementitious systems are compared to each other and compiled as a reference database for phase identification. The quantification of bound water, portlandite and calcium carbonate and different hydrates is described in detail. Special attention is given to the methods of stopping hydration and their effect on TGA measurements and a step-by-step guideline for analysis of hydrated cements is given.

Minerals and hydrates can undergo several thermal reactions: dehydration, dehydroxylation, decarbonation, oxidation, decomposition, phase transition or melting. These reactions are generally associated with weight changes or release of heat. The temperature at which these processes occur are typical for the mineral or hydrate. During TGA the sample is heated while the weight loss is recorded as shown in Figure 5.1a for portlandite. Differentiation of the thermogravimetric (TG) data allows a better resolution and identification of consecutive weight losses: derivative thermogravimetry or differential thermogravimetry (DTG) as shown in Figure 5.1.

In addition to TGA, differential scanning calorimetry (DSC) or differential thermal analysis (DTA) are also often used to analyse cements. DSC measures the difference in the heat needed to increase the temperature of a sample compared to a reference as a function of temperature. An alternative technique, which has much in common with DSC, is DTA. During DTA the sample and reference are heated identically; however, phase changes and other thermal processes cause a difference in temperature between the sample and an inert reference, which can be either exothermic or endothermic (Figure 5.1b). Both DSC and DTA provide similar information. The measured heat or temperature difference is related to the enthalpy changes caused by loss of water or CO_2, or redox or recrystallisation reactions. The deconvolution of DTA or DSC signals can be very difficult in hydrated cement pastes as different reactions, e.g. dehydroxylation of C-S-H and phase transformation of the glasses, appear in the same temperature regions as discussed in Section 5.6.

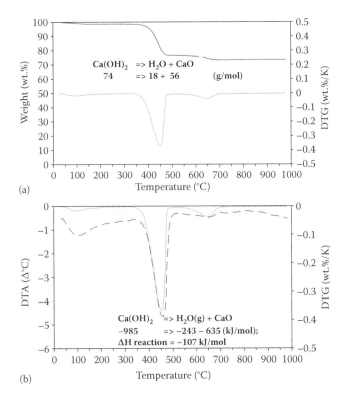

Figure 5.1 (a) Weight loss recorded by TGA (solid line) and differential thermogravimetry (DTG) data (dotted line) and (b) comparison of DTG and differential thermal analysis (DTA; dashed line) for portlandite. 17 mg of sample, heating rate 20°C/min.

DTG and DTA (or DSC) often show similar peaks (see Figure 5.1b) as many weight change reactions, such as the loss of water, are also associated with endothermic changes in the enthalpy.

Some instruments allow to record simultaneously with the weight loss the type and the amount of gaseous reaction products. For this purpose the thermogravimeter is coupled directly to a mass spectrometer or an infrared spectrometer. In cementitious systems the mass loss up to ≈600°C is generally related to the loss of water and above 600°C mainly to the release of CO_2, as shown in Figure 5.2.

In cementitious systems, TGA (or also DSC or DTA) is generally used to identify hydration phases, to confirm the presence of both crystalline and amorphous phases found by XRD, nuclear magnetic resonance (NMR) or scanning electron microscopy–energy-dispersive X-ray spectroscopy (SEM-EDX) analysis and/or to quantify the amount of solids present. The discussion presented in this chapter will mainly focus on TGA; only Section

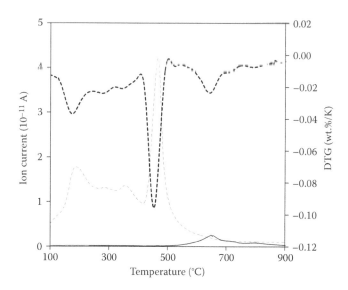

Figure 5.2 DTG (dotted line) and H_2O (dashed line) and CO_2 currents (solid line) measured by mass spectrometry during the TGA of a portland cement after a hydration time of 28 days. Further details are provided by De Weerdt et al. (2011b).

5.6 focuses on the potential use of DSC or DTA to determine the fraction of unreacted glass. Based on the recorded weight changes during the TGA measurement, the amount of mineral or hydrate can be easily calculated in systems where only one or two solids are present. However, in cementitious systems, where a multitude of minerals and hydrates are present, the reactions often overlap, which makes quantification difficult, as shown in Figure 5.3a and b.

5.2 FACTORS INFLUENCING THERMOGRAVIMETRIC MEASUREMENTS

The results of thermal analysis are strongly influenced by the architecture of the measuring device, the kind of vessel used, the heating rate, the amount of solid, the particle size, gas flow rate and kind of purging gas (e.g. O_2 or N_2) and the pretreatment of the sample. The different influences make it nearly impossible to transfer measurements done at one laboratory directly to another laboratory. The relative peak intensities and the sequence of the weight loss, however, follow the same rules independent of the details of the measurement protocol used. As the results are influenced differently by many factors, one of the main rules to do reliable TGA measurements is to stick as much as possible to the same procedure for all measurements.

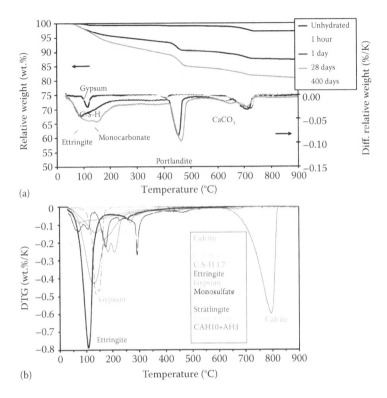

Figure 5.3 (a) TGA/DTG of a hydrating portland cement containing 4 wt.% of limestone (Data from Lothenbach, B. et al., *Cement and Concrete Research*, 38(6), 848–860, 2008.) and (b) DTG of solids typical for cementitious systems (for further details, see Section 5.3).

5.2.1 Heating rate, open and closed vessels and gas flow

The heating rate and the kind of vessel influence the position of dehydration peaks, as exemplified for gypsum ($CaSO_4 \cdot 2H_2O$) in Figure 5.4; similar findings have also been reported by Paulik et al. (1992). Higher heating rates lead to better-defined, narrower peaks but also to higher observed dehydration and dehydroxylation temperatures as the water vapour pressure over the sample is higher if the sample is heated faster. Under static conditions and in the presence of low vapour pressures, the dehydration can occur at much lower temperatures.

Gypsum dehydrates via hemihydrate to anhydrite (Hudson-Lamb et al. 1996) and at higher vapour pressures the dehydrations of gypsum and hemihydrate can be differentiated as the dehydration of hemihydrate ($CaSO_4 \cdot 0.5H_2O$) to anhydrite ($CaSO_4$) occurs at slightly higher temperatures. While under TGA conditions, the dehydration of gypsum is generally

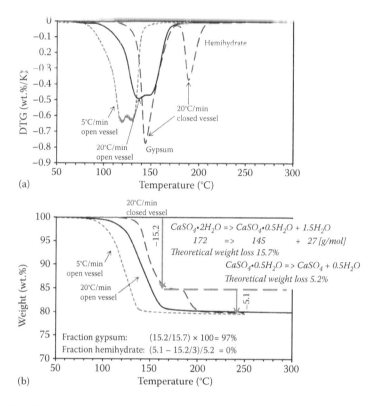

Figure 5.4 (a) DTG and (b) weight loss recorded by TGA for gypsum using ≈20 mg of sample. A heating rate of 5 or 20°C/min and open Al_2O_3 vessels or tightly closed aluminium vessels with a 50 μm hole in the lid were used.

observed at ≈140°C, the exact temperature can vary strongly, depending on the vapour pressure present. Dehydration of gypsum has been observed at around 50°C in equilibrium with only 1 Pa vapour pressure, while at higher vapour pressure the temperature of dehydration rises up to 150°C or even higher (Badens et al. 1998; Strydom et al. 1995). The same trend is also visible in Figure 5.4. If closed vessels with only a very small hole are used, the vapour pressure over the sample is increased as the H_2O can escape only slowly through the small opening. The higher partial pressure of H_2O under these conditions results in two distinct peaks for the dehydration of gypsum and of hemihydrate at clearly separated temperatures, as shown in Figure 5.4a. The associated weight losses from the TGA can be used to calculate the fraction of hemihydrate to gypsum in the sample; if more than one-fourth of the weight loss is associated with the hemihydrate peak, this difference indicates the initial presence of hemihydrate in addition to gypsum.

Also, the gas flow affects the main weight loss region (Paulik et al. 1992; Salvador et al. 1989). Higher gas flow lowers the temperature where

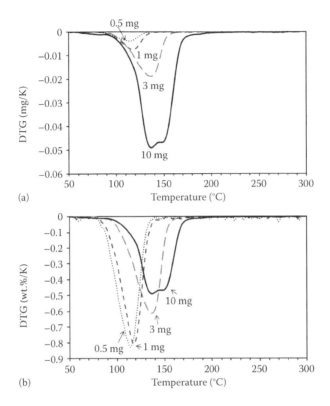

Figure 5.5 Influence of the sample weight on the peak position and shape of the dehydration peak of gypsum. Differential TG data normalised to (a) milligrams
of sample and (b) wt.%. A heating rate of 20°C/min and open Al_2O_3 vessels
were used.

the weight loss is observed but also leads to a not as good differentiation
between different peaks. Similarly, the size and shape of the pan used influences the vapour pressure and thus the temperature where the main weight
loss is observed.

5.2.2 Sample weight

Figure 5.5 illustrates that the sample weight is another important factor
affecting the temperature of the main weight loss, especially if the differential data are normalised to 100 wt.%* of the initial weight (Figure 5.5b). In
the normalised plots higher quantities of a phase lead to a seemingly higher
dehydration temperature and to broader peaks.

---

* In most cases the measured TGA and DTG data are normalised to 100 wt.% to ease the
comparison between the different samples as slightly different starting weights are used for
each sample.

If the measured data are plotted without normalisation, as in Figure 5.5a, it is visible that the weight loss caused by the dehydration of gypsum starts in all cases at ≈80°C. If little gypsum is present, the main weight loss occurs at 115°C and only anhydrite is present above 135°C. If more gypsum is present in the sample pan, the main weight loss occurs at 145°C and continues up to 180°C. The dehydration of a larger quantity of gypsum releases more water, resulting in higher vapour pressure in the environment of the solid; thus, higher temperatures are needed to dehydrate gypsum to anhydrite. The normalisation to 100 wt.% accentuates this difference in temperature and even seems to indicate (wrongly) that the starting temperature of the weight shifts from 80 to 100°C with increasing amount of gypsum present, as the relatively small weight loss below 100°C is not well visible in the presence of high quantities of gypsum in the normalised data shown in Figure 5.5b.

The effect of the sample weight on the temperature of the weight loss peak has also been reported by Paulik et al. (1992) for gypsum and by Salvador et al. (1989) for $CaCO_3$. Besides the sample size, heating rate and gas flow, the particle size also affects the peak temperature (Criado and Ortega 1992; Salvador et al. 1989). Hence, a fixed grinding procedure should be employed.

5.2.3 Pretreatment

One of the most important parameters affecting the TG analysis of hydrated cements is the sample preparation, i.e. how the hydration reaction is stopped and how long the sample is dried. The aim of the sample preparation is to remove the free water to preserve the hydrates in the sample in a specific state, e.g. after 24 h or 28 days of hydration. In literature a range of sample preparation methods for TG analysis of hydrated cement paste can be found as summarised recently by Zhang and Scherer (2011). For early-age samples the methods typically consist of a hydration stopping step by, e.g. quenching with liquid nitrogen or solvent exchange, followed by a drying step, e.g. freeze-drying, vacuum drying or oven drying. Further details on the drying techniques are discussed in Chapter 1. There are, however, some aspects of sample preparation important for TG analysis which are discussed in the following sections.

5.2.3.1 Effect of organic solvents

Hydration of cement paste can be stopped by diluting and removing the water present in the pores of the cement paste by a solvent. Finely ground cement paste or small discs are submerged in a relatively large amount of solvent for a certain period of time, which depends on the size of the sample (Zhang and Scherer 2011). Typical solvents that have been used to stop cement hydration are methanol, ethanol, isopropanol and acetone.

However, solvents have been observed to interact with the cement hydrates. Taylor and Turner (1987) reported that methanol, acetone and some other organic liquids are strongly sorbed by C_3S paste. The organics appear to replace some of the bound water and are not entirely removed, even by vacuum drying. A similar behaviour was observed on cement paste by Thomas (1989). Also, an interaction between methanol and portlandite has been reported by some authors (Beaudoin 1987; Day 1981), while others observed no such interaction (Parrott 1983; Thomas 1989). Besides possible interactions with C-S-H and portlandite, solvents can also affect the TGA signals of ettringite and AFm phases. Prolonged storage (up to 24 hours) in methanol, ethanol and isopropanol dehydrate monosulfate (C_4AsH_{12}) from 12 to 10 H_2O (C_4AsH_{10}) in the interlayer, while the alcohols intercalate in the interlayer (Khoshnazar et al. 2013a). Only little water (one to two H_2O) is removed from ettringite ($C_6As_3H_{32}$) during 24 hours of immersion in isopropanol or ethanol, while methanol leads to complete breakdown of the ettringite structure (Khoshnazar et al. 2013b). Methanol has clearly a damaging effect on the microstructure of ettringite and monosulfate, while the effect of ethanol and isopropanol is much weaker.

The sorption of alcohol on the hydrates increases indirectly the weight loss in the temperature range typical for carbonate decomposition in hydrated cements (Day 1981; Knapen et al. 2006; Thomas 1989; Zhang and Scherer 2011). The increased weight loss in the carbonate temperature range has been attributed to a reaction of the organic solvents liberated during the TG analysis with portlandite, resulting in the formation of calcium carbonate during TG analysis (Jenni et al. 2005; Knapen et al. 2006; Kriegel et al. 2003), as discussed further in Section 5.5. The increase in calcium carbonate content has been observed to be stronger at longer soaking times and also stronger in the case of methanol and acetone than for isopropanol and ethanol (Thomas 1989; Zhang and Scherer 2011). Figure 5.6 shows the effect of ethanol on the TGA and DTG curves for a portland cement sample hydrated for 91 days. An increased weight loss is observed above 600°C if the sample has been immersed in ethanol. The impact of this is, however, limited: there is a small decrease in portlandite content and an increase in the $CaCO_3$ content. It should be noted that the sample with ethanol was dried in the TGA to remove most of the solvent prior to the TG analysis.

Instead of using alcohols, which tend to sorb to the hydrates, either drying under N_2 (De Weerdt et al. 2011a,b; Lothenbach et al. 2008a; Figure 5.6) or the exchange of the alcohols used to remove the pore solution by a less polar solvent, such as diethyl ether, can help to minimise or even avoid the sorption of organics. A three-step procedure using isopropanol (exchange of pore water by isopropanol, followed by the removal of isopropanol by diethyl ether and short drying at 40°C to evaporate any remaining diethyl ether; for details see Deschner et al. [2012] and Schöler et al. [2015]) leads to no significant decrease in the portlandite

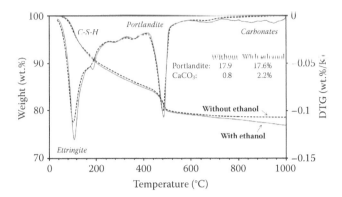

Figure 5.6 Effect of stopping hydration with ethanol (2 times 10 minutes) on the TGA and DTG curves for a 91-day-hydrated ordinary portland cement (De Weerdt et al. 2011b). The sample without ethanol was dried at 40°C under N_2 in the TG device prior to the measurements.

content or increased carbonate content compared to freeze-drying, as shown in Figure 5.7a.

5.2.3.2 Other drying procedures

The use of solvents to stop the hydration of early-age samples can be avoided by quenching the samples with liquid nitrogen, which is commonly combined with subsequent freeze-drying. Many authors recommend this method because it allows to stop hydration at a specific time without altering the composition and one is able to dry the samples to a specific equilibrium in a CO_2-free environment (Knapen et al. 2006; Taylor and Turner 1987; Zhang and Glasser 2000; Zhang and Scherer 2011). However, care should be taken as freeze-drying can lead to too severe drying, thereby suppressing the signal associated with ettringite upon prolonged drying as part of the water associated with ettringite can be removed ($3CaO \cdot 3CaSO_4 \cdot Al_2O_3 \cdot 32H_2O$) (Zhang and Glasser 2000) and X-ray amorphous metaettringite ($3CaO \cdot 3CaSO_4 \cdot Al_2O_3 \cdot 12H_2O$) can be formed (Baquerizo et al. 2015; Zhou et al. 2004), as also shown in Figure 5.7a for hydrated cement.

Other hydrates, particularly C-S-H (Figure 5.7c) and to a lesser extent monosulfate (Baquerizo et al. 2014; Khoshnazar et al. 2013a), are also sensitive to drying and lose a part of the water under such conditions. Hydrates such as hydrogarnets are less sensitive to the drying conditions as they lose most of their water at higher temperatures as discussed in Section 5.3. It should be noted that it is not solely the drying procedure itself but rather the duration of the drying which affects the TGA results. Drying for only a few minutes at 105°C or in the freeze-dryer does not necessarily destabilise

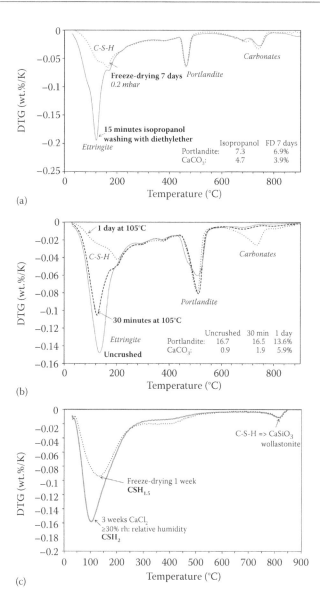

Figure 5.7 Effects of (a) stopping by isopropanol (15 minutes followed by washing with diethyl ether) or freeze-drying (ternary blend of 50 wt.% portland cement, 20% blast furnace slag, 25% fly ash and 5% fly ash after 196 days of hydration; further details in Schöler et al. [2015]); (b) drying at 105°C (portland cement hydrated for 28 days [De Weerdt et al. 2011b]) and (c) C-S-H (Ca/Si = 1.0) after freeze-drying or reequilibration over saturated CaCl₂ solution for 3 weeks. Details on C-S-H preparation are provided by L'Hopital et al. (2015).

the ettringite, in contrast to an extended drying period of some hours or days, which leads to dehydration of ettringite, as shown in Figure 5.7b for a hydrated portland cement dried at 105°C for 1/2 hour and 1 day. It should be noted that the short-term drying methods, as mentioned previously, have to be combined with solvent exchange as the short drying is not sufficient to remove all free water and thus to stop hydration.

5.2.3.3 Carbonation during storage

Another important parameter regarding the sample preparation for TGA is the possible carbonation of the hydrated cement during the sample preparation and storage before the measurement. The presence of some water and the large surface area of the crushed sample make the samples very susceptible to carbonation, as shown in Figure 5.8a for a hydrated portland cement exposed to air directly after crushing. The rate of carbonation of hydrated cement paste is at its highest around a relative humidity of 50%–60% and reduces towards lower and higher relative humidities. Hence, the optimal carbonation conditions are present during the drying step, and care needs to be taken to avoid contact with CO_2 during drying. Drying over longer time in an aerated oven can lead to accelerated carbonation, independent of whether solvent exchanges have been performed or not, and therefore it should be banned as a preparation method for TGA. The use of an organic solvent, such as isopropanol, to remove the free water and the storage of such samples in tightly closed containers under N_2 or with a slight vacuum slows down the carbonation, but it is not able to completely suppress it; see Figure 5.8b.

The suitability of the sample preparation method depends strongly on the sample to be analysed and the parameters one wishes to measure.

An additional factor for carbonation can also be the time the prepared sample is exposed to air on the auto sampler before the TG measurement. Both the peaks from 50 to 200°C, associated mainly with C-S-H, AFm and AFt phases, as well as the amount of portlandite, decrease, while the amount of carbonates increases, indicating both further drying of the sample and some carbonation. The drying and carbonation has been observed to be most distinct during the first 6 hours, while afterwards the changes are much smaller.

5.3 THERMOGRAVIMETRIC ANALYSIS DATA OF SOLIDS TYPICALLY OBSERVED IN CEMENTITIOUS SYSTEMS

As discussed previously, the peak position of DTG data depends on the measurement conditions and the device used; thus, reference measurements with pure minerals should be carried out using the respective measurement conditions and device. However, as the temperature and the peak shape are just

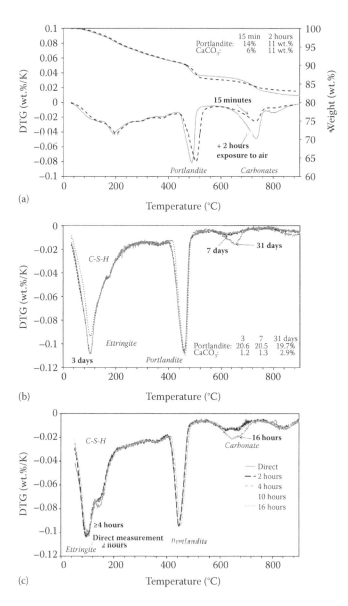

Figure 5.8 (a) Effect of the exposure of the wet powder (after crushing) to air of a port-land cement hydrated for 28 days before drying at 105°C (De Weerdt et al. 2011b); effects of (b) storage under slight vacuum (water pump) and (c) 'wait-ing' time in the auto sampler before TG measurement for a white portland cement hydrated for 28 days and after sample stoppage with isopropanol.

shifted or modified, the relative position of reference measurements from other devices such as those presented here can be used for comparison. All the measurements shown in this chapter have been done unless otherwise stated on a Mettler Toledo Thermogravimetric Analyser TGA/SDTA 851e using open 70 μL aluminium oxide vessels filled with approximately 20 mg solid, a heating rate of 20°C/min and using N_2 as protective gas with a flow of 30 mL/min.

In the following, TGA/DTG data for the most common solids present in cementitious systems are discussed.

5.3.1 Gypsum

Gypsum ($CaSO_4 \cdot 2H_2O$) loses its water in two steps, at around 100 to 140°C to hemihydrate ($CaSO_4 \cdot 0.5H_2O$) and at 140–150°C to anhydrite ($CaSO_4$). The heating rate used strongly influences the temperature at which the main weight loss occurs, as shown in Figure 5.4. Generally the two water loss peaks overlap strongly; closed vessels with a small hole in the lid allow to distinguish the dehydration of gypsum and hemihydrate, as shown in Figure 5.4.

5.3.2 Portlandite, brucite, calcium and magnesium carbonates

Portlandite ($Ca(OH)_2$) dehydroxylates ($Ca(OH)_2 \rightarrow CaO + H_2O$) at around 460°C; brucite $Mg(OH)_2$, at around 420°C, as shown in Figure 5.9. Between 500 and 600°C, magnesite ($MgCO_3$) decomposes to MgO and CO_2; calcite ($CaCO_3$), between 600 and 800°C. Aragonite and vaterite, two other crystalline calcium carbonate polymorphs, recrystallise without weight change at ≈450°C to calcite during the TGA measurements, as shown by Goto et al. (1995), while amorphous calcium carbonate decarbonates partially

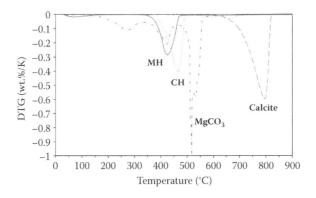

Figure 5.9 DTG of brucite, portlandite, magnesite and calcite.

between 400 and 600°C, forming CaO and calcite (Goto et al. 1995). Dolomite decomposes in two temperature steps; in the first step MgO and calcite are formed at ≈650°C, and in the second step the calcite decomposes (McCauley and Johnson 1991).

The exact position of the decarbonation peaks observed in cements depends strongly on the presence of hemi- and monocarbonates, the amount of calcium carbonate present and the fineness of the calcium carbonates, such that in hydrating portland cement two or more peaks associated with carbonates can be observed, as shown in Figure 5.3a or 5.8. A CO_2 weight loss peak is observed at ≈720°C because of the presence of relatively coarse calcite in the unhydrated cement; a second weight loss peak is present at 600 to 650°C because of the presence of mono- or hemicarbonates and due to the carbonation of portlandite and possibly C-S-H. Both the weight losses of portlandite and calcite in TGA can be used for quantification as they are generally well separated from the weight loss of other main hydrate phases observed in cements. Techniques for quantification are detailed in Section 5.4.

5.3.3 Calcium silicate hydrates

Calcium silicate hydrate (C-S-H) phases show water loss over a wide range of temperature (50 to 600°C) caused by the loss of the water present in the interlayer and dehydroxylation. An additional small dehydroxylation peak (≈0.1 H_2O/Si) at 800°C is related to the decomposition of C-S-H to wollastonite ($CaSiO_3$) (Mitsuda and Taylor 1978; Myers et al. 2015) as also visible by the exothermic DTA peak shown in Figure 5.10. At high Ca/Si ratio (Ca/Si$_{total}$ = 1.6; Ca/Si in C-S-H = 1.5), in addition a small amount of portlandite and carbonate is visible.

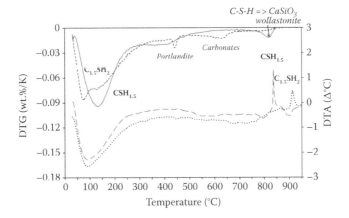

Figure 5.10 DTG of C-S-H with a Ca/Si = 1.0 and 1.5 after freeze-drying. Details on the preparation of C-S-H are given by L'Hopital et al. (2015).

The amount of water lost from C-S-H depends mainly on the drying conditions applied before the measurements (Snoeck et al. 2014). Drying progressively removes bulk and gel water; at 11% relative humidity (freeze-drying) the interlayer water is still present in the C-S-H (Gutteridge and Parrott 1976; Jennings 2008) and approximately 1.5 H_2O per Si are observed for Ca/Si = 1 and 2 H_2O for Ca/Si = 1.5 (Gutteridge and Parrott 1976; Jennings 2008; L'Hopital et al. 2015; Figure 5.10). This value is comparable to the 1.8 H_2O/Si interlayer water determined by ^{1}H-NMR for C-S-H in hydrated portland cement (Muller et al. 2013). Compared to the influence of the drying conditions, the effect of variation in the Ca/Si ratio on the amount of interlayer water present is relatively small. Note that the effect of Ca/Si on the amount of gel water is not captured under these drying conditions.

5.3.4 AFt phases

Ettringite ($3CaO \cdot Al_2O_3 \cdot 3CaSO_4 \cdot 32H_2O$) has a hexagonal prismatic shape. Columns of $[Al(OH)_6]^{3-}$ octahedra are linked together by calcium and hydroxide ions ($[Ca_6 \cdot Al_2(OH)_{12}]^{6+}[3 SO_4^{2-} \cdot 26H_2O]^{6-}$) and sulfate and water molecules are located on the outer surface of the columns. The water between the columns is lost around 100°C, as shown in Figure 5.11, while the water from the dehydroxylation of the aluminium hydroxide is removed between 200 and 400°C (main weight loss at ≈260°C, as observed also for aluminium hydroxide [AH_3], discussed in Section 5.3.6). Thaumasite ($CaSiO_3 \cdot CaCO_3 \cdot CaSO_4 \cdot 15H_2O$), which has a structure ($[Ca_3 \cdot Si(OH)_6]^{4+}[CO_3^{2-} \cdot SO_4^{2-} \cdot 12H_2O]^{4-}$) comparable to that of ettringite, similarly shows a loss of the water between the columns at around 130°C and a continued water loss due to the hydroxides associated with the silica

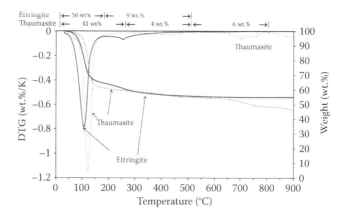

Figure 5.11 TGA and DTG of ettringite ($C_6As_3H_{32}$; continuous line) and thaumasite (C_3SscH_{15}; dotted line). The weight losses observed are indicated.

Table 5.1 Molecular weights (MW) of fully hydrated solid and calculated weight losses for H_2O and CO_2

Name	Formula[a]	MW (g/mol)	H_2O loss (g/mol)	H_2O (wt.%)	CO_2 loss (g/mol)	CO_2 (wt.%)
Gypsum	CsH_2	172	36	20.9		
Hemihydrate	$CsH_{0.5}$	145	9	6.2		
Anhydrite	Cs	136				
Portlandite	CH	74	18	24.3		
Calcite	Cc	100			44	44.0
Aragonite	Cc	100			44	44.0
Vaterite	Cc	100			44	44.0
Magnesite	Mc	84			44	52.2
Dolomite	$CcMc$	184			88	47.7
High-Ca C-S-H	$C_{1.5}SH_2$	180	36	20.0		
Low-Ca C-S-H	$CSH_{1.5}$	140	27	19.3		
Ettringite	$C_6As_3H_{32}$	1255	576	45.9		
Thaumasite	C_3SscH_{15}	623	270	43.4	44	7.1
Monosulfate	C_4AsH_{12}	623	216	34.7		
Monocarbonate	C_4AcH_{11}	568	198	34.9	44	7.7
Hemicarbonate	$C_4Ac_{0.5}H_{12}$	564	216	38.3	22	3.9
Friedel's salt	$C_4ACl_2H_{10}$	561	180	32.1		
Kuzel's salt	$C_4ACls_{0.5}H_{12}$	610	216	35.4		
Strätlingite	C_2ASH_8	418	144	34.5		
Hydrotalcite	M_4AH_{10}	443	180	40.6		
	C_4AH_{19}	669	342	51.2		
	C_4AH_{13}	560	234	41.8		
Katoite	C_3AH_6	378	108	28.6		
	$C_2AH_{7.5}$	349	135	38.7		
	CAH_{10}	338	180	53.3		
Aluminium hydroxide	AH_3	156	54	34.6		

[a] Cement short hand notation is used – A: Al_2O_3; c: CO_2; C: CaO; H: H_2O; M: MgO; s: SO_3; S: SiO_2.

up to 400°C. In addition, a peak associated with the weight loss due to CO_2 is observed between 600 and 750°C. For comparison a summary of the theoretical weight losses of the most important solids observed in cementitious systems is given in Table 5.1.

5.3.5 AFm phases

AFm (Al_2O_3-Fe_2O_3-monophase) phases have a layered crystal structure built by periodical stacking of positively charged $(M^{2+}, 2M^{3+})(OH)_6$ octahedral layers and negatively charged interlayers containing anions and water molecules, with the general formula $[Ca_4Al_2(OH)_{12}]^{2+}[A_{2/n}^{n-} \cdot mH_2O]^{2-}$, where A is a mono- or bivalent anion. There is a wide range of anions A^{n-} that can be observed in the interlayer: SO_4^{2-}, CO_3^{2-}, OH^-, $Al(OH)_4^-$, $AlSi(OH)_8^-$ and Cl^- or mixtures thereof are the most important in the cement matrix (Allmann 1977; Taylor 1997). The minerals form hexagonal platy crystals; the layer thickness varies with the nature of the anion and the amount of water molecules incorporated and thus the relative humidity.

The layered structure and the presence of loosely bound water are mirrored in the TGA measurements of the AFm phases, where typically several weight loss regions are observed (Pöllmann 1984; Figure 5.12). For monocarbonate ($4CaO \cdot Al_2O_3 \cdot CO_2 \cdot 11H_2O$), the loss of the five interlayer H_2O can be observed between 60 and 200°C; between 200 and 300°C the six H_2O from the octahedral layer are lost and CO_2 at around 650°C (Figure 5.12). In the case of monosulfate ($4CaO \cdot Al_2O_3 \cdot SO_3 \cdot 12H_2O$), three temperature steps are observed for the six interlayer waters; while the water loss from the octahedral layer occurs at slightly higher temperatures (250 to 350°C) than for monocarbonate. This difference in temperature can be used to differentiate

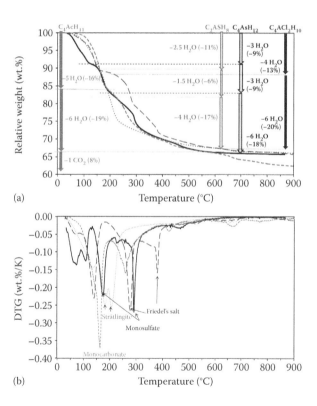

Figure 5.12 (a) TGA and (b) DTG of monosulfate (C_4AsH_{12}; continuous line), monocarbonate (C_4AcH_{11}; dashed line), strätlingite (C_2ASH_8; dotted line) and Friedel's salt ($C_4ACl_2H_{10}$; long-dashed line).

between monosulfate and monocarbonate (Lothenbach et al. 2008a). For Friedel's salt ($3CaO \cdot CaCl_2 \cdot Al_2O_3 \cdot 10H_2O$), the interlayer water is lost around 140°C, while for the main-layer water two temperature regions between 250 and 400°C are present. In the case of strätlingite, the losses of the interlayer and the main-layer water both occur around 200°C, which makes the deconvolution more difficult.

5.3.6 Calcium aluminium hydrates

Aluminium hydroxide ($Al(OH)_3$) loses its water at ≈270°C. CAH_{10} shows a distinct weight loss at 120°C; katoite (C_3AH_6), at 320°C; and hydrogarnets containing also silica, at slightly higher temperatures (≈340°C for C_3ASH_4) (Figure 5.13). The dehydroxylation temperature of aluminium hydroxide is in the same temperature range (260 to 320°C) as the dehydroxylation of calcium aluminate hydrate present in katoite, AFm (except Friedel's salts) and AFt phases.

5.3.7 M-S-H and hydrotalcite

Synthetic magnesium silicate hydrates (M-S-H) are gel-like solids and show a water loss related to loosely bound water up to 270°C (Nied et al. 2015) similarly as C-S-H, as shown in Figure 5.14. In contrast to C-S-H, a second weight loss between 270 and 700°C is visible, which is related to the dehydroxylation of hydroxyl groups, $2 \equiv OH^- \rightarrow O^{2-} + H_2O(g)$, in the M-S-H.

Hydrotalcite (M_4AH_{10}) has a layered structure, as the AFm phases shown in Figure 5.12, and thus loses in the first step around four interlayer waters up to 270°C and in the second step six H_2O in the main layer at around 400°C.

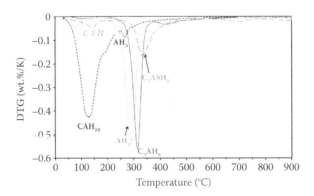

Figure 5.13 DTGs of CAH_{10}, AH_3, C_3AH_6, C_3ASH_4 and CH for comparison.

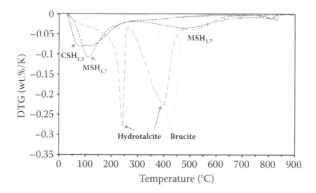

Figure 5.14 DTGs of M-S-H, hydrotalcite, brucite and C-S-H for comparison.

5.4 QUANTIFICATION OF BOUND WATER, PORTLANDITE AND CaCO₃

5.4.1 Bound water

A widely used technique to assess the degree of reaction of portland cements is to obtain the bound water content from the difference between the weight losses of samples dried to 105 and at 1000°C (Gómez-Zamorano and Escalante-García 2010; Mouret et al. 1997). This procedure assumes that only pore water is removed until 105°C. Based on the TGA results presented in Figures 5.8 through 5.14, it is evident that many of the cement hydrates – particularly C-S-H, ettringite and monosulfate – lose part of their chemically combined water below 105°C, especially during prolonged drying. In addition, the weight loss due to the decomposition of carbonates is included when the interval 105 to 1000°C is used. Another possibility to obtain bound water is to use the difference between the weight after solvent exchange, e.g. as done by De Weerdt et al. (2011a), Lothenbach and Winnefeld (2006) and Schöler et al. (2015), and the weight after drying at 500 or 550°C to differentiate chemically bound water (H_2O_{bound}) from the mass loss from decarbonation, based either on direct weighing or on TGA measurements.

5.4.2 Portlandite and calcite

Portlandite ($Ca(OH)_2$) decomposes generally between 400 and 500°C to CaO and H_2O, as shown in Figure 5.1. This weight loss ($WL_{Ca(OH)_2}$) due to the evaporation of water can be used to calculate the amount of portlandite present, using the molecular masses of portlandite ($m_{Ca(OH)_2} = 74$ g/mol) and water ($m_{H_2O} = 18$ g/mol):

$$Ca(OH)_{2,measured} = WL_{Ca(OH)_2} \times m_{Ca(OH)_2} / m_{(H_2O)} = WL_{Ca(OH)_2} \times \frac{74}{18}. \quad (5.1)$$

Calcium carbonate ($CaCO_3$) decomposes above 600°C to CaO and CO_2. Similarly as for portlandite, this weight loss (WL_{CaCO_3}) can be used to calculate the amount of calcium carbonate present, using the molecular masses of $CaCO_3$ (m_{CaCO_3} = 100 g/mol) and CO_2 (m_{CO_2} in 44 g/mol):

$$CaCO_{3,measured} = WL_{CaCO_3} \times m_{CaCO_3}/m_{CO_2} = WL_{CaCO_3} \times \frac{100}{44}. \quad (5.2)$$

As the sample weight of the solid fraction of the sample is changing during hydration (see Figure 4.16 in Chapter 4 about XRD), the results need to be rescaled, to either paste or anhydrous, similarly as for the results of XRD-Rietveld analysis.

- Per 100 g paste:

$$Ca(OH)_{2,paste} = Ca(OH)_{2,measured}/[(1-H_2O_{bound})(1+w/c)]$$

$$= \frac{Ca(OH)_{2,measured}}{\text{weight at } 600°C(1+w/c)} \quad (5.3)$$

- Per 100 g anhydrous:

$$Ca(OH)_{2,dry} = Ca(OH)_{2,measured}/(1-H_2O_{bound}) = \frac{Ca(OH)_{2,measured}}{\text{weight at } 600°C} \quad (5.4)$$

A normalisation to 100 g anhydrous cement increases the relative amount of portlandite or limestone, as illustrated in Figure 5.15 for the amount of portlandite observed during the hydration of a portland cement.

Also, the amount of other phases can be calculated using the weight losses given in Table 5.1 as, e.g. done by Gruskovnjak et al. (2011) for supersulfated blast furnace slags. In hydrated portland cement, however, there is generally too much overlap between the different weight loss regions (see Figure 5.3) to be able to make meaningful deconvolutions. In particular, the water loss region of C-S-H, the main hydrate phase in portland cements, is not well defined and spans from 40 to 600°C (Figure 5.10). Thus, quantification based on TGA is often limited to gypsum (during the first hours of hydration), portlandite and calcite. The quantification is further complicated by the shift in the DTG peak position depending on the amount of bound water, as shown in Figure 5.5.

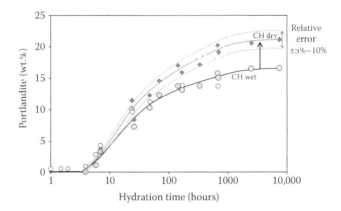

Figure 5.15 Amount of portlandite measured by TGA (filled circles; tangential method) and as free lime (empty circles; measured according to Franke [1941]); CH dry = CH wet normalised to dry weight (diamonds), i.e. to the weight at 650°C. The relative measurement error of ±5% to ±10% is indicated by the dotted lines. (Data from Lothenbach, B., and F. Winnefeld, *Cement and Concrete Research*, 36(2), 209–226, 2006.)

5.4.3 Measurement error

The relative error of measurement of bound water or the amount of hydrates between several independent preparations of hydrated cement paste is observed to be between ±5% and ±10% (Deschner et al. 2012; Gruskovnjak et al. 2011; Lothenbach and Winnefeld 2006). This error is related to the heterogeneity of the paste samples and the relative small amount weighted in, typically 10 to 50 mg and to the errors introduced by the deconvolution of the TGA as discussed in Section 5.4.4. Because of the heterogeneity even in paste samples, a minimal weight of approximately 50 mg is recommended for hydrated cement paste samples.

5.4.4 Tangential versus stepwise method for quantification

The large temperature region (40–600°C) where weight loss from C-S-H is observed makes the quantification of portlandite and other hydrates difficult in hydrated portland cements, as visible in Figure 5.3. For calcite, where generally the weight loss occurs only above 600°C, quantification is more simple except in cases where amorphous calcium carbonate has formed (e.g. during carbonation studies; amorphous calcium carbonate decarbonates at 400–600°C (Goto et al. 1995; Figure 5.10).

In literature, generally either the so-called stepwise or the tangential method is used to quantify portlandite. Figure 5.16 shows the effect of the stepwise and tangential methods on the quantification of portlandite in a

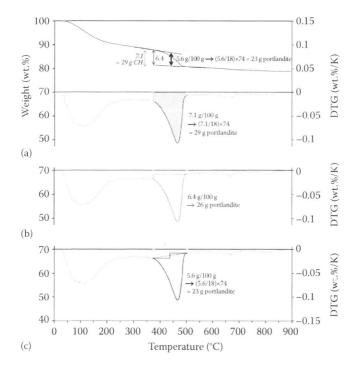

Figure 5.16 Effects of the tangential and stepwise methods on the quantification of portlandite in a mix containing 75% of C-S-H (with no portlandite) and 25% 'portlandite' (which corresponds to 23% CH and 2% Cc): (a) stepwise; (b) software and (c) tangential.

mix containing 75% of C-S-H (with no portlandite) and 23 wt.% portlandite (CH). The stepwise method is the simplest method and is often used to determine portlandite in hydrated cements. With a stepwise approach, the weight difference between the onset at 370°C (87.9 wt.%) and the end at 500°C (80.8 wt.%) of the weight loss related to portlandite is used, resulting in 29 g portlandite for the example shown in Figure 5.16. The data tend to agree reasonably well with XRD results as, e.g. shown by De Weerdt et al. (2011a) and Lothenbach et al. (2008a). The stepwise method, however, overestimates the real portlandite content as it includes not only the weight loss of the portlandite but also the weight loss of the C-S-H and any other phase which might lose water in this temperature region, as visualised on the derivative DTG curve in Figure 5.16a.

The tangential method assumes that the weight change due to the presence of C-S-H or other hydrates observed before and after the peak will continue (linearly) in the portlandite region. Such a correction results in the quantification of the portlandite weight loss only and is equivalent to

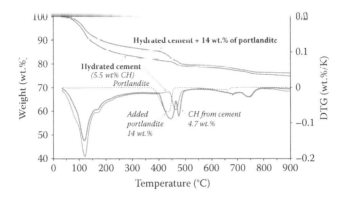

Figure 5.17 Amount of portlandite determined in a hydrated cement (blend of 50 wt.% portland cement, 20% slag, 25% fly ash and 5% limestone [Schöler et al. 2015] hydrated for 196 days) and in the hydrated cement after the addition of 14 wt.% of portlandite: tangential method (5.5 + 14 wt.%) and stepwise method (6.3 + 16.6 wt.%).

the integration of the peak area, as shown in Figure 5.16c. The amount of portlandite determined by the tangential method, 23 wt.%, agrees with the amount of portlandite added to the C-S-H. The correctness of the tangential method is also confirmed by the quantification of portlandite in a hydrated cement (5.5 wt.%), to which 14 wt.% of portlandite has been added (Figure 5.17) or by the good agreement with the amount of portlandite determined with free lime measurements, as shown in Figure 5.15.

Often the software of the instrument allows the direct quantification of the weight loss either by the stepwise or the tangential method or by integration of the derivative curve. The results of the tangential method, however, depend on the manufacturer, as different approaches have been chosen. The software of some manufacturers calculates under the label *tangential* the differences at the start temperature (Figure 5.16b) instead of at the median temperature (Figure 5.16c). The use of the weight at the initial temperatures results in hydrated cements generally in higher weight loss. Other software will take the difference at the temperature where the maximum weight loss occurs. As alternative to using the often not well-described tangential method in the software, the integration of the peak area offers a better traceable quantification method and the results correspond to those in Figure 5.16c.

5.5 OXIDATION REACTIONS DURING THERMOGRAVIMETRIC ANALYSIS

Not only the heating during TGA measurements can result in the loss of H_2O and CO_2, but also the loss of H_2S can be observed if blast furnace slags are heated. The reduced sulfur species (S^{2-}) present in blast furnace slag can react with water to H_2S and O^{2-}. Analysis of the gaseous reaction products confirms the evolution of H_2S gas and a corresponding weight loss of -1.06 g per g S^{2-} ($H_2S/S^{2-} = [2 + 32]/32 = 1.06$), as shown in Figure 5.18a. Above approximately 700°C a weight increase is observed ($+1.996$ g per g S^{2-}) as S^{2-} is oxidised to sulfate (SO_4^{2-}) ($S^{2-} + 2O_2 \rightarrow SO_4^{2-}$; $4 \times 16/32 = 1.996$) and solid sulfates, probably $CaSO_4$ (Montes-Morán et al. 2012), are formed. In addition, elemental sulfur might also form (Montes-Morán et al. 2012). Above 900°C, gaseous SO_2 is observed, resulting in a further weight loss of -0.5 g per g S^{2-} ($S^{2-} + 1.5O_2 \rightarrow SO_2 + O^{2-}$; $16/32 = 0.5$).

If blast furnace slags are heated under oxidising conditions, the presence of H_2S is observed below 500°C (Figure 5.18b and c). A small fraction of SO_2 is present at all temperatures because of the availability of O_2 and, as under N_2, its fraction increases above 900°C. The TGA data show above 600°C a more distinct weight increase under O_2 than under N_2, as shown, e.g. in Figure 5.18c or by Montes-Morán et al. (2012), as the presence of oxygen eases the formation of sulfates. In addition to the weight changes caused by the oxidation of sulfide, the oxidation of metallic iron or manganese in slags could lead also to weight increase.

The formation of gases (H_2S and SO_3) and of solids ($CaSO_4$) from sulfide present in the slag results in a complex pattern of weight changes, which is affected not only by the surrounding atmosphere but also by the heating rate in the TGA, as the heating rate affects the oxidation rate and the relative fraction of sulfur in the gas (H_2S and SO_2) and in the solid phase (SO_4^{2-}). Due to these possible oxidation reactions in blast furnace slags, the determination of the loss on ignition (LOI) of blast furnace slags needs special attention. If slag is heated to 950°C in air to determine the LOI, a weight increase instead of a weight loss is often observed (Ben Haha et al. 2011; Gruskovnjak et al. 2006) due to the formation of SO_4^{2-} during the heating. The LOI measured in air can be corrected to $LOI_{corrected} = LOI_{in\ air} + 1.996 \times S^{2-}$ as also recommended in EN 196-2. Alternatively, a determination of the LOI under N_2 gives comparable results as the correction (Matthes 2014), as the formation of sulfate is (partially) suppressed, especially if the heating occurs fast.

Also, in the case of fly ashes, oxidation reactions can be observed during TGA measurement, as fly ashes contain a small fraction of unburnt coal, which can oxidise to CO_2. This is confirmed by the TGA measurements of a fly ash under protective gas and under O_2 shown in Figure 5.19. The oxidation of unburnt coal to CO_2 ($C + O_2 \rightarrow CO_2$) in the presence of oxygen results in a weight loss of -1 g per g C (carbon). This corresponds to a weight loss of 2.5 wt.% for the fly ash shown in Figure 5.19, while the same fly ash heated under a protective helium atmosphere shows a very low mass loss. The release of minor amounts of CO_2 even under protective gas is probably related to redox reactions of the unburnt carbon with oxides such as Fe_2O_3 (Deschner et al. 2012).

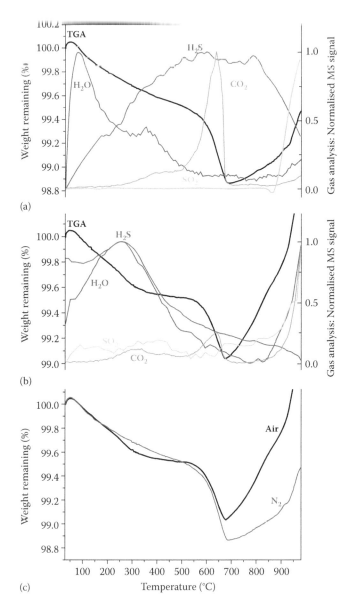

Figure 5.18 Weight loss and exhaust gas analysis using mass spectrometry during TGA of a blast furnace slag (containing 0.26 wt.% of S^{2-}) under (a) N_2 and (b) O_2 atmosphere. (Courtesy of S. Bernal.)

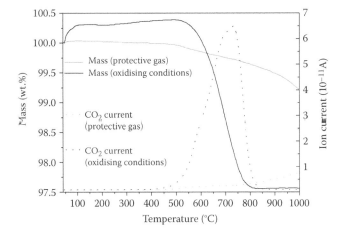

Figure 5.19 Weight loss and CO_2 in the exhaust gas analysed by mass spectrometry during TGA of a fly ash (containing 2.7 wt.% of unburnt coal) under He and oxidising atmosphere. (Data from Deschner, F. et al., *Cement and Concrete Research*, 42, 1389 1400, 2012.)

Also, organics present can oxidise to CO_2 during TGA measurements as shown by Kriegel et al. (2003). In hydrated cements where water and portlandite are present, a part of the CO_2 from the oxidation of the organics reacts with portlandite to calcium carbonate, as also discussed by Thomas (1989). This leads in the presence of high amounts (≥ 5 wt.%) of organics, such as in tile adhesives, to apparent too low portlandite and too high calcium carbonate contents determination by TGA (Kriegel et al. 2003; Jenni et al. 2005). In contrast to these cements with high amounts of organics, the presence of low quantities (<2%) of organics, e.g. plasticisers or retarders, has no significant influence on the amount of portlandite and calcium carbonate quantified by TGA (Lothenbach et al. 2008b; Möschner et al. 2009; Wieland et al. 2014). Similarly, as discussed above in Section 5.2.3.1, the use of organic solvents to stop the hydration can lead also to a (however minor) decrease in the amount of portlandite and the presence of more carbonates. Instead of using alcohols, which tend to sorb to the hydrates, either drying under N_2 (De Weerdt et al. 2011a,b; Lothenbach et al. 2008a; Figure 5.6) or the exchange of the isopropanol by diethyl ether seems to avoid the sorption of organics on hydrates, as shown in Figure 5.7a.

5.6 DIFFERENTIAL THERMAL ANALYSIS/ DIFFERENTIAL SCANNING CALORIMETRY

DTA or DSC measures the enthalpy changes related to glass transitions. The deconvolution of DTA or DSC signals is generally very difficult in hydrated cement pastes as different reactions, e.g. dehydroxylation of C-S-H, oxidation reactions and phase transformation of the glasses, appear in the same temperature regions. DTA and DSC measurements are important techniques to characterise pure phases or the glass fraction in blast furnace slags or fly ash. In hydrated cements, however, DTA and DSC are generally not used quantitatively but rather to assess the presence of glassy phases or certain hydrates with a characteristic heat development as, e.g. M-S-H (Zhang et al. 2014).

The devitrification of the glasses and slags to melilite and merwinite is visible by exothermic peaks between 800 and 1000°C (Fredericci et al. 2000; Schneider and Meng 2000). It has been suggested to use the heat generated by these crystallisation reactions to quantify the amount of unreacted slag in hydrating cement (Lommatsch 1956; Schrämli 1963). Recent DTA and DSC studies, however, indicate that in hydrated blast furnace slags or blends of portland cements with slag, a quantification of the heat generated by the crystallisation reaction is not possible, as the background is not constant (Gruskovnjak et al. 2011; Kocaba et al. 2012). The complex pattern of sulfur oxidation reactions in blast furnace slags, which occurs in the same temperature range, contributes significantly to the total heat measured, as shown in Figure 5.20. These enthalpy changes overlay the signal related to the heat due to the crystallisation reaction; the decrease in the amount of sulfur present in the glass with hydration and the contribution of the increasing amount of hydrates present make a meaningful deconvolution of the different signals quite impossible and the method is thus not recommended as already stated by Gruskovnjak et al. (2011) and Kocaba et al. (2012).

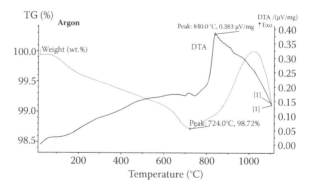

Figure 5.20 Weight loss and DTA of a blast furnace slag determined under argon atmosphere. (Courtesy of D. Hooton.)

5.7 GUIDELINES FOR ANALYSIS

The main rule of performing reliable TG analysis is to use the same procedure for all measurements. The temperature where the main weight losses is observed can vary strongly between different laboratories because of the large number of factors which can influence the resulting TG curve, e.g. type of instrument, sample holder, sample mass and fineness, pretreatment, heating rate and type and flow of purging gas.

In the following a step-by-step procedure for TG analysis of hydrated cement paste is recommended.

5.7.1 Step-by-step procedure for thermogravimetric analysis of hydrated cement

5.7.1.1 Boundary conditions

- Alumina crucibles with a volume of preferably 150 μL (for *50 mg of paste* sample) or more.
- *Heating from 40 to 1000°C at a rate of 20°C/min:* Heating rates in the range of 10 to 20°C per minute are a good compromise between the duration of the experiments and the ability to differentiate between the different weight losses.
- N_2 is used as a purging gas with a rate of 30 to 50 mL/min.
- A *blank curve* is determined for the selected crucible, heating procedure and purging gas before each series of measurement.

5.7.1.2 Pretreatment

To remove the pore solution in hydrating cement and to stop the reaction, either solvent exchange (focusing on the determination of bound water and identification of hydrates such as ettringite) or a freeze-drying procedure (for the minimisation of carbonation to determine portlandite content or for the characterisation of C-S-H) is recommended.

5.7.1.2.1 Solvent exchange

- For health and safety reasons one should perform the pretreatment in a fume cabinet to remove dust and fumes of the solvents.
- *Crush and grind* the cement paste sample roughly and fast using mortar and pestle; alternatively grinding under N_2 atmosphere will further minimise carbonation.
- *Mix ≈5 g of the ground cement paste with 50 mL isopropanol* in a glass beaker to remove the pore solution. Let the suspension rest for 10 to 15 minutes, which should be sufficient to replace the free water with isopropanol in the case of fine particles (see Zhang and Scherer

[2011] and Section 4.6.3). Isopropanol sorbs less to the hydrates than low-molecular weight alcohols, such as methanol or ethanol.

- *Filtrate the suspension* using, c.g. a Büchner funnel and an aspirator pump to remove the excess isopropanol.
- Wash the solid (still in the filter device) *with 5 to 10 mL diethylene ether* to remove the isopropanol and pump until the sample gets a lighter colour (or the pressure increases).
- *Dry the resulting solid* either by spreading it on a petri dish and drying it for 8 to 10 minutes in a 40°C aerated oven or by putting it for a short period in a vacuum desiccator. A third option is to dry until constant mass in the TG under a stream of N_2 without heating prior to the TG analysis.
- *Analyse the sample immediately or the same day if possible.* Alternatively, the sample can be stored for a few days in a closed vessel or under a light vacuum.

5.7.1.2.2 Freeze-drying

- *Immerse a piece of unground cement paste in liquid nitrogen* for ≈15 minutes.
- Dry the frozen piece of cement paste in the freeze-dryer.
- Crush the dried cement paste sample using a mortar and pestle.
- *Analyse the sample immediately or the same day if possible.* Alternatively, the sample can be stored for a few days in a closed vessel or under a light vacuum.

5.7.1.3 Thermogravimetric analysis

- *Weigh in the empty* crucibles.
- *Weigh in 50 mg* powder in 150 μL crucibles (or more in larger crucibles).
- Start the measurement (subtract the blank curve).
- Perform quantitative data analysis as described in Section 5.4.

For other purposes, different procedures might be better suited, such as the use of a closed aluminium vessel to determine the amount of hemihydrate and gypsum, as detailed in Section 5.2.1.

5.7.1.4 Guidelines for reporting

The following information should be included in the reporting:

- Sample properties
  - Composition and storage history of sample
  - Method to stop hydration and/or dry

- Storage history between stopping hydration and analysis, e.g. storage in desiccator and waiting time in auto sampler
- Amount and fineness of sample
- Boundary conditions
  - TG instrument
  - Crucible type and size
  - Temperature interval and heating rate
  - Type and rate of purging gas
- Results
  - Complete TG or DTG curves and/or the weight loss including temperature interval, the quantification method (as described in Section 5.4.4) and the error or quantification
  - Report whether carbonation is observed

5.7.2 Further points to consider

In the case of freeze-drying one has to consider that a part of the chemically bound water from the C-S-H and AFm and AFt phases may be removed. In the case of solvent exchange, sorption of alcohols can lead to a slight lowering of portlandite content, especially if the samples are not washed with diethyl ether.

The drying technique used and its duration depend on the aim of the analysis. For a detailed analysis of the AFm and AFt phases, a short gentle drying technique has to be used as these phases decompose at relatively low temperatures, while if only the amount of portlandite is of interest, a harsher drying technique can be applied.

To quantify bound water, it is recommended to remove the free water by solvent exchange and to avoid the often-used predrying at 105°C, as drying at 105°C removes water from the ettringite, C-S-H and AFm phases. The temperature range from 50 to 550 or 600°C is recommended to exclude carbonates.

For the quantification of portlandite the tangential method is suggested, as discussed in Section 5.4.

5.8 CONCLUSIONS AND OUTLOOK

This chapter on TG analysis discusses the main factors affecting TGA, such as heating rate, vapour pressure and sample weight, and discusses the effect of common pretreatment methods. In addition, the TGA signals of the most common minerals observed in cementitious systems are compared to each other and given as a reference database.

An important advantage of TGA is its sensitivity towards X-ray amorphous materials, such as C-S-H, M-S-H or AH_3, such that TGA is often used as a technique complementary to XRD to characterise cements. TGA is also more sensitive than XRD in detecting and quantifying low amounts of carbonates, portlandite or aluminium hydroxides. As any analytical method, TGA has limitations. The characteristic temperatures for weight losses vary depending on sample weight, the equipment, the vapour pressure and other parameters. C-S-H, the main hydrate in most cements, shows weight loss over a broad temperature range, from 50 to 600°C, the same temperature range where also most of the other hydrates show their characteristic signals. This strong overlap makes deconvolution in hydrated cements difficult to impossible with the notable exception of portlandite and calcium carbonates. Thus, the quantification of the different phases in hydrated cements by TGA alone is difficult but possible for well-defined peaks with little overlap as, e.g. for portlandite or calcium carbonates or aluminium hydroxides in calcium aluminate cements.

ACKNOWLEDGEMENTS

Special thanks to Luigi Brunetti, Belay Dilnesa, Astrid Gruskovnjak, Emilie L'Hôpital, Göril Möschner, Dominik Nied, Laure Pelletier and Thomas Schmidt for the preparation and measurements of the pure phases. Susan Bernal, Florian Deschner, Doug Hooton, Gwenn Le Saout, Winnie Matthes and Axel Schöler are acknowledged for providing measurements on hydrated cements and blast furnace slags and Marta Palacios, Karen Scrivener and Ruben Snellings for helpful discussions. This work was initiated through discussions in the RILEM TC238-SCM committee and its members are gratefully acknowledged for their constructive comments and discussion.

REFERENCES

Allmann, R. (1977). 'Refinement of the hybrid layer structure $(Ca_2Al(OH)_6)^+$ $(0.5SO_4 \cdot 3H_2O)^-$'. *Neues Jahrbuch für Mineralogie Monatshefte* 4: 136–143.

Badens, E., P. Llewellyn, J. M. Fulconis, C. Jourdan, S. Veesler, R. Boistelle and F. Rouquerol (1998). 'Study of gypsum dehydration by controlled transformation rate thermal analysis (CRTA)'. *Journal of Solid State Chemistry* 139(1): 37–44.

Baquerizo, L. G., T. Matschei and K. L. Scrivener (2015). 'Thermodynamics of the water content of ettringite'. *Cement and Concrete Research*: submitted.

Baquerizo, L. G., T. Matschei, K. L. Scrivener, M. Saeidpour, A. Thorell and L. Wadsö (2014). 'Methods to determine hydration states of minerals and cement hydrates'. *Cement and Concrete Research* 65: 85–95.

Beaudoin, J. J. (1987). 'Validity of using methanol for studying the microstructure of cement paste'. *Materials and Structures* 20(1): 27–31.

Ben Haha, M., G. Le Saout, F. Winnefeld and B. Lothenbach (2011). 'Influence of slag chemistry on the hydration of alkali activated blast-furnace slag – Part I: Effect of MgO'. *Cement and Concrete Research* **41**(9): 955–963.

Criado, J. M., and A. Ortega (1992). 'A study of the influence of particle size on the thermal decomposition of $CaCO_3$ by means of constant rate thermal analysis'. *Thermochimica Acta* **195**: 163–167.

Day, R. L. (1981). 'Reactions between methanol and portland cement paste'. *Cement and Concrete Research* **11**(3): 341–349.

De Weerdt, K., M. Ben Haha, G. Le Saout, K. O. Kjellsen, H. Justnes and B. Lothenbach (2011a). 'Hydration mechanisms of ternary Portland cements containing limestone powder and fly ash'. *Cement and Concrete Research* **41**(3): 279–291.

De Weerdt, K., E. J. Sellevold, H. Justnes and K. O. Kjellsen (2011b). 'Fly ash-limestone ternary cements: Effect of component fineness'. *Advances in Cement Research* **23**(4): 203–214.

Deschner, F., F. Winnefeld, B. Lothenbach, S. Seufert, P. Schwesig, S. Dittrich, F. Goetz-Neunhoeffer and J. Neubauer (2012). 'Hydration of a Portland cement with high replacement by siliceous fly ash'. *Cement and Concrete Research* **42**: 1389–1400.

Franke, A. (1941). 'Bestimmung von Calciumoxid und Calciumhydroxid neben wassertreiem und wasserhaltigem Calciumsilikat'. *Zeitschrift für Anorganische und Allgemeine Chemie* **247**: 180–184.

Fredericci, C., E. D. Zanotto and E. C. Ziemath (2000). 'Crystallization mechanism and properties of a blast furnace slag glass'. *Journal of Non-Crystalline Solids* **273**: 64–75.

Gómez-Zamorano, L. Y., and J.-I. Escalante-García (2010). 'Effect of curing temperature on the nonevaporable water in Portland cement blended with geothermal silica waste'. *Cement and Concrete Composites* **32**: 603–610.

Goto, S., K. Suenaga, T. Kado and M. Fukuhara (1995). 'Calcium silicate carbonation products'. *Journal of the American Ceramic Society* **78**(11): 2867–2872.

Gruskovnjak, A., B. Lothenbach, L. Holzer, R. Figi and F. Winnefeld (2006). 'Hydration of alkali-activated slag: Comparison with ordinary Portland cement'. *Advances in Cement Research* **18**(3): 119–128.

Gruskovnjak, A., B. Lothenbach, F. Winnefeld, B. Münch, S. C. Ko, M. Adler and U. Mäder (2011). 'Quantification of hydration phases in super sulphated cements: Review and new approaches'. *Advances in Cement Research* **23**(6): 265–275.

Gutteridge, W. A., and L. J. Parrott (1976). 'A study of the changes in weight, length and interplanar spacing induced by drying and rewetting synthetic CSH (I)'. *Cement and Concrete Research* **6**(3): 357–366.

Hudson-Lamb, D. L., C. A. Strydom and J. H. Potgieter (1996). 'The thermal dehydration of natural gypsum and pure calcium sulphate dihydrate (gypsum)'. *Thermochimica Acta* **282–283**: 483–492.

Jenni, A., L. Holzer, R. Zurbriggen and M. Herwegh (2005). 'Influence of polymers on microstructure and adhesive strength of cementitious tile adhesive mortars'. *Cement and Concrete Research* **35**: 35–50.

Jennings, H. M. (2008). 'Refinements to colloid model of C-S-H in cement: CM-II'. *Cement and Concrete Research* **38**(3): 275–289.

Khoshnazar, R., J. J. Beaudoin, L. Raki and R. Alizadeh (2013a). 'Solvent exchange in sulfoaluminate phases; Part II: Monosulfate'. *Advances in Cement Research* 25(6): 322–331.

Khoshnazar, R., J. J. Beaudoin, L. Raki and R. Alizadeh (2013b). 'Solvent exchange in sulphoaluminate phases; Part I: Ettringite'. *Advances in Cement Research* 25(6): 314–321.

Knapen, E., O. Cizer, K. Van Balen and D. Van Gemert (2006). Comparison of solvent exchange and vacuum drying techniques to remove free water from early age cement-based materials. Second International Symposium on Advances in Concrete through Science and Engineering. Quebec City, Canada.

Kocaba, V., E. Gallucci and K. Scrivener (2012). 'Methods for determination of degree of reaction of slag in blended cement pastes'. *Cement and Concrete Research* 42(3): 511–525.

Kriegel, R., R. Hellrung and A. Dimmig (2003). Qualitative and quantitative analysis of selected organic additives in hardened concrete by thermal analysis and infrared spectroscopy. 11th International Congress on the Chemistry of Cement, Durban, South Africa. 171–180.

L'Hopital, E., B. Lothenbach, D. Kulik and K. Scrivener (2015). 'Influence of the Ca/Si on the aluminium uptake in C-S-H'. *Cement and Concrete Research*: submitted.

Lommatsch, A. (1956). 'Untersuchung von Hochofenschlacke mit der Differentialthermoanalyse'. *Silikattechnik* 7(11): 468.

Lothenbach, B., G. Le Saout, E. Gallucci and K. Scrivener (2008a). 'Influence of limestone on the hydration of Portland cements'. *Cement and Concrete Research* 38(6): 848–860.

Lothenbach, B., T. Matschei, G. Möschner and F. P. Glasser (2008b). 'Thermodynamic modelling of the effect of temperature on the hydration and porosity of Portland cement'. *Cement and Concrete Research* 38(1): 1–18.

Lothenbach, B., and F. Winnefeld (2006). 'Thermodynamic modelling of the hydration of Portland cement'. *Cement and Concrete Research* 36(2): 209–226.

Matthes, W. (2014). Personal communication.

McCauley, R. A., and L. A. Johnson (1991). 'Decrepitation and thermal decomposition of dolomite'. *Thermochimica Acta* 185(2): 271–282.

Mitsuda, T., and H. F. W. Taylor (1978). 'Normal and anomalous tobermorite'. *Mineralogical Magazine* 42: 229–235.

Montes-Morán, M. A., A. Concheso, C. Canals-Batlle, N. V. Aguirre, C. O. Ania, M. J. Martín and V. Masaguer (2012). 'Linz–Donawitz steel slag for the removal of hydrogen sulfide at room temperature'. *Environmental Science & Technology* 46(16): 8992–8997.

Möschner, G., B. Lothenbach, R. Figi and R. Kretschmar (2009). 'Influence of citric acid on the hydration of Portland cement'. *Cement and Concrete Research* 39(4): 275–282.

Mouret, M., A. Bascoul and G. Escadeillas (1997). 'Study of the degree of hydration of concrete by means of image analysis and chemically bound water'. *Advanced Cement Based Materials* 6(3–4): 109–115.

Muller, A. C., K. L. Scrivener, A. M. Gajewicz and P. J. McDonald (2013). 'Densification of C–S–H measured by ^{1}H NMR relaxometry'. *The Journal of Physical Chemistry C* 117(1): 403–412.

Myers, R. J., E. L'Hôpital, J. L. Provis and B. Lothenbach (2015). 'Effect of temperature and aluminium on calcium (alumino)silicate hydrate chemistry under equilibrium conditions'. *Cement and Concrete Research* **68**: 83–93.

Nied, D., K. Enemark-Rasmussen, E. L'Hopital, J. Skibsted and B. Lothenbach (2015). 'Synthesis, structural characterisation and solubility of magnesium-silicate-hydrates (M-S-H)'. *Cement and Concrete Research*: submitted.

Parrott, L. J. (1983). 'Thermogravimetric and sorption studies of methanol exchange in an alite paste'. *Cement and Concrete Research* **13**(1): 18–22.

Paulik, F., J. Paulik and M. Arnold (1992). 'Thermal decomposition of gypsum'. *Thermochimica Acta* **200**: 195–204.

Pöllmann, H. (1984). Die Kristallchemie der Neubildung bei Einwirkung von Schadstoffen auf hydraulische Bindelmittel, PhD thesis, University of Erlangen, Germany.

Salvador, A. R., E. G. Calvo and C. B. Aparicio (1989). 'Effects of sample weight, particle size, purge gas and crystalline structure on the observed kinetic parameters of calcium carbonate decomposition'. *Thermochimica Acta* **143**(1): 339–345.

Schneider, C., and B. Meng (2000). Bedeutung der Glasstruktur von Hüttensanden für ihre Reaktivität. Ibausil, 14. Internationale Baustofftagung, Weimar, Bundesrepublik Deutschland, Bauhaus-Universität Weimar. 0455–0463.

Schöler, A., B. Lothenbach, F. Winnefeld and M. Zajac (2015). 'Hydrate formation in quaternary Portland cement blends containing blast-furnace slag, siliceous fly ash and limestone powder'. *Cement and Concrete Composites* **55**: 374–382.

Schrämli, W. (1963). Zur Charakterisierung von Hochofenschlacken mittels Differentialthermoanalyse. *Zement-Kalk-Gips* **16**(4): 140–147.

Snoeck, D., L. F. Velasco, A. Mignon, S. Van Vlierberghe, P. Dubruel, P. Lodewyckx and N. De Belie (2014). 'The influence of different drying techniques on the water sorption properties of cement-based materials'. *Cement and Concrete Research* **64**: 54–62.

Strydom, C. A., D. L. Hudson-Lamb, J. H. Potgieter and E. Dagg (1995). 'The thermal dehydration of synthetic gypsum'. *Thermochimica Acta* **269–270**: 631–638.

Taylor, H. F. W. (1997). *Cement chemistry*. Second edition. London, Thomas Telford.

Taylor, H. F. W., and A. B. Turner (1987). 'Reactions of tricalcium silicate paste with organic liquids'. *Cement and Concrete Research* **17**(4): 613–623.

Thomas, M. D. A. (1989). 'The suitability of solvent exchange techniques for studying the pore structure of hardened cement paste'. *Advances in Cement Research* **2**(1): 29–34.

Wieland, E., B. Lothenbach, M. Glaus, T. Thoenen and B. Schwyn (2014). 'Influence of superplasticizer on the long-term properties of cements and possible impacts on radionuclide uptake in cement based repository for radioactive waste'. *Applied Geochemistry* **49**: 126–142.

Zhang, J., and G. W. Scherer (2011). 'Comparison of methods for arresting hydration of cement'. *Cement and Concrete Research* **41**(10): 1024–1036.

Zhang, L., and F. P. Glasser (2000). 'Critical examination of drying damage to cement pastes'. *Advances in Cement Research* **12**(2): 79–88.

Zhang, T., L. J. Vandeperre and C. R. Cheeseman (2014). 'Formation of magnesium silicate hydrate (M-S-H) cement pastes using sodium hexametaphosphate'. *Cement and Concrete Research* **65**: 8–14.

Zhou, Q., E. E. Lachowski and F. P. Glasser (2004). 'Metaettringite, a decomposition product of ettringite'. *Cement and Concrete Research* **34**(4): 703–710.

Chapter 6

High-resolution solid-state nuclear magnetic resonance spectroscopy of portland cement-based systems

Jørgen Skibsted

CONTENTS

6.1 INTRODUCTION

Solid-state nuclear magnetic resonance (NMR) spectroscopy represents an important research tool for the characterisation and structural analysis of cement pastes and cement-based materials and has been increasingly employed in this field for more than the past 3 decades. A main advantage of the method is nuclear-spin selectivity, where one nuclear-spin isotope of the NMR periodic table (e.g. ^{1}H, ^{11}B, ^{19}F, ^{27}Al and ^{29}Si) is detected at the time, which often results in rather simple but informative spectra for complex multiphase systems, such as cementitious materials. In addition, the resonances from these spin nuclei are most sensitive to local structural ordering and/or dynamic effects, which permits studies of not only crystalline phases but also amorphous components, such as the calcium–silicate–hydrate (C-S-H) phase, produced during the setting and hardening of portland cement. Thus, NMR complements a number of other analytical techniques which probe the long-range order of crystalline phases (e.g. X-ray diffraction [XRD]; cf. Chapter 4) or bulk chemical features.

The first application of NMR to cement phases occurred in 1948, a few years after the discovery of NMR as a physical phenomenon in 1945, when G. Pake reported the ^{1}H-NMR spectrum of a single crystal of gypsum ($CaSO_4 \cdot 2H_2O$) (Pake 1948). He demonstrated that the splitting in the spectrum (the Pake doublet) reflects the ^{1}H–^{1}H dipolar interaction and thereby the distance between the hydrogen atoms in the crystal water molecules of gypsum, which was reported to be 1.5 Å. Several years later, ^{1}H-NMR was used in studies of portland cement hydration (Blaine 1960; Blinc et al. 1978; MacTavish et al. 1985; Schreiner et al. 1985; Seligmann 1968), utilising variations in ^{1}H spin–spin and spin–lattice relaxation times to identify different states of water in the hydrated systems. More recently and since 2000, these approaches and ^{1}H-NMR relaxometry have been further developed for studies of solid and mobile water in hardened cement systems (Barberon et al. 2003; Greener et al. 2000; Holly et al. 2007; Korb et al. 2007; McDonald et al. 2005), providing new knowledge about porosity, pore sizes, pore connectivities and water diffusion (see Chapter 7). An advantage of these low-resolution NMR experiments is that they are performed at low magnetic fields (roughly 0.5–1.4 T, corresponding to ^{1}H resonance frequencies of 20–60 MHz), implying low equipment costs compared to high-resolution NMR experiments, which are generally performed in superconducting magnets with magnetic field strengths in the range of 4.7–23.5 T (200–1000 MHz for ^{1}H).

The magic-angle spinning (MAS) technique (Andrew et al. 1958; Lowe 1959) was a breakthrough in high-resolution NMR of solids and technological improvements of this technique in the early 1980s opened the door for studies of nuclei other than ^{1}H and ^{13}C, principally ^{27}Al and ^{29}Si in the field of inorganic materials science. Pioneering work on ^{29}Si MAS NMR by Lippmaa et al. (1980) demonstrated that ^{29}Si chemical shifts for silicates reflect the degree of condensation of SiO_4 tetrahedra in studies of zeolites and aluminosilicate minerals. This relationship was also found for calcium silicates

(Mägi et al. 1984), and the amorphous C-S-H phase produced by portland cement hydration (Lippmaa et al. 1982). Furthermore, it has formed the basis for a wide range of structural and semiquantitative ^{29}Si MAS NMR studies of the hydration of portland cement and cement blends, including a variety of silicate- and aluminosilicate-rich additives. For ^{27}Al MAS NMR, Müller et al. (1977; 1981) showed that aluminium in tetrahedral and octahedral coordinations can be clearly distinguished by their 50–60 parts per million (ppm) difference in ^{27}Al chemical shift. This observation was subsequently utilised in ^{27}Al MAS NMR studies of the hydration of calcium aluminate (Müller et al. 1984), high-alumina cements (Luong et al. 1989; Skibsted 1988) and the tricalcium aluminate phase in portland cement (Hjorth et al. 1988), where the anhydrous aluminate phases include Al in tetrahedral coordination, whereas the calcium aluminate hydration products contain Al in octahedral environments. ^{27}Al ($I = 5/2$) is a quadrupole nucleus and for structurally distorted ^{27}Al sites this results in a strong quadrupolar broadening of the resonances that is only partly reduced by MAS. However, the second-order quadrupolar interaction is inversely proportional to the magnetic field strength; thus, improved resolution can be achieved at high magnetic fields and in combination with high-speed MAS, as illustrated earlier for tricalcium aluminate ($Ca_3Al_2O_6$) (Skibsted et al. 1991a, 1993). The development of the multiple-quantum (MQ) MAS radio-frequency (rf) pulse sequence (Frydman and Harwood 1995; Medek et al. 1995) in the mid-1990s represents a breakthrough in NMR studies of half-integer–spin quadrupolar nuclei, such as ^{17}O and ^{27}Al. This technique correlates MQ and single-quantum coherences under MAS and gives a two-dimensional (2D) MAS spectrum where the second-order quadrupolar interaction is removed in one isotropic dimension.

A range of other solid-state NMR rf pulse sequences have been used in studies of cement materials which utilise homo- or heteronuclear dipolar interactions or J couplings to probe connectivities and internuclear distances between spin pairs. For example, the double-quantum back-to-back pulse sequence (Feike et al. 1996) has been employed in a 2D ^{29}Si–^{29}Si correlation experiment to study tetrahedral SiO_4 connectivities for C-S-H samples (Brunet et al. 2004), whereas ^{1}H–^{29}Si connectivities for similar samples have been studied by 2D ^{29}Si{^{1}H} cross-polarisation (CP) experiments (Rawal et al. 2010). Furthermore, the rotational-echo double-resonance (REDOR) experiment (Gullion and Schaefer 1989) has been used for ^{19}F-^{29}Si spin pairs to clarify the incorporation of fluoride guest ions in the crystal structure of alite (Ca_3SiO_5) (Tran et al. 2009). Although the majority of applications of solid-state NMR in cement science have focused on ^{27}Al and ^{29}Si MAS NMR, a number of other spin isotopes (e.g. ^{1}H, ^{2}H, ^{11}B, ^{13}C, ^{17}O, ^{19}F, ^{23}Na, ^{25}Mg, ^{31}P and ^{43}Ca) have also been investigated. These nuclei can provide unique information, for example, about guest-ion incorporation in the principal cement minerals, interactions between cement and admixtures (e.g. superplasticisers) and the impact of exposure of hardened cements to ionic aqueous solutions, including waste materials.

Several textbooks can be found which give a basic introduction to NMR spectroscopy in general (Derome 1987; Harris 1986; Keeler 2010; Levitt 2007), to solid-state NMR spectroscopy (Apperley et al. 2012; Duer 2002; Harris et al. 2009; Wasylishen et al. 2012) and to applications of solid-state NMR in materials science (Bakhmutov 2012; Fyfe 1983; MacKenzie and Smith 2002), including monographs on zeolites (Engelhardt and Michel 1987) and polymers (Schmidt-Rohr and Spiess 1994). The importance of and general interest in solid-state NMR in studies of cementitious materials has been recognised by two international conferences on NMR spectroscopy of cement-based materials from which the proceedings (Colombet and Grimmer 1994; Colombet et al. 1998) illustrate the state-of-art of NMR in cement science in the 1990s. Since then, a couple of reviews have appeared (Skibsted and Hall 2008; Skibsted et al. 2002) which highlight some of the progress made in the field in the subsequent decade.

This chapter describes the basic features and common practise of solid-state NMR in studies of portland cement systems and illustrates how this tool can be used to derive structural and quantitative information about cement components and cement hydration for both pure portland cement and portland cements, including admixtures or supplementary cementitious materials (SCMs). Particular emphasis is given on the NMR techniques generally used in solid-state NMR studies of portland cement systems. However, a comprehensive overview of the NMR studies that have been performed so far or of the new chemical and physical knowledge derived from these studies will not be given.

6.2 BASIC THEORY AND NUCLEAR MAGNETIC RESONANCE INTERACTIONS

Several NMR techniques have been employed in studies of cement systems, ranging from multinuclear solid-state NMR investigations for chemical and structural analyses of anhydrous and hydrated cement phases to low-field ^{1}H-NMR studies of porosity and water transport in hydrated systems and to spatially resolved NMR (i.e. magnetic resonance imaging) for investigations on the micrometre to millimetre level of hydration, pores, cracks and water diffusion. Although this chapter deals only with high-resolution NMR, the background for these different applications is the same and relies on the fundamental NMR theory described in the following two sections.

6.2.1 Nuclear spin, magnetisation and the basic nuclear magnetic resonance experiment

When a nuclear-spin isotope, with spin quantum number I, is subjected to a strong, static magnetic field (B_0), its nuclear magnetic moment will orient

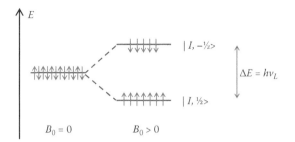

Figure 6.1 Energy-level diagram for a spin I = 1/2 nucleus, illustrating the polarisation of the spins in a static magnetic field and the resulting two spin eigenstates.

either parallel or antiparallel to the magnetic field direction, $B = (0, 0, B_0)$. For a spin $I = 1/2$ nucleus, this gives two different energy states (Figure 6.1), where the difference is the so-called Zeeman splitting, given by

$$\Delta E = h\nu_L = \gamma\hbar B_0, \tag{6.1}$$

Here, ν_L is the Larmor frequency, $\hbar = h/2\pi$ is the Planck constant and γ is the gyromagnetic ratio. The latter is constant for specific spin nuclei and its variation for different elements (Table 6.1) is the principal reason for the nuclear-spin selectivity of NMR experiments. For a macroscopic sample, the large number of identical spins (N) will be distributed over the two energy levels (for $I = 1/2$) with quantum numbers $m = 1/2$ and $m = -1/2$ according to a Boltzmann distribution,

$$\frac{N_{-1/2}}{N_{1/2}} = \exp\left(-\frac{\Delta E}{kT}\right), \tag{6.2}$$

where k is the Boltzmann constant, T is the kelvin temperature and $N = N_{1/2} + N_{-1/2}$. At thermal equilibrium, this gives a bulk magnetisation (M_0) along the magnetic field direction, which can be expressed as

$$M_0 = \frac{N\gamma\hbar I(I+1)}{3kT}\Delta E. \tag{6.3}$$

The net magnetisation can be rotated away from the z direction by applying a short rf pulse (magnetic field) perpendicular to B. After the pulse, the magnetisation, $M = (M_x, M_y, M_z)$, will relax towards the equilibrium state and during this process induce a current in the receiver coil, oriented perpendicular to the B field. A classic, phenomenological description of the equation of motion for the magnetisation after the rf pulse is provided by

Table 6.1 NMR properties of nuclear-spin isotopes relevant in cement chemistry

Nucleus	Spin (I)	Natural abundance	γ ($\times 10^{-7}$)[a] (rad T^{-1} s^{-1})	$\nu_L/\nu_L(^1H)$[b] (MHz/%)	Relative receptivity[c]	Reference sample[d]
^{1}H	1/2	99.99	26.752	100.0	1.0	Si(CH$_3$)$_4$ – neat
^{2}H	1	0.0115	4.106	15.35	1.11×10^{-6}	Si(CD$_3$)$_4$ – neat
^{11}B	3/2	80.1	8.584	32.08	0.132	BF$_3$·Et$_2$O – neat
^{13}C	1/2	1.07	6.728	25.15	1.70×10^{-4}	Si(CH$_3$)$_4$ – neat
^{17}O	5/2	0.04	-3.628	13.56	1.11×10^{-5}	D$_2$O – neat
^{19}F	1/2	100	25.181	94.09	0.834	CCl$_3$F – neat
^{23}Na	3/2	100	7.081	26.45	9.27×10^{-2}	NaCl – 0.1 M
^{25}Mg	5/2	10.00	-1.639	6.12	2.68×10^{-4}	MgCl$_2$ – 0.1 M
^{27}Al	5/2	100	6.976	26.06	0.21	Al(NO$_3$)$_3$ – 1.1 m
^{29}Si	1/2	4.68	-5.319	19.87	3.68×10^{-4}	Si(CH$_3$)$_4$ – neat
^{31}P	1/2	100	10.839	40.48	6.65×10^{-2}	H$_3$PO$_4$ – 85%
^{33}S	3/2	0.76	2.056	7.68	1.71×10^{-5}	CS$_2$ – neat
^{35}Cl	3/2	75.78	2.624	9.80	3.58×10^{-3}	NaCl – 0.1 M
^{39}K	3/2	93.26	1.250	4.67	4.76×10^{-3}	KCl – 0.1 M
^{43}Ca	5/2	0.14	-1.803	6.73	8.68×10^{-6}	CaCl$_2$ – 0.1 M

Source: Harris, R. K. et al., *Pure Appl. Chem.*, 73, 1795–1818, 2001.

[a]   Gyromagnetic ratio; cf. Equation 6.1.
[b]   Larmor frequency, $\nu_L = |\gamma/2\pi|B_0$, at a magnetic field of 2.35 T or relative to $\nu_L(^1H)$ in percentage.
[c]   Receptivity relative to ^{1}H, $D^H = \gamma^3 x I(I+1)$, where x is the natural abundance of the spin isotope.
[d]   The concentrations correspond to aqueous solutions in D$_2$O; M = molarity, m = molality.

the following Bloch equations, which allow the components parallel (M_z) and perpendicular (M_x and M_y) to the magnetic field to relax with different time constants:

$$\frac{dM_z(t)}{dt} = -\frac{M_z(t) - M_0}{T_1},$$

$$\frac{dM_x(t)}{dt} = -\frac{M_x(t)}{T_2}, \tag{6.4}$$

$$\frac{dM_y(t)}{dt} = -\frac{M_y(t)}{T_2}.$$

T_1 is the longitudinal or spin–lattice relaxation time and T_2 is the transverse or spin–spin relaxation time.

In the single-pulse experiment, the spin system is excited by applying a magnetic field (B_1) perpendicular to the z axis. The B_1 field is generated by a current in a coil, oriented along the x axis. When the current (rf pulse) is turned on, the magnetisation starts to rotate clockwise around the x axis, i.e. a rotation on the yz plane, starting from the z axis towards the y axis. When the current is turned off, the magnetisation will precess around the z axis and move towards its equilibrium state. This precession induces a current in the rf coil, whose magnitude is proportional to the projection of the magnetisation vector (M) onto the xy plane. In the absence of relaxation, this component is given by

$$M_y^0(\tau_p) = M_0 \sin(\gamma B_1 \tau_p),\qquad\qquad(6.5)$$

where τ_p is the duration of the pulse and B_1 is the strength of the magnetic field for the rf pulse, $B_1 = (B_1, 0, 0)$. As an example, Figure 6.2 shows an array of ^{29}Si-NMR spectra for Na_2SiF_6, incrementing the rf pulse width from $\tau_p = 1$–30 μs. These spectra reveal that maximum intensity is observed for $\tau_p \cong 6.5$ μs, corresponding to the magnetisation vector oriented along the y axis immediately after the pulse (i.e. a 90° pulse). A zero crossing from negative to positive intensities is observed for $\tau \cong 25.4$ μs, reflecting that the magnetisation has rotated 360°. The array of spectra shown in Figure 6.2 is generally acquired to calibrate the rf pulse width and to determine the corresponding rf field strength. The zero crossing observed

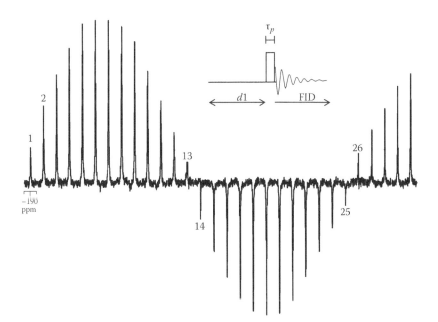

Figure 6.2 ^{29}Si MAS NMR spectra (4.7 T, $\nu_R = 7.0$ kHz) of Na_2SiF_6, acquired using the single-pulse rf sequence (inset) with a repetition delay $dl = 5$ s, for pulse widths in the range $\tau_p = 1$–30 μs (1 μs increments). The resonance from Na_2SiF_6 is observed at $\delta_{iso} = -188.9$ ppm.

for a 360° pulse corresponds to $\gamma B_1 \tau_p = 2\pi$ (Equation 6.5), and the rf field strength of the pulse is generally expressed as $\gamma B_1/2\pi = 1/\tau_{360}$, where τ_{360} is the pulse length for the 360° pulse. In the actual case, $\gamma B_1/2\pi = 39.4$ kHz, and it is noted that Na_2SiF_6 is a suitable sample for calibration of ^{29}Si rf fields, due to the short T_1 relaxation time for ^{29}Si in this compound ($T_1 = 0.46$ s at $B_0 = 7.1$ T and $T_1 = 0.63$ s at $B_0 = 9.4$ T).

The induced voltage in the receiver coil, $S(t)$, is proportional to the rate of change in magnetic flux ϕ following Faraday's law of electromagnetic induction, and it can be expressed as

$$S(t) = \frac{d\phi}{dt} = \frac{dM_x(t)}{dt}.$$

(6.6)

The maximum signal is generated by a 90° pulse, corresponding to $dM_x(t=0)/dt = \omega_L M_y^0(\tau_{90})$, where $\omega_L = 2\pi\nu_L$. Thus, immediately after the 90° pulse, the signal amplitude can be expressed as

$$S(t=0) = N \frac{\gamma^3 B_0^2 \hbar^2 I(I+1)}{3kT}.$$

(6.7)

The time evolution of this signal is denoted the free induction decay (FID), since there is no rf field at this stage perturbing the system. It is apparent from this expression that the signal intensity for a given resonance is proportional to the corresponding number of spins in the sample, which forms the basis for quantitative NMR studies. Moreover, the expression reveals a significant gain in signal intensity by increasing the external magnetic field. However, a more detailed analysis of the signal-to-noise ratio (S/N) as a function of the B_0 field, including other effects such as the performance of the NMR probe, reveals that the increase in S/N is closer to S/N $\propto |\gamma|^{5/2}(B_0)^{3/2}$ (Levitt 2007), which is still a very important gain in sensitivity, in particular for studies of low-γ nuclei.

The induced current in the receiver coil is typical in the microvolt range and it is seldom sufficient to generate a spectrum in the frequency domain after Fourier transformation of the time-domain FID with an acceptable S/N. Thus, the experiment is repeated several times (NS) and signal intensity is accumulated for each experiment (scan) until a satisfactory S/N is obtained. The signal from the receiver coil is a sum of the NMR signal (Equation 6.7) and noise in the form of random rf signals arising from thermal motions of electrons in the receiver coil and from other sources. If the frequencies of the noise signal are uncorrelated, it can be shown that the S/N is proportional to the square root of the number of scans, i.e. S/N $\propto \sqrt{NS}$. The process of signal averaging can be quite time consuming, in particular for spin isotopes in low natural abundance, such as ^{29}Si and ^{13}C. Moreover, the observation of quantitative reliable spectra requires that full spin–lattice relaxation is achieved for each component between the individual scans. The time evolution of the longitudinal magnetisation, obtained from the Bloch equation (Equation 6.4), takes the form

$$M_z(t) = M_0 + [M_z(0) - M_0]\exp(-t/T_1), \tag{6.8}$$

where $M_z(0)$ is the z magnetisation immediately after the pulse. For a 90° pulse ($M_z(0) = 0$), relaxation delays ($d1$) of five times the longest T_1 are generally required to obtain quantitative reliable intensities in the spectra. According to Equation 6.7, $t = 5T_1$ corresponds to the recovery of 99.3% of the equilibrium magnetisation (M_0) after a 90° pulse; however, in practise, spectra are often recorded with smaller flip angles (e.g. 45°), which lower the requirement of $d1 = 5T_1$ for the repetition time.

6.2.2 Spin–lattice and spin–spin relaxation

Relaxation is a result of transitions between different spin-energy levels, caused by fluctuations in the local interactions at the nucleus. Thus, relaxation processes depend on the strength of the spin interactions and the number of fluctuations, where the latter is related to dynamic processes such as atomic or molecular motions and vibrations. Thereby, the relaxation times are very sensitive to the local environment of the spins, temperature, molecular motions and impurity ions, and a determination of T_1 and T_2 can often be used to distinguish different species or environments. For rigid solids, T_1 and T_2 increase with decreasing temperature and, generally, T_1 increases and T_2 decreases on going from a mobile (liquid) to a rigid (solid) phase.

The spin–lattice relaxation time (T_1) can be determined by either the inversion-recovery (IR) or the saturation-recovery experiment. The IR experiment (Figure 6.3a) employs a 180° pulse, which inverts the magnetisation, $M_z(t = 0) = -M_0$, followed by a recovery time (τ) and a 90° observe

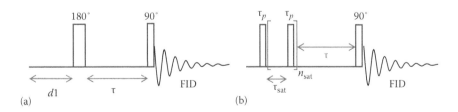

(a)   (b)

Figure 6.3 Radio-frequency pulse sequences for (a) the IR and (b) the saturation-recovery experiments for determination of spin–lattice (T_1) relaxation times. The recovery time (τ) is incremented in both experiments. The saturation-recovery sequence employs typically 10–20 saturation pulses (n_{sat}) with a length corresponding to 90° pulses ($\tau_p = 90°$).

pulse prior to detection of the FID (Vold et al. 1968). An array of spectra is acquired for typically 8–15 recovery times, where 3–5 of the values are chosen before the zero crossing ($\tau_0 = T_1 \ln 2$) and the remaining in the range, $T_1 \ln 2 < \tau < 3T_1$. For a single exponential relaxation process (e.g. for pure samples), the intensities in these spectra can be fitted by Equation 6.8, employing $M_z(t = 0) = -M_0$, which gives the expression

$$M_z(\tau) = M_0[1 - 2 \exp(-\tau/T_1)]. \tag{6.9}$$

A disadvantage of the IR experiment is that full relaxation ($d1 \sim 5T_1$) is required between each scan, which makes the experiment very time consuming for long T_1 values. A reduction in spectrometer time can be achieved by the fast IR experiment (Canet et al. 1975), where the relaxation delay is reduced to $d1 \sim 3T_1$ and the intensities fitted to the equation

$$M_z(\tau) = M_0[1 - \alpha \exp(-\tau/T_1)] \tag{6.10}$$

in a three-parameter fit (M_0, α, T_1). The new variable, $\alpha \leq 2.0$, takes into account effects from the nonideal relaxation delay and compensates also for minor imperfections of the 180° and 90° pulses. Thus, Equation 6.10 may preferentially be used even if a repetition delay corresponding to full spin–lattice relaxation is employed. For very long relaxation times, a significant reduction in instrument time can be achieved by the saturation-recovery experiment (Figure 6.3b), which saturates the spin system by applying a number of pulses (typically 5–20) with interpulse delays, $\tau_{sat} \ll T_1$, followed by a recovery period (τ) and a subsequent 90° observe pulse, prior to detection of the FID (Markley et al. 1971). The initial saturation of the magnetisation implies that $M_z(t = 0) = 0$, and thereby Equation 6.8 gives the relationship

$$M_z(\tau) = M_0[1 - \exp(-\tau/T_1)], \tag{6.11}$$

which allows determination of the relaxation time from a two-parameter (M_0, T_1) fit. This sequence has the advantage that a very small repetition delay can be used, $d1 \cong 0$; however, the analysis of the intensities is sensitive to the precision of the 90° observe pulse and the acquisition of at least one spectrum in the τ array that approaches the equilibrium magnetisation.

For portland cement systems, the spin–lattice relaxation is often governed by paramagnetic ions (e.g. Fe^{3+}) either in separate phases in the near vicinity or incorporated as impurity ions in the crystal lattices. The gyromagnetic ratio of the spin for the unpaired electron is about three orders of magnitude larger than the γ values for the nuclear spins, which results in a very efficient relaxation mechanism caused by the dipolar interaction between the nuclear spin of the detected nucleus and the electron spin of the paramagnetic ions. Considering this mechanism for low concentrations of paramagnetic ions, and assuming the absence of spin diffusion, allows derivation (Tse and Hartmann 1968) of the following relations between the observed intensity and the recovery time for the IR ($\alpha = 2$) and saturation-recovery experiments, respectively:

$$M_z(\tau) = M_0\left[1 - \alpha\exp\left(-\sqrt{\tau/T_1'}\right)\right],$$ (6.12)

$$M_z(\tau) = M_0\left[1 - \exp\left(-\sqrt{\tau/T_1'}\right)\right].$$ (6.13)

This behaviour is also known as a 'stretched exponential' relationship, with the associated relaxation time T_1', and it has been successfully employed in ^{29}Si MAS NMR studies of silicates containing small amounts of iron (Hartman and Sherriff 1991; Skibsted et al. 1995a,b). Moreover, it has been used to analyse the spin–lattice relaxation for ^{29}Si in alite and belite in portland cements (Poulsen et al. 2009) and to establish the incorporation of phosphorus in the alite and belite phases of anhydrous portland cement (Poulsen et al. 2010). A more detailed description of the effects of paramagnetic ions on ^{29}Si spin–lattice relaxation can be found in the first of these studies (Poulsen et al. 2009). An array of saturation-recovery ^{29}Si-NMR spectra for a mineral sample of belite (β-Ca$_2$SiO$_4$) is illustrated in Figure 6.4. Analysis of the intensities in these spectra reveals a stretched exponential relationship (Equation 6.13) and gives the relaxation time $T_1' = 65.7$ s. The fit shows a convincing agreement for all $M_z(\tau)$ and τ values and the

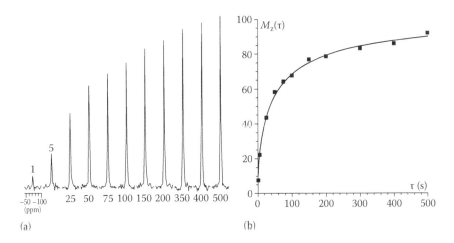

(a)   (b)

Figure 6.4 (a) Saturation-recovery ^{29}Si MAS NMR spectra (4.7 T, ν_R = 7.0 kHz) of a mineral sample of belite (β-Ca$_2$SiO$_4$; Scawt Hill, Northern Ireland), employing the recovery times (τ) indicated in seconds. (b) Plot of the observed ^{29}Si intensities as a function of τ along with a least-squares fit to these data using a stretched exponential relationship (Equation 6.13).

stretched exponential relation suggests that the relaxation process is governed by interactions with impurity ions in the present sample of mineral belite.

6.2.3 Chemical shift and dipolar and quadrupole interactions

The principal source of structural information in high-resolution NMR spectroscopy is the chemical shift, which originates from the shielding of the spin nucleus from the external magnetic field by the small magnetic field induced by the motions of the electrons surrounding the nucleus. Thus, the nucleus experiences an effective field, $B = (1 - \sigma)B_0$, where σ, the shielding constant, is a dimensionless number, usually listed in parts per million. This leads to a small modification of the resonance condition in Equation 6.1:

$$h\nu = \gamma\hbar(1-\sigma)B_0. \tag{6.14}$$

The chemical shielding is measured relative to the value for the nucleus in a standard sample (ref); the difference is expressed as the chemical shift:

$$\delta = \frac{\nu_{sample} - \nu_{ref}}{\nu_{ref}} \times 10^6 \, (\text{in ppm}). \tag{6.15}$$

Since the shielding constants are small, ν_{ref} in the denominator of Equation 6.15 may be replaced by the spectrometer operating frequency (ν_L), resulting in the following relation between chemical shifts (δ) and shielding constants (σ):

$$\delta = (\sigma_{ref} - \sigma_{sample}) \times 10^6 \, (\text{in ppm}). \tag{6.16}$$

The reference compounds recommended by the International Union of Pure and Applied Chemistry (Harris et al. 2001) are included in Table 6.1. Several of them are liquids; thus, a spectrum of the reference sample is acquired and referenced prior to the study of the sample under investigation. In practise, solid compounds with well-defined structures are often used as secondary reference materials, since it may be more convenient to test the setup of the experiment on the solid sample.

Generally, the spin state for a spin nucleus subjected to a strong magnetic field (B_0) and to rf field radiation (B_1) is governed by the spin Hamiltonian

$$H = H_Z + H_{rf} + H_\sigma + H_D + H_J + H_Q, \tag{6.17}$$

where H_Z and H_{rf} denote the Hamiltonians for the interactions with the external B_0 and B_1 fields, respectively, while the remaining terms are the

Hamiltonians for the spin interaction with internal fields. These include the chemical shift interaction (H_σ), homo- and heteronuclear direct (dipolar, H_D) and indirect (J coupling, H_J) spin–spin couplings and the quadrupole coupling interaction (H_Q) for $I > 1/2$ nuclei. The internal interactions result in characteristic shifts of the resonances, which are the principal source of structural information, and provide mechanisms for relaxation. Generally, the internal interactions are several orders of magnitude smaller than the Zeeman interaction, which allows calculation of the effects from these as perturbations to the Zeeman interaction. In the high-magnetic field approximation, this is performed by a transformation from the laboratory-fixed frame to a frame precessing with the Larmor frequency (ν_L) around the magnetic field direction B_0 (i.e. the Zeeman interaction frame). The periodical time dependence introduced by this transformation can be removed by average Hamiltonian theory (Maricq and Waugh 1979), resulting in an effective Hamiltonian of the type

$$\bar{H}_{\text{eff}} = \bar{H}_\sigma + \bar{H}_D + \bar{H}_J + \bar{H}_Q. \tag{6.18}$$

The resonance frequency (Equation 6.14) for the $(m, m - 1)$ transition, $h\nu = \hbar\omega_{m,m-1}$, in the secular approximation is calculated as

$$\omega_{m,m-1} = \langle I, m|\bar{H}_{\text{eff}}|I, m\rangle - \langle I, m-1|\bar{H}_{\text{eff}}|I, m-1\rangle, \tag{6.19}$$

where $|I, m\rangle$ and $|I, m - 1\rangle$ denote the Zeeman eigenstates for the spin system.

6.2.3.1 The chemical shift interaction

The Hamiltonian for the chemical shift interaction, describing the coupling of the spin $(\bar{I})$ with the external magnetic field, is given by

$$H_\sigma = \bar{I} \cdot \bar{\bar{\sigma}} \cdot \bar{B} \tag{6.20}$$

where $\bar{\bar{\sigma}}$ is the chemical shift anisotropy (CSA) tensor, reflecting the three-dimensional nature of the chemical shielding. The CSA tensor takes the following form in its principal axis system:

$$\bar{\bar{\sigma}} = \begin{pmatrix} \sigma_{xx} & 0 & 0 \\ 0 & \sigma_{yy} & 0 \\ 0 & 0 & \sigma_{zz} \end{pmatrix} = \begin{pmatrix} \sigma_{\text{iso}} - \frac{\delta_\sigma}{2}(1+\eta_\sigma) & 0 & 0 \\ 0 & \sigma_{\text{iso}} - \frac{\delta_\sigma}{2}(1-\eta_\sigma) & 0 \\ 0 & 0 & \sigma_{\text{iso}} + \delta_\sigma \end{pmatrix}. \tag{6.21}$$

The trace of the tensor is the isotropic chemical shift, $\delta_{iso} = -\sigma_{iso}$, whereas the anisotropy is described by the shift anisotropy (δ_σ) and the CSA asymmetry parameter ($0 \leq \eta_\sigma \leq 1$). These parameters are given by the principal tensor elements according to

$$\delta_{iso} = -\sigma_{iso} = \frac{1}{3}(\sigma_{xx} + \sigma_{yy} + \sigma_{zz}), \quad \delta_\sigma = \sigma_{zz} - \sigma_{iso}, \quad \eta_\sigma = \frac{\sigma_{yy} - \sigma_{xx}}{\delta_\sigma}, \quad (6.22)$$

following the convention $|\sigma_{zz} - \sigma_{iso}| \geq |\sigma_{xx} - \sigma_{iso}| \geq |\sigma_{yy} - \sigma_{iso}|$. Alternatively, the CSA can be characterised by the span ($\Omega = \sigma_{33} - \sigma_{11}$) and skew [$\kappa = 3(\sigma_{iso} - \sigma_{22})/\Omega$], where the principal elements are defined as $\sigma_{33} \geq \sigma_{22} \geq \sigma_{11}$ (Mason 1993).

For the static-powder NMR experiment, transformation from the CSA principal-axis system to the laboratory-fixed frame results in the following expression for the first-order Hamiltonian for the CSA interaction:

$$\bar{H}_\sigma^{(1)} = \omega_L \sigma_{iso} - \frac{1}{2}\omega_L \delta_\sigma \{3\cos^2\beta - 1 - \eta_\sigma \sin^2\beta\cos 2\alpha\}I_z \quad (6.23)$$

for a single crystallite with an orientation described by the Euler angles (α, β, γ) ($\gamma = 0$ as a result of the axial symmetry of the magnetic field). Thus, the chemical shift interaction is described by three parameters of which δ_σ and η_σ reflect the anisotropic part. This part is not observed in liquid-state NMR, since rapid tumbling of the molecules in solution efficiently averages out the dependency on the α and β angles in Equation 6.23. However, in NMR of rigid solids, the CSA may result in significant line broadening, depending on the anisotropic nature of the electronic field around the nucleus. A description of the full line width and shape is achieved by averaging the expression in Equation 6.23 over all uniformly distributed crystallite orientations in the powder sample. This gives line shapes of the form simulated in Figure 6.5 for a single site under static and MAS conditions and for different values of the CSA asymmetry parameter. The two upper simulations (Figure 6.5) show that a change in sign for δ_σ results in an inversion of the line shape, reflecting whether the unique element of the CSA tensor (σ_{zz}) is smaller or larger than the isotropic chemical shift.

As an experimental example, Figure 6.6 illustrates the experimental and simulated $^{29}Si\{^1H\}$ CP/MAS NMR spectra (14.1 T) of a synthetic sample of jaffeite ($Ca_6Si_2O_7(OH)_6$), acquired for a static sample and at slow spinning speeds of $v_R = 1.0$ and 4.0 kHz. The static and $v_R = 1.0$ kHz spectra clearly reveal that the Si site in jaffeite is in nearly axially symmetric environments ($\eta_\sigma = 0$), in accordance with the presence of $Si_2O_7^{6-}$ dimers in its crystal structure, each tetrahedron including three short and one long Si–O bond with nearly threefold symmetry around the Si–O–Si bond. A precise determination of the CSA parameters is achieved from simulations based on the powder

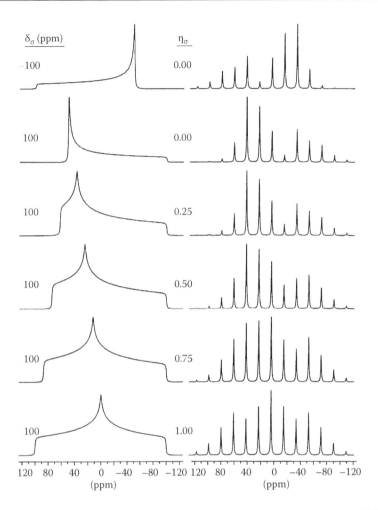

Figure 6.5 Simulated static-powder (left column) and MAS (right column; v_R = 1500 Hz) spectra for a spin I = 1/2 nucleus influenced only by the CSA interaction for the Larmor frequency v_L = 79.43 MHz (i.e. ^{29}Si at 9.4 T) and δ_{iso} = 0.0 ppm, illustrating the variation in line shape/spinning sideband intensities as a function of the CSA parameters, δ_σ and η_σ.

average of Equation 6.23 of the static and v_R = 1.0 kHz spectra (δ_{iso} = −82.2 ± 0.1 ppm, δ_σ = −51.2 ± 1.2 ppm and η_σ = 0.07 ± 0.03). On the other hand, the spectrum obtained with v_R = 4.0 kHz contains only a few spinning sidebands and it is therefore not suitable for a determination of the CSA parameters.

^{29}Si CSA parameters have been reported only for several silicate, calcium silicate and calcium silicate hydrate phases relevant to cement chemistry. These are summarised in Table 6.2 along with ^{29}Si isotropic chemical shifts reported for structurally well-defined cement constituent phases.

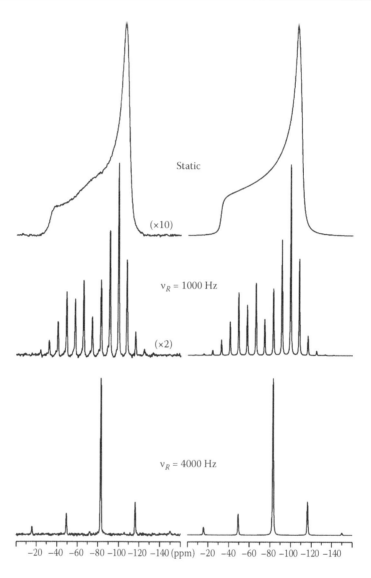

Figure 6.6 Experimental (left column) and simulated (right column) ^{29}Si-NMR spectra of a synthetic sample of jaffeite, acquired under static conditions and with spinning speeds of v_R = 1000 and 4000 Hz. The experimental spectra were obtained with the ^{29}Si{^{1}H} CP experiment (CP contact time of 5.0 ms and an 8 s relaxation delay) and the vertical expansion factors for the spectra are indicated relative to the v_R = 4000 Hz spectrum. The simulations employ the parameters δ_{iso} = −82.2 ppm, δ_σ = −51.2 ppm and η_σ = 0.07, as determined from analysis of the static and MAS (v_R = 1000 Hz) spectra.

Table 6.2 ^{29}Si isotropic chemical shifts (δ_{iso}), chemical shift anisotropies (δ_σ) and CSA asymmetry parameters (η_σ) for structurally well-defined silicate phases common in cement systems

Cement phase	Formula	Q^n site	δ_{iso} (ppm)	δ_σ (ppm)	η_σ	Ref.
Dicalcium silicate	α'_L-Ca$_2$SiO$_4$	Q^0	-70.1/-70.8			a
	β-Ca$_2$SiO$_4$	Q^0	-71.3 ± 0.1	16.6 ± 1.5	0.83 ± 0.07	b
	γ-Ca$_2$SiO$_4$	Q^0	-73.7 ± 0.1	25.5 ± 1.0	0.83 ± 0.04	b
Tricalcium silicate (triclinic, T1)	Ca$_3$SiO$_5$					c
	Si(1)	Q^0	-69.06 ± 0.01	7.2	0.84	
	Si(2)	Q^0	-69.18 ± 0.01	-6.8	0.93	
	Si(3)	Q^0	-71.85 ± 0.01	8.7	0.85	
	Si(4)	Q^0	-72.83 ± 0.01	9.5	0.53	
	Si(5), Si(6)	Q^0	-73.60 ± 0.01	4.6	1.0	
	Si(7)	Q^0	-73.80 ± 0.01	-5.3	0.81	
	Si(8)	Q^0	-74.05 ± 0.01	-5.7	0.74	
	Si(9)	Q^0	-74.66 ± 0.01	-8.8	0.67	
Calcium chlorosilicate	Ca$_3$SiO$_4$Cl$_2$	Q^0	-73.6 ± 0.1	11.0 ± 1.3	0.78 ± 0.04	b
Rankinite	Ca$_3$Si$_2$O$_7$					
	Si(1)	Q^1	-74.5 ± 0.1	-55.3 ± 1.4	0.69 ± 0.03	b
	Si(2)	Q^1	-75.9 ± 0.1	-40.5 ± 1.4	0.65 ± 0.03	b
Cuspidine	Ca$_4$Si$_2$O$_7$F$_2$	Q^1	-79.9 ± 0.1	-58.3 ± 1.3	0.61 ± 0.05	b
Gehlenite	Ca$_2$Al$_2$SiO$_7$	Q^1	-72.5	-52	0.92	d
Wollastonite	β-Ca$_3$Si$_3$O$_9$					
	Si(1)	Q^2	-89.0 ± 0.2	59.8 ± 2.0	0.62 ± 0.04	b
	Si(2)	Q^2	-89.5 ± 0.2	52.1 ± 2.0	0.68 ± 0.04	b
	Si(3)	Q^2	-87.8 ± 0.1	69.4 ± 1.5	0.60 ± 0.02	b

(Continued)

Table 6.2 (Continued) ^{29}Si isotropic chemical shifts (δ_{iso}), chemical shift anisotropies (δ_σ) and CSA asymmetry parameters (η_σ) for structurally well-defined silicate phases common in cement systems

Cement phase	Formula		Q^n site	δ_{iso} (ppm)	δ_σ (ppm)	η_σ	Ref.
Pseudowollastonite	α-Ca$_3$Si$_3$O$_9$		Q^2	−83.6 ± 0.2	88.9 ± 4.1	0.55 ± 0.06	b
Alpha-dicalcium silicate hydrate	α-Ca$_2$(SiO$_3$OH)OH		Q^0	−72.7 ± 0.1	26.0 ± 0.5	0.30 ± 0.03	b
Afwillite	Ca$_3$(HSiO$_4$)$_2$·2H$_2$O		Q^0	−71.3, −73.3			e
Calciochondrodite	Ca$_5$(SiO$_4$)$_2$(OH)$_2$		Q^0	−72.7			f
Jaffeite	Ca$_6$Si$_2$O$_7$(OH)$_6$		Q^1	−82.2 ± 0.1	−51.2 ± 1.2	0.07 ± 0.03	g
Lawsonite	CaAl$_2$Si$_2$O$_7$(OH)·2H$_2$O		Q^1	−81	−53	0.58	d
Hillebrandite	Ca$_2$SiO$_3$(OH)$_2$		Q^2	−85.8 ± 0.1	39.3 ± 0.9	0.71 ± 0.03	b
Xonotlite	Ca$_6$Si$_6$O$_{17}$(OH)$_2$	Si(1)	Q^2	−86.4 ± 0.1	38.1 ± 1.8	0.65 ± 0.04	b
		Si(2)	Q^2	−87.2 ± 0.1	39.7 ± 1.8	0.57 ± 0.04	
		Si(3)	Q^3	−97.6 ± 0.1	33.3 ± 1.1	0.02 ± 0.05	
Scawtite	Ca$_7$(Si$_6$O$_{18}$)CO$_3$·2H$_2$O	Si(1)	Q^2	−85.1 ± 0.1	49.1 ± 1.2	0.70 ± 0.04	b
		Si(2)	Q^2	−86.5 ± 0.1	61.1 ± 1.5	0.66 ± 0.06	
Strätlingite	Ca$_2$Al$_2$SiO$_7$·8H$_2$O		Q^2(2Al)	−82.2			h
			Q^2(1Al)	−85.6			
			Q^2	−87.1 ± 0.1			

(Continued)

Table 6.2 (Continued) ^{29}Si isotropic chemical shifts (δ_{iso}), chemical shift anisotropies (δ_σ) and CSA asymmetry parameters (η_σ) for structurally well-defined silicate phases common in cement systems

Cement phase	Formula	Q^n site	δ_{iso} (ppm)	δ_σ (ppm)	η_σ	Ref.
Kaolinite	$Al_2Si_2O_5(OH)_4$	Si(1) Q^3 Si(2) Q^3	-90.76 ± 0.02 -91.40 ± 0.02			i
Low-quartz	$c\text{-}SiO_2$	Q^4	-107.4	3.6 ± 0.3	0.6 ± 0.3	j
Thaumasite	$Ca_3[Si(OH)_6]CO_3SO_4 \cdot 12H_2O$	Si(VI)	-179.4 ± 0.1	-5.0 ± 0.5	0.85 ± 0.10	k

Note: The isotropic chemical shift is only reported for some of the samples. Uncertainty limits are not included in all studies. The definition of the CSA parameters follows Equation 6.22.

a Grimmer et al. (1985).
b Hansen et al. (2003).
c Grimmer and Zanni (1998).
d Smith et al. (2011).
e Engelhardt and Michel (1987).
f Bell et al. (1990).
g This book; Figure 6.6.
h Kwan et al. (1995).
i Hayashi et al. (1992). A clear splitting of the two resonances is observed only for very pure samples.
j Spearing and Stebbins (1989).
k Skibsted et al. (1995b).

6.2.3.2 The dipolar interaction

Spin–spin interactions are divided into *indirect* spin–spin couplings, generally denoted J couplings, which are through-bond couplings, and *direct* spin–spin couplings, denoted dipolar interactions, which are through-space couplings depending on the internuclear distances. J couplings and homonuclear dipolar interactions are seldom observed for light elements of oxidic systems, and they will not be considered here. Although they are not resolved in experimental spectra, they can be utilised in correlation NMR experiments to transfer magnetisation from one spin to another.

For an isolated heteronuclear-spin pair (I, S), the Hamiltonian for the dipolar coupling can be written as

$$H_D = \bar{I} \cdot \bar{\bar{D}} \cdot \bar{S}, \tag{6.24}$$

where $\bar{\bar{D}}$ is the dipolar coupling tensor. The unique element of this tensor (d_{zz}) is oriented along the internuclear vector $(\bar{r}_{IS})$ and the tensor is axially symmetric $(d_{xx} = d_{yy})$ as well as traceless. For a static-powder NMR experiment, the first-order dipolar Hamiltonian for a heteronuclear-spin pair takes the form

$$\bar{H}_D^{(1)} = \frac{Dh}{4\pi} \{3\cos^2 \beta - 1\} I_z S_z, \tag{6.25}$$

where β is the angle between the internuclear vector and the external magnetic field and D is the dipolar coupling constant defined by

$$D = \frac{\mu_0 \gamma_I \gamma_S h}{16\pi^3 r_{IS}^3}. \tag{6.26}$$

Here, μ_0 is the permittivity of free space and r_{IS} the internuclear distance between the two spins. The inverse dependency of the dipolar coupling on the cube of the internuclear distance $\left(D \propto r_{IS}^{-3}\right)$ implies that strong dipolar couplings are observed only for neighbouring spin nuclei. Moreover, the proportionality with the gyromagnetic ratios (γ_I and γ_S) results in line broadening from dipolar interactions being most critical for spins coupled to ^1H or ^{19}F in cement systems. However, dipolar couplings with these spins can efficiently be removed by MAS combined with high-power ^1H or ^{19}F decoupling, where these spins are radiated simultaneously during the detection of the FID for the observe nucleus. Heteronuclear dipolar couplings serve as the key interaction for transfer of magnetisation in the CP NMR experiment (see Section 6.2.4.2) for sensitivity enhancement or selective detection of dipolar coupled spins. Moreover, several rf pulse sequences, such as REDOR (Gullion and Schaefer 1989), transferred-echo double resonance (Hing et al. 1992) and rotational-echo adiabatic-passage double resonance (REAPDOR) (Gullion 1995), reintroduce the dipolar couplings in MAS NMR spectra for determination of the dipolar coupling constant (D) and thereby the internuclear distance (r_{IS}). This approach has been utilised in a study of the incorporation of fluoride ions in the alite phase of mineralised portland clinkers (Tran et al. 2009). Analysis of the $^{29}Si\{^{19}F\}$ REDOR NMR spectra in that work gave $D = 285$ Hz, corresponding to an average Si-F distance of $\langle r_{Si-F}\rangle = 4.29$ Å. This value demonstrates that fluoride substitutes only for the ionic interstitial oxygen atoms in the M_{III} form of alite ($\langle r_{Si-O}\rangle = 4.32$ Å) and not the covalently bonded Si-O oxygens ($\langle r_{Si-O}\rangle = 1.63$ Å).

6.2.3.3 The quadrupole coupling interaction

A nucleus with $I > 1/2$ has a nonspherically symmetric charge distribution, which leads to a strong coupling between the nuclear quadrupole moment and the electric field gradients (EFGs), produced by surrounding electrons at the nuclear site. This is the quadrupole coupling interaction and its Hamiltonian can be expressed as

$$H_Q = \bar{I}\cdot\bar{\bar{Q}}\cdot\bar{I}. \tag{6.27}$$

The elements of the quadrupole coupling tensor ($\bar{\bar{Q}}$) are related to the EFG tensor ($\bar{V}$) by

$$Q_{ij} = \frac{eQ}{2I(2I-1)h}V_{ij}, \tag{6.28}$$

where eQ is the nuclear quadrupolar moment. $\bar{\bar{Q}}$ is a symmetric and traceless tensor which in its principal axis system is given by

$$\bar{\bar{Q}} = \frac{eQ}{2I(2I-1)h}\begin{pmatrix} V_{xx} & 0 & 0 \\ 0 & V_{yy} & 0 \\ 0 & 0 & V_{zz} \end{pmatrix} = \frac{C_Q}{2I(2I-1)}\begin{pmatrix} -\frac{1}{2}(1+\eta_Q) & 0 & 0 \\ 0 & \frac{1}{2}(\eta_Q-1) & 0 \\ 0 & 0 & 1 \end{pmatrix},$$

$$\tag{6.29}$$

where $|V_{zz}| \geq |V_{yy}| > |V_{xx}|$ and $V_{zz} = eq$ is used for the unique EFG tensor element. Since the tensor is traceless, only two parameters are required to characterise the quadrupole interaction, i.e. the quadrupole coupling constant $C_Q = e^2qQ/h$ and the EFG asymmetry parameter $\eta_Q = (V_{yy} - V_{xx})/V_{zz}$, with $0 \leq \eta_Q \leq 1.0$. The energy-level diagram, for a spin $I = 5/2$ nucleus (e.g. ^{17}O and ^{27}Al) is shown in Figure 6.7 and illustrates the types of single-quantum

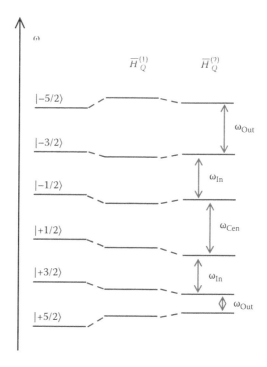

Figure 6.7 Energy-level diagram for a spin *I* = 5/2 nucleus, illustrating the perturbations from the first- and second-order quadrupolar interaction. $\Delta E = \hbar\omega$, and ω_{Cen} is the frequency for the central transition, whereas ω_{In} and ω_{Out} denote the inner and outer satellite transitions.

transitions that may be observed. The transition with the highest intensity (for *I* = 3/2, 5/2) is the so-called *central* transition (*m* = 1/2 ↔ *m* = −1/2), while transitions between the other energy levels normally are denoted the inner (*m* = ±1/2 ↔ *m* = ±3/2) and outer (*m* = ±3/2 ↔ *m* = ±5/2) satellite transitions. The quadrupole interaction requires consideration of both the first- and second-order terms of the Hamiltonian for half-integer–spin nuclei, which can be expressed, respectively, as

$$\overline{H}_Q^{(1)} = \omega_Q^{(1)}\left[3I_z^2 - I(I+1)\right], \tag{6.30}$$

$$\overline{H}_Q^{(2)} = \omega_Q^{(21)}\left[-8I_z^2 + 4I(I+1) - 1\right]I_z + \omega_Q^{(22)}\left[-2I_z^2 + 2I(I+1) - 1)\right]I_z, \tag{6.31}$$

where the factors $\omega_Q^{(i)}$ describe the spatial dependencies for the transformation from the EFG principal-axis system to the laboratory-fixed frame and include the parameters C_Q, η_Q and *I*. Explicit expressions for the $\omega_Q^{(i)}$ terms can be found elsewhere (Skibsted et al. 1991b). The resonance frequency for the $(m, m - 1)$ transition, employing Equation 6.19, takes the form

$$\omega_{m,m-1}^{(1),Q} = \omega_Q^{(1)}\left(m - \frac{1}{2}\right),\tag{6.32}$$

which shows that the central transition ($m = 1/2$) is not affected by the first-order quadrupole interaction, as indicated in Figure 6.7.

The second-order quadrupolar term (Equation 6.31) affects all transitions and has a spatial dependency, which cannot be averaged out by spinning the sample about a single axis. Thus, the central transition exhibits characteristic line shapes, depending on C_Q, η_Q and B_0, which are reduced only by a certain factor ranging from 2.43 ($\eta_Q = 0$) to 3.43 ($\eta_Q = 1.0$) by high-speed MAS. These line shapes are illustrated in Figure 6.8 by simulations of static-powder and MAS NMR spectra of the central transitions for different values of η_Q. Each line shape contains distinct singularities, shoulders and edges and the measurement of the frequency for three of these allows calculation of the C_Q, η_Q and δ_{iso} parameters for a given structural site (Samoson et al. 1982). In practise, this approach is often used to obtain an estimate of C_Q, η_Q and δ_{iso}, and these values are then used in a fit to the experimental spectrum by a full calculation based on the powder average of the second-order quadrupolar term in Equation 6.31. The width of the second-order line shapes (Figure 6.8) exhibits the proportionality $\bar{H}_Q^{(2)} \propto C_Q^2/\nu_L$, which implies that improved resolution can be obtained at higher magnetic fields as a result of the increased chemical shift dispersion and the reduced second-order quadrupolar broadening. For small and intermediate quadrupole couplings, the second-order quadrupolar line shape for the central transition may not be experimentally resolved to a level where C_Q, η_Q and δ_{iso} can be determined by line shape simulations. For such systems, these parameters can be obtained with high precision from analysis of the spinning sidebands observed for the satellite transitions (Jakobsen et al. 1989; Skibsted et al. 1991b).

This procedure is illustrated in Figures 6.9 and 6.10 by ^{27}Al MAS NMR spectra of $AlCl_3 \cdot 6H_2O$ and ettringite, respectively. The octahedrally coordinated Al^{3+} site in $AlCl_3 \cdot 6H_2O$ is highly symmetric, as reflected by the small quadrupole coupling ($C_Q = 116$ kHz) and the axially symmetric EFG tensor ($\eta_Q = 0$), which forms the basis for observation of nearly ideal line shapes for all single-quantum transitions in the spectrum of the static sample. Moreover, the MAS NMR spectra illustrate that the first order quadrupolar term $\bar{H}_Q^{(1)}$ is averaged by MAS, since the 'high-speed' spectrum includes only a few spinning sidebands with most intensity gathered in the centre-band resonance. The low-speed ^{27}Al MAS spectrum of ettringite (Figure 6.10) forms a valuable basis for a precise determination of C_Q, η_Q and δ_{iso}, as the spinning sideband intensities give a clear reflection of the overall line shape

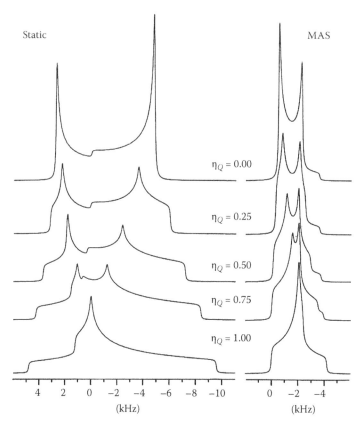

Figure 6.8 Simulated static-powder (left column) and MAS (right column) NMR spectra of the central transition for a spin I = 5/2 nucleus with ν_L = 104.2 MHz (i.e. ^{27}Al at 9.4 T), C_Q = 5.0 MHz and asymmetry parameters in the range η_Q = 0.0–1.0. An infinite spinning speed is assumed for the MAS NMR spectra, where the centre-band resonance contains all intensity for the central transition.

for the satellite transitions. In the absence of CSA, the spectrum of the satellite transitions should be symmetric around the central transition. However, the small asymmetry in the manifold of spinning sidebands around the centre band reflects that the individual inner and outer satellite transitions exhibit different T_2 relaxation times as a result of cross-correlation effects from the quadrupole coupling and the ^{27}Al-^{1}H dipolar interaction on the T_2 relaxation for the individual satellite transitions (Andersen et al. 2005).

Determination of the C_Q, η_Q and δ_{iso} parameters for the structurally different sites in a compound, preferentially by analysis of a pure sample, is generally a prerequisite for an unambiguous identification of the compound in complex mixtures. This requirement reflects that the centre of gravity of the centre band from the central transition as well as the centres of gravity

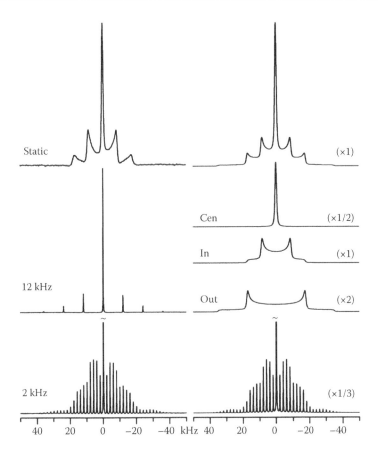

Figure 6.9 Left: Experimental ^{27}Al MAS NMR spectra (7.1 T) of the central and satellite transitions for AlCl$_3$·6H$_2$O obtained without spinning and with spinning speeds of ν_R = 12.0 kHz and 2.0 kHz. Right: simulations of the full static-powder spectrum along with subspectra of the central transition (Cen), the inner (In) and outer (Out) satellite transitions and the MAS spectrum including all transitions acquired with ν_R = 2.0 kHz. The vertical expansion factors are indicated at the right-hand side. The spectra were simulated with the parameters δ_{iso} – 0.0 ppm, C_Q = 0.116 MHz and η_Q = 0.0, as determined from the experimental spectra.

for the satellite transitions are shifted relative to δ_{iso}, since they depend on the second-order quadrupolar interaction, which is inversely proportional to the Larmor frequency. For spin I = 3/2 and I = 5/2 nuclei, the following expressions for the centres of gravity for the individual transitions can be derived (Samoson 1985):

$$\delta^{cg}_{1/2,-1/2} = \delta_{iso} - \frac{1}{40}\frac{C_Q^2\left(1+\eta_Q^2/3\right)}{\nu_L^2}\times 10^6 \text{(in ppm)} \quad \text{for } I = 3/2, \qquad (6.33)$$

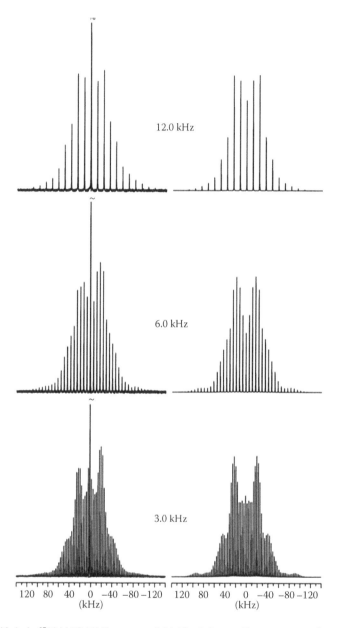

12.0 kHz

6.0 kHz

3.0 kHz

120 80   40   0 −40 −80 −120   120 80   40   0 −40 −80 −120
        (kHz)   (kHz)

Figure 6.10 Left: ^{27}Al MAS NMR spectra (14.1 T) of the satellite transitions for ettringite, acquired with spinning speeds of v_R = 3.0, 6.0 and 12.0 kHz, employing ^{1}H two-pulse phase-modulated (TPPM) decoupling ($\gamma B_2/2\pi$ = 50 kHz) during acquisition of the FID. The centre band from the central transition is cut off in all spectra. Right: simulated spectra of the satellite transitions, employing the parameters δ_{iso} = 13.4 ppm, C_Q = 0.36 MHz and η_Q = 0.19. (From Skibsted J. et al., *Inorg. Chem.* 32, 1013–1027, 1993.)

$$\delta^{cg}_{\pm1/2,\pm3/2} = \delta_{iso} + \frac{2}{40} \frac{C_Q^2\left(1+\eta_Q^2/3\right)}{v_L^2} \times 10^6 \text{(in ppm)} \quad \text{for } I = 3/2, \qquad (6.34)$$

$$\delta^{cg}_{1/2,-1/2} = \delta_{iso} - \frac{24}{4000} \frac{C_Q^2\left(1+\eta_Q^2/3\right)}{v_L^2} \times 10^6 \text{(in ppm)} \quad \text{for } I = 5/2, \qquad (6.35)$$

$$\delta^{cg}_{\pm1/2,\pm3/2} = \delta_{iso} + \frac{3}{4000} \frac{C_Q^2\left(1+\eta_Q^2/3\right)}{v_L^2} \times 10^6 \text{(in ppm)} \quad \text{for } I = 5/2, \qquad (6.36)$$

$$\delta^{cg}_{\pm3/2,\pm5/2} = \delta_{iso} + \frac{81}{4000} \frac{C_Q^2\left(1+\eta_Q^2/3\right)}{v_L^2} \times 10^6 \text{(in ppm)} \quad \text{for } I = 5/2. \qquad (6.37)$$

The different second-order quadrupolar shifts for the individual transitions allows determination of δ_{iso} and the quadrupolar product parameter, second-order quadrupolar effect or SOQE $= C_Q\sqrt{1+\eta_Q^2/3}$, from the centres of gravity for the central and inner satellite transitions, $\delta^{cg}_{1/2,-1/2}$ and $\delta^{cg}_{\pm1/2,\pm3/2}$, measured in a MAS NMR spectrum, or from the centre of gravity for the central transition, determined from spectra at different magnetic fields. The centre of gravity for the inner satellite transitions $\left(\delta^{cg}_{\pm1/2,\pm3/2}\right)$ can often be determined with high precision as the average frequency of the centres of gravity for symmetric pairs ($\pm n$) of spinning sidebands, e.g. $\delta^{cg}_{\pm1/2,\pm3/2} = \frac{1}{i}\sum_i\left(\delta^{cg}_{i=+n} + \delta^{cg}_{i=-n}/2\right)$. For example, this approach (Equation 6.35) has been utilised to determine the ^{27}Al parameters, δ_{iso} = 39.9 ppm and SOQE = 5.1 MHz, for pentacoordinated Al in the C-S-H phase from the centres of gravity of 33.5 and 37.1 ppm measured for the central transition at 14.09 and 21.15 T, respectively (Andersen et al. 2003).

The second-order quadrupolar interaction also affects the intensities observed in the NMR spectra of the central transition for long rf pulses, which has an impact on the quantification of centre-band intensities for sites experiencing different quadrupole couplings (Samoson and Lippmaa 1983a,b). This effect can be studied in a 2D nutation NMR experiment (Kentgens et al. 1987; Samoson and Lippmaa 1988) where the evolution of the spin system in the presence of an rf field is investigated in the rotating frame in the indirect dimension, incrementing the pulse length. The response in the indirect dimension depends on the quadrupole coupling parameters (C_Q and η_Q), the magnetic field strength and the rf field strength, and simulation of the excitation profile in this dimension can be used to determine C_Q and η_Q. Figure 6.11 shows calculated excitation profiles for the central transition of a spin I = 5/2 nucleus at v_L = 104.2 MHz (i.e. ^{27}Al

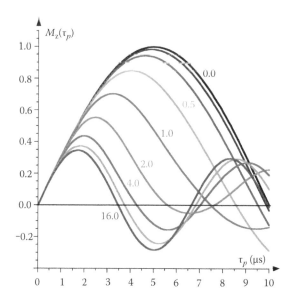

Figure 6.11 Simulated excitation profiles for the central transition of a spin $I = 5/2$ nucleus at 104.2 MHz for $\gamma B_1/2\pi = 50$ kHz, $\eta_Q = 0$ and $C_Q = 0$, 0.125, 0.25, 0.5, 1.0, 2.0, 4.0, 8.0 and 16.0 MHz. Effects of the spinning speed on the centre-band intensities are not taken into account.

at 9.4 T) for different C_Q values, assuming $\eta_Q = 0$ and $\nu_{rf} = \gamma B_1/2\pi = 50$ kHz. A pure sinusoidal intensity variation is found for $C_Q = 0$, corresponding to the response from ^{27}Al in a liquid sample (Equation 6.5), whereas increasing deviations from this behaviour are observed when the C_Q values are raised. Moreover, the pulse lengths for maximum intensity (τ_{max}) decrease with increasing C_Q value and it can be shown that $\tau_{max} = \tau_{90}^{liq}$ for $\nu_{rf} \gg C_Q$ and $\tau_{max} = \tau_{90}^{liq}/(I+1/2)$ for $\nu_{rf} \ll C_Q$ (Samoson and Lippmaa 1983a). The intensity variation of the central transition as a function of the quadrupole coupling (Figure 6.11) also shows that there is a linear regime for short pulse widths where the increase in intensity is nearly independent of the quadrupole coupling. Thus, if a pulse angle is chosen which fulfils $\tau_p \leq \pi/[2(I + 1/2)]$, then the observed central-transition intensities in the single-pulse MAS NMR experiment can be considered as independent of the quadrupole interaction, and quantitatively reliable spectra can thereby be obtained for sites with different quadrupole couplings (Samoson and Lippmaa 1983a). The pulse length has also an impact for the excitation profile of wideband spectra, and short rf pulses ($\tau_p = 0.5$–2 μs) are generally used for observation of the satellite transitions; for example, excitation of a 1.0 MHz spectral width requires $\tau_p \leq 1.0$ μs.

^{27}Al quadrupole coupling parameters and isotropic chemical shifts have been reported for the most common aluminate phases present in anhydrous and hydrated portland cement and high-alumina cement systems. These parameters have been determined from line shape simulations of the central transition or from simulations of the manifold of spinning sidebands, observed for the satellite transitions in ^{27}Al MAS NMR spectra. In addition, ^{27}Al MQ MAS NMR has been utilised in a few cases. The ^{27}Al parameters, δ_{iso}, C_Q and η_Q, reported for well-defined aluminate phases, are summarised in Table 6.3.

6.2.4 Line-narrowing and sensitivity-enhancement techniques

The orientational dependence of the internal spin interactions in the solid state, reflected by the transformations from the principal axes of the interactions to the laboratory-fixed frame (e.g. Equations 6.23, 6.25, 6.30 and 6.31), can result in line widths of several kHz (or MHz for $I > 1/2$) for powdered samples. To reduce or eliminate this broadening, several techniques have been developed which employ either mechanical rotation of the sample, to reduce the spatial dependent part, or multiple rf pulse schemes, to modify the spin-dependent part of the internal interactions. Of these experimental techniques, MAS (Andrew et al. 1958; Lowe 1959), heteronuclear X{^{1}H} decoupling and ^{1}H → X CP (Hartmann and Hahn 1962; Pines et al. 1972) represent the most fundamental methods for improving resolution and sensitivity in solid-state NMR spectra. In addition, for quadrupolar nuclei, the MQ MAS technique (Frydman and Harwood 1995; Medek et al. 1995) allows acquisition of high-resolution MAS NMR spectra of quadrupolar nuclei by removal of the second-order quadrupolar broadening for the central transition.

6.2.4.1 Magic-angle spinning

Mechanical sample spinning results in a modulation of the anisotropic interactions and in several cases, fast sample spinning may provide spectra which approach the high resolution observed for liquid samples. The first-order Hamiltonians for the CSA, heteronuclear dipolar couplings and the quadrupole interaction (Equations 6.23, 6.25 and 6.30) contain the geometrical factor, $(3 \cos^2 \beta - 1)$, where β describes the angle between a specific molecule-fixed vector and the static magnetic field. In liquids, rapid molecular tumbling averages this geometrical factor to zero, thus explaining that these interactions are not observed in liquid-state NMR spectra. The MAS technique mimics this molecular tumbling by rotating the sample mechanically around a fixed axis (θ) relative to the magnetic field. From employing the heteronuclear dipolar interaction as the illustrative example, the molecule-fixed vector is the internuclear vector (r) between the two spins (I and S). Rotation of the sample about an axis at the angle θ relative to the magnetic field causes the vector r to trace out a conical path described

Table 6.3 ^{27}Al isotropic chemical shifts and quadrupole coupling parameters for aluminate phases common in cement systems

Cement phase	Formula	Al coordination	Al site	δ_{iso}^{a} (ppm)	C_Q (MHz)	η_Q	Ref.
Calcium hexaluminate	$CaAl_6O_{10}$	Al(VI)	Al(1)	16.26 ± 0.05	0.15 ± 0.05		[b]
		Al(IV)	Al(2)	55.8 ± 0.1	21.40 ± 0.05	0.00 + 0.05	
		Al(IV)	Al(3)	68.1 ± 0.1	3.10 ± 0.05	0.00 + 0.05	
		Al(VI)	Al(4)	9.92 ± 0.05	1.60 ± 0.05		
		Al(V)	Al(5)	22.3 ± 0.5	4.8 ± 0.1	0.7 ± 0.1	
Grossite	$CaAl_4O_7$	Al(IV)	Al(1)	75.5 ± 0.5	6.25 ± 0.05	0.88 ± 0.02	[c]
		Al(IV)	Al(2)	69.5 ± 0.5	9.55 ± 0.05	0.82 ± 0.02	
Calcium monoaluminate	$CaAl_2O_4$	Al(IV)	Al(1)	81.9	2.50	0.20	[c]
			Al(2)	83.8	2.60	0.75	
			Al(3)	86.2	2.60	0.95	
			Al(4)	82.7	3.32	0.53	
			Al(5)	81.6	3.37	0.39	
			Al(6)	81.2	4.30	0.47	
C_4A_3	$Ca_4Al_6O_{13}$	Al(IV)	Al(1)	80.3 ± 1.0	2.4 ± 0.1	0.95 ± 0.05	[d]
$C_{12}A_7$	$C_{12}Al_{14}O_{33}$	Al(IV)	Al(2)	85.9 ± 1.0	9.7 ± 0.2	0.40 ± 0.10	[e]
				80.2 ± 0.3	3.8 ± 0.2	0.70 ± 0.10	
Tricalcium aluminate	$Ca_3Al_2O_6$	Al(IV)	Al(1)	79.5 ± 0.5	8.69 ± 0.05	0.32 ± 0.02	[f]
			Al(2)	78.3 ± 0.5	9.30 ± 0.05	0.54 ± 0.02	
Belite Al^{3+} guest ions	β-$Ca_2Si_{1-x}Al_xO_{4-(1/2)x}$	Al(IV)		96.1 ± 0.5	7.1 ± 0.2	0.33 ± 0.05	

(Continued)

Table 6.3 (Continued) ^{27}Al isotropic chemical shifts and quadrupole coupling parameters for aluminate phases common in cement systems

Cement phase	Formula	Al coordination, Al site		δ_{iso}^{a} (ppm)	C_Q (MHz)	η_Q	Ref.
Corundum	α-Al_2O_3	Al(VI)		16.0 ± 0.2	2.38 ± 0.01	$0.00 + 0.02$	g
Gibbsite	γ-$Al(OH)_3$	Al(VI)	Al(1)	10.4 ± 0.3	1.97 ± 0.07	0.73 ± 0.04	c
		Al(VI)	Al(2)	11.5 ± 0.3	4.45 ± 0.05	0.44 ± 0.03	
Monocalcium aluminate decahydrate	$CaAl_2(OH)_{12}\cdot 4H_2O$	Al(VI)		10.2 ± 0.3	2.4 ± 0.2^h	0.8	c
$C_4A_3H_3$	$Ca_4Al_6O_{13}\cdot 3H_2O$	Al(IV)	Al(1)	78 ± 1.0	1.8	0.5	i
		Al(IV)	Al(2)	79 ± 1.0	5.4	0.45	
C_2AH_8	$Ca_2Al_2(OH)_{12}\cdot 2H_2O$	Al(VI)		10.3	1.2	0.6	i
Hydrogarnet-katoite	$Ca_3Al_2(OH)_{12}$	Al(VI)		12.4 ± 0.1	0.688 ± 0.006	$0.00 + 0.02$	c,g
C_4AH_{13}	$Ca_4Al_2(OH)_{12}\cdot 7H_2O$	Al(VI)		10.2 ± 0.2	1.8 ± 0.2^h		c
Monosulfate	$Ca_4Al_2(OH)_{12}(SO_4)\cdot 6H_2O$	Al(VI)		11.8 ± 0.2	1.7 ± 0.2^h		c
Calcium monocarbonate hydrate	$Ca_4Al_2(OH)_{12}(CO_3)\cdot 5H_2O$	Al(VI)		8.7	1.7	0.8	i
Hydrotalcite	$Mg_4Al_2(OH)_{12}(CO_3)\cdot 6H_2O$	Al(VI)		10.1	1.2	0.8	i
Friedel's salt	α-$Ca_4Al_2(OH)_{12}Cl_2\cdot 4H_2O$ $T = 18°C$	Al(VI)		9.2 ± 0.3	1.42 ± 0.03	0.93 ± 0.03	k
	β-$Ca_4Al_2(OH)_{12}Cl_2\cdot 4H_2O$ $T = 38°C$	Al(VI)		9.1 ± 0.3	1.09 ± 0.03	0.04 ± 0.05	

(Continued)

Table 6.3 (Continued) ^{27}Al isotropic chemical shifts and quadrupole coupling parameters for aluminate phases common in cement systems

Cement phase	Formula	Al coordination, Al site		δ_{iso}^{a} (ppm)	C_Q (MHz)	η_Q	Ref.
Ettringite	$Ca_6Al_2(OH)_{12}(SO_4)_3 \cdot 26H_2O$	Al(VI)		13.4 ± 0.2	0.36 ± 0.01	0.19 ± 0.03	i,l
Strätlingite	$2CaO \cdot Al_2O_3 \cdot SiO_2 \cdot 8H_2O$	Al(IV)		8.4			m
		Al(VI)		61.7			
Kaolinite	$Al_2Si_2O_5(OH)_4$	Al(VI)	Al(1)	8.0 ± 1.0	3.0 ± 0.1	0.9 ± 0.1	n
			Al(2)	7.5 ± 1.0	3.4 ± 0.1	0.8 ± 0.1	

[a] Isotropic chemical shifts relative to 1.0 M AlCl$_3$·6H$_2$O(aq).
[b] Du and Stebbins (2004).
[c] Skibsted et al. (1993).
[d] Müller et al. (1986a).
[e] Skibsted et al. (1991a).
[f] Skibsted et al. (1994).
[g] Jakobsen et al. (1989).
[h] Quadrupolar product parameter, $SOQE = C_Q\sqrt{1+\dfrac{\eta_Q^2}{3}}$.
[i] Müller et al. (1986b).
[j] Faucon et al. (1998a).
[k] Andersen et al. (2002).
[l] Figure 6.10.
[m] Kwan et al. (1995).
[n] Paris (2014).

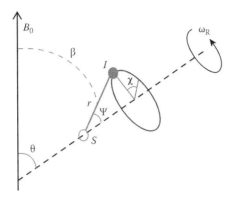

Figure 6.12 Illustration of the internuclear vector (*r*) for two dipolar coupled spins (*I* and *S*) in a rotating sample in a magnetic field (B_0) and the angles that describe its time dependence during rotation (ω_R).

by the angles ψ and χ, as shown in Figure 6.12. Thus, the orientation of *r* relative to B_0 becomes time dependent, implying that the time average of the geometrical factor must be evaluated. For a rotation frequency of ω_r ($= 2\pi\nu_r$), the following time average for $3\cos^2\beta - 1$ can be calculated:

$$\langle 3\cos^2\beta - 1 \rangle = \frac{1}{2}(3\cos^2\theta - 1)(3\cos^2\psi - 1)$$

$$+ \frac{1}{3}[\sin 2\theta \sin 2\psi \cos(\omega_r t + \chi) + \sin^2\Theta \sin^2\psi \cos(2\omega_r t + 2\chi)].$$

(6.38)

At high spinning speeds, the time-dependent terms, $\cos(\omega_r t + \chi)$ and $\cos(2\omega_r t + 2\chi)$, will average to zero and the time average will be proportional to $(3\cos^2\Theta - 1)$. The θ angle can be adjusted experimentally and for $\theta = \theta_m = \cos^{-1}(1/\sqrt{3}) \cong 54.736°$, the time average becomes $\langle 3\cos^2\beta - 1 \rangle = 0$. The angle θ_m is the so-called magic angle, and spinning the sample about this angle improves the resolution of the resonances by elimination of the above-mentioned first-order anisotropic interactions. However, a complete elimination requires that the applied spinning speed (ν_R) is larger than the width of the resonance in a static powder spectrum. For the CSA and heteronuclear dipolar interactions, this situation can be achieved in several cases by spinning the sample at spinning speeds of 5–40 kHz. If this condition is not fulfilled, the time-dependent terms of Equation 6.38 will not average completely to zero and result in spinning sidebands in the MAS NMR spectrum. The technical limits on increasing the maximum spinning speed are related to the fact that the speed on the surface of the rotors cannot exceed the speed of sound and to the manufacture of rotor materials that can withstand these speeds. Although the experiments could in principle be performed by spinning the sample in He gas, where the speed of sound is higher than in air, the most common approach has been to reduce the diameter of the NMR rotors. Currently (2015), MAS probes with rotor diameters of 0.7 mm are commercially available, capable of attaining spinning speeds of up to 110 kHz.

6.2.4.2 Cross polarisation

For dilute spin $I = 1/2$ nuclei (e.g. ^{13}C and ^{29}Si), single-pulse experiments often suffer from inherently low sensitivity and requirements of long repetition delays due to long spin–lattice relaxation times. These disadvantages can in several cases be overcome by the CP technique (Hartmann and Hahn 1962; Pines et al. 1972), which transfers magnetisation from an abundant spin (e.g. ^1H and ^{19}F) to the dilute spin via the heteronuclear dipolar interaction. The basic pulse sequence for the $I \rightarrow S$ CP NMR experiment is shown in Figure 6.13a, where I represents the abundant spin with a high gyromagnetic ratio (γ_I) and S is the dilute spin with a lower γ value. The first step is to excite the I spins by a 90°_y pulse and subsequently to spin lock the I magnetisation along the x axis in the rotating frame by a long x-phase I pulse. At this point the S rf channel is switched on and the amplitude of this field (B_{1S}) is adjusted to match the so-called Hartmann–Hahn condition, $\gamma_I B_{1I} = \gamma_S B_{1S}$. In MAS experiments, this condition is modified by the spinning frequency (Stejskal et al. 1977):

$$\gamma_I B_{1I} = \gamma_S B_{1S} \pm n\nu_R, \tag{6.39}$$

implying that polarisation transfer can be accomplished for different values of n, which is an integer (usually $n = -1$ or $+1$ gives the optimum transfer).

If the dilute spin is a quadrupolar nucleus, the Hartmann–Hahn match becomes

$$\gamma_I B_{1I} = \sqrt{S(S+1) - m(m-1)}\gamma_S B_{1S} \pm n\nu_R \tag{6.40}$$

for polarisation transfer to the $(m, m - 1)$ transition of the S nucleus. The Hartmann–Hahn match implies that the magnetisations for the I and S spins precess at equal frequencies around their spin-locking fields in their respective rotating frames. Thus, the energy required for spin flips is identical for both spins, allowing an efficient transfer of magnetisation via the dipolar interaction. The maximum enhancement of the S-spin magnetisation is given by the ratio of the gyromagnetic ratios for the two spins, γ_I/γ_S. However, in practise a lower enhancement factor is generally observed, partly as a result of relaxation of spin-locked magnetisation in the rotating frame for both the I and S spins. The time constants characterising

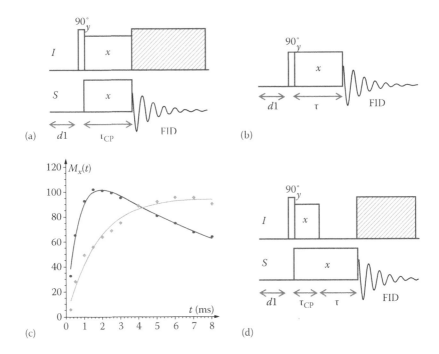

Figure 6.13 Radio-frequency pulse sequences for (a) the S{I} CP NMR experiment, (b) a spin-lock experiment for direct determination of $T_{1\rho}$ and (d) a spin-lock experiment combined with CP for determination of the $T_{1\rho}^S$ relaxation time. (c) Plot of the observed, transferred magnetisation $M_x(t)$ as a function of the CP contact time ($t = \tau_{CP}$) in ^{29}Si{^{1}H} CP/MAS NMR spectra (7.1 T, $\nu_R = 4.0$ kHz) of a mineral sample of kaolinite (circles) and a synthetic sample of α-dicalcium silicate hydrate (diamonds). The experiments employed $\gamma B_1/2\pi \approx \gamma B_2/2\pi = 50$ kHz for the CP contact time and $\gamma B_2/2\pi = 62$ kHz for the initial 90°(^{1}H) pulse and the ^{1}H TPPM decoupling during acquisition of the FID. The lines represent the result of least-squares fitting of the observed intensities to Equation 6.43; see text.

relaxation in the rotating frame, $T_{1\rho}^I$ and $T_{1\rho}^S$, can be determined from a spin-lock experiment or spin lock combined with a CP experiment (Figure 6.13b and d). For both experiments, the time constants can be determined by fitting the observed intensities to the expression

$$M_x^i(\tau) = M_x^i(0)\exp\left[-\left(\frac{\tau}{T_{1\rho}^i}\right)\right], \quad i = I, S,$$

(6.41)

where $M_x^i(0)$ is the transverse spin-locked magnetisation at $\tau = 0$. The observed magnetisation in the CP experiment as a function of the CP contact time

$(t = \tau_{CP})$ can be derived by the inverse spin-temperature approach (Mehring 1983) and for a dilute–abundant $(S–I)$ spin system it takes the form

$$M_x(t) = \frac{M^I(0)}{\lambda} \exp\left(-\frac{t}{T_{1\rho}^I}\right)\left[1 - \exp\left(\frac{-\lambda t}{T_{IS}}\right)\right],$$

$$\lambda = 1 + \frac{T_{IS}}{T_{1\rho}^S} - \frac{T_{IS}}{T_{1\rho}^I}.$$

(6.42)

$M^I(0)$ is the I magnetisation after the $90_y°$ pulse and T_{IS} (the CP time) is the time constant describing the buildup of S magnetisation, which is related to the $I–S$ dipolar coupling. As an example, Figure 6.13c shows the observed intensities as a function of the contact time (τ_{CP}) in a $^{29}Si\{^1H\}$ CP/MAS NMR experiment for kaolinite. A determination of the CP time (T_{SiH}) from these intensities requires a preknowledge of the two rotating-frame relaxation times ($T_{1\rho}^H$ and $T_{1\rho}^{Si}$), which can be obtained by spin-lock experiments (Figure 6.13b and d). However, for silanol sites in silicates and on silica surfaces, it is often observed that $T_{1\rho}^{Si}$ is more than an order of magnitude larger than $T_{1\rho}^H$. Assuming $T_{1\rho}^S \gg T_{1\rho}^I$ allows simplification of Equation 6.42 to

$$M_x(t) = \frac{M^I(0)}{1 - T_{IS}/T_{1\rho}^I} \exp\left(-\frac{t}{T_{1\rho}^I}\right)\left\{1 - \exp\left[\left(\frac{1}{T_{1\rho}^I} - \frac{1}{T_{IS}}\right)t\right]\right\}.$$

(6.43)

From this expression, the time constants T_{IS} and $T_{1\rho}^I$ can be determined from a three-parameter least-squares fit of the intensities observed in a variable–contact time CP/MAS NMR experiment. This is illustrated in Figure 6.13c for kaolinite ($Al_2Si_2O_5(OH)_4$) and α-dicalcium silicate hydrate (α-$Ca_2(SiO_3OH)OH$), where the least-squares fitting of Equation 6.43 to the intensities observed for kaolinite results in the parameters $T_{1\rho}^H = 11.2 \pm 0.8$ ms and $T_{SiH} = 0.64 \pm 0.04$ ms at 7.1 T. The mutual presence of two strong rf fields during the CP contact time is quite demanding for the NMR probe; thus, τ_{CP} may be limited to values below ~10 ms. In the present case, this implies that the decrease in transferred magnetisation caused by $T_{1\rho}^H$ relaxation is not observed for the synthesised sample of α-dicalcium silicate hydrate for contact times below 7 ms (Figure 6.13c). Thus, the variation in transferred magnetisation for this sample is analysed with the simplification of $T_{1\rho}^H \gg T_{SiH}$ in Equation 6.43, giving the value $T_{SiH} = 1.68 \pm 0.14$ ms for α-dicalcium silicate hydrate from a two-parameter ($M^I(0)$, T_{SiH}) least-squares fit. As noted elsewhere (Kolodziejski and Klinowski 2002), kaolinite represents a suitable sample for the initial setup and establishment of the Hartmann–Hahn matching condition (Equation 6.39) in $^{29}Si\{^1H\}$ CP/MAS NMR experiments. The latter is normally performed by acquiring CP/MAS spectra for a fixed contact time (e.g. τ_{CP} = 2.0 ms) where one of the rf field strengths, $B_1(^{29}Si)$ or $B_1(^1H)$, is varied and the observed intensity is optimised. For kaolinite, each experiment requires only a few scans ($NS \sim 8$–16) and a short repetition delay ($d1 \sim 2$–4 s).

In addition to the gain in intensity by the I–S magnetisation transfer, the CP NMR experiment has also the advantage that the repetition delay depends on the spin–lattice relaxation of the abundant I spins, which generally is much shorter than the T_1 values for the dilute spins. This difference may result in a significant reduction in instrument time. An extreme example has been reported for a natural sample of thaumasite ($Ca_6[Si(OH)_6]_2(CO_3)_2(SO_4)_2\cdot24H_2O$) where $T_1(^{29}Si)$ = 1003 s and $T_1(^1H)$ = 4.1 s along with a CP intensity-gain factor of ~2.5 were determined from $^{29}Si\{^1H\}$ CP/MAS NMR experiments (Skibsted et al. 1995a). For that sample, it was estimated that the instrument time could be reduced by roughly three orders of magnitude, employing CP rather than single-pulse NMR.

The clear detection of thaumasite in cementitious materials is illustrated in Figure 6.14 for a portland–limestone cement hydrated for 5 days at 40°C followed by 28 days at 20°C and then immersed in a 0.25 M $MgSO_4$ solution for 9 months (Skibsted et al. 2003). A disadvantage of the CP experiment is that the observed intensities depend on the I–S dipolar interaction along with the CP dynamics during the CP period. Thus, a strict quantification from CP/MAS NMR requires determination of the time constants $T_{1\rho}^I$, $T_{1\rho}^S$ and T_{IS} for both the sample under investigation and a suitable reference sample, which can be quite time consuming. This procedure has been used to quantify thaumasite in laboratory samples exposed to different sources of sulfates and carbonates (Skibsted et al. 1995a, 2003). For the portland–limestone cement exposed to 0.25 M $MgSO_4$ solution (Figure 6.14), this procedure gives a thaumasite content of 16.2 ± 1.5 wt.%, employing Equation 6.43 and the assumption of $T_{1\rho}^{Si} \gg T_{1\rho}^H$.

The CP NMR experiment can advantageously also be used to identify dipolar coupled I–S spins in samples containing different phases. In ^{29}Si MAS NMR this is illustrated in Figure 6.15 by ^{29}Si MAS and $^{29}Si\{^1H\}$ CP/MAS NMR spectra of a white portland cement hydrated for 28 days, where the CP experiment detects only the hydrated phases. In addition, it is shown that $^{29}Si\{^{19}F\}$ CP/MAS NMR can be used to locate fluoride guest ions in a mineralised white portland cement. The $^{29}Si\{^{19}F\}$ CP/MAS NMR spectrum (Figure 6.15) is very similar to the ^{29}Si MAS NMR spectrum observed for the M_1 form of alite, whereas the narrow resonance from belite is absent, thereby clearly showing that fluoride ions are incorporated as guest ions in alite only (Tran et al. 2009).

Further information about dipolar coupled I–S spins can be obtained from a 2D heteronuclear correlation (HETCOR) NMR experiment (Figure 6.16a), utilising CP to transfer magnetisation from the I to the S spins (Caravatti et

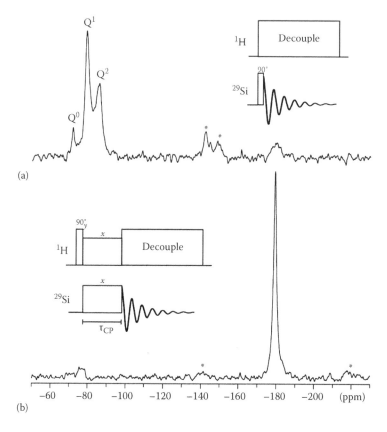

(a)

(b)

Figure 6.14 (a) ^{29}Si MAS and (b) ^{29}Si{^{1}H} CP/MAS NMR spectra (9.4 T) of a portland cement including 15 wt.% limestone (raw feed precipitator dust from cement production; ~85 wt.% CaCO$_3$), hydrated for 5 days at 40°C, 28 days at 20°C and then immersed in a 0.25 M MgSO$_4$ solution for 9 months. The ^{29}Si MAS spectrum employed a spinning speed of ν_R = 5.0 kHz and a repetition delay of dl = 15 s, whereas the CP experiment used ν_R = 3.0 kHz, dl = 8 s and a CP contact time of τ_{CP} = 0.8 ms. The spectra are shown with identical vertical scales. Asterisks indicate spinning sidebands. (Reprinted from *Cement and Concrete Composites*, 25, Skibsted, J., S. Rasmussen, D. Herfort and H. J. Jakobsen, ^{29}Si cross-polarisation magic-angle spinning NMR spectroscopy – An efficient tool for quantification of thaumasite in cement-based materials, 823–829, Copyright 2003, with permission from Elsevier.)

al. 1982). In this experiment the chemical shifts of the I spins (^{1}H) are allowed to evolve in the period t_1, which is incremented to give the second dimension of the spectrum. Various multiple-pulse decoupling sequences can be used to reduce the ^{1}H–^{1}H homonuclear dipolar couplings during the t_1 evolution period, resulting in an improvement of the resolution of the resonances in the ^{1}H dimension of the spectrum. For example, frequency-switched Lee–Goldberg (FSLG) decoupling (Bielecki et al. 1989) can be considered as a

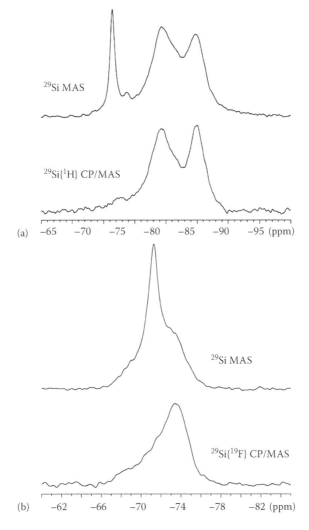

Figure 6.15 (a) ^{29}Si MAS and ^{29}Si{^{1}H} CP/MAS NMR spectra (9.4 T) of a white portland cement hydrated for 28 days, acquired with spinning speeds of ν_R = 5.0 and 3.0 kHz and relaxation delays of dl = 30 and 4 s, respectively. A CP contact time of τ_{CP} = 1.5 ms is used for the lower spectrum. (b) ^{29}Si MAS and ^{29}Si{^{19}F} CP/MAS NMR spectra (7.1 T) of a mineralised white portland cement. The single-pulse spectrum employs ν_R = 7.0 kHz and dl = 30 s, whereas the CP/MAS spectrum is obtained with ν_R = 5.0 kHz, dl = 4 s and τ_{CP} = 5.0 ms.

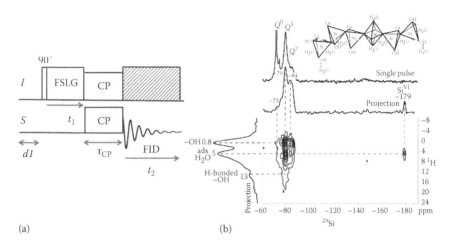

(a)   (b)

Figure 6.16 (a) rf pulse sequence for a 2D S{I} HETCOR CP NMR experiment where the t_1 period is incremented, allowing evolution of the I spin magnetisation, and the S spin magnetisation is detected as the FID during t_2. Frequency-switched Lee–Goldberg decoupling is used in the present experiment to reduce homonuclear I–I dipolar couplings. (b) 2D ^{29}Si{^{1}H} HETCOR CP/MAS NMR spectrum (11.7 T, ν_R = 6.5 kHz), acquired with an rf pulse sequence similar to the sequence in a, for a white portland cement hydrated for 4 weeks. One-dimensional projections of the ^{29}Si and ^{1}H dimensions of the 2D spectrum are shown along the horizontal and vertical axes, respectively. In addition to ^{1}H–^{29}Si connectivities for the ^{1}H and ^{29}Si sites of the C-S-H phase, a minor quantity of thaumasite ($\delta(^{29}$Si) = −179 ppm) is also detected. (Reprinted with permission from Rawal, A. et al., *J. Am. Chem. Soc.* 132, 7321–7337. Copyright 2010 American Chemical Society.)

spin-space variant of MAS where the ^{1}H magnetisation is forced to evolve about an effective field oriented at the magic angle in spin space by using a set of 2π pulses with carefully chosen offset frequencies. ^{29}Si{^{1}H} HETCOR CP/MAS NMR has been used to study ^{1}H–^{29}Si connectivities in C-S-H samples synthesised with different Ca/Si ratios (Brunet et al. 2004), allowing CP from interlayer water molecules and Ca-OH protons to be distinguished in the 2D spectra. Moreover, the technique has been employed in a study of the C-S-H structure, resulting from 4 weeks of hydration of white portland cement (Rawal et al. 2010). An illustrative ^{29}Si{^{1}H} HETCOR CP/MAS spectrum from that work is shown in Figure 6.16, where connectivities from the Q^1, Q^2(1Al) and Q^2 ^{29}Si peaks from the C-S-H phase and resonances from hydroxyl groups, water molecules and hydrogen-bonded Si-OH sites in the ^{1}H dimension are detected. Similar HETCOR spectra have also been reported from ^{27}Al{^{1}H} and ^{13}C{^{1}H} CP/MAS experiments in a study of the surface interactions between saccharides and hydrated portland cement (Smith et al. 2012).

The CP/MAS technique has also been employed in a few ^{13}C-NMR studies of cementitious materials, utilising that a significant sensitivity enhancement can be achieved by ^{13}C{^{1}H} CP/MAS NMR for ^{13}C present in a hydrous environment or in direct C–H bonds. Thus, this technique is convenient for studies of organic admixtures in cement systems, as demonstrated in a ^{13}C{^{1}H} CP/MAS NMR investigation of the interface between poly(vinyl alcohol) and the inorganic matrix of calcium silicate hydrates (Comotti et al. 1996). The method has also been used to study the interaction and stability of different latex polymers in portland cement mixes (Rottstegge et al. 2005), where the absence of ^{13}C resonances from calcium acetate and poly(vinyl alcohol) showed that the latex polymer was stable towards hydrolysis in the alkaline medium of the hydrating portland cement. The application of ^{13}C{^{1}H} CP/MAS NMR in studies of cement materials is illustrated by three examples in Figure 6.17, where the first spectrum shows the detection of the carbonate ion in calcium monocarbonate hydrate. [13]C

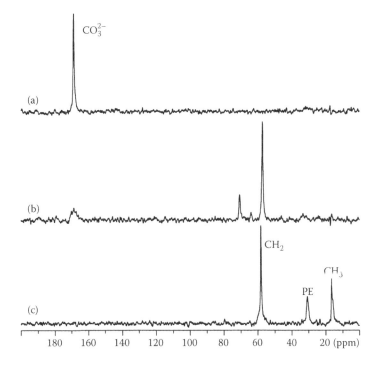

Figure 6.17 ^{13}C{^{1}H} CP/MAS NMR spectra (7.1 T, ν_L = 5.0 kHz and d1 = 4.0 s) of (a) a synthetic sample of calcium monocarbonate hydrate (τ_{CP} = 2.5 ms), (b) a C-S-H sample including X-Seed® (BASF) in the synthesis procedure (τ_{CP} = 2.5 ms) and (c) a synthesised magnesium-silicate-hydrate sample, shaken in a polyethylene (PE) container for 1 year, then washed with ethanol and freeze-dried (τ_{CP} = 1.5 ms). The spectra employed 20,000 to 40,000 scans.

chemical shifts and CSA parameters have recently been determined for a number of inorganic carbonates relevant to cement chemistry, utilising the spinning sideband patterns observed in low-speed ^{13}C MAS or CP/MAS NMR spectra (Sevelsted et al. 2013). The variation in these parameters (δ_{iso}, δ_σ and η_σ) are rather small; thus, it may be hard to distinguish different carbonate ions present in small amounts in portland cement systems by conventional ^{13}C-NMR techniques. However, comparison of the intensities in ^{13}C MAS and ^{13}C{^{1}H} CP/MAS NMR spectra can provide information about the relative amounts of carbonate ions in hydrous and anhydrous environments, as demonstrated for a hydrated portland–limestone cement (Sevelsted et al. 2013). ^{13}C resonances from C-H sites are strongly enhanced by ^{13}C{^{1}H} CP/MAS NMR, allowing detection of organic components in cement mixtures. This is illustrated by the ^{13}C{^{1}H} CP/MAS NMR spectrum acquired for a C-S-H phase synthesised in the presence of X-Seed (Figure 6.17b), which clearly illustrates the ^{13}C resonances of the polymer in the X-Seed product, and by a similar spectrum (Figure 6.17c) of a synthesised M-S-H sample with an initial Mg/Si ratio of 0.6, shaken in a polyethylene (PE) container for 1 year, then washed with ethanol and freeze-dried (Nied et al. 2015). In the latter spectrum, the resonance at 31 ppm originates from PE, whereas the peaks at 58 and 17 ppm can be assigned to the $-CH_2$ and $-CH_3$ groups from solid ethanol, demonstrating a small degree of abrasion of the PE container during shaking and that some ethanol is left from the washing, physically or chemically bound to the sample, as it would otherwise not be seen by ^{13}C{^{1}H} CP/MAS NMR.

6.2.4.3 Heteronuclear decoupling

A major source of line broadening in solid-state NMR spectra of dilute spins is heteronuclear dipolar couplings with abundant spins, usually ^{1}H or ^{19}F, which causes a modulation of the effective local field around the dilute spin nucleus. The interaction can be strongly reduced/eliminated in a double-resonance experiment where a strong rf field is applied to the abundant spins at their Larmor frequency during detection of the magnetisation from the dilute spins. Decoupling is a well-known concept from liquid-state NMR where it is used to remove scalar J couplings. These couplings are generally very small (<200 Hz), implying that only low-power decoupling (~5–10 W) is required. However, dipolar interactions in the solid state are much stronger than the scalar couplings; thus, high-power decoupling (~50–1.000 W) is generally required for rigid systems. As mentioned earlier (Section 6.2.3.2), the heteronuclear dipolar interaction is inversely proportional to the cube of the internuclear distance, implying that decoupling is most crucial for spins directly bonded to hydrogen, such as in polymers and rigid organic molecules. However, for most spin nuclei of interest in cement chemistry, hydrogen is normally present in the second- and further-distant coordination spheres, resulting in weaker dipolar couplings. For example, in most ^{29}Si-NMR studies of silicates and portland cements, the ^{29}Si–^{1}H dipolar couplings are efficiently removed by MAS at spinning speeds of roughly 3–5 kHz. In ^{27}Al-NMR experiments the dipolar couplings can be stronger, in particular for phases including Al(OH)$_n$ sites (such as ettringite; Figure 6.10). However, these ^{27}Al–^{1}H couplings can often be removed by ^{1}H decoupling using moderate rf fields of $\gamma B_2/2\pi = 40$–50 kHz. The ^{1}H decoupling field is generally subjected to a specific modulation scheme to make the decoupling more efficient, and several rf pulse sequences for ^{1}H/^{19}F decoupling have been developed, which consider specific spin systems or experimental conditions (e.g. offset and CSA effects at high magnetic fields). The simplest decoupling sequence is continuous-wave decoupling, where the I spins are continuously radiated with a constant pulse phase at their Larmor frequency. However, other decoupling sequences, such as wonderful alternating phase technique for zero residual splittings (WALTZ) (Shaka et al. 1983), TPPM (Bennett et al. 1995) and small phase incremental alternation (SPINAL) (Fung et al. 2000) are commonly used, and implemented on most modern spectrometer systems, and improve the decoupling efficiency for a given maximum ^{1}H rf field strength, defined by the probe and spectrometer.

To summarise, the effects of ^{1}H decoupling along with MAS and CP are illustrated by the ^{13}C-NMR spectra of camphor in Figure 6.18, which show that a combination of these methods can result in solid-state NMR spectra approaching the resolution observed in liquid-state NMR. Roughly speaking, ^{1}H decoupling reduces ^{1}H–^{13}C dipolar interactions, and MAS, the ^{13}C CSA; while CP leads to a sensitivity enhancement.

6.2.4.4 The multiple-quantum magic-angle spinning experiment

The principal limitation of MAS NMR of half-integer–spin quadrupolar nuclei is that the second-order quadrupolar broadening for the central transition cannot be removed by MAS alone. A complete averaging of this interaction can be achieved by double-rotation NMR (Llor and Virlet 1998; Wu et al. 1990), where the sample is spun simultaneously about two different angles, or by dynamic-angle spinning (DAS) NMR (Llor and Virlet 1998; Mueller et al. 1990), where the spinning axis is changed from one angle to another during the rf pulse experiment. However, these techniques are technically quite demanding; thus, the number of applications have been rather limited. A reduction of the second-order quadrupolar interaction can also be achieved by the MQ MAS NMR experiment, which has the advantage that spinning only around a single axis (the magic angle) is required, utilising the dependencies of the spin-dependent parts of the second-order Hamiltonian to average out this interaction (Frydman and Harwood 1995; Medek et al. 1995).

The possibility of eliminating second-order quadrupolar effects in a 2D experiment may be visualised from the energy-level diagram for a spin $I = 5/2$ shown in Figure 6.19a. This diagram indicates that the central (1/2 ↔

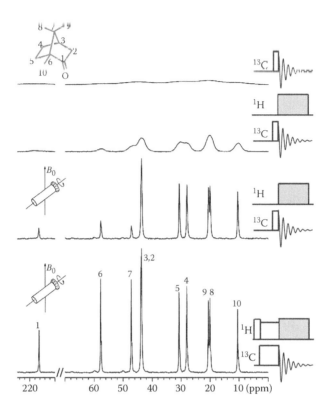

Figure 6.18 ^{13}C MAS NMR spectra (7.1 T) of camphor (C$_{10}$H$_{16}$O). Top to bottom: static single-pulse experiment, static single-pulse experiment with ^{1}H decoupling, static single-pulse experiment with ^{1}H decoupling and MAS and static single-pulse experiment with ^{1}H decoupling and CP/MAS.

$-1/2$), triple ($m = 3/2 \leftrightarrow m = -3/2$) and quintuple ($m = 5/2 \leftrightarrow m = -5/2$) transitions are unaffected by the first-order quadrupolar interaction. Furthermore, these transitions are equally affected by the second-order quadrupolar interaction which is proportional to $\Delta = \left\| H_Q^{(2)} \right\| / \left\| H_Z \right\|$. The MQ MAS technique utilises that evolution of either triple-quantum (3Q) or quintuple-quantum coherences under the effect from Δ for a time (t_1) can be refocused by conversion of these transitions into single-quantum coherences proportional to $-\Delta$. In practise the 3Q MQ MAS experiment is much more sensitive than the quintuple-transition MQ MAS NMR experiment; thus, only 3Q MAS ($p = 3$) is considered in the following part. The basic two-pulse sequence for the MQ MAS NMR experiment is shown in Figure 6.19b, where the 3Q coherences are created by the first rf pulse and converted into single-quantum coherence by the second pulse. The evolution of single-quantum coherence will result after a certain time ($t_2 = kt_1$) in an echo, where the effects from the second-order quadrupolar interaction have been refocused. Here, k

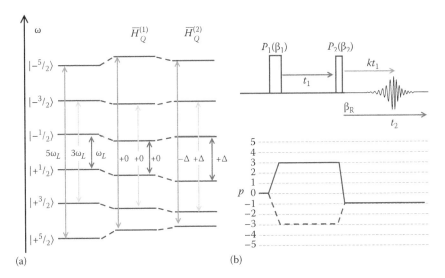

(a)

(b)

Figure 6.19 (a) Energy-level diagram for a spin $I = 5/2$ nucleus, illustrating the effects from the first- and second-order quadrupolar interactions on the single-quantum, triple-quantum and quintuple-quantum transitions. (b) rf pulse scheme for the basic two-pulse MQ MAS experiment, selecting the coherence transfer pathway shown below for a triple-quantum MQ MAS experiment for an $I = 5/2$ nucleus, where $p = 2m$ is the order of the coherence. The echo (solid lines) and antiecho (dashed lines) pathways are selected by phase cycling of the two pulses (P_1 and P_2) and of the receiver (β_R). The most simple phase cycle for the echo pathway is $\beta_1 = 0°, 60°, 120°, 180°$ and $240°$; $\beta_2 = 0°$ and $\beta_R = 180°, 0°, 180°, 0°, 180°$ and $0°$.

depends on the nuclear spin (I) and the type of MQ transition employed in the experiment. For $I = 3/2$ and $5/2$ nuclei, $k = 7/9$ and $19/12$ for the 3Q MAS experiment, respectively (Medek et al. 1995). The evolution time (t_1) is incremented in steps of $1/sw_1$, where sw_1 is the spectral width in the indirect dimension, and the pulse and receiver phases are cycled to obtain the desired coherence transfer pathways (Massiot et al. 1996; Medek et al. 1995). In addition, two FIDs are acquired for each increment with a 90°/p phase shift of the first pulse in the second FID to obtain phase sensitivity in the indirect dimension by the hypercomplex method (States et al. 1982). The array of FIDs obtained for each increment is subjected to a 2D Fourier transformation, which gives a correlation spectrum, where 3Q coherences are correlated with single-quantum coherences, corresponding to the isotropic (indirect, F1) and anisotropic (direct, F2) dimensions of the 2D spectrum. The 2D Fourier transformation needs to be combined with a shearing transformation in order to obtain 2D peak shapes that follow the F1 and F2 axes, since $k \neq 1$ (for $sw_1 = sw_2$) (Goldbourt and Madhu 2002; Medek et al. 1995). The shearing transformation is also used in 2D DAS

NMR experiments (Grandinetti et al. 1993) and it is often combined with Fourier transformation software on commercial spectrometers. The phase sensitivity in the indirect dimension of the basic two-pulse sequence (Figure 6.19b) may be improved by the z filter version (Amoureux et al. 1996a) which stores the single-quantum magnetisation along the z axis and includes a short delay before it is transferred into detectable magnetisation by a soft $90°/(I + 1/2)$ pulse. Several modifications of the MQ MAS experiment have been reported, which principally aim at improving the efficiency of the 3Q excitation and the conversion into single-quantum coherences. These processes depend strongly on the applied rf field strength and the magnitude of the quadrupole interaction, where high rf fields ($\gamma B_1/2\pi > 100$ kHz) gives an improved excitation for sites experiencing strong quadrupole interactions. For the 3Q MAS NMR experiment, the optimum pulse widths for the P_1 and P_2 pulses of the basic sequence (Figure 6.19b) are roughly 270° and 70° for $I = 3/2$ and 180° and 60° for $I = 5/2$, respectively (Amoureux et al. 1996b), where the pulse angles correspond to liquid-state pulses (i.e. pulses in the absence of the second-order quadrupolar interaction).

As an example, Figure 6.20 illustrates the ^{27}Al 3Q MAS NMR spectrum (14.1 T, ν_R = 40.0 kHz) of tricalcium aluminate, obtained with the

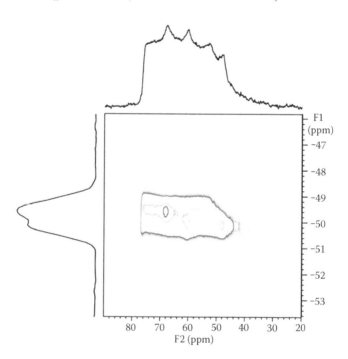

Figure 6.20 ^{27}Al MQ MAS NMR spectrum of Ca$_3$Al$_2$O$_6$ (14.1 T, ν_R = 40.0 kHz) acquired with the basic two-pulse sequence (Figure 6.19b), using $\gamma B_1/2\pi$ = 200 kHz, pulse widths of P_1 = 2.5 µs and P_2 = 0.84 µs and 96 increments in t_1 for sw$_1$ = sw$_2$ = 40 kHz.

two-pulse sequence, using $\gamma B_1/2\pi = 200$ kHz and pulse widths of $P_1 = 2.5$ μs and $P_2 = 0.84$ μs, and 96 increments in t_1 for $sw_1 = sw_2 = 40$ kHz. The contour plot along with the projections onto the F1 and F2 axes clearly illustrate the line-narrowing effect by MQ MAS for the two different tetrahedral AlO_4 sites. These two sites exhibit very similar ^{27}Al chemical shifts and quadrupole couplings (Skibsted et al. 1991a), which accounts for only a partial resolution of the two sites in the isotropic dimension of the MQ MAS spectrum. The isotropic shifts in the F1 dimension of the MQ MAS spectrum represent scaled values of the isotropic chemical shift and the isotropic second-order quadrupolar shift, following the expressions

$$\delta_{3Q}^{calc} = \frac{17}{8}\delta_{iso} + \frac{1}{32}\frac{C_Q^2\left(1+\eta_Q^2/3\right)}{v_L^2}\times 10^6 \text{ (in ppm)} \quad \text{for } I = 3/2, \quad (6.44)$$

$$\delta_{3Q}^{calc} = -\frac{17}{31}\delta_{iso} + \frac{3}{1550}\frac{C_Q^2\left(1+\eta_Q^2/3\right)}{v_L^2}\times 10^6 \text{ (in ppm)} \quad \text{for } I = 5/2 \quad (6.45)$$

for the 3Q MQ MAS experiment (Massiot et al. 1996; Goldbourt and Madhu 2002). For tricalcium aluminate, $\delta_{3Q}^{exp} = -49.9$ ppm and $\delta_{3Q}^{exp} = -50.3$ ppm are determined from the spectrum in Figure 6.19, which match well the calculated values, $\delta_{3Q}^{calc} = -49.8$ ppm and $\delta_{3Q}^{calc} = -50.4$ ppm, using the δ_{iso}, C_Q and η_Q data for tricalcium aluminate (Table 6.3) along with Equation 6.45. It is noted that referencing of the isotropic dimension in the MQ MAS spectrum is not straightforward and requires consideration of the type of experiment, the k factor and the difference between the transmitter frequency and the spectral reference (see Massiot et al. [1996] and Millot and Man [2002] for further details). Thus, several spectra have been published where the isotropic dimension is not properly referenced but only given in frequency units relative to the transmitter frequency. Moreover, it is noted that tricalcium aluminate does not represent a suitable sample for setting up and testing the MQ MAS experiment due to the strong quadruple couplings for the two Al sites. For such experiments, gibbsite (γ-Al(OH)$_3$) could alternatively be used (Faucon et al. 1998a; Skibsted et al. 2002), since this aluminate contains two octahedral Al sites with much smaller quadrupolar couplings. However, high-power ^1H decoupling is required for gibbsite both during the evolution (t_1) and detection (t_2) periods, since the heteronuclear dipolar coupling scales with the coherence order in the evolution period. Finally, it is noted that ^{27}Al MQ MAS NMR has been used to study the distribution of aluminium environments in calcium aluminate hydrates (Faucon et al. 1998a) as well as the types of Al sites present in synthesised C-A-S-H phases (Faucon et al. 1998b).

6.3 NUCLEAR MAGNETIC RESONANCE INSTRUMENTATION AND SIMULATION OF NUCLEAR MAGNETIC RESONANCE SPECTRA

High-resolution solid-state NMR spectrometers including superconducting magnets with field strengths (^{1}H frequencies) ranging from 4.7 T (200 MHz) to 23.5 T (1.0 GHz) are currently commercially available. The NMR experiments described in the preceding section can be applied on all systems. However, the equipment costs increase nearly exponentially with the magnetic field strength; thus, very high-field spectrometers are generally devoted to the most delicate experiments, for example, studies of low-γ nuclei in low natural abundance (e.g. ^{33}S and ^{43}Ca), admixtures in low concentration or surface interactions on cement particles where sensitivity is a major challenge. ^{29}Si is the most widely studied NMR nucleus in cement materials, which, in addition to its chemical importance, may reflect that ^{29}Si MAS NMR spectra of good quality may be obtained on a wide range of systems including 'low-field' (4.7–9.4 T) spectrometers. This section contains some general considerations regarding the experimental hardware, sample preparation and simulation tools for analysis of solid-state NMR spectra relevant to cement chemistry.

6.3.1 The magnet

Almost all high-resolution NMR spectrometers include a superconducting magnet (Figure 6.21), where the magnetic field is generated by a constant current in a superconducting coil, operating in persistent mode with no external interference. The wires are typically made from alloys of Nb or Sn (e.g. Nb_3Sn and NbTi) and superconductivity is achieved only at very low temperatures, which implies that the coil is immersed in liquid He ($T \sim$ 4.2 K). The helium can is covered by vacuum jackets and an outer larger can of liquid N_2, which serves to reduce the boil-off of He. Once established, the superconducting magnet can run for many years (>25 years), the only maintenance and running costs being the supply of liquid cryogens. The helium can needs typically refilling every 2–6 months (depending on the system), whereas N_2 is supplied more frequently (every 1–3 weeks), and both service operations take on the order of 1 hour. The magnet also includes superconducting shim coils, whose currents are adjusted during installation, and serve to generate magnetic field gradients that compensate for inhomogeneities in the main field, in order to obtain a very homogeneous magnetic field in the centre of the magnet. In addition, the magnet bore contains a set of room-temperature shim coils, whose currents can be adjusted externally to optimise the magnetic field homogeneity for specific probes and samples. The bore of the magnet has typically a diameter of 39–52 mm ('narrow bore') on liquid-state NMR spectrometers, whereas larger bore diameters (e.g. 89 mm ['wide bore']) often are preferred for

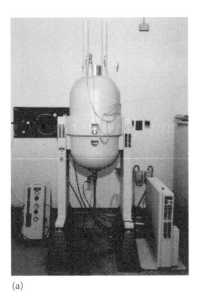

(a)   (b)

Figure 6.21 Superconducting NMR magnets with magnetic fields of (a) 4.7 T (200 MHz for ^{1}H) and (b) 22.3 T (950 MHz). The 4.7 T magnet in (a) represents Varian's 'R2D2' design from the mid-1980s and has a height and a weight of 160 cm and 136 kg, respectively. The 950 MHz magnet in (b) was produced by Bruker in 2014 and has a height of ~4 m and a weight of 7200 kg. The magnet is situated in a 1 m deep pit and it is surrounded by a staircase which is used for manual sample inject at the top of the magnet and the refill of cryogens.

solid-state NMR experiments since more space may be required for the mechanical rotation (MAS stators) and the electronic components for high rf-field operations. However, MAS probes have been designed to accommodate narrow-bore magnets, allowing the spectrometer to be used for both liquid- and solid-state applications.

The magnetic field homogeneity should be better than 1 part in 10^8–10^9 of the main field and the drift in magnetic field should be less than 1 part in 10^7 per day (MacKenzie and Smith 2002). In liquid-state NMR, the magnetic-field homogeneity is routinely optimised for the individual samples in a deuterated solvent, allowing spectral line widths <0.1 ppm for ^{1}H and ^{13}C to be observed. Moreover, small variations in the main magnetic field are continuously being accounted for by a ^{2}H lock system, where the ^{2}H frequency is measured and kept fixed during the experiment. For solid-state NMR experiments, the spectral lines are generally much broader than those observed in liquid-state NMR and shimming of the system is seldom performed for each sample but more likely from time to time for the individual MAS probes. Moreover, magnetic field drift is generally not considered in an interactive manner with a ^{2}H lock system, but corrections are made after acquisition of the NMR spectra.

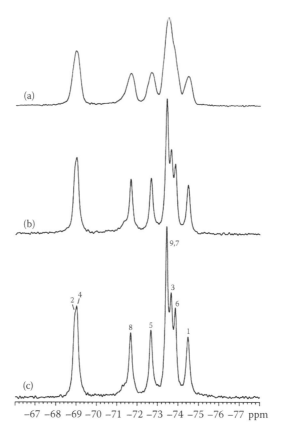

-67 -68 -69 -70 -71 -72 -73 -74 -75 -76 -77 ppm

Figure 6.22 ^{29}Si MAS NMR spectra (14.1 T, ν_R = 6.0 kHz and dI = 120 s) of a synthetic sample of triclinic (T1) tricalcium silicate (Ca$_3$SiO$_5$) obtained with (a) a standard set of room-temperature shim parameters and (b) after shimming of the probe for ^{1}H on a sample of H$_2$O:D$_2$O (1:10 volume/volume). Six hundred scans were accumulated for the spectrum in (b). (c) Same as (b) but for an array of five subspectra (each 120 scans), taking into account a constant drift in the magnetic field before summation (see text). The numbers indicate the assignment of the individual resonances to the nine inequivalent Si sites in triclinic Ca$_3$SiO$_5$. (From Skibsted J. et al., *Chem. Phys. Lett.* 172, 279–283, 1990.)

As an example, Figure 6.22 shows ^{29}Si MAS NMR spectra (14.1 T) of a synthetic sample of tricalcium silicate (Ca$_3$SiO$_5$, T1 polymorph), obtained with a standard set of shim parameters for the specific 7 mm CP/MAS probe and after shimming the probe for ^{1}H on a liquid sample (1:10 mixture of H$_2$O:D$_2$O) prior to acquisition of the MAS spectrum. The ^{29}Si spin–lattice relaxation times for synthetic Ca$_3$SiO$_5$ are quite long, requiring long acquisition times. Thus, effects from drift of the magnetic field may affect the observed line widths of the resonance. In the present case, the magnetic drift of the 14.1 T magnet is rather stable and corresponds to 56 Hz per day for ^{1}H at 599.5 MHz. In the absence of a ^{2}H lock system for the MAS probe, a correction for this drift can be performed by acquisition of an array of spectra with few scans whose frequencies are systematically corrected for the constant drift in magnetic field, prior to the summation that gives the overall signal averaging. Although the effect seems minimal in the present case, this procedure is illustrated in Figure 6.22c, where the total acquisition has been divided into five subspectra and each subjected to magnetic-field drift correction upon summation.

6.3.2 The nuclear magnetic resonance console

The NMR console includes all electronic parts for pulse programming, transmission of rf pulses, detection, signal amplification and data processing. Most systems contain two to four separate transmitter systems ('channels') which are restricted either to specific frequencies (e.g. ^{1}H, ^{13}C and ^{15}N) or are able to cover a broad band of low frequencies (e.g. roughly $v_L \leq v_L(^{31}\text{P})$). A diagram illustrating the main parts of a two-channel system (^{1}H and X) is shown in Figure 6.23, with a setup for detection of the X-channel nucleus. Each channel includes a frequency synthesiser that produces a low-power rf signal with a well-defined frequency (ω_0). This signal enters a pulse-gate unit, which is a fast switch that splits the continuous signal from the synthesiser into short time slots (approximately microseconds) of the rf waveform (i.e. pulses). These signals then enter a high-power rf amplifier that amplifies the rf amplitude of the pulses by several orders

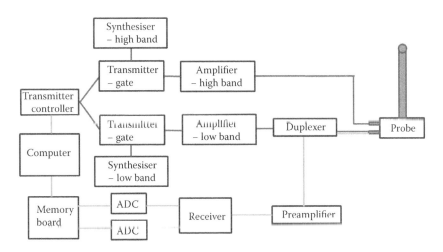

Figure 6.23 Schematic drawing of the basic elements in the NMR console of a two-channel system, operating in broadband (X)–high band (^{1}H/^{19}F) mode for detection of the X nucleus. Transmitted signals are indicated in red and detected signals are in green. ADC: analogue-to-digital converter.

of magnitude before the signals go into the NMR probe. The signal for the observe nucleus passes a duplexer before the probe, which provides a separation of the strong rf pulse that enters the probe from the very weak signal in the opposite direction, generated as the response of the pulse from the nuclear spins with the frequency $\omega_0 + \Delta\omega$. The latter signal is enhanced by a preamplifier before it enters the receiver, where the frequency of the signal is changed to an intermediate frequency range ($\omega_{if} + \Delta\omega$) by multiplying the observed signal with a generated signal of frequency, $\omega_{mix} = \omega_0 - \omega_{if}$, and filtering out the $\omega_{if} + \Delta\omega$ component. Thereby, signals corresponding to different Larmor frequencies can be detected with the same filters and analogue-to-digital converters (ADCs). The mixed-down signal ($\omega_{if} + \Delta\omega$) is now split into two parts, which are compared with an intermediate-frequency signal that is either in phase with ω_{if} or 90° out of phase with ω_{if}. The two different parts enter the filters, which ensure that only signals with the desired frequencies pass through. The in-phase– and the 90°-phase–shifted signals go then into two separate ADCs, which digitise the alternating current signal with a certain rate and transfers the digital signal to the computer memory for storage and signal accumulation. This detection procedure is called quadrature detection, where the two different outputs correspond to the real and imaginary parts of the complex signal (i.e. the FID). For spectra acquired with quadrature detection, the frequency in the centre of the spectrum corresponds to the transmitter frequency (ω_0) and the spectrum exhibits a spectral width (sw) defined by the dwell time (dw) of the ADCs, i.e. sw = 1/dw. Solid-state spectrometers often contain two different receiver/ADC systems with different digitisation rates, used for acquisition of spectra with normal spectral widths (sw = 100–500 kHz) and for wideband spectra with spectral widths in the range 1–5 MHz.

6.3.3 Magic-angle spinning probes

High-resolution NMR spectra of rigid solids can generally be achieved by MAS, where the sample is spun by typically 3000 to 20,000 revolutions per second (Hz) at the magic angle $\theta_m = 54.736°$ relative to the magnetic field. The sample, as a fine powder, is packed in a NMR rotor, which enters the stator of the NMR probe (Figure 6.24). Generally, the stator includes two supplies of air gas, where the bearing gas via a number of small nozzles/jets in the stator lifts the rotor from the inner walls of the stator. Spinning is obtained by the drive source, which through small jets at either the top or the bottom (or both) of the stator hits the turbine cap(s) of the rotor. The homebuilt Jakobsen probes (Daugaard et al. 1988), illustrated in Figure 6.23, employ only an upper end cap whose vanes drive the rotor, whereas other designs have pencil-like rotors with a turbine cap at the bottom of the rotor (Andrew 1981). The maximum attainable spinning speed is achieved when the peripheral velocity of the rotor/end cap approaches the speed of sound (346 ms^{-1} at 25°C in air). Obviously, this limit is achieved only for

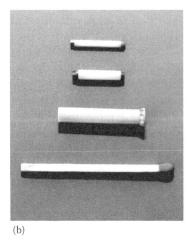

(a)   (b)

Figure 6.24 (a) Spinner housings of homebuilt MAS NMR probes for 4 and 7 mm outer-diameter rotors. (From Daugaard, P. et al., High speed spinner for NMR spectrometer. US Patent 4739270, April 19, 1988.) The outer shields of the probes have been removed. The cupper coil for rf excitation and detection can be seen in the centre of the 7 mm stator (left). (b) Rotors with a diameter of 4 (homebuilt), 2.0 (Bruker) and 1.6 mm (Varian) capable of spinning at 20, 35 and 45 kHz, respectively. The 4 mm rotor is made of Si_3N_4 and has a sample volume of 80 µL, whereas the 2 and 1.6 mm rotors are made of partially stabilised zirconia with sample volumes of 12 and 8.1 µL, respectively.

specific rotor, stator and end-cap materials capable of handling the stresses under such high spinning speeds. MAS rotors are often made of high-strength ceramics such as silicon nitride (Si_3N_4) or partially stabilised zirconia (PSZ), while the end caps are produced from polymers such as Vespel, Kel-F and poly(methyl methacrylate) (Doty 1996). Thus, in practise the upper spinning speed, provided by the probe manufacturer, is defined by the type of stator, the rotor material, the rotor wall thickness and, to some extent, the density of the sample under investigation. Moreover, the air-bearing and drive-gas settings for achieving different spinning speeds are normally provided by the manufacturer as well. For most NMR systems, the spinning speed can be controlled automatically via a MAS controller unit, often incorporated in the NMR console, by monitoring the spinning speed and regulating the pressure of the drive gas. A high and long-term stability in spinning speed (± 1 Hz) requires that both the air-bearing and drive-gas supplies are very stable prior to the pressure regulation. This can be achieved by incorporating two to three ballast tanks in series in the line between the air compressor and the air inlet of the MAS controller unit (Jakobsen et al. 2007).

Double- and triple-resonance MAS probes are commercially available and can be designed for the observation of specific NMR nuclei/spin systems.

The common double-tuned CP/MAS probes have one high-frequency channel ($^1H/^{19}F$) and a broadband channel for lower frequencies, both utilising one common coil surrounding the central part of the rotor, as an inner part of the stator. A range of low frequencies can generally be achieved for the broadband channel by introducing fixed capacitors or capacitor sticks into the double-resonance rf circuitry of the probe. Tuning and match for each frequency can be optimised by variable capacitors (located near the stator; Figure 6.24a), which are connected to adjustable rods at the bottom of the probe. Some MAS probes have a fixed (soldered) setting of the magic angle, whereas other probes have a mechanical adjustment of the stator for exact setting of the magic angle. For the latter type of probes, a highly accurate magic-angle setting (< ±0.001°) can be obtained spectroscopically by optimising the intensity of the rotational echoes in the time-domain signal (FID) or by minimising the spinning sideband line widths in the frequency-domain spectra. Suitable samples for this adjustment is the ^{79}Br MAS NMR spectrum of KBr (Frye and Maciel 1982) or the ^{23}Na MAS NMR spectrum of $NaNO_3$ (Jakobsen et al. 1989), both for which high S/N can be obtained with two to eight scans. Every manual handling of the probe and sample change may affect the setting of the magic angle; thus, a magic-angle setting is generally performed from time to time. A highly precise setting of the angle is particularly important for wideband experiments where manifolds of spinning sidebands are detected (cf. Figures 6.6 and 6.10).

The upper limit for the peripheral velocity of the rotors implies that higher spinning speeds can be obtained by lowering the rotor diameters. However, this results in a smaller sample volume and thereby a reduction in sensitivity. Thus, very high spinning speeds should be employed only for systems with strong internal spin interactions (e.g. C–H dipolar couplings), whereas lower MAS rates are sufficient for spin systems with weaker interactions. For example, the ^{29}Si CSA and $^{29}Si-^1H$ dipolar couplings are generally strongly reduced by spinning speeds of ν_R = 3–6 kHz, implying that these experiments are advantageously performed in large-diameter rotors (5–10 mm) with larger sample volumes.

Finally, it should be mentioned that the spinning of the sample has an impact on the actual sample temperature, as a result of frictional heating (Bjorholm and Jakobsen 1989), and sample temperatures of 30–50°C is common for high-speed spinning experiments, using bearing and drive gas at ambient temperature. The actual sample temperature for a specific setup and given spinning speed can be determined spectroscopically by measuring the change in ^{207}Pb chemical shift for $Pb(NO_3)_2$, $\Delta\delta$ = 1.32 ppm/K (Bielecki and Burum 1995), and the sample temperature can be adjusted by regulating the temperature of the bearing gas. Special designs of variable-temperature (VT) MAS probes are available where the most common probes can operate from roughly –150 to 200°C. However, special probes for VT MAS experiments at both higher (<600°C) and lower (> –196°C) temperatures have also been produced. Probably, the most critical issue of VT NMR experiments is not to damage the magnet, since neither the inner-bore room-temperature shim system or the fittings (O-rings) of the inner-bore dewars should be exposed to large temperature variations.

6.3.4 Sample handling and preparation

MAS NMR experiments are generally performed on dry powders and require sample volumes of typically 20–500 µL, depending on the size of the NMR rotors. A main advantage of the technique is that it is nondestructive as long as the samples are stable to temperature variations in the range of 20–50°C for ambient-temperature experiments. If this is not the case, MAS NMR experiments at a well-defined temperature can be performed with VT MAS equipment. The particle size and particle size distribution for the sample under investigation is not critical, the principal requirement being that the NMR rotors can be packed with the sample in a homogeneous manner, allowing stable spinning to be obtained. Usually, it is sufficient to grind hardened paste samples in an agate mortar, providing particles of 100 µm and smaller. The packing tools are rather simple, at least for rotor sizes of 4 mm and above; thus, the operation can easily be carried out in a glove box for air-sensitive samples. Moreover, most rotor end caps include one or more O-rings and can often be considered airtight at least for the time of a typical NMR experiment.

The low sensitivity, in particular for NMR studies of dilute spins such as ^{29}Si, implies that large rotors/sample volumes should be employed if possible. Moreover, low sensitivity for such systems results in the method being less amenable to studies of concrete samples with large amounts of inert fillers. For such experiments, paste samples should preferentially be used. In studies of hydration kinetics, the hydration process should be stopped chemically and the samples gently dried, prior to the NMR experiment. For the hydration kinetics of the principal silicate phases in portland cement, the hydration process can be stopped by removing the mobile (liquid) water from the hydrating system by suspending the ground paste sample for a certain time (e.g. 15–60 min) in an organic solvent, the most common being ethanol, isopropanol and acetone. After this process, the suspension is filtered, eventually further washed with the solvent and then vacuum dried or dried gently in an inert atmosphere in a desiccator (for further details on such procedures, see Section 1.7).

The large centrifugal forces induced by the sample spinning may have an impact on the results from in situ MAS experiments following the hydration of paste samples, since the sample spinning will result in an inhomogeneous distribution of water and solid material in the rotor. Moreover, it may also be difficult to obtain a stable probe tuning, in particular for the ^{1}H channel, for slurries or wet cement pastes. The hydration of a tricalcium silicate paste has been followed by in situ ^{29}Si MAS and ^{29}Si{^{1}H} CP/MAS NMR using a 7 mm rotor system and a spinning speed of ν_R = 1.5 kHz (Brough et al. 1994). A tricalcium silicate sample enriched to nearly 100% was used in that study to improve S/N and reduce the instrument time. Moreover, the paste was stored in rotor inserts made from Delrin™ with push fit lids and sealed after packing with a fast-setting epoxy resin. Comparison of spectra for static and spinning (ν_R = 1.5 kHz) pastes hydrated for 18 h revealed a slightly lower degree of tricalcium silicate reaction for the spinning sample and the observation of an additional resonance ($\delta(^{29}Si)$ = −82.5 ppm) in the $^{29}Si\{^1H\}$ CP/MAS experiment for the spinning sample. Although this resonance constituted only a small percentage of the total intensity of the ^{29}Si MAS NMR spectrum, it reflects that the large centrifugal forces induced by the MAS technique may have an impact on the cement hydration chemistry. This has also been observed in ^{27}Al MAS NMR spectra (ν_R = 4.2 kHz, 7.5 mm pencil rotors) of a hardened portland cement sample exposed to a 5.0 M NaCl solution for 6 months, where spectra acquired for 1 hour over a period of 20 hours showed that the strong centrifugal forces resulted in a gradual decomposition of the AFt phase, whereas the resonance from Friedel's salt remained unaffected by the sample spinning under these conditions (Jones et al. 2003).

Finally, attention should be paid to the content of paramagnetic species in the sample, which may affect the sample spinning and, for high contents, result in spinning failure or extinction of the rotors into the centre of the magnet for open probe systems (cf. Figure 6.24a). For iron-containing samples, it is recommended not to analyse samples with a bulk Fe_2O_3 content above roughly 10 wt.%. For heterogeneous samples such as fly ashes, the most iron-rich grains may be removed from the sample, prior to the NMR experiment, by exposing the sample to the surface of a strong permanent magnet.

6.3.5 Acquisition of basic magic-angle spinning nuclear magnetic resonance spectra

Before starting a single-pulse MAS NMR experiment, the experimentalist should consider the following basic parameters of the NMR experiment for a given sample and spin nucleus:

1. A calibration of the rf field strength should be performed for a given NMR probe and pulse amplification in order to determine $\gamma B_1/2\pi$ and the length of a 90° pulse (τ_{90}), see Figure 6.2. From this calibration, the desired flip angle (pulse length) can be chosen.
2. A spectral width, covering the full chemical shift range, including the spinning sidebands, should be chosen along with the number of data acquisition points that is sufficient to detect the full decay of the FID. For ^{29}Si MAS NMR experiments, spectral widths of 25–50 kHz are sufficient along with an acquisition of 1024–4096 data points.
3. A short delay (pulse dead time) between the rf pulse and the data acquisition is generally a part of the pulse sequence and it is introduced

to protect the receiver from any breakthrough of the rf pulse. This delay is typically on the order of 5–25 µs, depending on the probe, the spectral width and the observe nucleus.

4. The amplification of the signal prior to digitisation, generally denoted the receiver gain, should be optimised and then kept fixed if a series of identical experiments are acquired. A too high value will result in truncation of the FID after digitisation and a very low value will reduce the overall S/N of the spectrum.

5. The relaxation delay should be estimated, either from an array of spectra employing increasing delay times or from an estimate of the T_1 relaxation times for the different species in the sample, in order to obtain quantitatively reliable NMR intensities.

6. The number of signal accumulations (scans) should be estimated from the concentration of spins to be detected in the sample and the desired S/N. Generally, a large number of scans can be chosen, as the process of signal averaging can be followed during the acquisition process and the experiment can be stopped when an acceptable S/N is achieved.

7. The digital resolution of the frequency domain spectrum can be improved by zero filling, where a large number of zeros are appended to the decayed FID, prior to Fourier transformation. At the same time, the S/N of the spectrum can be improved by applying an exponential or Gaussian decay function to the FID before Fourier transformation. This artificial line broadening will obviously affect the line widths of the resonances and as a rule of thumb, the artificial line broadening (in hertz) should not exceed one-half of the natural line width for the most narrow line in the spectrum.

A number of additional initial steps must be performed for multiple-pulse and double-resonance NMR experiments, which will not be outlined here. When NMR spectra are reported in publications, the basic parameters and spectrometer characteristics listed in Table 6.4 should be given in the experimental section or in the figure captions.

6.3.6 Simulation of isotropic magic-angle spinning nuclear magnetic resonance spectra

In high-speed MAS NMR spectra of spin $I = 1/2$ nuclei, all intensity for a given structural site is gathered in the centre-band resonance. Ideally, these resonances exhibit a symmetric line shape with a certain line width that originates from imperfections in the local structure, inhomogeneities in the external magnetic field, T_2 relaxation and residual anisotropic interactions such as dipolar couplings. For example, this situation can be achieved in most ^{29}Si MAS NMR experiments, performed at moderate spinning speeds ($\nu_R = 5$–10 kHz) and magnetic field strengths (e.g. 7.1–14.1 T). For such a system and if the spectrum is acquired with full spin–lattice relaxation

Table 6.4 Guidelines for reporting NMR spectra in scientific documents

Data collection properties and settings		Examples
Spectrometer	Manufacturer	Bruker, JEOL, Varian
	Model	Avance, Infinity, Direct-Drive
	Magnetic field strength (B_0)	4.7, 9.4, 18.8 T
NMR probe	Manufacturer	Bruker, Chemagnetics, Doty
	Rotor diameter and material	2 mm PSZ, 7mm Si_3N_4
NMR experiment	Pulse sequence	Single-pulse, CP, REDOR
	Applied spinning speed	$\nu_R = 5.0$ kHz
	rf field strength	
	• observe nucleus	$\gamma B_1/2\pi = 50$ kHz
	• decoupling/CP	$\gamma B_2/2\pi = 75$ kHz
	Excitation pulse	
	• Pulse angle	45°, 90°
	or pulse width	2.5, 5.0 μs
	Relaxation delay	5, 30 s
	Number of accumulations	512 scans
	Chemical shift reference	Tetramethyl silane, 1.0 M aqueous AlCl$_3$·6H$_2$O
	Temperature	Ambient, VT (50°C)
Simulations	Software	WinFIT, STARS, IGOR

for the individual components in the sample, the observed intensities are proportional to the molar fractions of the individual components. For well-resolved spectra, information about the relative intensities can be obtained simply by spectral integration. However, in several cases a significant overlap of resonances is observed and deconvolution methods need to be employed. Generally, these methods employ summations of a number of peaks with either Gaussian or Lorentzian line shapes, or a combination of these peak shapes, where the frequency, line width, intensity and peak shape (e.g. fraction of Lorentzian/Gaussian line shape) are considered as variable parameters in an optimisation to the experimental spectrum. Routines for deconvolutions, including optimisation (e.g. least-squares fitting) to the experimental spectra, are often incorporated in the spectrometer software.

Alternatively, the experimental spectra are transferred to common software packages for data processing, which generally include features for peak summation and different optimisation routines. In all cases, scripts may be developed by the user, which ease the deconvolution process in the analysis of a multitude of similar spectra.

In most cases the overlap of resonances is so severe that an unbiased model, where the spectrum is considered to be composed of a variable number of peaks for which the frequencies, line widths, intensities and peak shapes are allowed to vary without any constraints, can hardly be used in a least-squares fit to the experimental spectrum. In such cases, the analyst needs to set up a model for the system under consideration, for example, a model that defines the number of different species in the sample and also potentially constrains the peaks to the same overall line shape and to a limited range of values for the line widths. The model defines the input to the optimisation of peaks to the experimental spectrum and, obviously, it can be modified by the analyst during the refinement, for example, if additional components need to be accounted for.

As a rather straightforward example, Figure 6.25 illustrates the ^{29}Si MAS NMR spectrum of an alkali-activated metakaolin–silica blend (Si/Al = 2.0 and Na/Al = 1.5 molar ratios) cured for 1 week at 80°C. The formation of a zeolitic phase in this system is anticipated, which will result in up to five distinct resonances from the $Q^4(nAl)$ $(n = 0, 1, 2, 3, 4)$ environments in the framework structure. Thus, the five resonances from –87 to –107 ppm

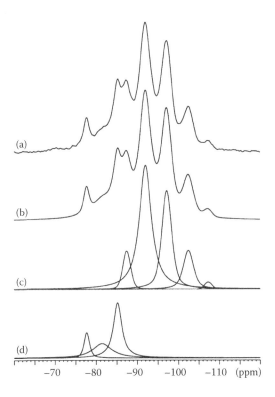

Figure 6.25 (a) ^{29}Si MAS NMR spectrum (7.1 T, ν_R = 7.0 kHz and 30 s relaxation delay) of an alkali-activated metakaolin–silica blend (molar ratios of Si/Al = 2.0 and Na/Al = 1.5) cured for 1 week at 80°C; (b) deconvolution of the experimental spectrum, assuming the presence of eight distinct Si sites and (c) optimised resonances for the five ^{29}Si originating from the zeolitic framework structure, corresponding to $Q^4(nAl)$; $n = 0, 1, 2, 3, 4$; (d) optimised resonances for the three ^{29}Si sites that are not a part of the framework structure.

are assigned to a zeolite component. In addition, two distinct peaks at –77.5 and –85 ppm are observed and an initial input to the simulation routine is established which includes seven peaks with fixed peak positions according to the experimental spectrum. A least-squares optimisation is then performed, optimising the line widths, peak shapes and intensities for the individual components, i.e. a fit with 21 variables. From the first optimisation, it is apparent that the simulation cannot account for the observed intensity in the range –79 to –83 ppm; thus, an eighth peak is added to the input file with a frequency in this region. A few additional optimisations are then performed, where the number of variables can be modified and restrictions on specific parameters can be introduced. For example, if the peak shapes from the first optimisation are acceptable, as judged from a visual comparison of the experimental and simulated spectra, they can be employed as fixed parameters in the subsequent optimisations. Moreover, an optimisation where the peak positions also are allowed to vary may improve the final result. The result from the deconvolution in Figure 6.25b is in agreement with the proposed model, since the peaks from the $Q^4(n\text{Al})$ sites exhibit very similar line widths and peak shapes, as expected since they are a part of the same basic structure. Thus, the deconvolution provides information about the molar fraction of the zeolite component in the alkali-activated sample and the relative intensities for the individual $Q^4(n\text{Al})$ peaks can be used to calculate the Si/Al ratio of the zeolite framework. An assignment of the three additional resonances in the range –77 to –85 ppm may be achieved from their ^{29}Si chemical shifts or, as in the present case, from additional $^{29}\text{Si}\{^1\text{H}\}$ CP/MAS and $^{29}\text{Si}\{^{27}\text{Al}\}$ REAPDOR NMR experiments (Tran 2011).

Distinct resonances from each crystallographic site in the crystal structure may hardly be resolved for the phases present in portland cement. For example, in ^{29}Si MAS NMR studies of anhydrous portland cements, it is not possible to resolve the resonances from the distinct SiO_4 tetrahedra in alite. The alite phase in commercial portland cements includes a number of impurity ions (e.g. Mg^{2+}, Al^{3+} and S^{6+}) which stabilise either the monoclinic M_I or M_{III} forms of alite, depending on their concentration. The crystal structure of the M_{III} form of alite includes 18 different SiO_4 tetrahedra in Q^0 environments and the resonances from these sites cannot be resolved, even for synthesised samples of M_{III} alite. However, ^{29}Si MAS NMR spectra of synthetic alites show that distinct and characteristic line shapes are observed for the different polymorphic forms (Bellmann et al. 2014; Skibsted et al. 1995b; Stephan et al. 2008), as illustrated in Figure 6.26 for the M_I and M_{III} forms. This can be utilised in the analysis of ^{29}Si MAS NMR spectra of commercial portland cements, using the following model: (1) The form of alite is first identified from a comparison of the overall line shape with those observed for the synthesised polymorphs. For the two white portland cements, studied in Figure 6.26, it is apparent that M_I alite is the dominating polymorph for cement wPc-I, whereas

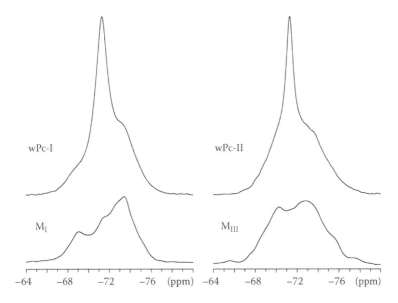

Figure 6.26 ^{29}Si MAS NMR spectra (9.4 T, ν_R = 6.0 kHz) of two anhydrous white portland cements, wPc-I and wPc-II, and of synthetic samples of alite in its monoclinic M_I and M_{III} forms.

the line shape for the M_{III} form matches well the line shape observed for cement wPc-II. (2) A deconvolution of the ^{29}Si MAS NMR spectrum for the synthesised alite is performed and the data from this procedure are stored as a subspectrum for the given form of alite. For the M_{III} form of alite a satisfactory deconvolution (Figure 6.26) can be achieved using nine different ^{29}Si peaks, although the crystal structure reveals the presence of 18 sites. This simplification is acceptable, since we need to account only for the intensity of the overall line shape. (3) A subspectrum for belite is constructed which consists of a single peak, following the crystal structure for this phase. (4) A routine is established which allows optimisation of the two subspectra to the experimental line shape for the portland cement. For the alite phase, this involves systematic changes in the intensities and line widths for the nine peaks constituting the subspectrum (i.e. a total of two variable parameters), whereas the frequency, line width, intensity and peak shape may be varied for belite. Thus, the applied model assumes that the spectrum is composed of contributions from belite and either the monoclinic M_I or M_{III} form of alite. The approach is illustrated for a white portland cement in Figure 6.27. The relative intensities, resulting from the deconvolution, give the molar fraction for the alite and belite phase, which can be converted into masses, assuming either stoichiometric Ca_2SiO_4 and Ca_3SiO_5 compositions or the phase compositions proposed by Taylor (1989) and employing the bulk SiO_2 content determined for the actual

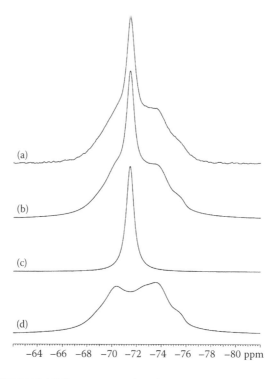

$$-64 \quad -66 \quad -68 \quad -70 \quad -72 \quad -74 \quad -76 \quad -78 \quad -80 \text{ ppm}$$

Figure 6.27 (a) ^{29}Si MAS NMR spectrum of a white portland cement acquired at 9.4 T, using a spinning speed of ν_R = 13.0 kHz and a 30 s relaxation delay; (b) optimised deconvolution of the spectrum in (a) employing the subspectra in (c) and (d); (c) subspectrum for belite and (d) subspectrum for M_{III} alite. The deconvolution in (b) gives the molar ratio I(belite)/I(alite) = 0.345. Combining this value with the bulk SiO_2 content of the cement (24.7 wt.%) gives quantities of 72.4 wt.% alite and 20.0 wt.% belite when Taylor's compositions (Taylor 1989) for these phases are employed.

cement from X-ray fluorescence spectroscopy or chemical analysis. More detailed descriptions of this approach to determining the quantities of alite and belite and correlations with alite and belite determinations from XRD-Rietveld refinements and Bogue calculations are given elsewhere (Poulsen et al. 2009; Skibsted et al. 1995b).

The deconvolution approach for anhydrous portland cement can be easily extended to hydrated portland cements and portland cement–SCM blends. For the latter systems, deconvoluted subspectra for the SCMs, e.g. slags, metakaolin and natural pozzolans, can be established and incorporated in the optimisation routine. As an example, Figure 6.28 illustrates the deconvolution of the ^{29}Si MAS NMR spectrum for a white portland cement hydrated for 2 weeks. The model used for this deconvolution includes subspectra of belite and M_{III} alite as well as single peaks for the Q^1, $Q^2(1Al)$ and Q^2 resonances of the C-S-H phase. For the C-S-H phase,

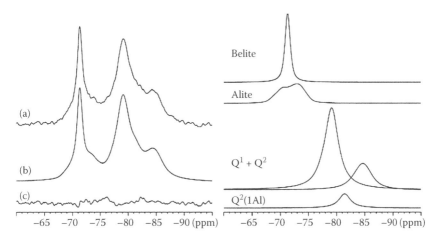

Figure 6.28 (a) ^{29}Si MAS NMR spectrum (9.4 T, ν_R = 6.0 kHz) of a white portland cement hydrated for 2 weeks; (b) optimised deconvolution, employing the subspectra shown in the right column; (c) difference plot between the ^{29}Si MAS NMR spectra in (a) and (b).

this is the simplest approach since different Q^1 sites (chain end groups and dimers) and Q^2 sites (paired and bridging tetrahedra) most likely are present in the structure. The resonance from these sites may heavily overlap; thus, the experimental spectrum does not give clear indications of more than three sites in the C-S-H region of the spectrum. Thus, this simple model is used in the analysis of the C-S-H intensities. The most difficult parameters in the present deconvolution are the parameters used for the Q^2(1Al) resonance, which overlaps strongly with the Q^1 peak. From deconvolution analyses for a range of white portland cement systems, we have found that a robust approach to coping with this problem is to constrain the values for the line width, peak shape and also, to some extent, the chemical shift for the Q^2(1Al) site only in the least-squares optimisations to the experimental spectra. In the present case, $\delta(^{29}$Si$)$ = −81.0 ppm, full width at half maximum = 2.5 ppm and a 50% Gaussian–50% Lorentzian line shape is used (Skibsted and Andersen 2013). Simulation of the spectra for anhydrous and hydrated portland cements, using the same set of subspectra, allows calculation of the degrees of reaction (H) for alite, belite and the SCMs, H = 1 − $I_x(t)/I_x(0)$, where $I_x(t)$ is the relative intensity for component x after the hydration time (t) and $I_x(0)$ is the corresponding relative intensity for the anhydrous mixture. This approach has the advantage that minor inaccuracies in the subspectra will be eliminated when H is calculated. For example, if the chosen subspectrum for alite results in an overestimation of this phase for the anhydrous mixture, this overestimation will be similar for all spectra; thereby, it will not affect the degree of hydration (H).

In the estimation of degrees of reaction, an alternative approach is the subtraction of a spectrum of the anhydrous component from the spectrum of the hydrated material. This approach has been utilised in studies of a hydrated portland cement–slag system (Dyson et al. 2007), for alkali-activated slags (Le Saout et al. 2011) and for synthetic glasses cured in saturated $Ca(OH)_2$ solutions (Moesgaard et al. 2010). In the last investigation, the broad and dominating component from the glass in the spectrum is removed, resulting in a difference spectrum that contains only resonances from the hydration products. The assumption of congruent dissolution was investigated for the hydrated portland cement–slag system, utilising selective dissolution of the silicate phases, leaving a residue assigned to unreacted slag (Dyson et al. 2007). The ^{29}Si MAS NMR spectrum of this residue was subsequently used in the deconvolution of the original spectrum of the hydrated blend. For the alkali-activated slags, it was found that a subspectrum based on the spectrum of the anhydrous slag gave the most satisfactory results in the deconvolutions (Le Saout et al. 2011).

6.3.7 Simulation of anisotropic magic-angle spinning nuclear magnetic resonance spectra

Supplementary structural information about different structural sites can be obtained from a determination of the anisotropic NMR parameters, characterising the CSA for spin $I = 1/2$ nuclei and the quadrupole coupling for spins with $I > 1/2$. These parameters may assist the assignment of the resonances to specific structural environments, for example, as observed for the ^{29}Si CSA parameters determined for calcium silicates and calcium silicate hydrates (Hansen et al. 2003), where the CSA parameters (δ_σ and η_σ) give an improved identification of Q^0, Q^1 and Q^2 sites as compared to the isotropic chemical shift. For quadrupolar nuclei, a determination of C_Q, η_Q and δ_{iso} for the individual sites is a prerequisite, since the centre of gravity of the centre band depends on these parameters and the applied magnetic field (cf. Equations 6.35 through 6.37). This should indeed be considered when spectra acquired at different magnetic fields are compared.

The CSA parameters for spin $I = 1/2$ nuclei can be determined either from analysis of the spinning sideband intensities in MAS NMR spectra or line shape analysis of static-powder NMR spectra, where both approaches are illustrated in Figure 6.6 for the ^{29}Si NMR spectra of jaffeite. For MAS NMR spectra, a simple approach is to evaluate the relative spinning sideband intensities using the Herzfeld and Berger approach (Herzfeld and Berger 1980) or to simulate the powder pattern using the expression for the first-order CSA interaction, averaged over all uniformly distributed crystallite orientations. Explicit expressions for these powder averages can rather easily be implemented in standard numerical optimisation packages. Moreover, several NMR software packages are available, in some cases even as a part of the spectrometer software, which allow a direct determination of the

CSA parameters by optimisation to intensities or line shapes observed in the experimental spectra. Common NMR simulation packages are WinFIT (Massiot et al. 1994), STARS (Bildsøe 1996) and SIMPSON (Bak et al. 2000). These packages can also be used in simulations of static and MAS NMR spectra of quadrupolar nuclei, utilising either the spinning sidebands from the satellite transitions (Figure 6.10) or the second-order quadrupolar line shape, observed for the central transition (Figure 6.8). However, in MAS NMR of less-ordered phases, it may be hard to extract a unique set of interaction parameters (e.g. C_Q, η_Q and δ_{iso}), since small variations in the local environments for the quadrupole nucleus leads to a dispersion in these parameters. For the satellite transitions, this results in bell-shaped envelopes of spinning sidebands, whereas the centre band from the central transition often exhibits an asymmetric line shape with a tail towards low frequency. For example, a bell-shaped manifold of spinning sidebands has been observed for the octahedral ^{27}Al site in monosulfate and ascribed to structural defects and disorder (Skibsted et al. 1993). This was justified by the construction of a similar overall envelope of spinning sidebands, obtained by an addition of 10 simulations, employing a small distribution in the C_Q and η_Q values. A more advanced simulation approach has been developed for the central-transition line shape, which implements a specific distribution of the EFG tensor elements (cf. Equation 6.29) using the Czjzek model (d'Espinose de Lecaillerie et al. 2008). This method allows deconvolution of the experimental spectra, providing values for the average isotropic chemical shift, the quadrupolar product parameter and a measure of the degree of structural disorder. The method has been demonstrated by the analysis of ^{27}Al MAS NMR spectra for a slag as well as the AFm and 'third aluminate hydrate' phases of a hydrated portland cement paste.

6.4 CONCLUSIONS AND PERSPECTIVES

The main advantage of solid-state NMR spectroscopy in studies of cement-based materials is its ability to detect and quantify amorphous and crystalline phases in an equal manner. This property and the fact that the detected NMR parameters are most sensitive to the local structure, roughly the nearest one to three coordination spheres, makes it a valuable and important supplement to XRD techniques that probe long-range order and crystalline phases. Moreover, the nuclear-spin selectivity often implies that unambiguous structural or quantitative information can be achieved for a subsystem of a complex multicomponent material. For example, ^{27}Al MAS NMR provides direct information about the aluminium coordination state and, for portland cement systems, it allows identification of aluminium in tricalcium aluminate, as guest ions in the alite and belite phases, in the AFm and AFt hydration products and as a part of the C-S-H structure, despite a low bulk Al_2O_3 content of 2–5 wt.% for such systems.

^{27}Al MAS NMR and ^{29}Si MAS NMR have made very important contributions over the last 3 decades to our current knowledge about the structure and composition of C-S-H phases. In that research, the full potential of MAS NMR spectroscopy has not been utilised yet. It is expected that future applications of a range of rf pulse sequences for probing homonuclear- or heteronuclear-spin connectivities and for distance measurements for dipolar coupled spins will bring new and more detailed information about the C-S-H structure.

Research in new SCMs and portland cement–SCM blends is increasing significantly and it is expected to further increase in the next decades, considering the global challenges that the cement industry faces for development of more sustainable concrete. The SCMs with the highest potential are all less-crystalline or amorphous phases and NMR spectroscopy will definitely play an important role in the exploration of these new SCMs and their reactivity in cementitious systems.

The continuous increase in magnetic field strength of commercial spectrometers implies that it will be possible to obtain new information for a range of low-γ spin nuclei, such as ^{33}S and ^{43}Ca, without isotopic enrichment. Higher magnetic field increases sensitivity and chemical shift dispersion and reduces quadrupolar effects for $I > 1/2$ nuclei, and all these features contribute to the possibility of acquiring ^{33}S and ^{43}Ca NMR spectra for real systems on an acceptable timescale. Another approach to significantly improve the sensitivity by up to two orders of magnitude is dynamic nuclear polarisation (DNP) (Maly et al. 2008), where a radical is introduced in the studied sample and magnetisation is transferred from the electron of the radical, excited by microwaves, to the nuclear spin under investigation. This technique is exposed to considerable research efforts at the moment since it will open the door for a range of new applications of NMR, for example, in studies of surface structures and reactions (Lelli et al. 2011). Although several applications of the DNP technique have already been published for inorganic oxidic materials, the technique is currently technically demanding as it requires MAS NMR operations at low temperatures, close to liquid nitrogen, and the successful incorporation of radical probe molecules in the material.

Limiting factors for wide and general applications of MAS NMR in studies of cement materials are still the rather high costs of the equipment, in particular for very high-field systems; the low sensitivity for the most important spin nuclei, resulting in long instrument times; and the fact that a rather high level of practical experience is required to maintain and operate the spectrometer. This prevents several companies from investing in MAS NMR equipment and most spectrometers are thus found at universities and research institutions. Although improvements of the spectrometers' interface for the users and of the equipment for sample spinning are still being made, it is not foreseen that a significant increase in the dissemination of the technique will occur during the next decade.

ACKNOWLEDGEMENTS

My former PhD students, Morten D. Andersen, Michael R. Hansen, Søren L. Poulsen, Tine F. Sevelsted, Thuan T. Tran and Kasper Enemark-Rasmussen, and lab technician Anne-Birgitte Johannsen are acknowledged for providing a major part of the experimental results shown in the figures in this chapter.

REFERENCES

Amoureux, J.-P., C. Fernandez and L. Frydman. 1996a. Optimised multiple-quantum magic-angle spinning NMR experiments on half-integer quadrupoles. *Chem. Phys. Lett.* 259, 347–355.

Amoureux, J.-P., C. Fernandez and S. Steuernagel. 1996b. Z filtering in MQMAS NMR. *J. Magn. Reson. Ser. A.* 123, 116–118.

Andersen, M. D., H. J. Jakobsen and J. Skibsted. 2002. Characterisation of the α–β phase transition in Friedels salt $(Ca_2Al(OH)_6Cl \cdot 2H_2O)$ by variable-temperature ^{27}Al MAS NMR spectroscopy. *J. Phys. Chem.* 106, 6676–6682.

Andersen, M. D., H. J. Jakobsen and J. Skibsted. 2003. Incorporation of aluminum in the calcium silicate hydrate (C-S-H) of hydrated Portland cements: A high-field ^{27}Al and ^{29}Si MAS NMR investigation. *Inorg. Chem.* 42, 2280–2287.

Andersen, M. D., H. J. Jakobsen and J. Skibsted. 2005. Effects of T_2 relaxation in MAS NMR spectra of the satellite transitions for quadrupolar nuclei: A ^{27}Al MAS and single-crystal NMR study of alum $KAl(SO_4)_2 \cdot 12H_2O$. *J. Magn. Reson.* 173, 209–217.

Andrew, E. R. 1981. Magic angle spinning in solid-state n.m.r. spectroscopy. *Phil. Trans. R. Soc. Lond. A.* 299, 505–520.

Andrew, E. R., A. Bradbury and R. G. Eades. 1958. Nuclear magnetic resonance spectra from a crystal rotated at high speeds, *Nature* 182, 1659–1659.

Apperley, D. C., R. K. Harris and P. Hodgkinson. 2012. *Solid-state NMR: Basic principles and practice.* New York: Momentum Press.

Bak, M., J. T. Rasmussen and N. C. Nielsen. 2000. SIMPSON: A general simulation program for solid-state NMR spectroscopy. *J. Magn. Reson.* 147, 296–330.

Bakhmutov, V. L. 2012. *Solid-state NMR in materials science – Principles and applications.* Boca Raton: CRC Press, Taylor & Francis.

Barberon, F., J.-P. Korb, D. Petit, V. Morin and E. Bermejo. 2003. Probing the surface area of a cement-based material by nuclear magnetic relaxation dispersion. *Phys. Rev. Lett.* 90, 116103.

Bell, G. M. M., J. Bensted, F. P. Glasser, E. E. Lachowski, D. R. Roberts and M. J. Taylor. 1990. Study of calcium silicate hydrates by solid state high resolution ^{29}Si nuclear magnetic resonance. *Adv. Cem. Res.* 3, 23–37.

Bellmann, F., J. Leppert, M. Görlach, M. Krbetschek, D. Damidot and H.-M. Ludwig. 2014. Analysis of disorder in tricalcium silicate by ^{29}Si NMR spectroscopy and additional methods. *Cem. Concr. Res.* 57, 105–116.

Bennett, A. E., C. M. Rienstra, M. Auger, K. V. Lakshmi and R. G. Griffin. 1995. Heteronuclear decoupling in rotating solids. *J. Chem. Phys.* 103, 6951–6958.

Bielecki, A., and D. P. Burum. 1995. Temperature dependence of ^{207}Pb MAS spectra of solid lead nitrate: An accurate, sensitive thermometer for variable temperature MAS. *J. Magn. Reson. Ser. A.* 116, 215–220.

Bielecki, A., A. C. Kolbert and M. M. Lewitt. 1989. Frequency-switched pulse sequences: Homonuclear decoupling and dilute spin NMR in solids. *Chem. Phys. Lett.* 155, 341–346.

Bildsøe, H. 1996. STARS User's Guide, Spectrum Analysis of Rotating Solids. Varian Assoc., Pub. No. 87-195233-00, Rev. A0296.

Bjorholm, T., and H. J. Jakobsen. 1989. ^{31}P MAS NMR of P_4S_3: Crystalline-to-plastic phase transition induced by MAS in a double air-bearing stator. *J. Magn. Reson.* 84, 204–211.

Blaine, R. L. 1960. Proton magnetic resonance (N. M. R.) in hydrated Portland cements. *Natl. Bur. Stand. Monogr.* 43, 501–511.

Blinc, R., M. Burgar, G. Lahajnar, M. Rozmarin, V. Rutar, I. Kocuvan and J. Ursic. 1978. NMR relaxation study of adsorbed water in cement and C_3S pastes. *J. Am. Ceram. Soc.* 61, 35–37.

Brough, A. R., C. M. Dobson, I. G. Richardson and G. W. Groves. 1994. In situ solid-state NMR studies of Ca_3SiO_5: Hydration at room temperature and at elevated temperatures using ^{29}Si enrichment. *J. Mater. Sci.* 29, 3926–3940.

Brunet, F., P. Bertani, T. Charpentier, A. Nonat and J. Virlet. 2004. Application of ^{29}Si homonuclear and ^{1}H–^{29}Si heteronuclear NMR correlation to structural studies of calcium silicate hydrates. *J. Phys. Chem. B* 108, 15494–15502.

Canet, D., G. C. Levy and I. R. Peat. 1975. Time saving in ^{13}C spin-lattice relaxation measurements by inversion-recovery. *J. Magn. Reson.* 18, 199–204.

Caravatti, P., G. Bodenhausen and R. R. Ernst. 1982. Heteronuclear solid-state correlation spectroscopy. *Chem. Phys. Lett.* 89, 363–367.

Colombet, P., and A.-R. Grimmer (eds.). 1994. *Application of NMR spectroscopy to cement science.* Amsterdam: Gordon & Breach Science.

Colombet, P., A.-R. Grimmer, H. Zanni and P. Sozzani (eds.). 1998. *Nuclear magnetic resonance spectroscopy of cement-based materials.* Berlin: Springer-Verlag.

Comotti, A., R. Simonutti and P. Sozzani. 1996. Hydrated calcium silicate and poly(vinyl alcohol): Nuclear spin propagation across heterogeneous interfaces. *Chem. Mater.* 8, 2341–2348.

d'Espinose de Lacaillerie, J.-B., C. Fretigny and D. Massiot. 2008. MAS NMR spectra of quadrupolar nuclei in disordered solids: The Czjzek model. *J. Magn. Reson.* 192, 244–251.

Daugaard, P., V. Langer and H. J. Jakobsen. 1988. High speed spinner for NMR spectrometer. US Patent 4739270, April 19, 1988.

Derome, A. E. 1987. *Modern NMR techniques for chemistry research.* Oxford: Pergamon Press.

Doty, F. D. 1996. Solid state probe design. In: *Encyclopedia of NMR spectroscopy*, Grant, D. M., and R. K. Harris (eds.). Vol. 7, 4475–4485. Chichester: Wiley.

Du, L.-S., and J. F. Stebbins. 2004. Calcium and strontium hexaluminates: NMR evidence that 'pentacoordinate' cation sites are four coordinated. *J. Phys. Chem. B* 108, 3681–3685.

Duer, M. J. (ed.). 2002. *Solid-state NMR spectroscopy – Principles and applications.* Oxford: Blackwell Science.

Dyson, H. M., I. G. Richardson and A. R. Brough. 2007. A combined ^{29}Si MAS NMR and selective dissolution technique for the quantitative evaluation of hydrated blast furnace slag cement blends. *J. Amer. Ceram. Soc.* 90, 598–602.

Engelhardt, G., and D. Michel. 1987. *High-resolution solid-state NMR of silicates and zeolites*. Chichester: John Wiley & Sons.

Faucon, P., T. Charpentier, D. Bertrandie, A. Nonat, J. Virlet and J. C. Petit. 1998a. Characterisation of calcium aluminate hydrates and related hydrates of cement pastes by ^{27}Al MQ-MAS NMR. *Inorg. Chem.* 37, 3726–3733.

Faucon, P., T. Charpentier, A. Nonat and J. C. Petit. 1998b. Triple-quantum two-dimensional ^{27}Al magic-angle nuclear magnetic resonance study of the aluminum incorporation in calcium silicate hydrates. *J. Am. Chem. Soc.* 120, 12075–12082.

Feike, M., D. E. Demco, R. Graf, J. Gottwald, S. Hafner and H. W. Spiess. 1996. Broad-band multiple-quantum NMR spectroscopy. *J. Magn. Reson. Ser. A* 122, 214–221.

Frydman, L., and J. S. Harwood. 1995. Isotropic spectra of half-integer quadrupolar spins from bidimensional magic-angle spinning NMR. *J. Am. Chem. Soc.* 117, 5367–5368.

Frye, J. S., and G. E. Maciel. 1982. Setting the magic angle using a quadrupolar nuclide. *J. Magn. Reson.* 48, 125–131.

Fung, B. M., A. K. Khitrin and K. Ermolaev. 2000. An improved broadband decoupling sequence for liquid crystals and solids. *J. Magn. Reson.* 142, 97–101.

Fyfe, C. A. 1983. *Solid-state NMR for chemists*. Guelph: C.F.C. Press.

Goldbourt, A., and Madhu, P. K. 2002. Multiple-quantum magic-angle spinning: High-resolution solid-state NMR spectroscopy of half-integer quadrupolar nuclei. *Monatshefte für Chemie*. 133, 1497–1534.

Grandinetti, P. J., J. H. Baltisberger, A. Llor, Y. K. Lee, U. Werner, M. A. Eastman and A. Pines. 1993. Pure-absorption-mode lineshapes and sensitivity in two-dimensional dynamic-angle spinning NMR. *J. Magn. Reson. A* 103, 72–81.

Greener, J., H. Peemoeller, C. Choi, R. Holly, E. J. Reardon, C. M. Hansson and M. M. Pintar. 2000. Monitoring of hydration of white cement paste with proton NMR spin-spin relaxation. *J. Am. Ceram. Soc.* 83, 623–627.

Grimmer, A.-R., F. von Lampe, M. Mägi and E. Lippmaa. 1985. High-resolution solid-state ^{29}Si NMR of polymorphs of Ca_2SiO_4. *Cem. Concr. Res.* 15, 467–473.

Grimmer, A.-R., and Zanni, H. 1998. ^{29}Si NMR study of chemical shift tensor anisotropy of tricalcium silicate. In: *Nuclear magnetic resonance spectroscopy of cement-based materials*, Colombet, P., A.-R. Grimmer, H. Zanni and P. Sozzani (eds.). Berlin: Springer-Verlag, 57–68.

Gullion, T. 1995. Measurement of dipolar interactions between spin-1/2 and quadrupolar nuclei by rotational-echo, adiabatic-passage, double-resonance NMR. *Chem. Phys. Lett.* 246, 325–330.

Gullion, T., and J. Schaefer. 1989. Rotational-echo double-resonance NMR. *J. Magn. Reson.* 81, 196–200.

Hansen M. R., H. J. Jakobsen and J. Skibsted. 2003. ^{29}Si chemical shift anisotropies in calcium silicates from high-field ^{29}Si MAS NMR spectroscopy. *Inorg. Chem.* 42, 2368–2377.

Harris, R. K. 1986. *Nuclear magnetic resonance spectroscopy: A physicochemical view*. Essex: Longman.

Harris, R. K., E. D. Becker, S. M. Cabral de Menezes, R. Goodfellow and P. Granger. 2001. NMR nomenclature: Nuclear spin properties and conventions for chemical shifts. *Pure Appl. Chem.* 73, 1795–1818.

Harris, R. K., R. E. Wasylishen and M. J. Duer (eds.). 2009. *NMR crystallography*. Chichester: John Wiley & Sons.

Hartmann, S. R., and E. L. Hahn. 1962. Nuclear double resonance in the rotating frame. *Phys. Rev.* 128, 2042–2053.

Hartman, J. S., and B. L. Sherriff. 1991. ^{29}Si MAS NMR of the aluminosilicate mineral kyanite: Residual dipolar coupling to ^{27}Al and non-exponential spin-lattice relaxation. *J. Phys. Chem.* 95, 7575–7579.

Hayashi, S., T. Ueda, K. Hayamizu and E. Akiba. 1992. NMR study of kaolinite: 1. ^{29}Si, ^{27}Al and ^{1}H spectra. *J. Phys. Chem.* 96, 10922–10928.

Herzfeld, J., and A. E. Berger. 1980. Sideband intensities in NMR spectra of samples spinning at the magic angle. *J. Chem. Phys.* 73, 6021–6030.

Hing, A. W., S. Vega and J. Schaefer. 1992. Transferred-echo double-resonance NMR. *J. Magn. Reson.* 96, 205–209.

Hjorth, J., J. Skibsted and H. J. Jakobsen. 1988. ^{29}Si MAS NMR studies of Portland cement components and effects of microsilica on the hydration reaction. *Cem. Concr. Res.* 18, 789–798.

Holly, R., E. J. Reardon, C. M. Hansson and H. Peemoeller. 2007. Proton spin-spin relaxation study of the effect of temperature on white cement hydration. *J. Am. Ceram. Soc.* 90, 570–577.

Jakobsen, H. J., A. R. Hove, H. Bildsøe, J. Skibsted and M. Brorson. 2007. Long-term stability of rotor-controlled MAS frequencies to 0.1 Hz proved by ^{14}N MAS NMR experiments and simulations. *J. Magn. Reson.* 185, 159–163.

Jakobsen, H. J., J. Skibsted, H. Bildsøe and N. C. Nielsen. 1989. Magic-angle spinning NMR spectra of satellite transitions for quadrupolar nuclei in solids. *J. Magn. Reson.* 85, 173–180.

Jones, M. R., D. E. Macphee, J. A. Chudek, G. Hunter, R. Lannegrand, R. Talero and S. N. Scrimgeour. 2003. Studies using ^{27}Al MAS NMR of AFm and AFt phases and the formation of Friedel's salt. *Cem. Concr. Res.* 33, 177–182.

Keeler, J. 2010. *Understanding NMR spectroscopy*, Second ed. Chichester: John Wiley & Sons.

Kentgens, A. P. M., J. J. M. Lemmens, F. M. M. Geurts and W. S. Weeman. 1987. Two-dimensional solid-state nutation NMR of half-integer quadrupolar nuclei. *J. Magn. Reson.* 71, 62–74.

Kolodziejski, W., and J. Klinowski. 2002. Kinetics of cross-polarisation in solid-state NMR: A guide for chemists. *Chem. Rev.* 102, 613–628.

Korb, J.-P., L. Monteilhet, P. J. McDonald and J. Mitchell. 2007. Microstructure and texture of hydrated cement-based materials: A proton field cycling relaxometry approach. *Cem. Concr. Res.* 37, 295–302.

Kwan, S., J. LaRosa and M. W. Grutzeck. 1995. ^{29}Si and ^{27}Al MASNMR study of strätlingite. *J. Am. Ceram. Soc.* 78, 1921–1926.

Le Saout, G., M. Ben Haha, F. Winnefeld and B. Lothenbach. 2011. Hydration degree of alkali activated slags: A ^{29}Si NMR study. *J. Amer. Ceram. Soc.* 94, 4541–4547.

Lelli, M.; D. Gajan, A. Lesage, M. A. Caporini, V. Vitzthum, P. Miéville, F. Héroguel, F. Rascón, A. Roussey, C. Thieuleux, M. Boualleg, L. Veyre, G. Bodenhausen, C. Copéret and L. Emsley. 2011. Fast characterisation of functionalised silica materials by silicon-29 surface-enhanced NMR spectroscopy using dynamic nuclear polarisation. *J. Am. Chem. Soc.* 133, 2104–2107.

Levitt, M. H. 2007. *Spin dynamics: Basics of nuclear magnetic resonance*, Second ed. Chichester: John Wiley & Sons.

Lippmaa, E., M. Mägi, A. Samoson, G. Engelhardt and A.-R. Grimmer. 1980. Structural studies of silicates by solid-state high-resolution ^{29}Si NMR. *J. Am. Chem. Soc.* 102, 4889–4893.

Lippmaa, E., M. Mägi, M. Tarmak, W. Weiker and A.-R. Grimmer. 1982. A high-resolution ^{29}Si NMR study of the hydration of tricalcium silicate. *Cem. Concr. Res.* 12, 597–602.

Llor, A., and J. Virlet. 1988. Towards high-resolution NMR of more nuclei in solids: Sample spinning with time-dependent spinner axis angle. *Chem. Phys. Lett.* 152, 248–253.

Lowe, I. J. 1959. Free induction decays of rotating solids. *Phys. Rev. Lett.* 2, 285–287.

Luong, T., H. Mayer, H. Eckert and T. I. Novinson. 1989. In situ ^{27}Al NMR studies of cement hydration: The effect of lithium-containing setting accelerators. *J. Am. Ceram. Soc.* 72, 2136–2141.

MacKenzie, K. J. D., and Smith, M. E. 2002. *Multinuclear solid-state NMR of inorganic materials*. Oxford: Pergamon, Elsevier Science.

MacTavish, J. C., L. Miljkovic, M. M. Pintar, R. Blinc and G. Lahajnar. 1985. Hydration of white cement by spin grouping NMR. *Cem. Concr. Res.* 15, 367–377.

Mägi, M., E. Lippmaa, A. Samoson, G. Engelhardt and A.-R. Grimmer. 1984. Solid-state high-resolution silicon-29 chemical shifts in silicates. *J. Phys. Chem.* 88, 1518–1522.

Maly, T., G. T. Debelouchina, V. S. Bajaj, K. N. Hu, C. G. Joo, M. L. Mak-Jurkauskas, J. R. Sirigiri, P. C. A. van der Wel, J. Herzfeld, R. J. Temkin and R. G. Griffin. 2008. Dynamic nuclear polarisation at high magnetic fields. *J. Chem. Phys.* 128, 052211.

Maricq, M. M., and J. S. Waugh. 1979. NMR in rotating solids. *J. Chem. Phys.* 70, 3300–3316.

Markley, J. L., W. J. Horsley and M. P. Klein. 1971. Spin–lattice relaxation measurements in slowly relaxing complex spectra. *J. Chem. Phys.* 55, 3604–3605.

Mason, J. 1993. Conventions for the reporting of nuclear magnetic shielding (or shift) tensors suggested by participants in the NATO ARW on NMR Shielding Constants at the University of Maryland, College Park, July 1992. *Solid State Nucl. Magn. Reson.* 2, 285–288.

Massiot, D., H. Thiele and A. Germanus. 1994. WinFit – A Windows-based program for lineshape analysis. *Bruker Report* 140, 43–46.

Massiot, D., B. Touzo, D. Trumeau, J. P. Coutures, J. Virlet, P. Florian and P. J. Grandinetti. 1996. Two-dimensional magic-angle spinning isotropic reconstruction sequences for quadrupolar nuclei. *Solid State Nucl. Magn. Reson.* 6, 73–83.

McDonald, P. J., J.-P. Korb, J. Mitchell and L. Monteilhet. 2005. Surface relaxation and chemical exchange in hydrating cement pastes: A two-dimensional NMR relaxation study. *Phys. Rev. E* 72, 011409.

Medek, A., J. S. Harwood and L. Frydman. 1995. Multiple-quantum magic-angle spinning NMR: A new method for the study of quadrupolar nuclei in solids. *J. Am. Chem. Soc.* 117, 12779–12787.

Mehring, M. 1983. *Principles of high resolution NMR in solids*. Berlin: Springer-Verlag.

Millot, Y., and P. P. Man. 2002. Procedures for labeling the high-resolution axis of two-dimensional MQMAS NMR spectra of half-integer quadrupole spins. *Solid-State Nucl. Magn. Reson.* 21, 21–43.

Moesgaard, M., D. Herfort, J. Skibsted and Y. Yue. 2010. Calcium aluminosilicate glasses as supplementary cementitious materials. *Glass Technology: European Journal of Glass Science and Technology A* 51, 183–190.

Mueller, K. T., B. Q. Sun, G. C. Chingas, J. W. Zwanziger, T. Terao and A. Pines. 1990. Dynamic-angle spinning of quadrupolar nuclei. *J. Magn. Reson.* 86, 470–487.

Müller, D., W. Gessner, H. J. Behrens and G. Scheler. 1981. Determination of the aluminium coordination in aluminium-oxygen compounds by solid-state high-resolution ^{27}Al NMR. *Chem. Phys. Lett.* 79, 59–62.

Müller, D., W. Gessner and A.-R. Grimmer. 1977. Determination of the coordination number of aluminum in solid aluminates from the chemical shift of aluminum-27. *Zeitschrift für Chemie* 17, 453–454.

Müller, D., W. Gessner, A. Samoson, E. Lippmaa and G. Scheler. 1986a. Solid-state ^{27}Al NMR studies on polycrystalline aluminates of the system $CaO-Al_2O_3$. *Polyhedron* 5, 779–785.

Müller, D., W. Gessner, A. Samoson, E. Lippmaa and G. Scheler. 1986b. Solid-state aluminium-27 nuclear magnetic resonance chemical shift and quadrupole data for condensed AlO4 tetrahedra. *J. Chem. Soc. Dalton Trans.* 1277–1281.

Müller, D., A. Rettel, W. Gessner and G. Scheler. 1984. An application of solid-state magic-angle spinning ^{27}Al NMR to study of cement hydration. *J. Magn. Reson.* 57, 152–156.

Nied, D., K. Enemark-Rasmussen, E. L'Hopital, J. Skibsted and B. Lothenbach. 2015. Properties of magnesium silicate hydrates (M-S-H). *Cem. Concr. Res.* (submitted).

Pake, G. E. 1948. Nuclear magnetic resonance absorption in hydrated crystals: Fine structure of the proton line. *J. Chem. Phys.* 16, 327–336.

Paris, M. 2014. The two aluminum sites in the ^{27}Al MAS NMR spectrum of kaolinite: Accurate determination of isotropic chemical shifts and quadrupole interaction parameters. *Amer. Mineral.* 99, 393–400.

Pines, A., M. G. Gibby and J. S. Waugh. 1972. Proton-enhanced nuclear induction spectroscopy: A method for high resolution NMR of dilute spins in solids. *J. Chem. Phys.* 56, 1776–1777.

Poulsen, S. L., H. J. Jakobsen and J. Skibsted. 2010. Incorporation of phosphorus guest ions in the calcium silicate phases of Portland cement from ^{31}P MAS NMR spectroscopy. *Inorg. Chem.* 49, 5522–5529.

Poulsen, S. L., V. Kocaba, G. Le Saoût, H. J. Jakobsen, K. L. Scrivener and J. Skibsted. 2009. Improved quantification of alite and belite in anhydrous Portland cements by ^{29}Si MAS NMR: Effects of paramagnetic ions. *Solid State Nucl. Magn. Reson.* 36, 32–44.

Rawal, A., B. J. Smith, G. L. Athens, C. L. Edwards, L. Roberts, V. Gupta and B. F. Chmelka. 2010. Molecular silicate and aluminate species in anhydrous and hydrated cements. *J. Am. Chem. Soc.* 132, 7321–7337.

Rottstegge, J., M. Arnold, L. Herschke, G. Glasser, M. Wilhelm, H. W. Spiess and W. D. Hergeth. 2005. Solid state NMR and LVSEM studies on the hardening of latex modified tile mortar systems. *Cem. Concr. Res.* 35, 2233–2243.

Samoson A. 1985. Satellite transition high-resolution NMR of quadrupolar nuclei in powders. *Chem. Phys. Lett.* 119, 29–32.

Samoson, A., E. Kundla and E. Lippmaa. 1982. High resolution MAS-NMR of quadrupolar nuclei in powders. *J. Magn. Reson.* 49, 350–357.

Samoson, A., and E. Lippmaa. 1983a. Central transition NMR excitation spectra of half-integer quadrupole nuclei. *Chem. Phys. Lett.* 100, 205–208.

Samoson, A., and E. Lippmaa. 1983b. Excitation phenomena and line intensities in high-resolution NMR powder spectra of half-integer quadrupolar nuclei. *Phys. Rev. B* 28, 6567–6570.

Samoson, A., and E. Lippmaa. 1988. 2D NMR nutation spectroscopy in solids. *J. Magn. Reson.* 79, 255–268.

Schmidt-Rohr, K., and H. W. Spiess. 1994. *Multidimensional solid-state NMR of polymers.* London: Academic Press.

Schreiner, L. J., J. C. MacTavish, L. Miljkovic, M. M. Pintar, R. Blinc, G. Lahajnar, D. Lasic and L. W. Reeves. 1985. NMR line-shape spin-lattice relaxation correlation study of Portland cement hydration. *J. Am. Ceram. Soc.* 68, 10–16.

Seligmann, P. 1968. Nuclear magnetic resonance studies of the water in hardened cement paste. *J. PCA Res. Dev. Lab.* 10, 52–65.

Sevelsted, T. F., D. Herfort and J. Skibsted. 2013. ^{13}C chemical shift anisotropies for carbonate ions in cement minerals and the use of ^{13}C, ^{27}Al, and ^{29}Si MAS NMR in studies of Portland cement including limestone additions. *Cem. Concr. Res.* 52, 100–111.

Shaka, A. J., J. Keeler and R. Freeman. 1983. Evaluation of a new broadband decoupling sequence: WALTZ-16. *J. Magn. Reson.* 53, 313–340.

Skibsted, J. 1988. High-speed ^{29}Si and ^{27}Al MAS NMR studies of Portland and high alumina cements: Effects of microsilica on hydration reactions and products. In: *Geopolymer '88: Proceedings of the First European Conference on Soft Mineralurgy*, Davidovits, J., and J. Orlinski (eds.), 2, 179–196. Université de Technologie de Compiègne.

Skibsted, J., and M. D. Andersen. 2013. The effect of alkali ions on the incorporation of aluminum in the calcium silicate hydrate (C-S-H) phase resulting from Portland cement hydration studied by ^{29}Si MAS NMR. *J. Am. Ceram. Soc.* 96, 651–656.

Skibsted, J., H. Bildsøe and H. J. Jakobsen. 1991a. High-speed spinning versus high magnetic field in MAS NMR of quadrupolar nuclei: ^{27}Al MAS NMR of $3CaO \cdot Al_2O_3$. *J. Magn. Reson.* 92, 669–676.

Skibsted, J., and C. Hall. 2008. Characterisation of cement minerals, cements and their reaction products at the atomic and nano scale. *Cem. Concr. Res.* 38, 205–225.

Skibsted, J., C. Hall and H. J. Jakobsen. 2002. Nuclear magnetic resonance spectroscopy and magnetic resonance imaging of cements and cement based materials. In: *Structure and Performance of Cements*, Second ed., Bensted, J., and P. Barnes (eds.), 457–476. London: Spon Press.

Skibsted, J., E. Henderson and H. J. Jakobsen. 1993. Characterisation of calcium aluminate phases in cements by ^{27}Al MAS NMR spectroscopy. *Inorg. Chem.* 32, 1013–1027.

Skibsted, J., J. Hjorth and H. J. Jakobsen, 1990, Correlation between ^{29}Si NMR chemical shifts and mean Si-O bond lengths for calcium silicates. *Chem. Phys. Lett.* 172, 279–283.

Skibsted, J., H. J. Jakobsen and C. Hall, 1994. Direct observation of aluminium guest ions in the silicate phases of cement minerals by ^{27}Al MAS NMR spectroscopy. *J. Chem. Soc. Faraday Trans.* 90, 2095–2098.

Skibsted, J., L. Hjorth and H. J. Jakobsen. 1995a. Quantification of thaumasite in cementitious materials by ^{29}Si{^{1}H} cross-polarisation magic-angle spinning NMR spectroscopy. *Adv. Cem. Res.* 7, 69–83.

Skibsted, J., H. J. Jakobsen and C. Hall. 1995b. Quantification of calcium silicate phases in Portland cements by ^{29}Si MAS NMR spectroscopy. *J. Chem. Soc. Faraday Trans.* 91, 4423–4430.

Skibsted, J., N. C. Nielsen, H. Bildsøe and H. J. Jakobsen. 1991b. Satellite transitions in MAS NMR spectra of quadrupolar nuclei. *J. Magn. Reson.* 95, 88–117.

Skibsted, J., S. Rasmussen, D. Herfort and H. J. Jakobsen. 2003. ^{29}Si cross-polarisation magic-angle spinning NMR spectroscopy – An efficient tool for quantification of thaumasite in cement-based materials. *Cem. Concr. Comp.* 25, 823–829.

Smith, B. J., L. R. Roberts, G. P. Funkhouser, V. Gupta and B. F. Chmelka. 2012. Reactions and surface interactions of saccharides in cement slurries. *Langmuir* 28, 14202–14217.

Spearing, D. R., and J. F. Stebbins. 1989. The ^{29}Si shielding tensor in low quartz. *Amer. Mineral.* 74, 956–959.

States, D. J., R. A. Haberkorn and D. J. Ruben. 1982. A two-dimensional nuclear Overhauser experiment with pure absorption phase in four quadrants. *J. Magn. Reson.* 48, 286–292.

Stejskal, E. O., J. Schaefer and J. S. Waugh. 1977. Magic-angle spinning and polarisation transfer in proton-enhanced NMR. *J. Magn. Reson.* 28, 105–112.

Stephan, D., S. N. Dikoundou and G. Raudaschl-Sieber. 2008. Influence of combined doping of tricalcium silicate with MgO, Al_2O_3 and Fe_2O_3: Synthesis, grindability, X-ray diffraction and ^{29}Si NMR. *Mater. Struct.* 41, 1729–1740.

Taylor, H. F. W. 1989. Modification of the Bogue calculation. *Adv. Cem. Res.* 2, 73–77.

Tran, T. T. 2011. Fluoride mineralisation of Portland cement: Applications of double-resonance NMR spectroscopy in structural investigations of guest ions in cement phases. PhD thesis, Interdisciplinary Nanoscience Center, Aarhus University, Denmark.

Tran, T. T., D. Herfort, H. J. Jakobsen and J. Skibsted. 2009. Site preferences of fluoride guest ions in the calcium silicate phases of Portland cement from ^{29}Si{^{19}F} CP-REDOR NMR spectroscopy. *J. Am. Chem. Soc.* 131, 14170–14171.

Tse, D., and Hartmann, S. R. 1968. Nuclear spin-lattice relaxation via paramagnetic centers without spin diffusion. *Phys. Rev. Lett.* 21, 511–514.

Vold, R. L., J. S. Waugh, M. P. Klein and D. E. Phelps. 1968. Measurement of spin relaxation in complex systems. *J. Chem. Phys.* 48, 3831–3832.

Wasylishen, R. E., S. E. Ashbrook and S. Wimperis (eds.). 2012. *NMR of quadrupolar nuclei in solid materials*. Chichester: John Wiley & Sons.

Wu, Y., B. Q. Sun, A. Pines, A. Samoson and E. Lippmaa. 1990. NMR experiments with a new double rotor. *J. Magn. Reson.* 89, 297–309.

Chapter 7

Proton nuclear magnetic resonance relaxometry

Arnaud C. A. Muller, Jonathan Mitchell and Peter J. McDonald

CONTENTS

7.1 INTRODUCTION

This chapter focuses on nuclear magnetic resonance (NMR) of hydrogen (^{1}H) nuclei, also known as proton NMR. We are concerned specifically with ^{1}H-NMR relaxometry studies of water mobility (as a probe of the physical environment), where the data analysis is performed in the time domain. This needs to be distinguished from NMR spectroscopy (see Chapter 6), where the data analysis is performed in the frequency domain to determine atomic structures (local bonding environment). An NMR spectrometer can be used for many applications. The scope of this chapter is limited to common applications of one-dimensional (1D) ^{1}H-NMR relaxometry for characterising the physical structures and properties of cementitious materials. Basic experiments such as setting up and carrying out pulse sequences to gain longitudinal (T_1) and transverse (T_2) relaxation times are treated. Other, more complicated NMR experiments exist such as two-dimensional (2D) relaxation time correlations, molecular self-diffusion, spatially resolved NMR (imaging) and cryoporometry, for example; these experiments are not discussed here. For more information, a recent review describes the complete scope of ^{1}H-NMR experiments already applied to cement-based materials (Valori et al. 2013).

^{1}H-NMR has come a long way from the first experiments on cement pastes in 1978 (Blinc et al. 1978). Because there is a wide variety of NMR equipment with different operating characteristics, and because the signal analysis is rather complicated, many scattered results and interpretations can be found in the literature. In consequence, an exhaustive review of all NMR studies on cementitious materials can be confusing for newcomers to the field. In addition, the use of alternatively T_1 or T_2 to probe water in pores further complicates the picture. A brief history of progress made over the years can be found in several papers (Korb 2011; Valori et al. 2013) and in PhD theses on the topic (Gajewicz 2014; Muller 2014). There is no such historical section in this chapter. Here, pulse sequences, NMR phenomena and signal analysis/interpretation as currently used and understood are described.

^{1}H-NMR has been extensively used to study water in porous media. While NMR has been used by physicists for a long time, it has now become accessible to engineers and material scientists. Low-field NMR spectrometers suitable for cement samples have become affordable and are sold as benchtop and portable systems. Basic pulse sequences are supplied as part of the operating software. All conditions are now met for ^{1}H-NMR to become a routine laboratory technique for characterising the state and location of water in cementitious materials.

^{1}H-NMR directly interrogates the hydrogen nuclei of water in a sample. This gives NMR major advantages over other pore characterisation techniques. First, it allows water-filled porosity to be probed without the need for sample drying or procedures for stopping hydration that can modify the pore structures of interest (Fonseca and Jennings 2010; Gallé 2001; Hunt et al. 1960). Second, it allows quasicontinuous measurements while preserving the pore structure. ^{1}H-NMR is a nondestructive and noninvasive characterisation technique. In principle, time-domain NMR can distinguish water confined on length scales ranging from isolated molecules up to large pores that are microns in size. In comparison, mercury intrusion porosimetry is restricted to pore diameters greater than 3–4 nm (Van Brakel et al. 1981) (see also Chapter 9), while gas adsorption, for instance, is restricted to a maximum pore size of the order of 100 nm (Abdel-Jawad and Hansen 1989; Diamond 1971) (Chapter 10).

^{1}H-NMR uses oscillating magnetic fields to manipulate hydrogen protons in such a way that they return a measurable signal. The intensity of this signal indicates how much hydrogen (water) there is in the sample. The lifetime of the decaying NMR signal, known as the T_2 relaxation time, provides information about water mobility and its degree of physical confinement in pores: bulk water has a long T_2 relaxation time of a few seconds, water in capillary pores has T_2 relaxation time of a few milliseconds and water in crystalline solids has a very short T_2 relaxation time of a few microseconds. For cement materials, ^{1}H-NMR enables us to distinguish chemically combined water and liquid water in different pores by the study of characteristic T_2 or T_1 relaxation time. Water within portlandite, water in C-S-H interlayers, water in C-S-H gel pores and water in capillary pores can all be distinguished and quantified.

Despite the clear advantages that ^{1}H-NMR offers to describing pore structures and water interactions within cement samples, special care must be taken to properly carry out NMR experiments. At first, NMR might appear challenging for nonspecialists as there are many adjustable parameters. The way these parameters impact results is not always trivial. This chapter explains how to carry out basic 1D NMR experiments to assess the porosity and the state of water in cement samples. Guidelines on choosing an appropriate spectrometer, calibration using test samples, cement sample preparation, pulse sequences, acquisition parameters and data processing and interpretation are included.

7.2 INTRODUCTION TO BASIC PRINCIPLES OF NUCLEAR MAGNETIC RESONANCE

As NMR uses the quantum physical properties of particles, it is highly recommended to follow lectures on the topic. However, a basic knowledge is sufficient to carry out ^{1}H-NMR experiments and to interpret NMR signals. In this chapter, for simplicity, the description of NMR phenomena will be based on a classic depiction in terms of nuclear magnetic moments. For more advanced and extensive literature, the reader is referred to the following texts:

- *The Principles of Nuclear Magnetism*, A. Abragam, *International Series of Monographs on Physics*, Oxford Science Publications, Clarendon Press, 1961.
- *Principles of Nuclear Magnetic Resonance Microscopy*, P. T. Callaghan, Oxford Science Publications, Clarendon Press, 1993.
- *Translational Dynamics and Magnetic Resonance: Principles of Pulsed Gradient Spin Echo NMR*, P. T. Callaghan, Oxford University Press, 2011.
- *NMR: Tomography, Diffusometry, Relaxometry*, R. Kimmich, Springer, 1997.

Magnetic resonance and relaxation phenomena are discussed and schematically explained in this section in the context of transverse T_2 and longitudinal T_1 relaxation times.

7.2.1 Magnetic properties of ^{1}H nuclei

The basis of NMR is the fact that moving elementary particles generate magnetic fields. As a result of the spinning motion of individual protons, neutrons and electrons, magnetic dipoles are created along spin axes that can be envisioned as subatomic bar magnets (Figure 7.1a). This gives a magnetic dipole moment to each particle, or spin. The ^{1}H nucleus is composed of a single proton and has a *nuclear-spin* quantum number $I = 1/2$.* A *nuclear magnetic moment* μ is associated with ^{1}H, characterised by

---

* I is a quantum property of spins. I relates to the nuclear magnetic moment μ (in conventional units) of the spin through the nuclear gyromagnetic ratio of ^{1}H, a constant involving the *g*-factor and the nuclear magneton.

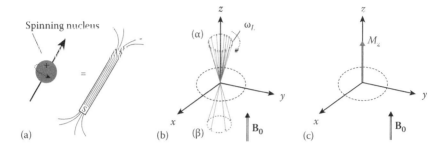

Figure 7.1 (a) Spinning ^{1}H nucleus compared to a little bar magnet; (b) individual spins in a static magnetic field $\mathbf{B_0}$; (c) the effective net magnetisation along z, M_z.

a magnitude and a direction. When exposed to an external static magnetic field $\mathbf{B_0}$, the field direction defines an axis about which the individual nuclear spins precess (Figure 7.1b). The spins precess about $\mathbf{B_0}$ at the Larmor (angular) frequency ω_L defined by

$$\omega_L = \gamma B_0, \tag{7.1}$$

where B_0 is the magnetic field strength (in tesla) and $\gamma/2\pi$ is the gyromagnetic ratio of ^{1}H nuclei divided by 2π equal to 42.57 MHz/T. At modest magnetic field strengths (e.g. $B_0 \sim 1$ T), the Larmor frequency is of the order of 42 MHz and corresponds to the radio frequency (rf) portion of the electromagnetic spectrum.

NMR measurements detect an ensemble of spins associated with a large number of molecules. A spin 1/2 nucleus can exist in one of two states, commonly called spin up (α, I_z or m_s = +1/2; a low-energy state parallel to $\mathbf{B_0}$) and spin down (β, I_{-z} or m_s = –1/2; a high-energy state antiparallel to $\mathbf{B_0}$). Normally, all spins are randomly oriented. When an external magnetic field $\mathbf{B_0}$ is applied, the low-energy state is slightly preferred. Hence, there are slightly more spins up in the direction of the field than spins down (Figure 7.1b), following a Boltzmann distribution. This gives rise to a net nuclear magnetisation parallel to $\mathbf{B_0}$ expressed as a vector sum $\mathbf{M}$ of all nuclei (M_z,* shown in Figure 7.1c). The net magnetisation projected onto the x–y plane remains at zero as the spins are precessing incoherently. The application of the external field is called polarisation and it is the first step of an NMR experiment.

---

* The $\mathbf{B_0}$ field direction is usually defined by convention along the z axis when representing and discussing spin states. Spins are aligned at an angle to $\mathbf{B_0}$ and because the individual spins are incoherently distributed around z, only the z components are invariant.

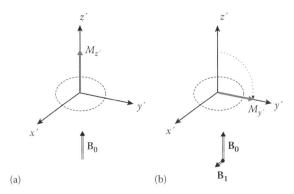

Figure 7.2 In the rotating frame of reference: (a) net magnetisation $M_{z'}$ of spins in a static magnetic field $\mathbf{B_0}$; (b) spins excited by an rf field $\mathbf{B_1}$ applied as a 90° rf pulse.

7.2.2 Magnetic resonance

In the static magnetic field $\mathbf{B_0}$, the nuclei can be further manipulated by applying a comparatively weak, time-dependent magnetic field that has some component orthogonal to the z axis. Typically, an electromagnetic field $\mathbf{B_1}$, oscillating at the Larmor frequency, is generated perpendicular to the existing static field $\mathbf{B_0}$ by an rf transmitter coil. As the oscillating field is rotating with the spins at the Larmor frequency, it appears static in the spins' frame of reference: this point of view is referred to as 'the rotating frame' and provides a convenient simplification to the description of spin dynamics as we can ignore spin precession occurring at ω_L. By convention, axes of the rotating frame are notated with prime symbols: x', y' and z' (as in Figure 7.2). The spins then precess about both $\mathbf{B_0}$ and $\mathbf{B_1}$. In the rotating frame, this motion is observed as a coherent precession of the nuclei on the z'–y' plane (assuming $\mathbf{B_1}$ is aligned along x'). In this way we achieve the *resonance* condition and this is the second step of a NMR experiment. The duration of $\mathbf{B_1}$ determines the ultimate position of the coherent spin ensemble. For example, a 90° rf pulse (designated P_{90}) applied along the x' axis of the rotating frame rotates the net magnetisation vector $\mathbf{M}$ from z' to y' (resulting in the so-called transverse magnetisation; Figure 7.2). This precessing coherent magnetisation induces a current in the rf detector coil and is the origin of the NMR signal. In modern NMR systems, the same rf coil is typically used for both transmission and detection.

7.2.3 Relaxation phenomena

After the spins have been excited away from their equilibrium state by the applied $\mathbf{B_1}$ field, the reverse process occurs; thereby the spins return to their original equilibrium state: this is called 'relaxation'. The measurement of the relaxation processes is the third step in our NMR experiment and is the underlying basis of NMR relaxometry.

The rate of relaxation is affected by magnetic field fluctuations in the local environment of excited hydrogen spins. For simple liquids such as water, the diffusion of H_2O molecules by Brownian motion allows ^1H spins to encounter and interact with other dipole magnetic moments as described by the Bloembergen–Purcell–Pound, or BPP, theory (Bloembergen et al. 1948). In cementitious materials, there are two main sources of relaxation:

1. The primary relaxation mechanism is the interaction of nuclear magnetic moments with other nearby nuclei. For cement samples, ^1H nuclei mainly interact with surrounding water molecules in pores and their interaction depends on the distances r between pairs of hydrogen nuclei and the θ angles between the vectors $\mathbf{r}$ and the magnetic field $\mathbf{B_0}$.
2. Paramagnetic impurities in the cement (e.g. Fe^{3+}) provide ^1H relaxation centres at pore surfaces. These particles have strong dipole moments due to unpaired electrons, which causes very fast spin relaxation. Barberon et al. (2003), Godefroy et al. (2001), Korb et al. (1997) and McDonald et al. (2005) showed how the surface relaxivity of water on pore surfaces may be calculated given the surface density of paramagnetic Fe^{3+} impurities. A measure of the pore volume then develops from the fast-exchange model of relaxation of Brownstein and Tarr (1979) and Zimmerman and Brittin (1957).

Transverse (T_2) and longitudinal (T_1) relaxation times

Two relaxation times exist. They are called transverse (or spin–spin) (T_2) and longitudinal (or spin–lattice) (T_1) relaxation times. Even though both time constants describe rates of nuclear-spin relaxation, they are driven by different mechanisms and are measured by different pulse sequences.

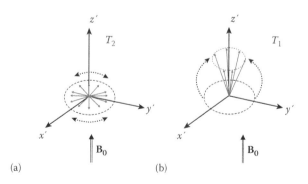

Figure 7.3 (a) T_2 loss of coherence among spins on the x'–y' plane; (b) T_1 recovery of the spins along the z' axis in equilibrium with $\mathbf{B_0}$.

The T_2 relaxation time characterises the rate at which the spin ensemble loses coherence on the x'–y' plane (Figure 7.3a). T_2 relaxation occurs due to energy exchange between neighbouring spins. The observed signal (magnetisation) decays with a time constant T_2^* due to loss of rotational phase coherence, a process known as a *free induction decay* (FID) (Hahn 1950a). When using benchtop magnets, the observed magnetisation decay, characterised by T_2^*, will be determined by inhomogeneities in the $\mathbf{B}_0$ field, not local spin interactions. To recover the transverse T_2 relaxation time, which is determined only by properties of the sample, additional $\mathbf{B}_1$ pulses need to be applied which negate loss of coherence due to field inhomogeneities on the x'–y' plane. This T_2 measurement is achieved with the Carr–Purcell–Meiboom–Gill (CPMG) pulse sequence (see Section 7.7.1).

The T_1 relaxation time is the time constant for magnetisation to recover on the z' axis (Figure 7.3b), restoring the equilibrium Boltzmann distribution of spins. This relaxation process occurs by energy loss to the surrounding 'lattice' (any nearby molecules) and the efficiency of the process is determined by molecular reorientations occurring near the Larmor frequency (e.g. molecular rotations and internal motions). T_1 is measured with the inversion-recovery (IR) or saturation-recovery (SR) pulse sequence (see Section 7.7.2).

T_1 recovery is equal to or slower than T_2. For bulk liquids, it is generally equal, while for solids it is generally orders of magnitude slower. For pore water in hydrated cement samples, it is in between: McDonald et al. (2005) reported that $T_1 = 4T_2$ at 20 MHz NMR frequency. During FID experiments, T_2 and T_1 processes occur simultaneously. Figure 7.4

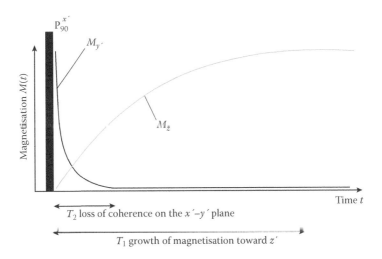

Figure 7.4 Schematic rates of relaxation T_1 and T_2 for water in pores of cement pastes. T_2 is a resonance signal decay with loss of coherence on the x'–y' plane. T_1 is a recovery of alignment of **M** along z'. The recovery of $M_{z'}$ cannot be monitored directly.

illustrates schematically T_1 and $T_\perp$ relaxation rates by showing the $M_{x'}$ and $M_{y'}$ magnetisation. Note that the recovery of $M_{z'}$ magnetisation cannot be monitored directly.

It is important to realise that new experiments cannot be conducted until the spins have returned to thermal equilibrium; i.e. the net magnetisation is realigned along the z' direction, a process that takes a few T_1 to complete fully. Therefore, it is necessary to wait approximately $5T_1$ between experiments (or repeat scans) in order to acquire quantitative signal intensities that can be interpreted in terms of water volumes.

7.2.4 Interpretation of relaxation times as pore sizes in porous media

The physics of NMR relaxation suggests that water in each different pore size should contribute to a separate $T_{1,2}$ component. Different pores can be distinguished either if they are isolated from each other (Figure 7.5a) or if the diffusive coupling between them is very weak (Figure 7.5b). If spins diffuse rapidly between two pores, one average relaxation time for the two pore populations is observed. On the other hand, if no exchange of water occurs during the measurement time, two distinct relaxation rates are recorded that can be related to the two pore sizes. The situation involves more complicated averaging if the exchange of water between pores is on the timescale of the experiment. For hydrating cement pastes, the exchange time between pores is slow (Monteilhet et al. 2006) in comparison to the slowest T_2 relaxation component. This scaling prevents interpretation problems associated with water diffusion between pores during the timescale of the measurement.

Researchers have studied the relationship between relaxation rates obtained by ¹H-NMR relaxometry and pore size. The fast-exchange model (also called the fast-diffusion model) developed by Brownstein and Tarr (1979), and Cohen and Mendelson (1982) relates T_1 and T_2 relaxation

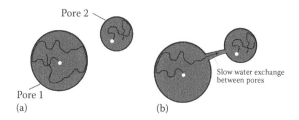

Figure 7.5 Conditions for observing two distinct relaxation components for two pores with different sizes: (a) two isolated pores and (b) two connected pores showing no exchange of water on the timescale of the experiment.

H$_2$O – bulk water (slow relaxation rate)

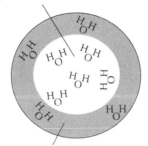

H$_2$O – surface water (fast relaxation rate)

Figure 7.6 Illustration of surface water molecules and bulk water molecules with two different relaxation rates. This hypothesis is the basis of the fast-exchange model of relaxation in pores (Brownstein and Tarr 1979; Cohen and Mendelson 1982). Only a few molecules of each population are shown.

times to the surface-to-volume ratio of pores. This model describes the existence of two distinct spin populations within a pore: a surface layer of water with a fast relaxation rate and bulk water* with a much slower relaxation rate (Senturia and Robinson 1970) (Figure 7.6). If the diffusion path length of a water molecule on the timescale of the NMR experiment is large compared to the pore size, then the water molecules are considered to be in fast exchange between the surface and bulk regions. The rate of relaxation which is experimentally observed (T_2 or T_1) is, according to the fast-exchange model, an average between the fast surface relaxation rate and the slow bulk relaxation rate as

$$\frac{1}{T_2} = \frac{1}{T_2^{\text{surf}}} + \frac{1}{T_2^{\text{bulk}}} . \tag{7.2}$$

A similar expression holds for T_1. For water in pores of cement samples, $T_2^{\text{bulk}} \gg T_2^{\text{surf}}$. Hence, the rate of relaxation of the bulk liquid can be considered negligible compared to the rate of relaxation of the surface liquid. Equation 7.2 can then be expressed as

$$\frac{1}{T_2} \approx \frac{S\varepsilon}{V} \cdot \frac{1}{T_2^{\text{surf}}} \approx \frac{S}{V} \cdot \lambda \qquad \text{with} \quad \lambda = \frac{\varepsilon}{T_2^{\text{surf}}} \Rightarrow T_2 \approx \frac{d}{2} \cdot \lambda^{-1}, \tag{7.3}$$

* Bulk water is the water in the middle of the pore and does not include surface water layers.

where S is the surface of the pore and V is its volume and ε represents the thickness of the surface water layer. For planar pores of area A and ignoring edge effects, $S = 2 \cdot A$ and $V = A \cdot d$, where d is the pore width or 'size'; hence, $V/S = d/2$. Other relationships between V/S and pore size exist, depending on the chosen geometry. The observed relaxation time T_2 is related to d through the parameter λ, the surface relaxivity. This parameter is assumed to be constant throughout the sample as a consequence of the supposed homogeneity of surface properties. The surface layer ε is assumed to be the thickness of a water monolayer (i.e. 0.28 nm). The characteristic relaxation T_2^{surf} of surface water depends on many parameters, such as Fe^{3+} surface density, temperature, pore fluid composition and magnetic field strength. For various reasons, determining λ is not straightforward. A robust technique for estimating λ experimentally, appropriate for use in hydrated cements, consists of a drying of the sample out to monolayer coverage. T_2^{surf} can also be obtained theoretically by ab initio calculations (Barberon et al. 2003; Godefroy et al. 2001; Korb et al. 1997). The fast-exchange model corresponds to the condition that $\lambda d/D \ll 1$, where D is the self-diffusion coefficient of bulk water. Finally, McDonald et al. (2010) showed an alternate analysis leading to the pore size through measurement of relaxation in a progressively dried sample that does not depend on λ.

7.3 NUCLEAR MAGNETIC RESONANCE EQUIPMENT

7.3.1 Description of a nuclear magnetic resonance instrument

Before carrying out NMR experiments, it is useful to recognise the essential components of an NMR instrument. The main components are shown in Figure 7.7 and briefly discussed in the following sections. These components represent a basic spectrometer used in material laboratories for relaxometry experiments. A standard benchtop slab-format magnet is illustrated. Although not discussed here, other magnet designs are available for special applications, such as the GARField (McDonald et al. 2007) or the Tree Hugger (Jones et al. 2012). Halbach magnets (Halbach 1980) are also popular for benchtop relaxometry, where the magnetic field is generated by a cylindrical array of permanent magnet blocks.

7.3.1.1 Permanent magnet (static magnetic field)

As discussed in the previous section, a static magnetic field is required to carry out NMR experiments. NMR instruments are defined by the strength, homogeneity and size of their permanent magnetic field. There are different ways to generate the static magnetic field:

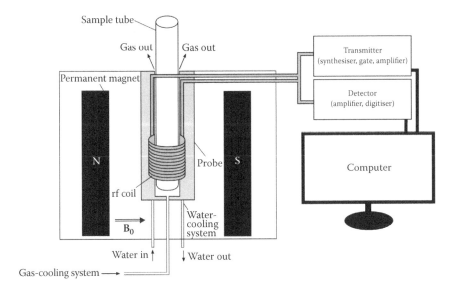

Figure 7.7 Schematic of a basic NMR instrument showing the primary components. The transmitter, detector and computer (i.e. the electronics) form the 'spectrometer' and can be used with a wide variety of magnet configurations.

1. The first one is to use a ferromagnetic material (such as a neodymium–iron–boron alloy) which has been permanently magnetised to a constant field. These types of magnets cannot be turned off. Most benchtop spectrometers used for studying water in porous media are of this type and provide relatively weak magnetic fields (cf. superconducting magnets, in number 3).
2. The second category uses resistive electromagnets that are made of a coil of wire (commonly of copper or aluminium) with current flowing through it. There are fewer and fewer magnets of this type as the weight, material cost and running conditions (energy consumption, thermal regulation and field stability) cannot compete with modern permanent magnetic materials. Although it is possible to design a 'strong' electromagnet, resistive electromagnets are impractical for general use in the laboratory.
3. The third category is superconducting magnets. In this case alloys such as niobium–titanium or niobium–tin are used and cooled down to a very low temperature using liquid helium. Superconducting magnets provide the most powerful magnetic fields. They are also the most expensive and are mainly used in fundamental physics, molecular chemistry, biochemistry and clinical imaging applications.

To achieve high-resolution NMR experiments, the static magnetic field must be extremely homogeneous. In practise, the 'perfect' magnetic field conditions can never be achieved. Even when a high-strength superconducting magnet is used, the field must vary in space due to the magnet's finite geometry and surrounding magnetic disturbances. To cope with this, some spectrometers are fitted with shim coils. The process of correcting field inhomogeneities is called 'shimming the magnet'. Shim coils are placed on both sides of the NMR probe and generate appropriate magnetic field gradients to compensate for the inhomogeneities in the active region of the permanent magnetic field. Shim coils are helpful but expensive as every shim coil comes with an associated high-stability current supply. Most benchtop instruments are equipped with basic shim coils, often with a fixed supply current that is factory set to remove gross distortions in the magnetic field.

7.3.1.2 Nuclear magnetic resonance probe

The rf probe is placed in the middle of the static magnetic field. This is the heart of the spectrometer and it contains the coil that generates the rf pulses. This coil is usually made of copper and is placed as close as possible around the sample. In most common spectrometers, the transmitter coil also fulfils the function of the receiver coil. This gives rise to a signal dead time during which the rf electronics switch from one mode to the other. Any signal arising during the dead time (which is important for studying fast-relaxing protons) is unfortunately lost. Even though manufacturers aim to reduce the dead time, there are still delays of the order of a few microseconds. The dead time is related to the Larmor frequency, probe size, design of the tuned rf circuit and the preamplifier recovery time. It is not uncommon to have a spectrometer with hundreds of microseconds of dead time for large coils in large magnets or spectrometers operating at low rf frequencies (e.g. < 10 MHz).

In the middle of the NMR probe is the sample compartment. Mostly cylindrical, it is made to accommodate a long glass NMR tube which contains the sample.

Some NMR spectrometers are equipped with temperature-controlling devices for the probe head sample compartment. Temperature control is achieved either with a cooling liquid circulation system, remote from the NMR coil, or with a gas line for passing high-purity gas (e.g. nitrogen or dry air) over the sample tube.

7.3.1.3 Transmitter

The transmitter generates the rf pulses sent into the coil. Considering the short time domain of NMR relaxometry experiments, pulse sequences are first loaded from the computer into what is called a 'pulse programmer'. It is this device that controls a synthesiser, which transforms the digital signal from the computer into rf waves, a 'gate', which allows pulses through, and an amplifier that generates the high rf power level of the pulses.

7.3.1.4 Receiver (or detector)

The receiver is made of high-capacity, low-noise amplifiers, which first boost received signals to a reasonable size. An analogue-to-digital converter (ADC) converts the continuous signal into digital data that can be stored in a computer. Modern digital spectrometers are able to sample the NMR signal directly, often performing the analogue-to-digital conversion at a frequency at least twice the Larmor frequency to reduce loss of information. A digital filter is then applied to downsample the digital data to the required sampling frequency. Overall, this process samples the continuous wave into discrete points. The amplitude resolution (gain parameter) and the time resolution (dwell time parameter) of digital-made signals are further discussed in Sections 7.5.4 and 7.5.5, respectively.

7.3.1.5 The operating software environment

A computer manages and coordinates all components of the spectrometer with a software interface provided by the manufacturer. It contains a pulse programmer and all tools to set the different NMR parameters. The programming language depends on the software used and on the brand of the spectrometer. Even if standard pulse sequences are commonly provided with the spectrometer, it is generally not complicated to manually code basic pulse sequences.

7.3.2 Magnet field strength (frequency)

NMR spectrometers are available over a wide range of magnetic field strengths from the earth's field (40 μT) to around 24 T (or higher in pulsed mode) for molecular structure determination. Benchtop permanent magnets are usually designed in the range of 50 mT to 1.5 T, depending on the application. For cement relaxometry studies, 0.5 T is considered standard. It is more common to refer to spectrometers by their ^{1}H resonance frequency, notated v_L* (see Figure 7.8).

A change in the magnetic field strength impacts the magnitude of the two quantum energy levels and the ratio of nuclei that align with the external magnetic field. The energy difference between spin states up and down (ΔE) is directly proportional to the external magnetic field strength B_0 and to v_L as

$$\Delta E = v_L \cdot h = \gamma \cdot B_0 \cdot \hbar = \omega_L \cdot \hbar , \tag{7.4}$$

with $\gamma/2\pi = 42.57$ MHz/T, $h = 6.6260 \times 10^{-34}$ J s and $\hbar = h/2\pi$. It follows that the higher the B_0 magnetic field strength, the larger is the difference in quantum energy levels (meaning more of the spins are aligned with the field) and the higher is the Larmor frequency. This relation is illustrated in

---

* v_l is related to the angular Larmor frequency of spins (ω_L) as $\omega_L = 2\pi \cdot v_l$.

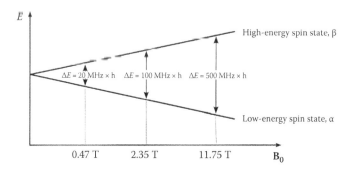

Figure 7.8 Effect of the magnetic field strength on the spin quantum energy levels and on the resonance frequency.

Figure 7.8. For example, a B_0 field strength of 0.47 T results in a ^1H resonance frequency of 20 MHz; a field strength of 2.35 T results in a ^1H resonance frequency of 100 MHz; a field strength of 11.75 T results in a ^1H resonance requency of 500 MHz and so on.

As more spins are aligned with the field, a stronger field gives a stronger signal. The inherent signal-to-noise ratio (S/N) of the NMR experiment increases with increasing field strength (between $S/N \propto B_0^{1.50}$ [Fukushima and Roeder 1993] and $B_0^{1.75}$ [Callaghan 1993], depending on the analysis, although at modest fields some experimentalists consider the relationship to be almost linear for practical considerations). This effect is the main reason for the current drive in NMR to build stronger magnets. Having a strong magnetic field becomes particularly important for nuclei that have weak magnetic moments or low abundance. It is important to realise that the S/N is a property of the *measurement* and not the *instrument*. Numerous factors influence the actual S/N of the detected signal, including rf coil performance and filling factor, filtering and environmental noise (i.e. coherent interference emitted by computers and other electrical devices that happens to coincide with the Larmor frequency).

7.3.3 Choosing the spectrometer

7.3.3.1 Spectrometer strength (or frequency)

Higher spectrometer strength results in more stimulated nuclei and consequently higher S/N. This is the reason for the current drive for researchers and manufacturers to go to higher NMR frequencies. Spectrometers with higher magnetic field strength also tend to have shorter dead time. This is advantageous as it gives more signals at the beginning of the decay.

There still are downsides to high magnetic fields. The magnetic susceptibility contrast between solid and liquid in porous materials distorts the static magnetic field B_0 at the pore scale. These pore-scale inhomogeneities, also called 'internal gradients', scale with magnetic field strength and can become very large, preventing accurate measurements of T_2 relaxation time (which is influenced by diffusion in field gradients) and self-diffusion coefficients, as well as limiting the maximum image resolution. It is therefore advisable to avoid using high-field magnets (> 60 MHz) to study porous materials, especially when the solid structure contains strong paramagnetic species such as Fe^{3+} ions.

For hydrogen protons with a high gyromagnetic ratio and almost 100% natural abundance, low-field magnets working at room temperature are sufficient to obtain satisfactory signal strength. Low- to intermediate-field* magnets (i.e. < 60 MHz resonance frequency) are suitable for the characterisation of cement samples.

7.3.3.2 Spectrometer bore size

From an *NMR point of view*, the bore size (size of the sample compartment) is a technological and financial constraint. Most spectrometers have small bore sizes (in the range of 10–25 mm) because it is easier and less expensive to build a magnet with a uniform magnetic field over a small volume: for a given mass of magnetic material, the field strength increases proportionally as the magnet poles are brought closer. The second advantage of small bore sizes is that there are fewer spins to stimulate. As a consequence, pulse lengths decrease (Valori et al. 2013).

From a *material point of view*, the bore size determines the degree to which NMR results are representative of the material studied. For cement samples, it obviously depends on the examined materials: concrete, mortar or cement paste. For cement pastes, bore sizes in the range of 5–15 mm are sufficient. For studying mortar, the bore size should be > 15 mm. For concrete, it depends on the size of the aggregates.

The choice of an appropriate bore size first depends on the type of application and the type of material studied. As short pulse lengths are required to study short T_2 components such as for solid-like hydrogen in crystalline phases (Valori et al. 2013), small bore sizes are preferable. Bore sizes in the range of 10–25 mm are suitable for the characterisation of cement samples.

Open-access and one-sided magnet configurations also exist that overcome the size constraint. However, they are prone to other difficulties, primarily much reduced magnet homogeneity.

7.3.4 Safety considerations

The obvious hazard associated with any NMR instrument is the static magnetic field, which in most instruments (including superconducting magnets) is

---

* The classification of field strengths into different categories is not officially established. The range of 10–60 MHz is usually considered 'intermediate field' and 'low field' corresponds to magnetic fields < 10 MHz. In this chapter, the term *low field* encompasses magnetic fields < 40 MHz.

impractical (if not impossible) to deactivate. For that, the 5G line (5×10^{-4} T) is a concept in NMR which defines the normally considered safe working distance from a magnet for pacemakers and ferrous metal objects, such as watches and those including metal implants or screws. Although supercon-ducting magnets tend to have a large stray field, where the 5G line is some considerable distance away from the magnet (~2 m), benchtop spectrome-ters have typically the 5G line within (or very close) to the 'box' around the instrument. Therefore, the instrument only presents a danger if the box is opened. Those safety considerations are always (or should always be) indi-cated at the entrance of NMR laboratories and on the machine themselves.

The rf radiation is another potential hazard. In medical magnetic reso-nance imaging applications, there are strict guidelines on patient dosage. The incident radiation, although transmitted at low power and radio fre-quencies, still imparts energy to the nuclei. Through T_1 processes, this energy is lost as heat to the lattice and can result in a localised rise in sample temperature. It is well known that rf pulse sequences with high duty cycles cause sample heating. It is worth considering that cement samples might be heated up to high temperature during NMR measurements.

7.4 GETTING STARTED WITH NUCLEAR MAGNETIC RESONANCE: OBTAINING A FREE INDUCTION DECAY ON TEST SAMPLES

Before starting NMR measurements on cement samples (which requires complicated experiments and data analyses), it is important to ensure the proper functioning of the spectrometer by first proceeding with simple tests and calibrations. This is usually done operating a single 90° pulse experi-ment on reference test samples.

7.4.1 Finding a test sample

The first step is to find an adequate test sample. It needs to be homoge-neous and to have, preferably, a single and relatively short relaxation time response. There are different options for this reference sample:

1. Many NMR users use copper sulfate ($CuSO_4$) solutions. The reasons are twofold. First, it is easy and cheap to make. Second, copper sulfate solutions have single relaxation times that can be adjusted by chang-ing the ionic concentration of copper sulfate. As for Fe^{3+} in cement materials, the presence of any transition metal ions with a strong magnetic moment affects relaxation times. Alternatively to copper sulfate, manganese chloride ($MnCl_2$) or gadolinium chloride ($GdCl_3$) solutions can also be used (e.g. as done by Pykett et al. [1983]). Note that both T_1 and T_2 will be reduced by the presence of metal ions, but

depending on the choice of ion and concentration, the ^{1}H relaxation times may differ even in bulk solutions.
2. The second most popular option as a reference sample for ^{1}H-NMR relaxometry is a rubber bung or any soft elastomer material. However, and contrary to copper sulfate solutions, their relaxation times are fixed. Also, note that elastomers can be structurally heterogeneous and may exhibit many different relaxation time components.

7.4.2 Turning on the spectrometer

In general, it is recommended to keep spectrometers permanently on if they are thermostatically regulated. From a cold start, it requires as much as 24 hours for the magnet to reach its operating temperature and to provide a stable magnetic field. Benchtop magnets are often heated to superambient temperatures, e.g. 25–35°C depending on the ambient conditions, to provide a stable magnet temperature.

7.4.3 Setting the sample position

It is very important to ensure the proper positioning of the sample in the probe head such that it remains in the homogeneous region of the magnetic field. Spectrometers are usually delivered with the optimum sample position preset (and specified in the instruction manual). If the correct sample position is not specified/set, it can be manually found/adjusted by locating where the sample returns the highest signal amplitude. A small test sample must be used for this purpose. In instruments designed for imaging applications, the sample position can be determined using a profile or image.

The probe head cavity bottom can be closed or open. In the latter case, a sleeve clamp around the NMR tube can be used to maintain the sample position.

7.4.4 Calibration of the spectrometer

The next step is to calibrate the spectrometer. Most commercial software have automatic procedures for calibrating the parameters of the spectrometer, such as resonance frequency, receiver dead time, receiver gain and pulse length/amplitude. More advanced systems may come with automated shim programmes. Calibrations can be made using test samples.

The calibration of the resonance frequency needs to be done frequently and for each experiment on some spectrometers. If there is no automatic calibration procedure, the user can manually set the correct resonance frequency by acquiring an FID and determining the centre frequency of the spectrum (in the Fourier-transformed frequency domain of the FID). The offset from resonance is usually calculated from the average phase shift between adjacent points in a FID.

7.4.5 Single 90° pulse (free induction decay experiment)

The easiest NMR experiment to carry out is a single 90° pulse aligned along the x' axis of the rotating frame (Figure 7.9), also called FID experiment. In the external magnetic field $\mathbf{B}_0$, hydrogen nuclear spins are thus rotated to the y' axis that initiates the free precession of individual spins on the x'–y' plane. The coherent precession of spins in the x'–y' plane induces a current in the receiver coil (y' direction) at the origin of the NMR signal. This FID signal decreases as spins dephase progressively (lose coherence among themselves). The speed and the direction of the dephasing vary as a function of local magnetic disturbances, for which there are two main contributions:

1. Magnetic influences of surrounding nuclei: spins dephase due to this mechanism in a time constant T_2. T_2 is the true spin–spin relaxation time T_2 of the sample.
2. Small $\mathbf{B}_0$ inhomogeneities related to the magnet quality: Spins dephase due to this mechanism in a time constant T_2^{inhom}.

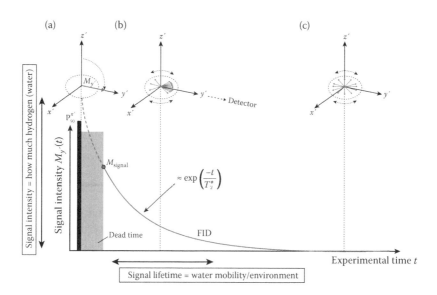

Figure 7.9 Pulse diagrams for a single 90° pulse experiment. The detected resonance signal, called FID, decreases exponentially following $1/T_2^*$. Spin states as seen on the x'–y' plane are reported at specific times: (a) spins are flipped onto the transverse plane for signal detection; (b) partial loss of coherence among spins and (c) total loss of rotational coherence among spins.

The total loss of coherence among spins is by convention defined as T_2^* and relates to T_2 and T_2^{inhom} as

$$\frac{1}{T_2^*} = \frac{1}{T_2} + \frac{1}{T_2^{\text{inhom}}}.$$

(7.5)

Most unimodal pore size porous materials as well as copper sulfate solutions exhibit a monoexponential FID following

$$M_{y'}(t) = M_0 \cdot \exp\left(\frac{-t}{T_2^*}\right),$$

(7.6)

where t is the experimental time (from the end of the $P_{90}^{x'}$), M_0 is the initial intensity of the signal at $t = 0$ and T_2^* is the characteristic relaxation time of the decaying NMR signal. The fitting of the monoexponential decay function to determine M_0 and T_2^* is therefore easy and can be done in Excel® or any other software such as Origin® or MATLAB® that have predefined least-squares fitting procedures.

The value of M_0 indicates the total hydrogen content of the sample. T_2^* is often dominated by the contribution of the magnet inhomogeneity $\left(T_2^{\text{inhom}}\right)$, which therefore often 'hides' the sample characteristics. For most benchtop systems with low-field magnetic fields, T_2^{inhom} is of the order of 0.5 to 1 ms (Valori et al. 2013) (depending on the availability and quality of the shim). Single 90° pulse experiment and FID acquisition is useful for the calibration of the spectrometer, the setting of the sample position or other NMR parameters using test samples. FID experiment is also often used to determine the amount of hydrogen in the sample. However and because the FID signal is influenced by magnetic field inhomogeneities, the study of cement samples by T_2^* is limited. To study the different water populations in hydrated cementitious materials, more sophisticated pulse sequences are required (see Section 7.7).

Even if the principle of the 90° pulse experiment is simple, different NMR parameters need to be adjusted to obtain proper results (see next section). Among them are the following:

- P_{90} pulse length
- Number of scans
- Recycle delay (RD)
- Receiver gain
- Dwell time

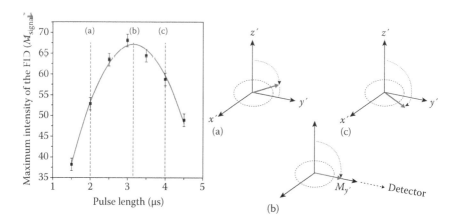

Figure 7.10 Influence of the pulse length on the intensity of the FID recorded during a 90° pulse experiment on a 7.5 MHz NMR spectrometer. In the experimental configurations, the correct P_{90} pulse length is 3.2 μs, value for which the spin rotation coincides with the detector direction. (a), (b) and (c) show the position of the net magnetisation for different pulse length values.

7.5 EXPERIMENTAL PARAMETERS

7.5.1 90° pulse length (P_{90})

During NMR experiments, the field B_1 is applied as pulses for which the pulse length is the time B_1 is maintained. B_1 induces the rotation of the spins about x' on the $z'-y'$ plane and a 90° pulse (P_{90}) is by convention the time required for B_1 to achieve rotation of the net magnetisation by 90°, from the z' direction to the y' direction (which is the direction of the detector; Figure 7.10b). The length of P_{90} varies from one spectrometer to another and depends on many parameters (bore size, NMR frequency, etc.). Note that pulse lengths change by changing the amplitude of the magnetic field B_1, i.e. the pulse power.

Pulse lengths are important NMR parameters as misset values result in incomplete (or too long) rotation and wrong signal amplitudes. As an example, Figure 7.10 shows the maximum intensity recorded during FID experiments (notated M_{signal}; see Figure 7.9) as a function of pulse length at a 7.5 MHz NMR frequency. The P_{90} pulse length was varied from 1.5 to 4.5 μs. In this experimental configuration, the correct 90° pulse length is $P_{90} = 3.2$ μs, the value for which the highest signal is detected.

7.5.2 Number of scans (s)

The number of scans is the number of times the experiment is repeated. The FID acquisition must be repeated multiple times and the data from each

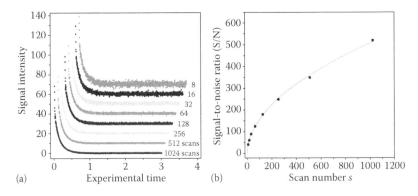

Figure 7.11 (a) The influence of the number of scans on FID results. These experiments were carried out on a portland cement paste after 7 days of sealed hydration on a 7.5 MHz NMR spectrometer. The different curves have been shifted on both axes for clarity, the signal decay being similar in all experiments. (b) Evolution of the S/N as a function of the number of scans s, fitted by a $\sqrt{s}$ scaling.

scan added in order to improve the S/N, which increases as the square root of the number of scans (Figure 7.11b). The S/N is calculated by

$$S/N = M_{signal}/\sigma_{noise}, \tag{7.7}$$

where M_{signal} is the maximum intensity of the received signal and σ_{noise} is the standard deviation of the noise, sometimes determined from the decay tail (part of the FID where the signal has decayed to zero). Alternatively, the noise level can be determined from the imaginary channel after the complex data have been correctly phase rotated (see Section 7.8.2).

It is preferable to have the number of scans by powers of 2. This helps to accommodate a phase cycle, for complex pulse sequences (such as 2D experiments) and for a regular improvement of the S/N through the cancellation of systematic errors. Figure 7.11a shows the influence of the number of scans on the recorded FID probing a portland cement paste after 7 days of sealed hydration. The number of scans s was increased by a factor of 2 from 8 scans to 1024 scans. Figure 7.11b shows in parallel the evolution of the S/N, which increases from 45:1 to 500:1.

7.5.3 Recycle delay

The recycle delay (RD) is usually defined as the time from the start of a pulse experiment to the start of its next repeat (or scan). Note, however, that RD is sometimes defined as the time between repeated pulse sequences excluding the time of the measurement itself. In any case, RD should allow all spins to recover their thermal equilibrium in the static magnetic field B_0 before a new measurement is conducted. A general and safe rule for the value of the RD for most cement samples is

$$RD = 5 \cdot T_1^{\text{longest}}, \tag{7.8}$$

which requires you to first measure, or know, the relaxation time(s) T_1 of the probed sample. For benchtop spectrometers (commonly < 40 MHz frequency), T_1 of pure bulk water is about 2–3 s at 20°C (Krynicki 1966) (limited by paramagnetic oxygen in solution). In comparison, T_1 of 1.25 g/L of $CuSO_4$ in distilled water ($\approx$ 8 mM) is about 0.1 s (Pykett et al. 1983) and the longest T_1 component in mature portland cement pastes is usually found in the range of 0.3–0.5 s (Korb 2009; Kowalczyk et al. 2014; Schreiner et al. 1985).

Too short an RD can result in inconsistent signal amplitudes from slow-relaxing protons. If a short RD is required (e.g. for monitoring rapid early hydration in cement pastes), then several dummy scans (rf pulses with no signal acquisition) should be used to drive the slowly relaxing spins to an equilibrium magnetisation state, acknowledging that the signal amplitude will not be proportional to water volume for any long relaxation time component.

Too long RD does not affect the signal but significantly increases the measurement time.

7.5.4 Receiver gain

The gain sets the amplification given to the FID received signal. For optimum results, the value of the gain needs to be adjusted to match the signal strength to the range of the digitiser (ADC). The S/N increases with increasing receiver gain following a logarithmic function. This implies high benefits in the short range of the gain and much smaller benefits once the gain has reached a certain value (Figure 7.12b).

If the amplification (the gain) is set too low, only a fraction of the digitisation levels is used, which leads to 'digital noise' and low S/N in the digital data.

If the amplification (the gain) is set too high, top signal of high intensities (at the beginning of the FID) will exceed the limits of the digitiser and the signal will be clipped (Figure 7.12a; 110 dB). Such data cannot be used.

The correct value of the gain varies from sample to sample depending on how weak the sample signal is. As an example, Figure 7.12a shows FID results obtained for different gain values measuring a cement paste. The gain is expressed here in decibels. The gain was varied between 50, 70, 90 and 110 dB for which a signal clip is observed. The S/N increases from 200:1 for gain = 50 dB to 370:1 for gain = 70 dB and further to 400:1 for gain = 90 dB (Figure 7.12b). It is important to realise that the gain value influences only the S/N for noise generated after the ADC. Noise (or

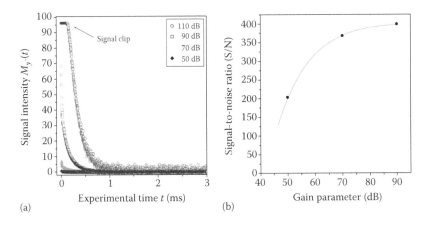

Figure 7.12 (a) Effect of the gain parameter on output digital FID results probing a portland cement paste after 7 days of sealed hydration. Digital FID results are for 50, 70, 90 and 110 dB as the gain parameter. A signal clip is observed for gain = 110 dB. (b) Evolution of the S/N as a function of the gain parameter.

interference) detected by the rf coil or introduced by the first stage of pre-amplification will be amplified at the same level as the signal by this gain parameter.

7.5.5 Dwell time (Δt)

The so-called dwell time, as for the gain, also concerns the digitiser. It represents the time interval into which the FID is digitally discretised. The dwell time parameter is often expressed in microseconds and corresponds to real acquisition time; it can also be set by a number of data points over a certain time range, or defined as a frequency range (inverse of dwell time). It is recommended to have at least two points on each wave cycle (Nyquist–Shannon theorem) which relates the dwell time to spectrometer frequency. The smaller the dwell time, the better it is but it mainly depends on how fast the digitiser can convert analogue data to digital data. The setting of the dwell time is always accessible to the user. Examples of dwell time samplings are shown in Figure 7.13 for an FID signal. If the dwell time is too long, the analysis of the data might suffer from the lack of information.

For modern spectrometers, the choice of dwell time is important as it will determine the performance of the digital filter stage (dead time and group delay). Normally, the dwell time determines the frequency bandwidth of the filter. However, expert users may wish to design oversampling filters to suppress noise (or interference) whilst retaining a short dwell time (e.g. a dwell time of 1 μs sets a detection bandwidth of ±500 kHz either side of the resonance frequency, but the digital filter may be set to pass only signals within ±50 kHz of the resonance frequency, equivalent to a dwell

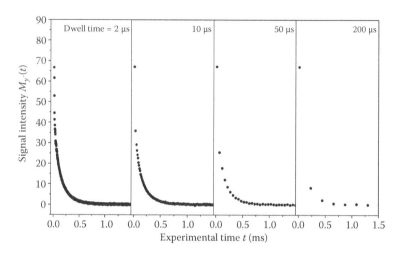

Figure 7.13 Effect of the dwell time parameter on FID digital results. These are 90°-pulse experiments carried out on a portland cement sample after 7 days of sealed hydration. The dwell time parameter was varied between 2, 10, 50 and 200 μs.

time of 10 μs). This type of oversampling filter allows a large number of data to be acquired, say in the FID of a fast-relaxing sample, without adding in the noise from a wide frequency range. In other words, the digital filter can be optimised for fast data acquisition without the S/N penalty.

7.6 PRACTICAL ASPECTS BEFORE STARTING ¹H NUCLEAR MAGNETIC RESONANCE EXPERIMENTS ON CEMENTITIOUS MATERIALS

This section discusses important details relating to ¹H-NMR experiments and the necessary precautions to be taken before starting measurements on cement samples. The following considerations are valid for cement samples measured with low-field spectrometers and do not apply to all materials and spectrometers.

7.6.1 Type of sample

7.6.1.1 White cement versus grey cement

Most NMR studies of cement samples are done using white cement. The reason is that the low iron content of such binders minimises the amount of paramagnetic impurities that causes magnetic disturbances and fast pore water relaxation (so difficult to measure).

Grey binders have more iron in comparison to white binders. In principle, an excess of Fe^{3+} should shorten NMR relaxation. In practise, white portland cements and grey portland cements have comparable relaxation times. The reason is that C-S-H, the main portland cement hydrate, has a limited number of surface sites to accommodate paramagnetic ions. In consequence, the iron density at the surface of C-S-H hydrates does not change significantly between white and grey binders and thus does not change relaxation times of water within pores. The extra iron in grey binders favours, however, the creation of iron-rich phases, such as ettringite (Buhlert and Kuzel 1971), AFm phases (Dilnesa et al. 2011) or hydrogarnet (Dilnesa et al. 2014), which locally distort the magnetic field. This worsens T_2^* performance and leads to lower S/N when using grey binders. Additionally, the susceptibility broadening in grey cements can lead to diffusive attenuation in CPMG pulse sequence, with insufficiently short pulse gaps.

If possible, white cement binders are preferable for ¹H-NMR relaxometry experiments. They provide a higher sensitivity and more accurate results. NMR on grey binders can still be carried out, however, with the knowledge that results have lower sensitivity and more noise.

7.6.1.2 Conventional binders versus pure C₃S or alite

Paramagnetic ions at surfaces of cement-based hydrates are an important cause of pore water relaxation. For conventional binders, this causes fast water relaxation within pores. In contrast, pure C_3S or alite samples, because of the lack of paramagnetic ions, have longer relaxation times compared to portland cement samples (up to a factor of 2–5). As a consequence, times of acquisition and RDs increase significantly. In these conditions, it becomes difficult to characterise C_3S or alite samples during the early stages of hydration when the material evolves rapidly. Long measurement times become less problematic for mature C_3S or alite pastes when the material evolves much more slowly.

7.6.1.3 Synthetic C-S-H: sample considerations

Preparation methods of synthetic C-S-H have been developed over the last 20 years. They aim to better understand the morphology and the growth of this material. C-S-H may be produced in various ways; for example, from a mixture of sodium silicate and calcium nitrate, mechanochemically from silica and calcium hydroxide (Garbev et al. 2008) or by the hydration of C_3S under controlled lime concentration (Nonat and Lecoq 1998). In all cases, high water-to-solid ratios are used and the resulting C-S-H hydrates are often in a form of white gels. This makes ¹H-NMR experiments to characterise synthetic C-S-H difficult to carry out because of the large amount of free water surrounding C-S-H hydrates, so the NMR signal of interest becomes very small (< 2% of the total signal).

To adequately characterise synthetic C-S-H by ^{1}H-NMR, it is neces-sary to increase the volume of C-S-H per sample volume and remove as much as possible of the surrounding free water. This can be done using controlled relative humidity (RH) environments. However, the choice of an appropriate RH is not straightforward. The RH should be low enough to remove most of the free water but high enough not to empty C-S-H pores.

Even so, relaxation times of water in synthetic C-S-H are longer than those for cement pastes, comparable to those for alite or pure C_3S pastes, because of the lack of paramagnetic impurities. Care must be taken to use appropriate RDs and appropriate pulse sequence parameters.

7.6.2 Choosing the sample size

Sample volume/height in the NMR tube should be restricted, if possible, to the homogeneous part of the magnetic field. Most of the time, absolute measurements are carried out where the sample volume is adapted to the tube diameter and its position confined to the middle of the magnetic field (Figure 7.14a). This configuration gives the best S/N. Oversized (or dominating) samples in the probe head should be avoided as the signal does not originate from the entire sample (Figure 7.14b). In addition, it leads to different rotation angles across the sample (B_1 heterogeneity): while spins experience a P_{90} in the middle of the rf coil, spins from outside the coil undergo, for example, a P_{75}. This is known as a pulse length arte-fact, which creates very reproducible systematic errors in experiments.

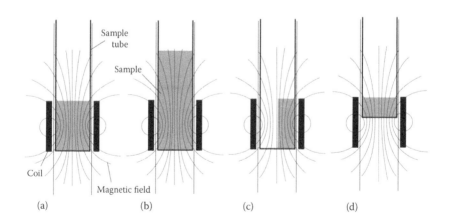

Figure 7.14 Schematics of different sample sizes/positions in the probe head: (a) abso-lute position giving the best S/N: the sample volume is maximised while restricted to the homogeneous part of the magnetic field; (b) dominating sample; (c) lateral displacement; (d) vertical displacement. Size (b) and posi-tions (c) and (d) all lead to unreliable results.

Off-centre samples also must be avoided. Figure 7.14c and d respectively show a lateral and a vertical displacement in the solenoid, the latter being the worse configuration for NMR experiments.

7.6.3 Sample casting

After mixing, there are two ways to cast NMR samples:

1. In cylindrical moulds that have a diameter slightly smaller to that of the NMR sample tube: specimens are then unmoulded and inserted into the NMR tube at the time of the measurement.
2. Directly into NMR tubes: This allows measurements to be made during the early stages of hydration when the paste is not solid enough to be unmoulded. In addition, samples cast in NMR tubes remain undisturbed throughout the hydration and provide slightly higher sample-to-volume ratios within the probe compared to samples cast in separated moulds.

Casting a cement sample directly into deep NMR tubes of a few millimetres diameter is easily done. The sample must be homogeneous and well placed at the bottom of the tube. For this, a normal plastic pipette can be used. In case a pipette cannot be used, homemade syringes can be made from a normal pumping device on top of which a long metallic tube is mounted (Figure 7.15).

NMR tubes need to be manipulated with gloves to prevent any grease from contaminating the probe and disrupting measurements. It is recommended to clean the external part of NMR tubes with isopropanol or ethanol by wiping with tissue before insertion into the probe.

7.6.4 Sample cure

After casting, the NMR tube or the mould in which the paste is placed should be tightly sealed to prevent any water loss. This can be done using

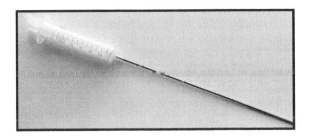

Figure 7.15 Homemade syringe for inserting cement pastes at the bottom of NMR tubes.

Parafilm® material for instance. A glass rod can be additionally included in the tube to reduce the free-air volume, into which the sample might release water in a gas phase.

For underwater curing, the samples can be unmoulded after 1 day and then immersed in saturated limewater up to the time of measurement. When the pastes are cast directly into the tubes, water solution can be added on top. The curing water surrounding the specimen must be removed before the ^{1}H-NMR experiment.

7.6.5 Sample mass and hydrogen content

There are two ways to know the hydrogen content of a sample:

1. Weigh the fresh paste inserted into the NMR tube. The amount of hydrogen within the sample can thus be calculated with the knowledge of the w/c of the material. Note that this estimation of the hydrogen content will remain valid only if the sample is kept tightly sealed from early after mixing.
2. If the w/c and/or the sample mass is unknown or if the sample has dried out, the amount of water can be obtained by removing the sample and by heating it in an oven at 550°C (not higher to avoid measurement of sample CO_2). In NMR experiments, the sample mass is small so great care must be taken in weighing the sample before and after ignition. Because cement samples have high specific surface areas, water can be quickly adsorbed by the sample once removed from the oven. To prevent measuring adsorbed water on the ignited sample, specimens must be weighed quickly.

7.6.6 Temperature control of the sample (if available)

If possible, it is recommended to maintain the sample at constant temperature in the NMR spectrometer. The reasons are twofold. First, it prevents the sample from being heated up by repeated pulses at short intervals. Second, it provides a temperature cure during continuous measurements of hydrating cementitious materials. Some spectrometers are equipped with temperature-controlling devices to maintain the sample at a defined temperature (most cement studies are done at 20 or 25°C). There are 2 options:

1. Circulating fluid systems using distilled water, mixtures of ethylene glycol or fluorinated heat transfer oils.
2. Gas flow systems: The gas must be free from impurities. It often is nitrogen or a mix of nitrogen and oxygen in appropriate proportions.

Either circulating fluid or blowing gas option is sufficient. A combination of both systems is unnecessary.

7.6.7 Sensitivity to the environment

The NMR technique is extremely sensitive to its environment as the signal received from ^{1}H spins is extremely small. Following the Boltzmann distribution, only 1 in 10^6 nuclei is detected. Many different factors influence NMR measurements. Computers or other electric equipment should be plugged into a separate electrical circuit. Ideally, the entire NMR system should be powered through an isolation transformer with an independent grounding stake to prevent noise injection through the building electrical earth/ground line. Wi-Fi networks always work at high frequencies and are not critical to NMR relaxometry measurements.

7.7 T_1 AND T_2 PULSE EXPERIMENTS SUITABLE FOR CEMENT SAMPLES

FID acquisition on test samples was described in Section 7.4 for the purpose of spectrometer calibration and of setting the primary NMR parameters. FID experiments can also be used to obtain the total hydrogen content of the sample or to determine T_2^*. However and because the FID signal is dominated by the inhomogeneity of the magnetic field, it is not suitable for detailed studies of cement samples. To properly characterise water in cement-based materials, more sophisticated experiments are required to obtain T_1 and T_2 relaxation times.

Similar information is contained in T_1 and T_2 decays, even though the underlying relaxation mechanisms are different (see Section 7.2.3). The advantages and disadvantages of studying T_1 or T_2 are as follows:

- Transverse (or spin–spin) T_2 relaxation time
  - Advantages: single-shot (fast) measurement; averages over short timescales (microseconds, so very little pore-to-pore exchange); robust solid/liquid water separation
  - Disadvantages: distorted by diffusion in internal gradients
- Longitudinal (or spin–lattice) T_1 relaxation time
  - Advantages: insensitive to diffusion in internal gradients
  - Disadvantages: slower measurement (although fast alternatives exist); averages over long timescales (milliseconds to seconds, so potential for pore-to-pore exchange)

For *liquid-like* hydrogen in cement samples (in capillary pores and in C-S-H pores), $T_1 \approx 4T_2$ at an NMR frequency of 20 MHz (McDonald et al.

2005). This T_1/T_2 ratio is frequency dependent and increases with increasing frequency. For *solid-like* hydrogen within crystalline phases (portlandite, ettringite, etc.), T_1 is long (seconds), while T_2 is invariably much shorter (a few microseconds). These two facts were/are the cause of much confusion when identifying cement water populations from T_1 and T_2 decays either separately or together. T_2 is obtained by the use of the CPMG pulse sequence. T_1 is obtained by the use of the IR or SR pulse sequence.

7.7.1 T_2: The Carr–Purcell–Meiboom–Gill pulse sequence

The CPMG pulse sequence is one of the most useful methods to characterise pore structures by ^{1}H-NMR relaxometry. It gives access to the transverse (or spin–spin) relaxation time T_2 of liquid water within the sample and allows different liquid water populations to be distinguished according to their mobility (degree of confinement in pores). However, T_2 CPMG experiments suffer from the fact that the solid signal from chemically combined water has substantially decayed when the first signal is recorded. The study of the solid signal is facilitated using solid-echo methods described in Section 7.7.3.

The CPMG pulse sequence was first proposed by Carr and Purcell (1954) and modified by Meiboom and Gill (1958). It is composed of a P_{90} pulse and a train of P_{180} pulses as $P_{90}^{x'} \tau P_{180}^{y'} 2\tau P_{180}^{y'} 2\tau P_{180}^{y'} 2\tau$, etc. (Figure 7.16). The τ value is, by NMR convention, the time between the middle of the initial P_{90} and the middle of the next pulse of the sequence, as in Figure 7.16.*

The CPMG pulse sequence has the advantage over the single 90° pulse experiment (cf. Section 7.4.5) to overcome the influence of magnetic field inhomogeneity and to measure directly the true sample T_2. This is achieved by the series of 180° pulses which cause spins to reverse their instantaneous phase angles and to precess back into phase. Since the magnet inhomogeneity is coherent, inversions of spin precessions cancel these variations out when spins rephase, leaving only the influence of other nuclei, i.e. the true sample T_2. During the CPMG pulse sequence, the best in-phase states are achieved at time 2τ from the start of the pulse sequence ($+1/2$ P_{90}^{length} and thereafter every 2τ. These in-phase states are called echoes. The signal envelope of the echoes is the T_2 spin–spin relaxation decay of the sample (Hahn 1950b). The analysis of T_2 CPMG decays is discussed in Section 7.8.

7.7.1.1   The τ value(s)

The pulse gap τ of the CPMG pulse sequence is an important parameter. It can be fixed to a constant value or be increased at each pulse interval.

---

* In some other cases, τ may be defined strictly as the pulse gap time excluding pulse lengths.

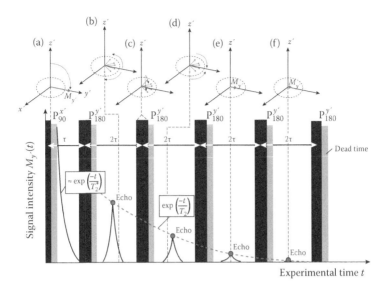

Figure 7.16 Pulse sequence diagram for T_2 CPMG experiment. The dashed line on the echoes shows the rate at which spins lose coherence among themselves. Spin states are reported at specific times: (a) net magnetisation flipped onto the transverse plane for signal detection; (b) spins dephasing from the y' direction; (c) $P_{180}^{y'}$ pulse which reverse spins instantaneous phase angles; (d) spins rephasing towards the y' direction. Panels (e) and (f) show spins' in-phase states (signal echoes).

7.7.1.1.1 Constant pulse gap τ

Cement samples contain different liquid water populations with different mobilities (100 μs $< T_2 <$ 100 ms). For this reason, it is sometimes not possible to capture all population decays with a constant pulse gap τ. Short τ (< 50 μs) might cause slow-relaxing protons, such as for water in large capillary pores, not to relax within the time of the pulse sequence (another issue is that too many pulses at short intervals of time might overheat the spectrometer or cause spin locking). On the other hand, large τ (> 0.2 ms) make experiments more sensitive to diffusion and might cause fast-relaxing protons, such as for water within C-S-H hydrates, to be missed out between the initial P_{90} and the following P_{180} pulse. The aforementioned considerations are critical because their effects might not be easily noticeable.

7.7.1.1.2 Varying pulse gap τ

An alternative to constant τ values is to increase τ over the course of the CPMG pulse sequence. Logarithmic spacing is common in cements work.

There are two essentially equivalent ways to calculate the necessary τ values:

1. Using a multiplication factor x:

$$\tau_{i+1} = \tau_i \cdot x. \tag{7.9}$$

   A common value for x is 1.02.

2. Stating an initial value (τ_{min}) and a final value (τ_{max}) for a given number of echoes (n_{echoes}) and using

$$\tau_i = \exp\left[\frac{n_i \cdot \ln\left(\dfrac{\tau_{max}}{\tau_{min}}\right)}{n_{echoes}} + \ln(\tau_{min})\right], \tag{7.10}$$

   where τ_i is the τ value used for the echo i.

The logarithmic spacing of τ keeps the total number of pulses low while covering approximately five orders of magnitude of relaxation times. The role of internal gradients caused by different magnetic susceptibilities between small and large pores has been raised (Hurlimann et al. 1995; Mitchell et al. 2013). This is, however, not critical to the measurement of water in cement samples because the different T_2 populations are still sufficiently close to each other that 'diffusive attenuations' do not occur.

The use of appropriate pulse gaps is important. To properly capture/quantify short liquid component(s) in cement pastes, i.e. the C-S-H interlayer water with T_2 of the order 80–120 μs (at 20 MHz NMR frequency), sufficiently short τ values must be used. The choice of τ_{min} depends on many parameters, such as spectrometer pulse lengths and spectrometer dead time. It is recommended to use $\tau_{min} \leq 50$ μs for which the first echo is located at $100 \, \mu s + 1/2 \, P_{90}^{length}$ from the beginning of the pulse sequence. On the other hand, τ_{min} should not be too short to prevent having a signal from solid-like protons that could make the analysis more difficult. Common values when measuring mature cement samples at NMR frequency of 5–20 MHz are $\tau_{min} = 25$ μs, $\tau_{max} = 6$ ms and $n_{echoes} = 256$. The resulting logarithmically spaced τ values are summarised in Table 7.1 along with the actual times of the echoes. Only the beginning and the ending τ values are shown in the table.

7.7.1.2 The number of echoes (n_{echoes})

The number of echoes to be used in the CPMG pulse sequence depends on the T_2 relaxation time components of the sample. It is important to have enough echoes to discretise short T_2 components and to ensure that the signals of long T_2 components decay entirely. The number of echoes also

Table 7.1 Logarithmically spaced τ values and actual
echo times

n_i	$τ_i$ (μs)	Actual echo time (μs)
1	25.00	50
2	25.54	101
3	26.10	153
4	26.67	206
5	27.24	261
252	5505.73	517733
253	5625.34	528986
254	5747.55	540483
255	5872.42	552229
256	6000.00	564229

Note: τ values are calculated with Equation 7.10 using $τ_{min}$ =
25 μs, $τ_{max}$ = 6 ms and n_{echoes} = 256 as parameters.
Echoes are spanning relaxation from 50 μs to 0.5 s.

depends on whether fixed or logarithmically spaced τ values are used. In
the second case, 128 or 256 echoes are usually sufficient for most white
cement pastes using the τ parameters given in the previous section. Note
that the end part of the decay with no residual signal is called the *tail*. The
tail is important to eventually fit a baseline or for the calculation of the
S/N of the data (Equation 7.7). On the other hand, having a long tail might
render inverse Laplace transform (ILT) analysis more difficult because long
relaxation time components might be fitted to the noise.

7.7.1.3 Number of scans, recycle delay and total measurement time

The number of scans s and the RD are other parameters to be adjusted in
the CPMG pulse sequence.

7.7.1.3.1 Recycle delay

As a reminder, the RD is usually defined as the time from the start of a
pulse experiment to the start of its next repeat (or scan). The RD is also,
sometimes, defined as the time between repeated pulse sequences excluding the time of the measurement itself (notated T_m). The RD should be
long enough to allow spins to recover the Boltzmann distribution in thermal equilibrium with B_0. This depends on the longest relaxation component (see Equation 7.8) and then on the sample type/age. For conventional
cement samples, RD ≈ 1.0 s is most of the time sufficient for low-frequency
spectrometers (< 1.0 T or 42 MHz frequency). This value must be adjusted,
however, for early-age measurements or for pure C_3S or synthetic C-S-H
samples, in which relaxation times are longer (see Section 7.6.1).

7.7.1.3.2 Number of scans

The number of scans is the number of times the pulse sequence is repeated to obtain satisfactory data. Increasing the number of scans increases the S/N (see Figure 7.11). However, it also increases the time of the experiment, T, which can be calculated by

$$T = s \cdot \text{RD or } T = s \cdot (T_m + \text{RD}), \tag{7.11}$$

depending on how the RD is defined. For an easy analysis of CPMG decays, the number of scans must be adjusted to achieve S/N of at least 400:1 (Kowalczyk et al. 2014) to 500:1 (Gajewicz 2014). For early-age samples for which the RD must be long and the experimental time must be short ($T <$ 15–20 min), fewer scans are possible (compared to mature cement pastes). Hence, the S/N is lower when measuring early-age samples. As the sample matures, the RD can be shortened and the experimental time can be made longer. This allows an increase in the number of scans with hydration time and consequently an increase in the S/N.

7.7.2 T_1: The inversion-recovery and the saturation-recovery pulse sequences

The IR or SR pulse sequences give access to the longitudinal (or spin–lattice) relaxation time T_1 of the different hydrogen components within the sample.

7.7.2.1 Inversion-recovery pulse sequence

The IR pulse sequence consists of a $P_{180}^{x'}$ and a $P_{90}^{x'}$ separated by a time interval τ_{rec} (Figure 7.17) (Roeder 1979). The first $P_{180}^{x'}$ inverts the initial magnetisation to the $-z'$ direction, resulting in a vector $M_{-z'}$. Subsequent to the $P_{180}^{x'}$, spins relax progressively because of interaction with the lattice and recover towards the low-energy state $M_{z'}$ in equilibrium with $\mathbf{B}_0$. A $P_{90}^{x'}$ pulse is used to flip spins that have partially or fully recovered towards the y' direction for signal detection. The IR pulse sequence therefore measures the spin recovery in the time interval τ_{rec}. The relaxation rate $1/T_1$ is obtained by varying the time interval τ_{rec} over a suitable range to capture the regain of $M_{z'}$ magnetisation.

7.7.2.2 Saturation-recovery pulse sequence

The SR pulse sequence is another way to measure T_1. It destroys any coherence among spins and saturates the bulk magnetisation to zero before it recovers progressively towards equilibrium. The SR experiment is similar to the IR experiment except that it starts with $P^{x'}$ pulses so that it saturates the

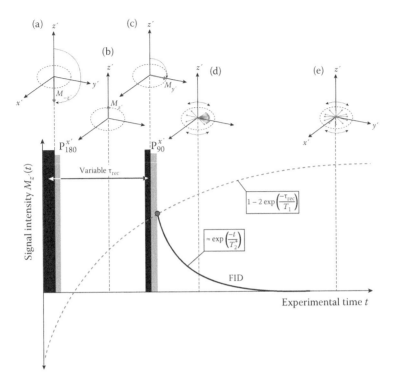

Figure 7.17 Pulse sequence diagram for T_1 IR experiment. The thick dashed line shows the T_1 recovery rate of the longitudinal magnetisation with $\mathbf{B_0}$. Spin states are reported at specific times: (a) the magnetisation is reversed; (b) partial T_1 recovery of spins in alignment with the $\mathbf{B_0}$ magnetic field; (c) spins that have partially or fully recovered are flipped onto the transverse plane for signal detection; (d) partial loss of coherence among spins and (e) total loss of rotational coherence among spins.

magnetisation which recovers from the $x'-y'$ plane ($M_{z'} = 0$) (Figure 7.18) rather than along z' ($M_{z'} = -M_0$).

7.7.2.3 T_1 pulse sequence parameters

Some parameters for the IR and the SR pulse sequences are similar to those for the CPMG pulse sequence such as the gain and the dwell time, for instance. The number of scans depends on the material examined and should provide a sufficient S/N for a robust data analysis (see Section 7.8). For the IR pulse sequence, the RD should allow spins to recover back to equilibrium with $\mathbf{B_0}$. For the SR pulse sequence, this precaution is no longer pertinent as the magnetisation is saturated to zero at the beginning of each experiment.

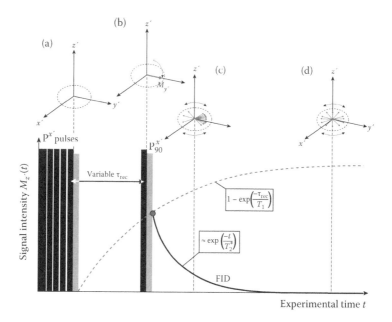

Figure 7.18 Pulse sequence diagram for T_1 SR experiment. The thick dashed line shows the T_1 recovery rate of the longitudinal magnetisation with $\mathbf{B_0}$. Spin states are reported at specific times: (a) the magnetisation is saturated; (b) spins that have partially or totally recovered are flipped onto the transverse plane for signal detection; (c) partial loss of coherence among spins and (d) total loss of rotational coherence among spins.

7.7.3 T_2: The solid signal and the solid-echo (quadrature-echo) pulse sequence

The solid-echo pulse sequence (Powles and Strange 1963) (Figure 7.19; also known as quadrature-echo, quad-echo or QE pulse sequence) allows very fast-relaxing T_2 components to be measured and quantified. In cement samples, it corresponds to hydrogen atoms strongly bound within phases such as portlandite, ettringite or AFm. Such solid-like hydrogen have a very short time constant $T_2 \approx 10\text{–}20$ µs (Gajewicz 2014; Greener et al. 2000; Holly et al. 2007; McDonald et al. 2010; Miljkovic et al. 1988; Muller et al. 2013a,b) with a signal decay mainly located within most spectrometers' dead time.

7.7.3.1 Pulse sequence

The QE pulse sequence is composed of two P_{90} pulses separated by a short time τ, so the sequence is $P_{90}^{x'} \tau P_{90}^{y'}$. While hydrogen atoms in liquid water are not sensitive to the second P_{90} (because it is aligned with the magnetisation), protons with static dipolar interactions become refocused, showing a Gaussian type of echo shape (McDonald et al. 2010). The second P_{90}, therefore, allows

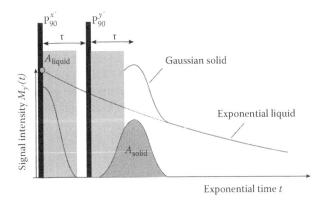

Figure 7.19 Pulse sequence diagram for solid-echo experiment. The detected signal (red line) is composed of an exponential decay part and a Gaussian part. (Adapted from Valori et al. 2013.)

the signal of solid-spins to partially recover (Powles and Strange 1963). To carry out informative QE experiments, τ must be varied over a certain range in order to capture the decay of the signal from solid-spins.

7.7.3.2 Parameters

7.7.3.2.1 τ value

The value of τ is, as for the CPMG, IR or SR pulse sequence, an important parameter. To successfully carry out QE experiments, τ must be similar to the relaxation time of the spins of interest. For solid-like protons in portlandite and ettringite having a T_2 time constant of about 10–20 µs, τ is best varied in the range of 0–60 µs.

7.7.3.2.2 Number of scans

The QE pulse sequence requires less scans than for the CPMG, IR or SR pulse sequence because the analysis of QE data is less sensitive to the noise. The choice of an optimum scan number is at the discretion of the experimentalist to have adequate S/N for a reliable analysis.

7.7.3.2.3 Recycle delay

The RD parameter should be comparable to that used for other pulse sequences for the same material. The RD should allow all spins to fully recover to thermal equilibrium with $\mathbf{B}_0$ between successive (repeated) scans (see Section 7.5.3 for more details).

7.7.3.3 Analysis of quadrature-echo signals

Literature is available on the shape of the solid signal from rigid coupled hydrogen which depends on the molecular configuration (Boden et al. 1974; Powles and Strange 1963). For cement samples, a Gaussian echo shape seems the most suitable following the work of McDonald et al. (2010). The liquid water contribution can be simplified as one single exponential decay. Hence, QE data are best fitted as the sum of a Gaussian part and an exponential decay part using

$$M(t,\tau) = A_{\text{solid}} \cdot \exp\left[-\left(\frac{t-t_c}{\sigma}\right)^2\right] + A_{\text{liquid}} \cdot \exp\left(\frac{-t}{T_2^*}\right), \tag{7.12}$$

where t is the experimental time counted from the end of the $P_{90}^{x'}$, t_c is the time of the centre of the Gaussian and σ is its width. The value of t_c depends on how the programme gives raw data. If the data are given from the end of the $P_{90}^{x'}$, $t_c = 2\tau + P_{90}^{\text{length}}$. If the data are given from the end of the $P_{90}^{y'}$, $t_c = \tau$ (see Figure 7.19). For solid-like protons in cement pastes with relaxation time $T_2 \approx 10$–20 µs, the width of the Gaussian echo is $\sigma \approx 10$–20 µs.

In Equation 7.12, the exponential amplitude A_{liquid} is the signal intensity of liquid water within the sample and is only very weakly, nearly not, pulse gap dependent. The Gaussian amplitude A_{solid} is the signal intensity of 'solid water' and decays with increasing pulse gap interval τ. Figure 7.20a shows different QE raw signals for a mature cement paste when $\tau = 15$, 30 and 45 µs. While the exponential part of the signal is equivalent for all τ values, there is a decrease in the Gaussian intensity part when increasing τ and a shift of echo centre to maintain that centre at $2\tau + P_{90}^{\text{length}}$ from the end of the $P_{90}^{x'}$. QE experiments are carried out for several τ values. The fitting of QE data using Equation 7.12 gives the associated A_{solid} and A_{liquid} amplitudes as a function of pulse gap τ, as presented in Figure 7.20b. To quantify the fraction of solid-like and liquid-like hydrogen in the sample, the Gaussian fraction and the exponential fraction must be extrapolated back to zero pulse gap free from relaxation phenomena:

1. Liquid signal: As the mobile part of QE signals is constant when varying τ, $A_{\text{liquid}}(\tau = 0)$ can be taken as an average of $A_{\text{liquid}}(\tau)$ or from linear back extrapolation (dashed line in Figure 7.20b). This value is, most of the time, equivalent to the total signal intensity obtained from the fitting of parallel T_2 CPMG decays.
2. Solid signal: The back extrapolation of the solid signal intensity to its initial value, $A_{\text{solid}}(\tau = 0)$, was (Boden and Mortimer 1973), and still is (McDonald et al. 2010) the subject of debate. In recent work on white cement pastes (Muller et al. 2013b), a Gaussian fit (as presented in Figure 7.20b, solid grey line) was used, giving good results in

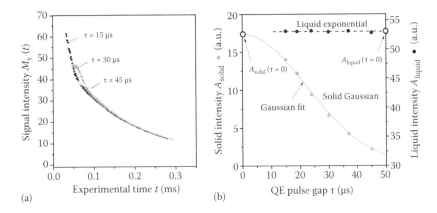

Figure 7.20 (a) Example of QE raw data for a mature cement paste when τ is 15, 30 and 45 μs obtained on a 7.5 MHz NMR spectrometer. (b) Fitting results of QE data using Equation 7.12. The variation in A_{solid} and A_{liquid} amplitudes is shown as a function of pulse gap τ. The dashed and solid grey lines respectively show the back extrapolation of liquid and solid signals to τ = 0, free from relaxation phenomena.

comparison to the estimation of that solid water by X-ray diffraction/ thermogravimetric analysis (XRD/TGA). On the other hand, an exponential law was used with success in other studies (McDonald et al. 2010). As it remains uncertain, the use of quantitatively known samples (calcium hydroxide and water, for instance) can help to choose the adequate fitting function. Spectrometers with short pulse lengths and short dead times are recommended to measure short signal decays.

The fraction of hydrogen (hence water) in solid phases, such as ettringite, portlandite and AFm, I_{solid}, is calculated as

$$I_{solid} = \frac{A_{solid}(\tau = 0)}{A_{solid}(\tau = 0) + A_{liquid}(\tau = 0)}. \tag{7.13}$$

The liquid fraction of the signal, I_{liquid}, being the fraction of liquid-like water in the sample, is calculated by $I_{liquid} = (1 - I_{solid})$.

7.7.3.4 Possible issues with quadrature-echo analysis

7.7.3.4.1 Pake doublet

In QE experiments, a 'wiggle' can be noticed due to the classic 'Pake doublet' of dipolar coupled spin pairs (Pake 1948). This mainly occurs when a high amount of solid-like protons is probed or for specific time-domain data. Figure 7.21 shows an example of this phenomenon that has occured

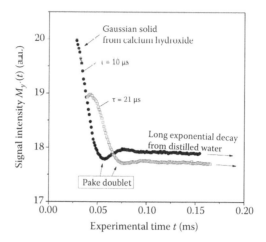

Figure 7.21 Illustration of the Pake doublet phenomenon in QE experiments. Signals were acquired on a 23.5 MHz spectrometer measuring simultaneously distilled water and calcium hydroxide held in two separated NMR tubes. Data for pulse gaps τ = 10 µs and τ = 21 µs are shown.

when probing a combination of calcium hydroxide and distilled water (for which the relaxation time is very long). The Pake doublet phenomenon is visible at the bottom of the Gaussian shape. This can be corrected using a sinc function as a factor within the fitting functions.

7.7.3.4.2 Sensitivity of quadrature-echo experiments and fitting

As QE signal decay occurs over relatively short times, there are always some uncertainties when measuring quadrature echoes. Errors might arise from the timing of the pulse sequence, the fact that the pulse lengths are finite, the proximity of the dead time, some possible group delays of filters and the amplification of short time data. One should be careful if fitting short τ value data, for which the top of the Gaussian might be located within the dead zone of the spectrometer (τ = 15 µs curve in Figure 7.20a is an example of this).

7.7.3.4.3 Long experimental time

QE experiments require running pulse sequences for several τ values. In consequence, the time of QE experiments can become long. For measuring cement pastes during the first hours of hydration, it is recommended to keep a total experimental time T < 15–20 minutes. This constraint limits the number of QE measurement scans, which consequently lower the S/N.

It is highly recommended to code the QE pulse sequence to run all required τ experiments consecutively.

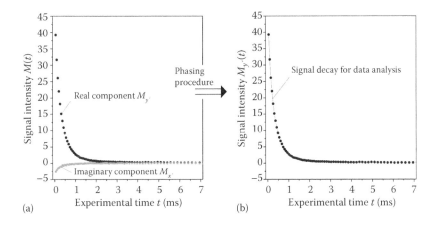

Figure 7.22 (a) Example of CPMG decay for a mature white cement paste showing real and imaginary envelope components of the signal and (b) phased data using Equation 7.14 ready for analysis.

7.8 DATA ANALYSIS FOR CEMENT SAMPLES

7.8.1 Data file properties

The way NMR results are given to the user will vary depending on the programme used, the brand of the spectrometer, the way the pulse sequence is coded or the user preferences. The signal intensity might be given as an average signal of all scans or as the sum of all scan intensities. As NMR signal amplitudes are not physical quantities, both types of data can be analysed as they are. The signal time (as shown in ms in Figure 7.22) may be given to the user from the beginning of the first pulse, from the end of the first pulse, from the end of the following dead time, from the first recorded echo (in case of CPMG), etc. In some cases, the signal is discretised with data point numbers and the time line must be explicitly calculated. The digital filtering of the raw signal, although necessary, also creates dead time and group, that is, filter, delays that should be accounted for when expressing decay times. For data analysis, the signal will be extrapolated back to its initial value; therefore, the decay time must be expressed as starting from the beginning of relaxation processes. This is, most of the time, the end of the first pulse of the sequence.

7.8.2 Raw data and phasing process

The coherent precession of spins in the x'–y' plane induces a current in the receiver coil. From the precessing motion of coherent spins, a signal oscillating at the Larmor frequency is recorded. The direct quadrature (or phase-sensitive) detection of most NMR spectrometers records the signal as orthogonal y' and x' components. The signal is represented as complex data made of 'real' (cosine wave) and 'imaginary' (sine wave) components, related to $M_{y'}$ and $M_{x'}$, respectively (Figure 7.22a). Both real and imaginary signals are normally provided to the user.

For further data analysis, a phase rotation is applied to minimise the imaginary channel and leave only a corrected real component. The phasing is done using

$$M(t) = \cos(\theta) \cdot M_{y'}(t) + \sin(\theta) \cdot M_{x'}(t), \tag{7.14}$$

with

$$\theta = \arctan\left[\frac{M_{x'}(0)}{M_{y'}(0)}\right],$$

for which the phase angle θ is calculated from the early data with maximum amplitude. This process is called phasing the data and can be used for all types of experiments. Another way to combine real and imaginary data is by the magnitude mode:

$$M(t) = \sqrt{[M_{y'}(t)]^2 + [M_{x'}(t)]^2}. \tag{7.15}$$

Note that the phasing mode is preferred for FID or CPMG T_2 analysis because the magnitude mode sums the white noise that would otherwise average to zero. This leads to a baseline offset at $t = \infty$. In the same way, the magnitude mode is not appropriate to T_1 analysis using IR because it implicitly has negative and positive data. Figure 7.22 shows typical real and imaginary components of a CPMG signal decay. In the case where there is only one positive component in the output file, the user must ensure that it consists of phased data. NMR software often displays magnitude data or in some cases only the real component of the signal.

7.8.3 Multiexponential fitting

CPMG T_2 and IR/SR T_1 decays of cement samples are usually composed of the sum of several exponential components characteristic of different water relaxations. In principle, each water population i (water type and pore type) is characterised by a signal amplitude M_0^i and a relaxation time $T_{1,2}^i$. The deconvolution of these T_2 and T_1 multiexponential responses into individual components remains the critical part of CPMG, IR and SR experiments:

For the transverse relaxation time T_2, CPMG data can be fitted using

$$M_{y'}(t) = \sum_{i=1}^{N} M_0^i \cdot \exp\left(\frac{-t}{T_2^i}\right) \qquad (7.16)$$

where t is the experimental time counted from the end of the $P_{90}^{x'}$, N is the number of water populations, i is each individual population and M_0^i is each associated signal intensity at $t = 0$.

For the longitudinal relaxation time T_1, IR data can be fitted using

$$M_{z'}(t) = \sum_{i=1}^{N} M_0^i \cdot \left[1 - 2 \cdot \exp\left(\frac{-\tau_{rec}}{T_1^i}\right)\right] \qquad (7.17)$$

and SR data fitted with

$$M_{z'}(t) = \sum_{i=1}^{N} M_0^i \cdot \left[1 - \exp\left(\frac{-\tau_{rec}}{T_1^i}\right)\right], \qquad (7.18)$$

where τ_{rec} is the pulse gap used in T_1 experiments.

To extract the different T_2 or T_1 water populations using Equations 7.16, 7.17 or 7.18, three different methods exist:

1. A least-squares fitting to a known number of components. However, the results have significant dependence on the 'guess values' for signal amplitudes and relaxation times if they are all left variable. The results might also change according to whether the data are fitted as real data to exponential curves or as logarithmic data to straight lines as the noise weights differently.
2. A discrete method known as *curve peeling* consisting of a decomposition of the signal by an iterative fit of the longest $T_{1,2}$ component at data point times sufficiently long to capture a signal at that decay time but none shorter. The result of each iterative fit is subtracted from the remaining signal until the shortest component has been accounted for. Critical to this type of analysis is the lower boundary assigned to each successive fit. As for the least-squares fitting method, successful use of this method really requires prior knowledge of the number of components.

3 An approximate numerical inversion of the Fredholm integral equation that describes the expected form of the relaxation data. Numerical inversion analyses of NMR data allow more consistent deconvolutions of multiexponential curves, where the number of components is not known, compared to the other two fitting options (assuming that the data are composed of only exponential decays). It allows the raw data to be fitted without preassumptions regarding the number of exponential components that contribute to the global decaying NMR signal. However, the results are still open to misinterpretation as the fitting process will always return a solution. Note that a robust inversion analysis requires S/N of at least 400:1 (Kowalczyk et al. 2014) to 500:1 (Gajewicz 2014). Numerous numerical inversion algorithms have been developed to solve multiexponential data problems. One particular method that has been used successfully for NMR signal analysis is the one developed by Venkataramanan et al. (2002) in 2002 (inaccurately referred to as an 'inverse Laplace transform' in much of the literature). This inversion algorithm contains a regularisation parameter (Tikhonov type) which defines the *smoothing* of the distribution (Godefroy et al. 2008) and which can be varied according to the sample response and the noise in the data. Special care must be taken to identify the influence and the correct value of the regularisation parameter. ILT limitations of this particular method are discussed in Section 7.10.1.

Typical T_1 and T_2 results for cement samples are presented in the next section.

In some cases, the empty probe can yield a signal. The former probe may contain hydrogen (e.g. a small amount of adsorbed water in polytetrafluoroethylene material) or it may have been inadvertently contaminated or the pulse may have sufficient bandwidth to just detect ^{19}F. Whatever the cause, in many instances, an appropriate filter bandwidth should be able to suppress what appears as an off-resonant signal. The exception is when working at very low frequency where ^1H and ^{19}F resonances are close or when very short T_2 components are detected since short times equate to large bandwidths. Note that in principle, good NMR instruments do not give background signals.

7.9 TYPICAL RESULTS FOR CEMENT PASTE SAMPLES

In this section, relaxation times are discussed. The given values are characteristic for relaxation times of water in mature white cement pastes probed with low-field spectrometers (< 1.0 T or 40 MHz NMR frequency). For spectrometers with comparable frequencies, the ratios of the relaxation times should remain roughly similar if portland binders are used at conventional *w/c*. For other samples or the use of high-field spectrometers, relaxation times are different. Some of these specific cases are discussed in Sections 7.3.2 (spectrometer field strength) and 7.5.1 (type of sample).

Cement samples are often mixed with distilled water; here the fresh cement paste initially exhibits one single and long relaxation component. This rapidly evolves towards shorter relaxation times (as cement grains dissolve into water) and into different relaxation components (as cement reactions create hydrates). Numerous experiments on hydrated white cements have led to the classification of hydrogen protons in two main categories:

1. *Solid-like protons with solid (static) dipole–dipole interactions:* These are characteristic of hydrogen strongly (often qualified as chemically) bound within phases such as portlandite, ettringite and AFm. In white cement pastes probed with low-field spectrometers, these solid-like protons have the shortest T_2 relaxation time with $T_2 \approx 10$–20 μs (Gajewicz 2014; Holly et al. 2007; McDonald et al. 2010; Muller et al. 2013a,b) and the longest T_1 relaxation time with $T_1 \approx 300$–500 ms (Korb 2009; Kowalczyk et al. 2014; Schreiner et al. 1985) (at 20 MHz NMR frequency). Solid-like protons can be quantified only with spectrometers having short pulse lengths and short dead time. For T_1, IR or SR pulse sequences are used. For T_2, the study of the solid signal is facilitated using solid-echo methods (see Section 7.7.3).

   Solid-like protons, in the way that they can be measured with the methods described in this chapter, appear as one relaxation entity. The distinction between solid water in the different phases (portlandite, ettringite, etc.) has not been established yet by relaxometry experiments. Valori et al. (2013) reported that hydrogen in portlandite and ettringite, for instance, may have identifiably different relaxation signatures; but this difference seems not to be sufficiently large to be distinguished when studying them embedded in a cement paste. Additionally, the degree to which all hydrogen atoms in alumina-containing phases (ettringite and AFm) have solid-like NMR properties is still a matter of debate and remains to be investigated more deeply. Ettringite, for instance, contains 32 H_2O molecules in total, among which 30 H_2O molecules are considered fixed and 2 H_2O molecules are reported to be zeolitic water loosely bound in channels (Skoblinskaya and Krasilnikov 1975). Recent NMR work on synthetic ettringite equilibrated at 85% RH (i.e. without the zeolitic water) showed no liquid-like NMR response (Muller 2014).

2. *Liquid-like protons with liquid (mobile) dipole–dipole interactions:* These hydrogen atoms are located within liquid water with some mobility in pores. The physics of NMR relaxation suggests that each different pore size might contribute to a separate T_1 or T_2 component (as discussed in Section 7.2.4). For liquid water in hydrated white

cement samples, four relaxation components are usually identified. From short to long, two components with relaxation times T_2 of the order 80–120 µs ($T_1 \approx$ 320–480 µs) and of the order 250–500 µs ($T_1 \approx$ 1–2 ms) arise from water in C-S-H interlayer pores and from water in C-S-H gel pores, respectively. A third and a fourth component usually appear from water in fine capillary pores (called interhydrates [Muller et al. 2013b] with $T_2 \approx 1$ ms and $T_1 \approx 4$ ms) and from water in large capillary pores with longer $T_{1,2}$. The aforementioned relaxation times might be somewhat different when studying other binders or using high-field spectrometers, for instance. Relaxation of liquid-like hydrogen is measured by the CPMG pulse sequence (for T_2) or the IR or SR pulse sequence (for T_1).

The degree to which there is hydrogen in silanol groups of C-S-H hydrates remains a matter of contention. Regarding NMR relaxometry, such OH groups at pore surfaces may be associated with relatively (rotationally?) mobile Ca ions in solution (Nonat 2004; Richardson 2008) and therefore appear within (contribute to) the relaxation of the pore water.

Note that only water-filled pores can be detected. If cement pastes are kept sealed, empty voids develop due to chemical shrinkage and self-desiccation (at around 8 vol.% for conventional cements [Bentz 2008]) which are not detected by NMR experiments.

7.9.1 Typical T_2 Carr–Purcell–Meiboom–Gill results

Figure 7.23 shows an example of T_2 CPMG decay for a mature white cement paste deconvoluted into its different exponential components. Figure 7.23a shows the results of the curve peeling method and Figure 7.23b shows the results of the ILT algorithm from Venkataramanan et al. (2002). Both methods give similar results and separate the CPMG T_2 decay into three water populations assigned here to C-S-H interlayer water, C-S-H gel water and capillary water. The different M_0^i amplitudes are the signal intensity of each individual population i at $t = 0$. The values of M_0^i obtained by the curve peeling method are comparable to peak areas of the ILT result. While the former is shown in the experimental time domain, ILT results are usually given in the relaxation time domain.

Figure 7.24 shows a typical evolution of CPMG T_2 results for a white cement paste as a function of hydration time. The raw data were analysed using the ILT method. At the beginning, the signal is dominated by that of 'free water' with long T_2. During the first 3 days of hydration, relaxation times are shifted towards smaller values and different water populations appear. Taking the 3 days data as an example, the map shows four discrete components. The first and the second component with relaxation times of 80–120 and 250–500 µs are signals from

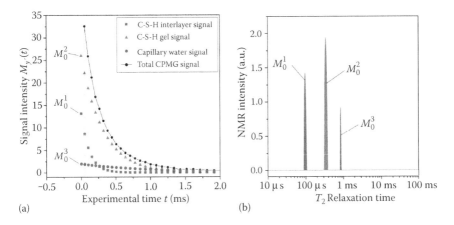

Figure 7.23 Deconvolution of a CPMG T_2 decay for a white cement paste after 28 days of sealed hydration into its different components: (a) using the curve peeling method – red squares are for the C-S-H interlayer water signal; green triangles are for the C-S-H gel water signal and blue circles are for the capillary water signal; (b) using the ILT method – an 'aggressive' regularisation parameter was used, leading to narrow peaks. Signal intensities of the peeling method at $t = 0$ are equivalent to peak areas of the ILT result. The signal assignment was done following the work of Muller et al. (2013a,b).

water in C-S-H interlayers and water in C-S-H gel pores, respectively. The third component rapidly displays a constant T_2 of about 1 ms and is attributed to fine capillary pores (called interhydrate pores [Muller et al. 2013b]). The fourth component is assigned to water in larger capillary pores and microcracks. Results of this type are expected when measuring portland cement pastes with low-field spectrometers. Relaxation times can be, however, different when moving outside common binders, outside common mix proportions and to high-frequency NMR spectrometers.

As previously mentioned, the decay of solid-like protons is too fast to be seen using the CPMG pulse sequence and must be measured separately by the QE pulse sequence (Section 7.9.3).

7.9.2 Typical T_1 inversion-recovery result

Figure 7.25 shows a T_1 distribution map obtained from an IR experiment on a white cement paste after 28 days of underwater cure. Compared to the T_2 distribution map obtained at the same NMR frequency (cf. Figure 7.24), relaxation times are roughly four times longer for each individual component. In addition, the solid water signal from crystalline phases appears with long relaxation time at around 300–400 ms.

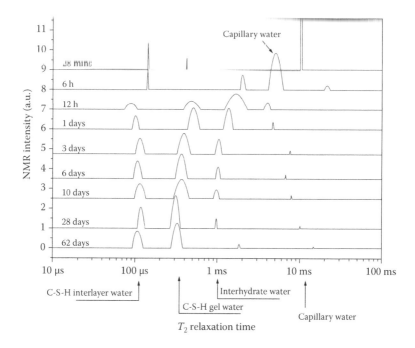

Figure 7.24 CPMG T_2 results for a white cement paste mixed at *w/c* = 0.40 and cured sealed throughout the hydration at 20°C; measured with a 7.5 MHz spectrometer. Data are presented from 28 minutes after mixing up to 62 days of sealed hydration. Distinct water populations are rapidly observed being, from short to long relaxation times, C-S-H interlayer water, C-S-H gel water, interhydrate water and water in large capillary pores. T_2 times can be translated into pore size using the fast-exchange model described in Section 7.2.4. For this, the surface relaxivity parameter is required. (Adapted from Muller 2014.)

7.9.3 Typical solid-echo (quadrature-echo) results

The QE pulse sequence quantifies the fraction of solid-like and liquid-like protons within the sample. Figure 7.26 shows, as an example, the evolutions of the signal fractions I_{solid} and I_{liquid} as a function of hydration time obtained by QE experiments (see Section 7.7.3) for a white cement paste mixed at *w/c* = 0.40 and cured sealed. There is a rapid increase in the solid signal during the first 2 days of hydration, after which it increases much more slowly. In this case, the rise in solid signal is associated with water used in the formation of ettringite and portlandite phases. An estimation of the quantity of water bound in portlandite and ettringite phases quantified by XRD and TGA shows that the solid signal fully accounts for the water bound in those two crystalline phases (Muller et al. 2013a,b). As the white cement studied contains a low amount of C_3A, there was no AFm formation.

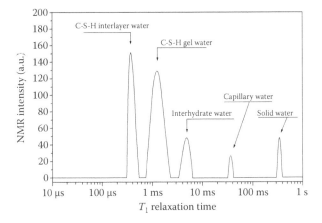

Figure 7.25 T_1 IR result for a white cement paste mixed at *w/c* = 0.40 and cured under water for 28 days at 20°C. The peak assignment is shown. Contrary to T_2 results, the solid signal appears with the longest relaxation time T_1.

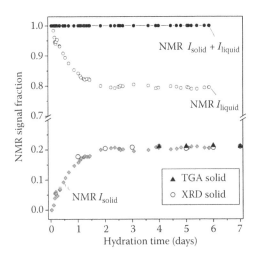

Figure 7.26 Example of QE results throughout the hydration of a white cement paste mixed at *w/c* = 0.40 and hydrated sealed at 20°C. There is an important rise in the solid signal fraction during the first 2 days of hydration because of water bound in crystalline phases. The QE liquid signal component follows the inverse trend. An estimation of portlandite and ettringite phases by XRD and TGA, converted into water mass fractions, shows good agreement with the NMR solid signal.

7.9.4 Evolution of the different nuclear magnetic resonance signals throughout hydration

By combining the QE and CPMG pulse sequences, all the water in a cement paste can be measured by T_2 and quantified in their different environments: water in crystalline phases (portlandite and ettringite) from QE and C-S-H interlayer water, C-S-H gel pore water and capillary pore water from CPMG. Figure 7.27 shows the evolution of these different water signals over the course of hydration. Note that the same information can be extracted from T_1 experiments.

During the first 2 days, there is a fast consumption of capillary water and a rise in signal from crystalline phase water (solid signal), C-S-H interlayer water and C-S-H gel water. Beyond 2 days, C-S-H gel water stops forming, while C-S-H layers and solid phases continue to be created. This type of quantitative analysis allows many aspects of the hydration process to be analysed. For example, equations for mass and volume balance allow the C-S-H density and chemical composition to be calculated (Muller et al. 2013a,b), in plain pastes as well as in blended systems (Muller et al. 2015).

The associated relaxation times are shown in Figure 7.28 for C-S-H interlayer water, C-S-H gel water and capillary water. The fast-exchange model of water relaxation in pores leads to average pore sizes of 0.85 nm (C-S-H interlayer spacing), 2.5 nm (gel pores) and 8.0 nm (interhydrate pores beyond 2 days of hydration) (Muller et al. 2013b). It needs to be specified

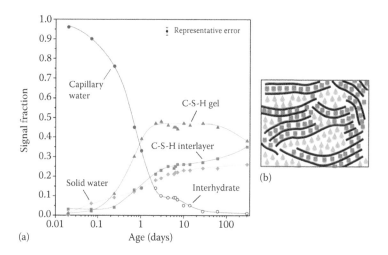

Figure 7.27 (a) Evolution of NMR signal fractions of different water populations in a hydrating white cement paste. The cement paste was mixed at w/c = 0.40 and kept sealed throughout the hydration at 20°C. These data were obtained by combining the CPMG and QE pulse sequences. (b) Schematic of C-S-H as viewed by ¹H-NMR relaxometry. (Adapted from Muller 2014.)

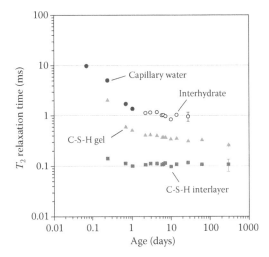

Figure 7.28 Associated relaxation times for the data presented in Figure 7.27. C-S-H gel water signal and C-S-H interlayer water signal are shown from the time they first appear. While the interlayer signal displays a constant T_2, the T_2 of gel water decreases with hydration time. Capillary water with initially high T_2 asymptotes to $T_2 = 1$ ms beyond 2 days of hydration.

that the paste was kept sealed, so larger capillary pores exist because of chemical shrinkage and self-desiccation, which are not seen by NMR.

7.10 SOME EXPERIMENTAL ISSUES

7.10.1 Inverse Laplace transform: Considerations and limitations

Numerical inversion methods are useful tools for discriminating between the different exponential components of a decaying NMR signal. It can be used to analyse CPMG T_2 and IR or SR T_1 decays. One particular method that is widely used by the NMR community is the ILT method from Venkataramanan et al. (2002). In this section we discuss ILT parameters and precautions that must be taken in interpreting the output results.

7.10.1.1 Regularisation parameter

In the ILT algorithm, a Tikhonov regularisation parameter (called α) is used to adapt the analysis to the quality (S/N) and the type of data. Figure 7.29 shows the influence of this α parameter on ILT results for a mature white cement paste. The parameter α was varied between what is considered as

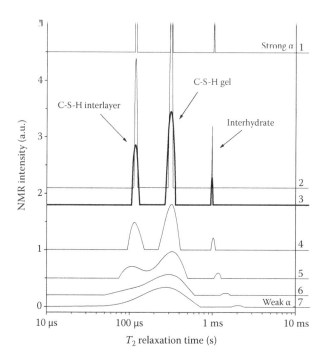

Figure 7.29 Effect of the regularisation parameter α implemented in the ILT developed by Venkataramanan et al. (2002) on the inversion results. A CPMG T_2 decay of a white cement paste mixed at w/c = 0.40 after 28 days of sealed hydration was analysed.

a strong α and a weak α, numbered here from 1 to 7. Note that the exact value of this parameter varies according to different software versions of the algorithm.

Here it can be seen that the α parameter influences the width of peaks and, to some extent, the way the different water populations become resolved as separated peaks. If α is too weak, the distribution is broad and peaks are not separated (they merge). This might be related to the 'pearling' effect (Callaghan 2011), described as a tendency of closely neighbouring peaks to move towards each other (Gajewicz 2014; Kowalczyk et al. 2014). On the other hand, if α is too strong, the distribution becomes very narrow (and peaks from fitting of the noise might appear). For inexperienced operators, automated α search methods exist, such as the BRD method and L-curve. The chosen α should be penalty weighted to match the noise figure or quality of the data. In the end, it remains the decision of the experimentalist to appreciate the over- or undersmoothing of the distribution. However, it is important to mention that peak areas (i.e. the signal intensity associated with each water population) show little dependence on the α parameter.

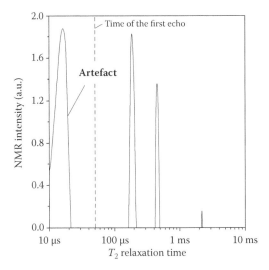

Figure 7.30 Example of T_2 CPMG deconvolution having artefact peak. This type of data is not reliable and should not be considered as a true sample result.

7.10.1.2 Artefact peak

Figure 7.30 shows the type of result that may come out in appropriate ILT analyses of cement samples. Peaks located at the limits of the T_2 fitting range are usually artefacts: CPMG, IR or SR decays are unlikely to have much information before the first recorded signal. Artefact peaks such as the one presented in Figure 7.30 might arise from baseline characteristics, from artificially high first few echoes or from the measurement of residual solid signals (if short pulse gaps are used). It is not recommended, however, to remove the first echo(es) as a lot of information is contained within them (see Figure 7.23a). Whether the data can be corrected or not depends on the source of the problem. However, to be on the safe side, data showing artefact peaks are better not used.

7.10.2 Troubleshooting and benchmarking

^{1}H-NMR is a sensitive technique where small disturbances, difficult to iden-tify, might affect the results. When the user has some doubts about the results, these are some possibilities to test the proper functioning of the spectrometer

7.10.2.1 Probing pulses by oscilloscope

It is possible to inspect pulses by the use of an oscilloscope. An oscilloscope can often be connected to the transmitter, before and after amplification as

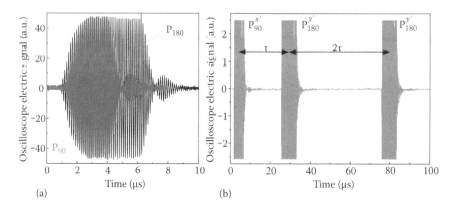

Figure 7.31 (a) P_{180} and P_{90} as seen by an oscilloscope on a 7.5 MHz spectrometer with 10 mm bore size; (b) the first three pulses of a CPMG pulse sequence where pulse gaps are visible.

well as at the spectrometer coil. To record pulses as seen by sample, a separated coil needs to be made, inserted into the NMR probe and connected to the oscilloscope. Figure 7.31a shows the length of P_{90} and P_{180} pulses as seen by an oscilloscope on a 7.5 MHz spectrometer with a 10 mm bore size. Figure 7.31b shows, in the same way, the beginning of a CPMG pulse sequence. Only the first three pulses are shown here; lengths of pulse gaps can be checked.

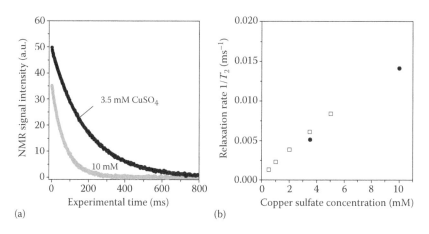

Figure 7.32 (a) CPMG results for 3.5 and 10 mM $CuSO_4$ in distilled water. (b) Squares: literature data for the influence of $CuSO_4$ concentration on T_2 relaxation time; circles: results of the experiments presented in (a). (From Pykett 1983.)

7.10.2.2 Spectrometer testing using known reference materials

Samples with known relaxation time or known porosity are useful in verifying the proper functioning of the spectrometer. Figure 7.32a shows CPMG results for 3.5 and 10 mM concentrations of copper sulfate in distilled water. The fit of one exponential decay for each curve led to T_2 = 193 ms and T_2 = 71 ms for 3.5 and 10 mM $CuSO_4$, respectively. The results are compared to the data from Pykett et al. (1983) in Figure 7.32b.

7.10.3 Measurement precision/accuracy

Estimating the error of NMR results is not a trivial task as this might depend on the type of experiment, the spectrometer and the type of material studied. For T_1 and T_2 analysis, Kowalczyk et al. (2014) reported three contributions to the error:

1. Error within the analysis: One of the most critical steps in characterising cementitious material by ^{1}H-NMR is the deconvolution of T_1 and T_2 decays into their individual exponential components. Most of the time, numerical inversion methods are used. Such analyses can be extremely sensitive to the noise in the data and some features of the results can be dependent on the parameters of the method (see Section 7.10.1).

2. Error from measurement uncertainty: NMR measurements are carried out over short time frames with finite pulse lengths and short pulse gaps. The accuracy of results can suffer therefore from the quality of the magnetisations. A way to estimate this type of error is to carry out several measurements of the same sample and to quantify the reproducibility. One might also evaluate the effect of recalibration. This, however, does not cover the cases where the experimental deficiency is systematic. Another way to evaluate the uncertainty in the measurement sequence is to measure the same sample on different spectrometers

3. The third source of error arises from sample reproducibility. This is not strictly inherent to NMR measurements but since the samples are necessarily small, they are affected by sample inhomogeneity of mixing, curing, etc. This variability can be tested by doing multiple castings and multiple samplings.

Knowing the aforementioned sources of errors, it is estimated that the uncertainty of NMR results should not exceed in total ±3% (Kowalczyk et al. 2014). However, NMR users should keep in mind that ^{1}H-NMR results depend on the current understanding of NMR mechanisms and the different tools available for signal acquisition and data analysis.

7.11 SUMMARY AND REPORTING

NMR quantitative analysis of water within a cementitious sample can be done in situ. The technique is nondestructive and noninvasive. It is possible to measure water in different pore (as discussed here); the specific surface area of cement pastes (Barberon et al. 2003; Halperin et al. 1994), the specific surface area of C-S-H including and excluding gel pores (Muller et al. 2013a) and the density and water fraction of C-S-H hydrates (Muller et al. 2013a,b), all in never-dried materials.

With the different ¹H-NMR experiments described in this chapter, all the water within a cement paste can be identified and quantified. NMR relaxometry experiments at frequencies in the range of 5–20 MHz measure the following:

1. Water in crystalline solid phases such as portlandite and ettringite: This water has short $T_2 \approx 10$–20 µs and long T_1 of the order 300–500 ms. The relaxation of this water can be detected only by spectrometers having short pulse lengths and short dead time.
2. Water in C-S-H interlayer pores with a time constant T_2 in the range of 80–120 µs ($T_1 \approx 320$–480 µs).
3. Water in C-S-H gel pores with T_2 in the range of 250–500 µs ($T_1 \approx 1$–2 ms).

Table 7.2 Reporting ¹H-NMR parameters for scientific publications

Data collection properties and setting		Examples
Equipment	Manufacturer	Bruker, Oxford Instruments, Kea
	Operating frequency	10, 20, 40 MHz
	Spectrometer dead time	15, 50 µs
	90° pulse length	3, 7, 15 µs
Method used	Type of study	T_1 or T_2
	Pulse sequence	CPMG, IR, SR, QE
Input parameter	Number of scans	512, 1024
	RD	1.0 s
	Number of echoes	128, 256
	Pulse gaps	Spacing rule: logarithmic spacing
		Range of values: from 15 to 45 µs, from 25 µs to 6 ms
	Typical S/N	600:1, 1000:1
Sample	Sample volume	0.35 cm³
	Iron content of the binder	1% C_4AF wt.%
	Mixing water	Distilled water
		w/c = 0.40, 0.48
	Curing	Underwater, sealed

4. Water in capillary spaces: Water in fine capillary pores called interhydrate with $T_2 \approx 1$ ms ($T_1 \approx 4$ ms) and water in larger capillary pores with longer T_2 and T_1.

When reporting NMR results in scientific publications, the different parameters, equipment characteristics and methods used need to be specified. Table 7.2 summarises the information that must be reported in the method section of scientific publications.

The progress made with ^{1}H-NMR allows a wide range of materials and mix parameters to be studied regarding hydrate's water and the surrounding capillary pores. These possibilities open up perspectives to improve our understanding of the fundamental hydration mechanisms.

ACKNOWLEDGEMENTS

The authors gratefully acknowledge the help, expert advice and comments of Agata Gajewicz, Karen Scrivener and Ruben Snellings. Arnaud Muller acknowledges the support of the European Union Seventh Framework Programme (FP7/2007-2013) under grant agreement 264448.

REFERENCES

Abdel-Jawad, Y., and W. Hansen (1989). 'Pore Structure of Hydrated Cement Determined by Various Porosimetry and Nitrogen Sorption Techniques'. In *Symposium Proceedings Vol. 137: Pore Structure and Permeability of Cementitious Material*, edited by Robert, L. R., and J. P. Skalny, 105–118. Warrendale, Pennsylvania: Materials Research Society.

Barberon, F., J. P. Korb, D. Petit, V. Morin and E. Bermejo (2003). 'Probing the Surface Area of a Cement-Based Material by Nuclear Magnetic Relaxation Dispersion'. *Physical Review Letters* 90 (11): 116103.

Bentz, D. P. (2008). 'A Review of Early-Age Properties of Cement-Based Materials'. *Cement and Concrete Research* 38 (2): 196–204.

Blinc, R., M. Burgar, G. Lahajnar, M. Rožmarin, V. Rutar, I. Kocuvan and J. Uršič (1978). 'NMR Relaxation Study of Adsorbed Water in Cement and C3S Pastes'. *Journal of the American Ceramic Society* 61 (1–2): 35–37.

Bloembergen, N., E. M. Purcell and R. V. Pound (1948). 'Relaxation Effects in Nuclear Magnetic Resonance Absorption'. *Physical Review* 73: 679–712.

Boden, N., Y. K. Levine and R. T. Squires (1974). 'NMR Dipolar Echoes in Solids Containing Spin-12 Pairs'. *Chemical Physics Letters* 28 (4): 523–525.

Boden, N., and M. Mortimer (1973). 'An NMR 'Solid' Echo Experiment for the Direct Measurement of the Dipolar Interactions between Spin-Pairs in Solids'. *Chemical Physics Letters* 21 (3): 538–540.

Brownstein, K. R., and C. E. Tarr (1979). 'Importance of Classical Diffusion in NMR Studies of Water in Biological Cells'. *Physical Review A* 19 (6): 2446–2453.

Buhlert, R., and H.-J. Kuzel (1971). 'The Replacement of Al^{3+} by Cr^{3+} and Fe^{3+} in Ettringite'. *Zement Kalk Gips* (2): 83–85.

Callaghan, P. T. (1993). *Principles of Nuclear Magnetic Resonance Microscopy*. Clarendon Press, Oxford, U.K.

Callaghan, P. T. (2011). *Translational Dynamics and Magnetic Resonance: Principles of Pulsed Gradient Spin Echo NMR*. Oxford University Press, Oxford, U.K.

Carr, H. Y., and E. M. Purcell (1954). 'Effects of Diffusion on Free Precession in Nuclear Magnetic Resonance Experiments'. *Physical Review* 94 (3): 630–638.

Cohen, M. H., and K. S. Mendelson (1982). 'Nuclear Magnetic Relaxation and the Internal Geometry of Sedimentary Rocks'. *Journal of Applied Physics* 53 (2): 1127–1135.

Diamond, S. (1971). 'A Critical Comparison of Mercury Porosimetry and Capillary Condensation Pore Size Distributions of Portland Cement Pastes'. *Cement and Concrete Research* 1: 531–545.

Dilnesa, B. Z., B. Lothenbach, G. Le Saout, G. Renaudin, A. Mesbah, Y. Filinchuk, A. Wichser and E. Wieland (2011). 'Iron in Carbonate Containing AFm Phases'. *Cement and Concrete Research* 41 (3): 311–323.

Dilnesa, B. Z., B. Lothenbach, G. Renaudin, A. Wichser and D. Kulik (2014). 'Synthesis and Characterization of Hydrogarnet $Ca_3(Al_xFe_{1-x})2(SiO_4)_y(OH)_{4(3-y)}$'. *Cement and Concrete Research* 59: 96–111.

Fonseca, P. C., and H. M. Jennings (2010). 'The Effect of Drying on Early-Age Morphology of C–S–H as Observed in Environmental SEM'. *Cement and Concrete Research* 40 (12): 1673–1680.

Fukushima, E., and S. B. W. Roeder (1993). *Experimental Pulse NMR: A Nuts and Bolts Approach*. Westview Press, Boulder, Colorado.

Gajewicz, A. M. (2014). 'Characterisation of Cement Microstructure and Pore–Water Interaction by 1H Nuclear Magnetic Resonance Relaxometry'. *PhD thesis*, Department of Physics, University of Surrey, Guildford, U.K.

Gallé, C. (2001). 'Effect of Drying on Cement-Based Materials Pore Structure as Identified by Mercury Intrusion Porosimetry: A Comparative Study between Oven-, Vacuum-, and Freeze-Drying'. *Cement and Concrete Research* 31 (10): 1467–1477.

Garbev, K., G. Beuchle, M. Bornefeld, L. Black and P. Stemmermann (2008). 'Cell Dimensions and Composition of Nanocrystalline Calcium Silicate Hydrate Solid Solutions; Part 1: Synchrotron-Based X-Ray Diffraction'. *Journal of the American Ceramic Society* 91 (9): 3005–3014.

Godefroy, S., J. P. Korb, M. Fleury and R. G. Bryant (2001). 'Surface Nuclear Magnetic Relaxation and Dynamics of Water and Oil in Macroporous Media'. *Physical Review E* 64: 021605.

Godefroy, S., B. Ryland and P.T. Callaghan (2008). '2D LaPlace Inversion Instruction Manual'. Victoria University of Wellington, New Zealand.

Greener, J., H. Peemoeller, C. Choi, R. Holly, E. J. Reardon, C. M. Hansson and M. M. Pintar (2000). 'Monitoring of Hydration of White Cement Paste with Proton NMR Spin–Spin Relaxation'. *Journal of the American Ceramic Society* 83 (3): 623–627.

Hahn, E. (1950a). 'Nuclear Induction Due to Free Larmor Precessio'. *Physical Review* 77 (2): 297–298.

Hahn, E. (1950b). 'Spin Echoes'. *Physical Review* 80 (4): 580–594.

Halbach, K. (1980). 'Design of Permanent Multipole Magnets with Oriented Rare Earth Cobalt Material'. *Nuclear Instruments and Methods* 169 (1): 1–10.

Halperin, W. P., J.-Y. Jehng and Y.-Q. Song (1994). 'Application of Spin-Spin Relaxation to Measurement of Surface Area and Pore Size Distributions in a Hydrating Cement Paste'. *Magnetic Resonance Imaging* 12 (2): 169–173.

Holly, R., E. J. Reardon, C. M. Hansson and H. Peemoeller (2007). 'Proton Spin–Spin Relaxation Study of the Effect of Temperature on White Cement Hydration'. *Journal of the American Ceramic Society* 90 (2): 570–577.

Hunt, C. M., L. A. Tomes and R. L. Blaine (1960). 'Some Effects of Aging on the Surface Area of Portland Cement Paste'. *Journal of Research of the National Bureau of Standards A* 64: 163–169.

Hurlimann, M. D., K. G. Helmer, T. M. Deswiet and P. N. Sen (1995). 'Spin Echoes in a Constant Gradient and in the Presence of Simple Restriction'. *Journal of Magnetic Resonance, Series A* 113 (2): 260–264.

Jones, M., P. S. Aptaker, J. Cox, B. A. Gardiner and P. J. McDonald (2012). 'A Transportable Magnetic Resonance Imaging System for in Situ Measurements of Living Trees: The Tree Hugger'. *Journal of Magnetic Resonance (San Diego, Calif.: 1997)* 218 (May): 133–140.

Korb, J. P. (2009). 'NMR and Nuclear Spin Relaxation of Cement and Concrete Materials'. *Current Opinion in Colloid & Interface Science* 14 (3): 192–202.

Korb, J. P. (2011). 'Nuclear Magnetic Relaxation of Liquids in Porous Media'. *New Journal of Physics* 13 (3): 035016.

Korb, J. P., M. Whaley-Hodges and R. G. Bryant (1997). 'Translational Diffusion of Liquids at Surfaces of Microporous Materials: Theoretical Analysis of Field-Cycling Magnetic Relaxation Measurements'. *Physical Review E* 56 (2): 1934–1945.

Kowalczyk, R. M., A. M. Gajewicz and P. J. McDonald (2014). 'The Mechanism of Water-Isopropanol Exchange in Cement Pastes Evidenced by NMR Relaxometry'. *RSC Advances* 4 (40): 20709–20715.

Krynicki, K. (1966). 'Proton Spin-Lattice Relaxation in Pure Water between 0°C and 100°C'. *Physica* 32 (1): 167–178.

McDonald, P. J., P. S. Aptaker, J. Mitchell and M. Mulheron (2007). 'A Unilateral NMR Magnet for Sub-Structure Analysis in the Built Environment: The Surface GARField'. *Journal of Magnetic Resonance (San Diego, Calif.: 1997)* 185 (1): 1–11.

McDonald, P. J., J. P. Korb, J. Mitchell and L. Monteilhet (2005). 'Surface Relaxation and Chemical Exchange in Hydrating Cement Pastes: A Two-Dimensional NMR Relaxation Study'. *Physical Review E* 72 (1): 11409.

McDonald, P. J., V. Rodin and A. Valori (2010). 'Characterisation of Intra- and Inter-C–S–H Gel Pore Water in White Cement Based on an Analysis of NMR Signal Amplitudes as a Function of Water Content'. *Cement and Concrete Research* 40 (12): 1656–1663.

Meiboom, S., and D Gill (1958). 'Modified Spin-Echo Method for Measuring Nuclear Relaxation Times'. *Review of Scientific Instruments* 29 (8): 688–691.

Miljkovic, L., D. Lasic, J. C. MacTavish, M. M. Pintar, R. Blinc and G. Lahajnar (1988). 'NMR Studies of Hydrating Cement: A Spin-Spin Relaxation Study of the Early Hydration Stage'. *Cement and Concrete Research* 18 (6): 951–956.

Mitchell, J., T. C. Chandrasekera and L. F. Gladden (2013) 'A General Approach to Measurements in the Presence of Internal Gradients'. *Microporous and Mesoporous Materials* 178: 20–22.

Monteilhet, L., J. P. Korb, J. Mitchell and P. J. McDonald (2006). 'Observation of Exchange of Micropore Water in Cement Pastes by Two-Dimensional T(2)-T(2) Nuclear Magnetic Resonance Relaxometry'. *Physical Review E* 74 (6): 61404.

Muller, A. C. A. (2014). 'Characterization of Porosity & C-S-H in Cement Pastes by 1H NMR'. *PhD thesis*, Ecole Polytechnique Fédérale de Lausanne, Switzerland.

Muller, A. C. A., K. L. Scrivener, A. M. Gajewicz and P. J. McDonald (2013a). 'Use of Bench-Top NMR to Measure the Density, Composition and Desorption Isotherm of C-S-H in Cement Paste'. *Microporous and Mesoporous Materials* 178: 99–103.

Muller, A. C. A., K. L. Scrivener, A. M. Gajewicz, and P. J. McDonald (2013b). 'Densification of C-S-H Measured by 1H NMR Relaxometry'. *Journal of Physical Chemistry C* 117: 403–412.

Muller, A. C. A., K. L. Scrivener, J. Skibsted, A. M. Gajewicz and P. J. McDonald (2015). 'Influence of Silica Fume on the Microstructure of Cement Pastes: New Insights from 1H NMR Relaxometry'. *Cement and Concrete Research* 74: 116–125.

Nonat, A. (2004). 'The Structure and Stoichiometry of C-S-H'. *Cement and Concrete Research* 34 (9): 1521–1528.

Nonat, A., and X. Lecoq (1998). 'The Structure, Stoichiometry and Properties of C-S-H Prepared by C3S Hydration Under Controlled Condition'. In *Nuclear Magnetic Resonance Spectroscopy of Cement-Based Materials*, edited by Colombet, P., H. Zanni, A.-R. Grimmer and P. Sozzani, 197–207. Springer, Berlin, Heidelberg, Germany.

Pake, G. E. (1948). 'Nuclear Resonance Absorption in Hydrated Crystals: Fine Structure of the Proton Line'. *The Journal of Chemical Physics* 16 (4): 327–336.

Powles, J. G., and J. H. Strange (1963). 'Zero Time Resolution Nuclear Magnetic Resonance Transient in Solids'. *Proceedings of the Physical Society* 82 (1): 6.

Pykett, I. L., B. R. Rosen, F. S. Buonanno and T. J. Brady (1983). 'Measurement of Spin-Lattice Relaxation Times in Nuclear Magnetic Resonance Imaging'. *Physics in Medicine and Biology* 28 (6): 723–729.

Richardson, I. G. (2008). 'The Calcium Silicate Hydrates'. *Cement and Concrete Research* 38 (2): 137–158.

Roeder, S. B. W. (1979). 'Principles of Magnetic Resonance (Slichter, C. P.)'. *Journal of Chemical Education* 56 (1): A38.

Schreiner, L. J., J. C. Mactavish, L. Miljković, M. M. Pintar, R. Blinc, G. Lahajnar, D. Lasic and L. W. Reeves (1985). 'NMR Line Shape-Spin-Lattice Relaxation Correlation Study of Portland Cement Hydration'. *Journal of the American Ceramic Society* 68 (1): 10–16.

Senturia, S. D., and J. D. Robinson (1970). 'Nuclear Spin-Lattice Relaxation of Liquids Confined in Porous Solids'. Society of Petroleum Engineers.

Skoblinskaya, N. N., and K. G. Krasilnikov (1975). 'Changes in Crystal Structure of Ettringite on Dehydration'. *Cement and Concrete Research* 5 (4): 381–393.

Valori, A., P. J. McDonald and K. L. Scrivener (2013). 'The Morphology of C–S–H: Lessons from 1H Nuclear Magnetic Resonance Relaxometry'. *Cement and Concrete Research* 49: 65–81.

Van Brakel, J., S. Modry and M. Svata (1981). 'Mercury Porosimetry: State of the Art'. *Powder Technology* 29 (1): 1–12.

Venkataramanan, L., Y.-Q. Song and M. D. Hurlimann (2002). 'Solving Fredholm Integrals of the First Kind with Tensor Product Structure in 2 and 2.5 Dimensions'. *IEEE Transactions on Signal Processing* 50 (5): 1017–1026.

Zimmerman, J. R., and W. E. Brittin (1957). 'Nuclear Magnetic Resonance Studies in Multiple Phase Systems: Lifetime of a Water Molecule in an Adsorbing Phase on Silica Gel'. *The Journal of Physical Chemistry* 61 (10): 1328–1333.

Chapter 8

Electron microscopy

Karen Scrivener, Amélie Bazzoni,
Berta Mota and John E. Rossen

CONTENTS

8.1 INTRODUCTION TO SCANNING ELECTRON MICROSCOPY

Electron microscopy is one of the most powerful techniques for studying the microstructure of cementitious materials. Since the publication of the first fracture surface imaged via scanning electron microscopy (SEM) (Chatterji and Jeffery 1966), many advances have been made, perhaps, most notably, the imaging of polished sections by backscattered electrons (BSEs) (Scrivener 2004). Nowadays SEMs are widespread and even benchtop machines are available at relatively low cost. Transmission electron microscopy (TEM) has also made a significant contribution to understanding these materials, although this is much less widely used due to more limited availability of instrumentation and difficult sample preparation.

Nevertheless, unfortunately, SEM and TEM are probably the most widely misused techniques in cement science and a very high proportion of published images provide no useful information, representing countless hours of wasted research time. In this chapter we try to provide guidelines to the effective use of electron microscopy (both SEM and TEM) and explain how these techniques can be best used to study cementitious microstructures.

8.1.1 Interaction of electrons with matter

There are many books on electron microscopy which describe the physics and operation of both SEM (e.g. Goldstein et al. 2003) and TEM (e.g. Williams and Carter 2009) in detail. Here we summarise the most relevant essentials.

To understand how electron microscopy works, we need to first look at what happens when a beam of electrons hits a material. The electrons undergo a series of elastic and inelastic collisions with the atoms of the material. These collisions generate signals which are detected in the microscope to form the image; this process is shown schematically in Figure 8.1.

In SEM the sample is usually thick and many collisions will occur before the energy of the incoming electrons is completely dissipated. The volume in which these collisions occur is known as the 'interaction volume'. For cementitious materials, which are made up of elements with fairly low atomic number, the size of this interaction volume is a few microns. The three most important signals generated are secondary electrons (SEs), BSEs and characteristic X-rays, which will be described in detail later. In all cases, images are formed by scanning the electron beam over the surface in a raster and using the signal detected at each point to give the intensity of the corresponding point in the image. The specimen is magnified simply by the difference in scale between the raster of the incident beam and the raster of the image.

In TEM, different types of image formation are possible, based either on a scanning arrangement as described above or on the use of electromagnetic lenses to focus the electrons and produce images or diffraction

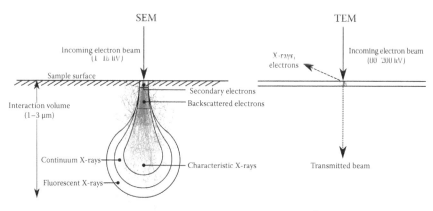

Electrons simulated using CASINO®

Figure 8.1 Schematic representation of the interaction of electrons with matter and the signals generated.

patterns in a manner analogous to a light microscope. The most important difference is that now the sample is very thin so the most important signal is the transmitted beam of electrons which have lost little energy through collisions with atoms on their passage through the sample. The thinness of the sample also means that the electrons are less spread by their interaction with the sample, which means the resolution is much higher. Further details about TEM are given in Section 8.7.

8.1.2 Main signals and imaging methods in scanning electron microscopy

8.1.2.1 Secondary electrons

SEs arise from inelastic collisions; for example, an incident electron may knock an electron out of the shell of an atom in the sample. These electrons have much lower energy than the incident electrons, so, although they are generated throughout the interaction volume, they can escape only from the near surface of the specimen, where the electrons have not yet spread much. For this reason they have the highest resolution of the signals discussed here and this resolution increases as the energy of the incident electrons decreases. Resolutions of 1–2 nm are now possible at operating voltages of a few kilovolts, but this is not particularly relevant for cementitious materials.

The main factor which determines the intensity of SEs emerging, and hence the brightness of the image, is the inclination of the surface to the incoming beam (see Figure 8.2). In addition, edges of the specimen have greater amounts of surface from which the weak SEs can escape, so edges

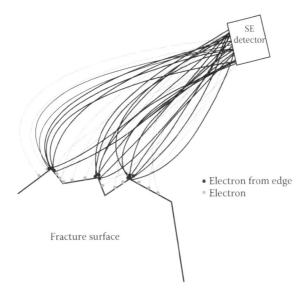

Figure 8.2 Schematic SE generation and detection.

or points also appear brighter in the image. The electrons emerging from the surface of the specimen are collected by a charged detector. So the signal is collected from most surfaces, whatever their inclination is. The result is an image of the surface topography analogous to what we see with the naked eye and visually easy for us to relate to in a qualitative sense. For example, Figure 8.3 shows the formation of C-S-H on the surface of cement grains after a few hours' hydration.

As will be discussed in detail later, the main limitation of SE images is their purely qualitative nature.

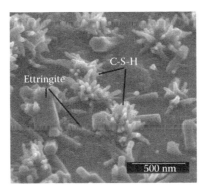

Figure 8.3 Surface of a cement grain after 4 hours' hydration.

8.1.2.2 Backscattered electrons

BSEs are the result of elastic collisions, like a ball bouncing back from a wall. They have energies similar to that of the incident electrons and so can escape from greater depth in the specimen and the images will be of lower resolution than SE images. The detector for BSEs is generally placed around the incident beam. This arrangement increases the resolution as the electrons most directly backscattered (greater scattering angle) come from the regions closest to the surface. Other arrangements, which collect electrons over a wider angle, are used for techniques such as electron back-scattered diffraction, but these are of very limited relevance to cementitious materials.

The most important aspect about BSEs is that their intensity, and so the brightness in the image, is primarily a function of the atomic number of the atoms in the sample. Bigger atoms have more electrons and the chance of the incident electron 'bouncing' on this electron cloud is greater. This relationship between BSE intensity (η) and atomic number has been well quantified:

$$\text{Element } \eta = -0.0254 + 0.016Z - 1.86 \times 10^{-4}Z^2 + 8.3 \times 10^{-7}Z^3 \quad (8.1)$$

The backscatter intensity of a compound (η) can be easily calculated from the weighted average of the elements in the compound:

$$\text{Compound } \eta_{cmp} = w_1\eta_1 + w_2\eta_2 + w_3\eta_3 + \cdots = \Sigma w_i\eta_i. \quad (8.2)$$

This leads to the types of image shown in Figure 8.4, where the different components of a cement paste can be easily identified, and most importantly, quantified by image analysis as discussed later.

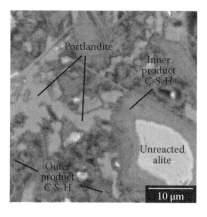

Figure 8.4 BSE image of a common cement paste.

However, in order to have good compositional contrast, it is extremely important to minimise other factors affecting contrast, particularly topography (surface roughness), by polishing the sample. If the sample is not well polished, compositional features will be obscured as discussed in Section 8.4.

8.1.2.3 Characteristic X-rays

Characteristic X-rays arise when an incident electron knocks an electron out of an inner shell of an atom. An electron from an outer shell falls back to fill the place of the ejected electron and, in doing so, emits an X-ray characteristic of the energy difference between the inner and outer shell electrons and so of the atom concerned (Figure 8.5). As the interaction of X-rays with matter is much less than that of electrons, the characteristic X-rays can escape from the whole of the interaction volume. This and the details of the X-ray detectors means that the images formed from characteristic X-rays have the lowest resolution of the techniques discussed here.

8.1.3 Environmental microscopy and other techniques to avoid prior drying

Conventional electron microscopes operate under high vacuum. This is because of the strong interaction of electrons with matter. Any gas molecules in the path of the incident beam will scatter and diffuse the electrons, at best lowering the resolution of the image. In order to place a sample in

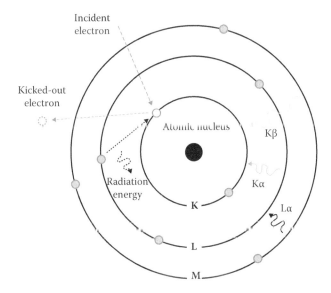

Figure 8.5 Generation of characteristic X-rays.

the high vacuum, the free water must be removed (methods for this are presented in Section 8.2). In addition, as cement-based materials are non-conducting, if samples are not coated, electrons will accumulate in the specimen, causing charge buildup which will distort the image. For this reason samples are usually coated with a thin layer of metal (usually gold) or carbon.

Recently 'environmental' microscopes have emerged on the market in which gas at low pressure is allowed into a small area just above the specimen. Some claim that such instruments have great advantages for cementitious materials. The authors of this chapter do not share this view for the following reasons. The holy grail in the study of cement hydration would be to 'see' what is happening in the real system in real time; that is to say, a cement grain surrounded by the solution in which it is hydrating. This is not possible with electron microscopy, again due to the strong interaction of electrons with matter. If a cement grain is surrounded by water, all that will be seen is the surface of the water. Environmental microscopes may image liquids, but because of the limitation of pressure in the environmental chamber, water will condense only at low temperatures. Studies have been done in which water is condensed, hydration allowed to occur, then the water removed, the sample imaged, water recondensed and so on. But this situation is very far removed from normal hydration as each time the water is removed, ions will be precipitated and then redissolved, the water-to-cement ratio (w/c) is not controlled and the temperature is quite different.

While it is now generally accepted that environmental microscopes cannot be used to follow hydration in situ, they are still claimed to have the advantage to be able to look at fracture surfaces without prior drying or other preparation. However, nowadays it is possible to look at specimens directly after fracture in normal SEMs, most of which now have low-vacuum or low-voltage modes. In low-vacuum mode the evaporation of remaining water from a sample can be tolerated and most samples do not need to be coated as the gas molecules can allow electrons to escape from the surface, avoiding charge buildup. In low-voltage mode the number of incident electrons is much reduced, so the charge can be dissipated more easily. In any case the samples must still be free of substantial liquid water.

Another possibility is to use a cryomicroscope. In this case the sample (e.g. cement paste) is frozen in liquid nitrogen. The frozen sample is then fractured and inserted into the microscope chamber still in the frozen state. The frozen water is sublimated from the surface to reveal the solid microstructure beneath.

These new methods, low-vacuum, cryomicroscopy and environmental microscopy, minimise damage to the delicate hydrate in cementitious materials (notably C-S-H). However, as illustrated in Section 8.3, the differences in morphology compared to carefully prepared conventional samples are not dramatic and support the view that conventional SEM can give images which are representative of the real microstructures existing in cement pastes.

One area where environmental microscopes can be more valuable is for the study of cementitious materials containing large amounts of organic compounds, such as latexes or oils, which suffer much more damage under vacuum or simply cannot be put in such conditions.

8.1.4 Samples for microscopy

One of the advantages of SEM is the huge range of samples which can be studied. Most of the techniques presented in this book can be, or are by far the best, applied only to cement pastes. SEM may also be applied to mortars and concretes and has the unique advantage of being able to isolate the paste part of these composites, avoiding the dilution effect of the aggregates. It is also possible to study the interface between paste and aggregates – the so-called interfacial transition zone or ITZ (Scrivener et al. 2004). A typical SEM sample is a few centimetres in size, but today many SEMs have chambers capable of taking samples up to the size of a 150 mm silicon wafer or more. For polished sections, this opens the possibility to look at how features of the microstructure vary in space – for example, from the surface of a real concrete element. However, care has to be taken about representativity. At the typical lowest magnification (e.g. nominal magnification of ×100) the field size is 2.6×1.9 mm^2, and at a magnification sufficient to resolve important details the area observed may be only a few hundred microns across or less. These matters are discussed further in Section 8.6.1.

8.2 STOPPING HYDRATION

A very common use of SEM in cement science is to look at the way the microstructure evolves over time during the hydration process, which transforms a fluid paste or concrete into a rigid solid. To look at samples at different times, first one has to stop the process of hydration by removing the free water. Even for hardened samples, free water must be removed before exposing samples to the high vacuum of the microscope chamber or as a first step in the polishing process.

8.2.1 Methods for early ages (less than 1 day)

The growth of hydrates takes place very fast during the first hours. Therefore, the chosen method to stop the hydration has to be fast in order to keep the time needed for stopping hydration short, compared to the interval between observations; typically this means a few minutes. Three different methods are compared here to illustrate their efficiency at stopping hydration.

Pastes of C_3S were studied at a water-to-solid ratio of 0.4. The hydration was stopped at 1 and 6 hours of hydration with the following three methods.

8.2.1.1 Method 1: Freeze-drying

This entails immersing the sample in liquid nitrogen (–196°C) for 15 min. This causes the water to vitrify without significant change in volume, minimising damage to the microstructure. The samples should be small to ensure that the pore water is instantly frozen. It is recommended to use plastic cylindrical moulds of diameter < 1 cm. The sample is then removed from the mould and placed in a freeze-dryer, where frozen water is removed by sublimation. For small samples at early ages, where the porosity is high, the time of freeze-drying is about 1 day. For older samples or larger sample sizes or longer freeze-drying times should be used. After this, the product is a powder which has to be stored in a desiccator, excluding moisture and CO_2, to avoid any modification of the sample.

8.2.1.2 Method 2: Filtration with isopropanol

Approximately 0.5 g of paste is taken from the sealed hydrating sample at the chosen times and immediately put in a filtrating funnel with two filter papers (retention diameter ≥ 5 μm). The setup is shown in Figure 8.6. To

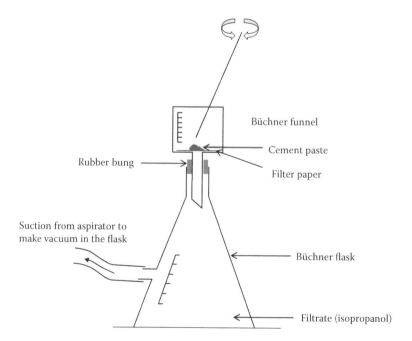

Figure 8.6 Setup for filtration with isopropanol.

minimise the exposure of the paste to air, the funnel should be filled immediately with isopropanol. Then the sample is continuously stirred in the funnel to maximise the removal of water. Afterwards, the powder from the filter is collected and further stored for 2 days under vacuum to remove the isopropanol. Part of the paste will remain stuck to the filter paper and should be discarded because traces of fibres from the paper could go into the powder and create artefacts in the microscope images.

8.2.1.3 Method 3: Immersion in isopropanol (during 24 hours)

For this method, approximately 0.5 g of the paste is taken out from the sealed sample at the required times and put in a polystyrene cylinder (e.g. 35 mm diameter × 50 mm high) filled with isopropanol (Figure 8.7). To ensure an optimum replacement of water by isopropanol in the sample, the solvent is exchanged after 1 and 15 hours followed by an agitation by hand. After 24 hours, the isopropanol is removed. The powdered sample is then collected and further stored for 1 day under vacuum drying to remove the remaining isopropanol.

Figure 8.8 compares images of powders prepared with these three techniques. It can be seen that there is no significant difference in the morphology or amount of clusters of C-S-H and/or portlandite precipitates on the surface of C_3S for the same time of hydration. Therefore, despite the samples being at early ages, isopropanol exchange (by filtration or immersion) is as suitable as freeze-drying regarding the precision of stopping the hydration in time. Moreover, the C-S-H forming on the surface of C_3S is not significantly damaged by any of the methods and looks very similar in all cases. Other phases, such as ettringite, are not present in this study of C_3S hydration. Although the stopping methods affect the crystallinity of this phase, and so detection by X-ray diffraction (XRD), they do not affect the appearance of ettringite crystals either.

Figure 8.7 Setup for immersion in isopropanol (during 24 hours).

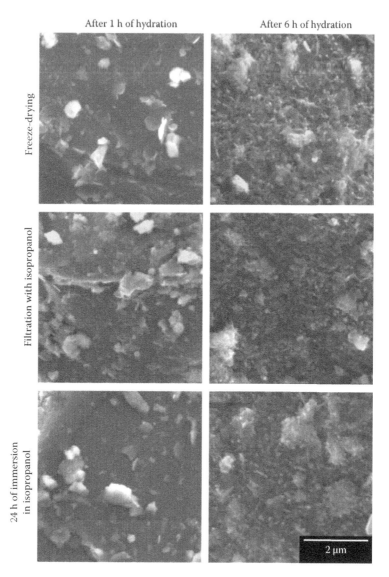

After 1 h of hydration   After 6 h of hydration

Freeze-drying

Filtration with isopropanol

24 h of immersion in isopropanol

2 µm

Figure 8.8 Comparison of the hydrated surface of C₃S with different methods for stopping hydration.

8.2.2 Stopping hydration for bulk samples (later ages from 24 hours)

For later-age samples (after about 24 hours), hydration is usually stopped by immersion during 5–7 days of a slice of cement paste approximately 3 mm thick × Ø 2.5 cm in a container filled with isopropanol. This is followed by vacuum drying in a desiccator for at least 48 hours. The volume of isopropanol should be at least 50–100 times that of the paste to facilitate solvent exchange. For an optimal removal of the water, the solvent should be exchanged twice during the first 24 hours. Mortar and concrete samples can be prepared in a similar way, but the samples are inevitably thicker and will require therefore a longer time in isopropanol.

8.3 MORPHOLOGICAL STUDIES (EARLY-AGE FRACTURE SURFACES)

After stopping the hydration, the samples (in powder form) should be gently crushed after drying in order to study single grains. The dried powder is dispersed on an adhesive carbon tab on a microscope stub. The final step is to coat with a conductive material to avoid charging. Traditionally coating is with gold, but this may give a grainy appearance on the surface, which may be visible in high-resolution images. Alternatives are carbon or osmium. Carbon coating is widely available, but it may sometimes be difficult to obtain a conductive thin coating. Osmium is the best to give a very fine coating, but because of its toxicity, is not widely available and must be used with strict attention to health and safety rules. Alternatively, as mentioned previously, samples can be studied uncoated in certain microscopes in low-vacuum or low-voltage modes. For example, the images in Figures 8.9 and 8.10 were taken in a microscope operating at 2 keV (Zeiss Merlin). Such a low voltage avoids charging effects. This microscope is designed to operate at low probe current (in this case about 200 pA) and possesses a high-sensitivity detector which permits the observation of beam-sensitive, nonconductive samples without damaging and charging effects.

A series of images with different coating techniques is shown in Figure 8.9. The first image shows the kind of damage which may occur if the coating apparatus is not used properly or a too thick coating is applied. The remaining three images show images of C-S-H in samples coated with carbon, coated with osmium and uncoated. It is clear that if the samples are coated correctly, there is very little impact on the morphology of the products. Optimum coating for carbon is 10 to 15 nm.

Throughout the whole process of specimen preparation, strict care should be taken to limit the exposure of the samples to air as this may allow carbonation to occur. Samples should always be stored in a desiccator. Ideally, fracture surfaces should be prepared and coated immediately before observation. In order to have a representative study of a single sample, the surface of at least six grains should be analysed. Any feature observed only once should not be considered representative of the sample as it is likely an artefact.

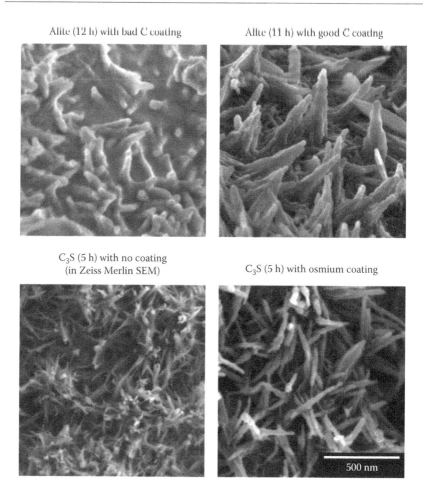

Alite (12 h) with bad C coating

Alite (11 h) with good C coating

C₃S (5 h) with no coating
(in Zeiss Merlin SEM)

C₃S (5 h) with osmium coating

500 nm

Figure 8.9 Hydrated samples coated with different techniques.

Fracture surfaces are very useful to look at the evolution of microstructure up to about 24 hours. Figure 8.10 shows the evolution in time of the surface of C_3S paste hydrated at 20°C. During this period the main feature is the growth of hydrates on the surface of the grains. For all typical commercial cements this evolution will be qualitatively similar, but fracture surfaces can give useful information about different hydration conditions, for example, the effects of the alkalinity of the solution. Contrary to the convergent needled morphology in the case of plain alite shown in Figure 8.9 (upper right), Figure 8.11a shows a planar/foil-like morphology of C-S-H observed in the presence of alkali salts and Figure 8.11b shows a divergent needle-like morphology observed in the presence of sulfate.

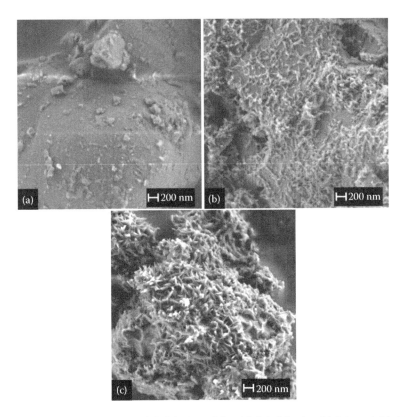

Figure 8.10 SEM observation of C₃S hydrated for (a) 1 h 30 min, (b) 3 h and (c) 6 h. Images are acquired on uncoated samples at 2 keV.

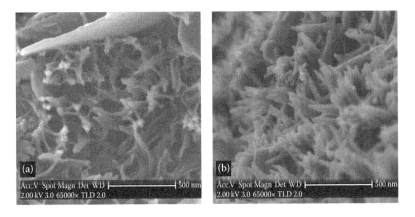

Figure 8.11 Different morphologies of C-S-H in alite pastes in the presence of (a) alkali salts and (b) sulfate.

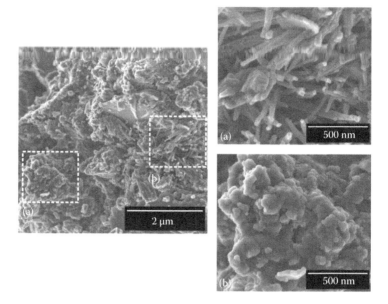

Figure 8.12 SE image of a cement sample hydrated for 7 days. Zones (a) and (b) look very different, although they are both from the same grain.

8.3.1 Later ages

Morphological studies are useful only within the first few days or so. When the degree of reaction becomes significant, the surface of the grains becomes covered and such observations have little value as everything tends to look the same, as shown in Figure 8.12, which shows a cement paste at 7 days. At the magnification needed to resolve well the morphology of the hydrates, the area observed is very small, such that the variation between images from the same sample (both framed in Figure 8.12) can be very different, even more than that of two images from different samples. It also has to be born in mind that a fracture surface is a path of weakness. Above all, the main drawback of SE imaging of cement samples is that it is not possible to obtain quantitative information so interpretation is often subjective.

It must also be noted that fracture surfaces are *not* suitable for making quantitative chemical analysis with characteristic X-rays.

8.4 BACKSCATTERED ELECTRON IMAGING OF POLISHED SECTIONS

Polished sections overcome the main limitation of fracture surfaces, in that they can provide representative images of a cross section of the microstructure.

BSE images reveal the different phases because of the atomic number contrast described earlier. Polished sections examined by BSE and characteristic X-rays can provide a wealth of information on length scales from millimetres to less than a micron. Nevertheless, the essential prerequisite is a well-polished sample. This is often challenging, particularly when the cementitious material contains aggregates (mortars and concretes). Aggregates are generally much harder than cement paste, so the paste component (which is the main subject of study) is frequently eroded between well-polished aggregates.

8.4.1 Sample preparation

This section describes the procedure to prepare a slice of paste, mortar or concrete for observation as a polished section. First, the free water must be removed as discussed previously. This is usually most conveniently done by immersing slices in isopropanol. The next stage is to impregnate the sample with resin to support the microstructure. The penetration of resin into hardened samples is quite limited, so it is useful to have a sample already flat (sawn but also lightly ground on SiC paper to improve flatness). Obtaining a good polish requires experience so it is best to learn from a trained person who will know whether any step is satisfactory or not. Regular observation of the quality of the surface under a light microscope is essential to check that polishing is proceeding correctly and has not been overdone. If relief (where some phases, such as sand grains, stand out from the surface) occurs, then the whole process should be started from the beginning with a new sample. Believe us, you will waste more time trying to rescue an overpolished sample than you will starting again.

8.4.1.1 Impregnation

The impregnation of the sample is done with a low-viscosity epoxy resin (e.g. EPO-TEK 301). The sample is placed in a cylindrical mould, with the side slightly ground with SiC paper against the bottom of the mould. A label can be placed on top of the sample, as shown in Figure 8.13, but should be written using a pencil as ink will dissolve in the resin. The impregnation should be done under vacuum, ideally with about 1–10 mbar (or less) to remove most air in accessible porosity and thus better allow infiltration of the resin.

For fluorescent light microscopy (not treated here), the resin can be mixed with a special dye (fluorescein) prior to impregnation. The choice of the resin depends on the intended usage. Most resins may be considered 'chemically transparent'. This is suitable for chemical analysis for all elements except carbon. Chlorine is often present in epoxy resins and should be accounted for prior to chemical analyses. If several samples are polished together, it is important to have a similar height of resin in all samples. Once the sample

Height of the resin

1. Write the name of the sample on a sticker and place it in a cylindrical mould. The surface of the sample to be studied has to face the bottom of the mould and the face with the sticker in the upper surface.

2. Fill the mould with resin under vacuum until the sample is fully covered. Wait for the resin to harden and take the sample out of the mould.

Figure 8.13 Schematic representation of the sample preparation for the impregnation process.

is impregnated, the resin is left to polymerise. At room temperature the waiting time is about 24 h for the resin to harden. This can be accelerated at higher temperature, but it is preferable to do it at room temperature and certainly not above 40°C and if possible to wait longer than the minimal polymerisation time to ensure the resin is completely hardened.

8.4.1.2 Polishing

Once the sample is removed from the mould, both the top side of the embedded sample (opposite the surface of the sample) and the bottom (where the sample is) are prepolished at 150 rpm using SiC paper and isopropanol as lubricant. Water should not be used at any time as it will lead to etching of any unhydrated grains. Care should also be taken that the polishing solutions are dry. Isopropanol will pick up water if not properly stored. The top side is prepolished using a 500 grade SiC paper until the surface is flat. The bottom side should be polished using a finer 1200 grade SiC paper and only for a few seconds because the resin does not penetrate very deeply below the surface. A satisfactory prepolishing should reveal the border of the sample when looking at the surface in grazing light and should have removed the pattern from the bottom of the mould. The top and bottom surfaces should be as parallel as possible to ensure a good distribution of the force during polishing. To check that the prepolishing is correct, it is useful to put the sample under an optical microscope and verify that the polishing scratches are not continuous from the resin to the sample, but it is still possible to see some resin in the sample itself. If no resin is left on the sample, the prepolishing was probably done for a too long time and it is best to prepare another sample rather than proceed. It is crucial that resin is present to stabilise the grains and products during the polishing.

Next, the sample is rinsed for a few minutes with isopropanol in an ultra-sonic bath and then polished with sprays of diamond powder of different sizes. Deodorised petrol is used as a lubricant. Figure 8.14 shows a common polishing machine. The polishing discs have to be washed between steps (or a dedicated disc kept for each grade of polishing media): the polishing disc is cleaned with soap and water using a dedicated brush (i.e. one for each size of diamond spray) to avoid contamination from larger grains. The sample must be rinsed each time with isopropanol in an ultrasonic bath to remove any residue from the surface. Polishing is best achieved going through the progressively smaller diamond grains with increasingly longer times. The optimum protocol varies from sample to sample and from machine to machine, so only general guidelines can be given and it is important to observe carefully how the polishing is proceeding under a light microscope. The typical sizes used are 9, 3 and 1 μm and sometimes 1/4 μm. The time spent at 9 μm should be low as it removes a lot of mate-rial and can easily produce relief. For most of the samples, the polishing at 1/4 μm does not lead to any improvement and it is seldom used. A general remark is when changing to a diamond spray which has particles with a smaller diameter than the one used previously, the polishing time should be at least doubled. The force applied is typically 15–25 N for pastes but can be 40 N for mortars and concretes. When decreasing the diamond grain size, the force is usually slightly increased. Here is an example of a protocol for a paste sample for our machine. However, for early-age samples, the

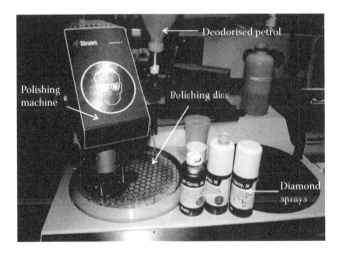

Figure 8.14 Polishing machine with the polishing disc and the diamond sprays.

times are usually shorter and require extra care because they have hardened less and have fewer hydrates to 'support' the anhydrous grains:

- 9 µm: 5 15 min, 15 N
- 3 µm: 1 h 30 minutes–2 h, 20 N
- 1 µm: 3 h or more, 25 N

Once it is polished, the sample has to be stored in a desiccator for at least 2 days in order to evaporate all the chemical products which could contaminate the SEM chamber and also limit the vacuum inside it. For the same reason, the polished samples should not be handled with bare hands but with nonpowdered gloves to keep the embedded sample clean from particles and grease, both of which can also contaminate the SEM. *These points are very important as contamination drastically reduces the performance of SEM and may result in you being banned from using shared facilities.* Polished samples should always be stored in a desiccator, avoiding moisture and CO_2, and should not be left lying around.

A well-polished sample is shown in Figure 8.15 for cement hydrated for 7 days. It contains only minor defects. The BSE image (Figure 8.15a) looks satisfactory, with only a faint scratch visible. Figure 8.15b shows the SE image, where minor defects such as small holes, dust particles and a small scratch are revealed by the topological contrast of the SE. Bright edge effects (accumulation of electrons) are visible for both the holes and the particles of the SE image.

Examples of bad polishing are shown in Figure 8.16. Often erosion of the sample occurs and it may become impossible to identify clearly the different hydrates as shown in Figure 8.16a. Figure 8.16b shows a young paste where damage of anhydrous grains has occurred because the volume of hydrates

Figure 8.15 Example of a well-polished cement sample hydrated for 7 days. The region shown contains very minor defects. (a) The image by BSE; (b) the same image by SE showing small holes, dust particles and a small scratch.

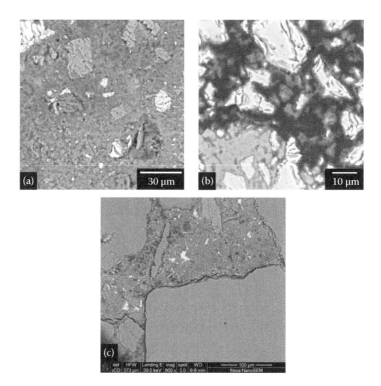

Figure 8.16 Examples of bad polishing. (a) The whole microstructure is destroyed. (b) In early-age pastes it is almost impossible to avoid damage as there is little support from hydrates. (c) In mortars there is often an important erosion of the ITZ zone, as observed in the bottom left of the image. (Courtesy of Moses Kiliswa, University of Cape Town, Cape Town, South Africa.)

is too low to support the grains during polishing. It is very difficult to prepare polished sections of pastes hydrated for less than 1 day. Figure 8.16c shows a mortar, which is generally well polished but there is severe erosion of the ITZs. This is very difficult to avoid due to the difference in hardness between pastes and aggregates and makes study of the ITZ particularly difficult. It is possible to check the quality of the polishing in the SEM by switching to SE imaging mode to ensure little damage has been caused by polishing (see Figure 8.15b). Edge charge effects indicate the absence of resin in that region and that visible cracks were not present prior to polishing.

8.4.1.3 Coating

The last step is the application of a carbon coating. For SEM in polished sections, it should be approximately 15–20 nm thick and not much more than 30 nm (for a high voltage of 15 kV). It is very important to keep

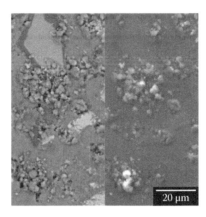

Figure 8.17 Carbonated sample showing calcium carbonate crystals on the surface.
Left: BSE image; right: SE image.

the sample in the desiccator to avoid carbonation. In Figure 8.17 there is
an example of carbonate crystals formed on the surface (portlandite was
transformed into calcium carbonate) of a sample which was not properly
stored. If a previously polished sample is carbonated, it can be prepared
once again for observation by polishing at the lowest grain size (usually
1 μm) for about 30 minutes, dried for 48 hours in a desiccator and carbon
coated again.

8.4.2 General appearance of backscattered electron images

The uses of BSE images were discussed by Scrivener (2004). A major
strength of the technique is the wide range of magnifications which can be
studied. The lower end of Figure 8.18 shows a composite image of a concrete
with a real size of 2 × 2 cm². This image was made by taking 144 images
at a nominal magnification of ×200 and then electronically 'stitching'
them together to give a composite image. In this image it is also possible
to see the very high quality of the polishing by the detail of the interfaces
between paste and aggregate. Such images are useful for studying concrete,
even from real structures, such as in the study of alkali aggregate reaction
detailed later. This technique can also be used to study other degradation
processes, which involve the ingress of species from the surrounding environment which will create gradients in microstructure at the surface, as
also discussed in detail later.

At the other end of the scale Figure 8.19 shows images at a nominal magnification of up to ×20,000. Early-age samples can show ettringite needles
or C-S-H fibrils at high magnification (Figure 8.19a). In Figure 8.19b the
resolution of a BSE image is compared to that of a fracture surface showing

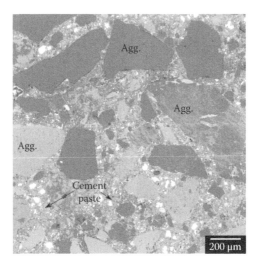

Figure 8.18 Concrete sample in a region containing cement paste and small aggregates (Agg.).

Figure 8.19 High-magnification images showing ettringite needles and C-S-H in a similar young cement paste sample. (a) BSE image of a polished paste. (b) High-magnification SE image of fracture surface. (Courtesy of Elise Berodier.)

similar features (C-S-H and ettringite needles). Because the space between grains gets filled with time, such features are less apparent at later ages.

SEM has the distinct advantage of allowing the user to focus on particular regions of a sample, e.g. the aggregates in concrete, the inner product) or the outer product regions of C-S-H at a later age and the ITZ in mortars and concretes (Scrivener et al. 2004). A further advantage of BSE imaging is that the contrast is reproducible as shown in the histograms in Figure 8.20a and b and can be used for quantitative analysis of images, as discussed further later.

However, it is important to bear in mind that we still see only 2D images of a three-dimensional (3D) microstructure. As discussed by Scrivener (2004), one should be aware about some bias in 2D sections of 3D microstructures.

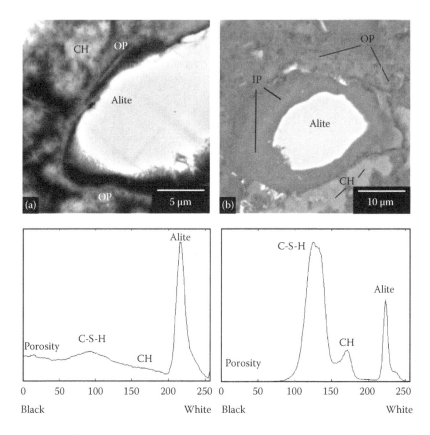

Figure 8.20 BSE images of a cement sample hydrated for (a) 1 and (b) 90 days, at 20°C. As illustrated in the histograms, it is possible to extract information from the grey levels which correspond to different phases. CH: portlandite; IP: inner product C-S-H; OP: outer product C-S-H.

They include the underestimation of the fraction of small particles and overestimation of the thickness of features such as inner product rims of C-S-H. It is also impossible to know the connectivity of the 3D structure in the volume below the observed surface. Attempts have been made to build up 3D images by serial sectioning, either by mechanical polishing (Scrivener 1988) or at very high resolution by focussed ion beam (FIB) techniques (Holzer and Münch 2009; Holzer et al. 2004, 2006, 2007; Trtik et al. 2011; Zingg et al. 2008) as presented at the end of this chapter. Unfortunately, even with modern automated FIB techniques, this process is too tedious, expensive and time consuming to be used routinely.

8.4.2.1 Accelerating voltage and resolution

Cementitious materials contain mainly light elements. Consequently, the interaction volume is larger than for heavier materials, such as metals commonly studied in the SEM. Many centrally located SEMs in materials science departments commonly operate at accelerating voltages of 20–40 kV. At such high accelerating voltages the resolution in BSE images of cementitious materials is far from optimal. In the 1980s one of the authors (Scrivener) found that 15 kV gave a good compromise between reducing the size of the interaction volume while still giving a high enough intensity of electrons. Improvements in microscope performance mean that good BSE contrast can now be obtained at even lower accelerating voltages. However, very low voltages will mean that electrons only come from the very near surface, which will usually be damaged from the polishing process, and if BSE imaging is to be combined with study by characteristic X-Ray (Section 8.5), a minimum voltage of about 12 kV is needed to excite iron, which is typically the heaviest element of interest. For general use 15 kV remains a good accelerating voltage.

8.4.3 Appearance of common cementitious materials by backscattered electron imaging

Here several examples of cementitious systems are shown to show the breadth of materials which can be studied by SEM.

8.4.3.1 Portland cement

A typical example of a plain cement paste is shown in Figure 8.21. It was hydrated for 5 years at 20°C with w/c = 0.4. Typical features can be seen.

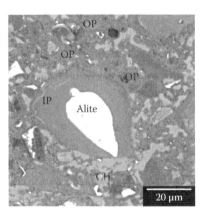

Figure 8.21 SEM image of a portland cement (5 years, 20°C, w/c = 0.4). CH: portlandite; IP: inner product C-S-H; OP: outer product C-S-H.

The water-filled space is now full of outer product C-S-H (OP) and portlandite (CH), while alite grains (and other clinker phases) have reacted. Around alite, inner product C-S-H (IP), has formed a rim broadly within the original alite grain boundaries. This image also demonstrates how as humans our attention is drawn to the 'feature' of the unhydrated alite in the centre of the image. We often conclude subjectively that there is 'a lot' of unhydrated materials left, whereas the degree of hydration (measureable by image analysis) is greater than 90% overall and close to 100% for alite.

8.4.3.2 Blends of portland cement with supplementary cementitious materials

Several binary blends of cement with a supplementary cementitious material (SCM) are shown in Figure 8.22. All were hydrated at 20°C, with a water/binder of 0.4. In these systems IP_{Cement} refers to the inner product

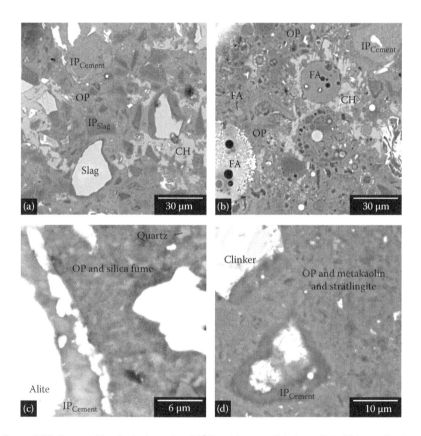

Figure 8.22 Several blends, hydrated at 20°C, with water/binder = 0.4. CH: portlandite; FA: fly ash; IP: inner product C-S-H; OP: outer product C-S-H.

from the cement grains, while IP_{SCM} refers to the inner product of the SCM in question. Unreacted slag is noted as Slag and unreacted fly ash particles are noted as FA (e.g. IP_{Slag}).

A cement–40% slag blend (Figure 8.22a) illustrates the strong changes in the microstructure caused by the presence of reactive ground granulated blast furnace slag grains. The chemical composition of a slag is usually homogeneous; thus, unreacted grains have a fixed grey level. Slag grains react similarly to alite to form a rim (IP_{Slag}) composed of an intimate mixture of C-S-H and hydrotalcite-like phases. The concentration of the magnesium containing hydrotalcite-like phases in the rims gives them a distinct dark appearance.

A cement–30% fly ash blend is shown in Figure 8.22b. Here the highly heterogeneous fly ash completely changes the appearance of the microstructure. Grains of fly ash of various sizes and of different compositions are distributed in the matrix. Several subclasses exist but are very difficult to differentiate based on grey level in the images alone. Some fly ash grains appear hollow and may even be filled with other hydrates (see the middle of Figure 8.22b). It is very difficult to determine if a fly ash grain has reacted. Some appear to have a sort of inner product rim, while most seem to remain unreacted.

Figure 8.22c and d show a blend of cement with 25% silica fume and a blend of cement with 40% metakaolin, respectively. Silica fume and metakaolin particles here have a mean diameter well below a micron and are almost impossible to distinguish even at the higher magnification shown here. While clinker and IP_{Cement} remain distinct in the microstructure, the OP region is filled with small particles of unreacted SCM or other phases such as strätlingite in the metakaolin system. Due to their size, they are very finely intermixed within the interaction volume of the SEM as discussed more in Section 8.5.

8.4.3.3 Calcium aluminate cement

Figure 8.23 shows an image of a calcium aluminate cement (CAC) after 1 day of hydration (Gosselin 2009). On the left of the image there is an anhydrous CAC grain. The hydrated matrix is composed of various hydrates. CAH_{10} and AH_x have an ill-defined morphology and seem to fill the available space. C_2AH_8 is a lamellar AFm phase and can be identified by its more even grey level and cracks formed by drying.

Figure 8.24 shows the microstructure of a CAC–gypsum blend as commonly used in ready-to-use specialist mortars. Here the matrix is composed of a mixture of ettringite (in light grey, full of cracks) and microcrystalline aluminium hydroxide (AH_3, in dark grey). Because ettringite contains a large amount of water, cracks appear due to the drying of the phase under high vacuum. Aside from those cracks, the porosity is hardly observable,

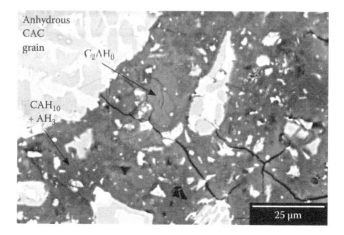

Figure 8.23 CAC after 1 day of hydration. (Courtesy of Christophe Gosselin.)

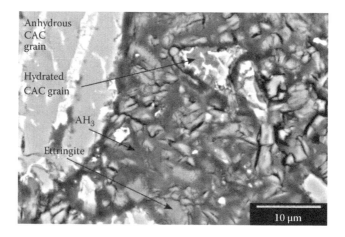

Figure 8.24 CAC with gypsum after 14 days of hydration. (Courtesy of Julien Bizzozero.)

especially when compared to images from portland cement systems in Figures 8.4 and 8.15.

8.4.3.4 Calcium sulfoaluminate cement

Figure 8.25 shows an example of a calcium sulfate cement blended with gypsum. The hydrates, ettringite and amorphous AH_3 are similar to the CAC–gypsum system (Figure 8.24) but more finely intermixed. Cracks again arise from the dehydration of ettringite in the vacuum of the SEM.

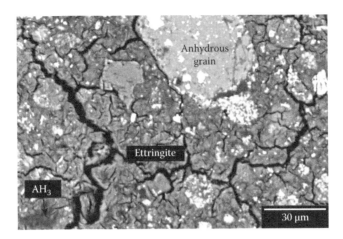

Figure 8.25 Blend of calcium sulfoaluminate cement with gypsum hydrated for 14 days. (Courtesy of Julien Bizzozero.)

8.5 CHARACTERISTIC X-RAY AND CHEMICAL ANALYSIS

More detailed information on SEM and chemical analysis can be found in many textbooks. A highly recommended read is *Scanning Electron Microscopy and X-Ray Microanalysis* by Goldstein et al. (2003).

It is very important to note that *quantitative chemical analysis by characteristic X-rays can only be made on polished samples in high-vacuum conditions*. Fracture surfaces and environmental conditions do not allow accurate analysis.

The range of detectable elements depends on the accelerating voltage. In cementitious samples, we are primarily interested in, but not limited to, detecting O, Na, Mg, Al, Si, P, S, Cl, K, Ca, Ti and Fe. Hydrogen is not detectable because it has no characteristic X-ray. The consequences of this are discussed further. The overvoltage U/U_0 (U is the SEM beam energy in kilo-electron-volts and U_0 is the considered X-ray energy in kilo-electron-volts) should be about 2.0 to properly measure the element with the highest energy X-ray (Kα). As iron Fe (Kα = 6.405 keV) is usually the element in cementitious materials with the highest energy X-ray, fifteen kilovolts still remains the recommended voltage (Harrisson et al. 1987) in order to have standard conditions for all types of samples.

It is common to use energy-dispersive spectrometry (EDS) or wavelength-dispersive spectrometry (WDS) to make chemical analyses in SEM. While WDS can have a lower detection threshold and signal-to-noise ratio (S/N) compared to EDS, it requires longer acquisition times and higher currents

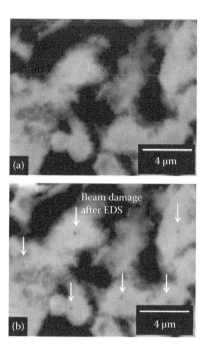

Figure 8.26 Alite paste at early age (a) before and (b) after the EDS analysis, where arrows indicate the sample damage (holes).

than for EDS to operate properly. Such conditions cause more damage to the beam-sensitive cement samples, as shown in Figure 8.26. In our experience both these reasons favour the choice of EDS in order to carry out shorter and more numerous analyses in a same amount of time. This section will deal exclusively with EDS.

As discussed by Rossen (2014) and Rossen et al. (2013), point analyses can be carried out on a few hundred points in only a few hours and give general information on the phases present. The optimal conditions should be determined for each combination of SEM and EDS detector because cement hydrates are highly susceptible to beam damage. General guidelines for the setup of the microscope include the following (Rossen 2014; Rossen et al. 2013):

- Use low currents (less than 1 nA in a FEI Quanta 200 with a tungsten source).
- Care should be taken with field emission gun (FEG) sources as they may easily have beam currents damaging to hydrates.
- Counting time should be short (but sufficient to obtain quantifiable spectra).
- For the use of mapping mode, short dwell times are preferred (100–200 μs/pixel) to limit damage.

Under some conditions, alkali ions such as Na migrate under the electron beam and should then be measured using a rastered (nonstationary) beam.

The quantification of the EDS analyses should be done using ZAF matrix correction schemes. The matrix effects arise from the fact that, in impure samples, there are differences in the elastic and inelastic scattering processes of X-rays and in the way they propagate through the specimen before reaching the detector. These effects are separated into the atomic number Z_i, the X-ray absorption A_i and the X-ray fluorescence (XRF) F_i. The most common form of the equation to correct for matrix effects (Goldstein et al. 2003) is

$$\frac{C_i}{C_{(i)}} = [\text{ZAF}]_i \cdot \frac{I_i}{I_{(i)}}, \tag{8.3}$$

where C_i is the mass fraction of the element i of interest in the sample, $C_{(i)}$ is the mass fraction of i in the standard, I_i and $I_{(i)}$ are the above-background intensities of the element in the sample and in the standard, respectively. Equation 8.3 is applied separately for each element i inside the sample and is the basis for both EDS and WDS. Each effect can be estimated from a model or experimental approach. The X-ray generation with depth, $\varphi(\rho z)$ (phi rho z), is often available in software packages to improve the calculation of the ZAF factors and should be used when possible.

Calibration should be done using a proper set of standards. These standards are typical oxides and metals relevant to cementitious materials which include compounds such as wollastonite, jadeite, silica and alumina. It is extremely important that the spectra from the standards be recorded in the same conditions used for analysis, most importantly the beam current. This is because the principle of quantification relies on the comparison between the elemental intensities of the unknown spectrum and those from a set of standards. For convenience, it is useful to define a preset list of elements such as O, Na, Mg, Al, Si, P, S, Cl, K, Ca, Ti and Fe but this can be adjusted as needed.

Oxygen is not reliably quantified as it is a peak at low energy and which can be sensitive to the thin EDS detector window and not perfectly deconvoluted during quantification. It is therefore usually quantified by stoichiometry by defining a list of oxides (e.g. SiO_2 and CaO), calculating the atomic percent (at.%) of oxygen according to the oxides considered and then by converting the atomic percent of oxygen back to mass percent. When representing the composition of different phases, one should use atomic ratios which minimise errors in quantifying oxygen and also minimise changes due to slight variations in beam current.

The quantification of oxygen by stoichiometry can also be used to differentiate between hydrates and anhydrous material. Because of the presence of water, analysis totals in hydrates never reach 100% because hydrogen is not taken into account. In the vacuum of a microscope, it is estimated from thermal dehydration and intensive drying at room temperature that the water content in hydrates is around 15–25 wt.% (Taylor 1997). In addition, porosity and irradiation damage may also lead to low analysis totals (Harrisson et al. 1987; Kjellsen and Helsing Atlassi 1998). So while anhydrous phases such as alite and belite will give analysis totals of ≈95%–105%, the total from hydrates may range from 65% to 85%.

To minimise errors in quantification, it is recommended for EDS analyses to use atomic ratios as results from chemical analyses (e.g. Ca/Si and Al/Si) and report analysis totals with them (Harrisson et al. 1987).

The most important factor limiting the interpretation of microanalyses is the intermixing of phases which often occurs within the interaction volume. For this reason the user should manually place the beam on regions to analyse in order to focus on the important phases. In doing so, one should avoid interfaces by keeping a distance of at least a micron from them. These efforts may not be sufficient with small particles, such as ground metakaolin and silica fume particles, which have mean diameters less than a micrometre.

8.5.1 C-S-H: Scatter plots and average composition

To deal with this problem of intermixing, atomic ratios for EDS point analyses are commonly plotted on 2D (or 3D) scatter plots, which can be interpreted to give the compositions of the phases present and the intermixing between them. This is particularly useful for estimating the average C-S-H composition (or C-A-S-H composition when aluminium is present).

Examples of scatter plots are given in Figure 8.27a and b, with Si/Ca versus Al/Ca and Al/Ca versus Si/Ca representations, respectively. Analyses of

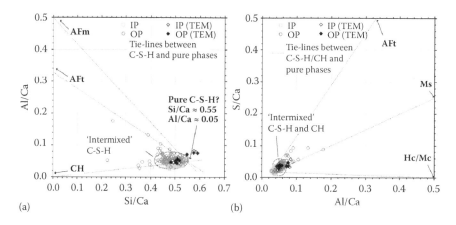

Figure 8.27 (a) The 2D scatter plot of EDS analyses of the inner product (IP) and outer product (OP) C-S-H of a 5-year-old cement paste hydrated at 20°C with w/c = 0.4; (b) same plot as a with different atomic ratios. AFt: ettringite; AFm and Ms: monosulfate; Hc: hemicarbonate; Mc: monocarbonate.

the inner product and the outer product C-S-H of a matured cement paste (5 years old and hydrated at 20°C with w/c = 0.4) were done separately here and are shown as two sets of data.

In such plots, there is the C-S-H region with typical Ca/Si between 1.50 and 2.00 for plain cements, alite and C_3S (Si/Ca between ≈0.65 and 0.5) or in a larger range of Ca/Si ≈ 0.8–2.0 for blended systems. Other phases of known theoretical compositions can be placed on the plot. They include portlandite (CH), ettringite (AFt), monosulfate (AFm or Ms), hemicarbonate (Hc) and monocarbonate (Mc). Their locations are indicated on the plots, e.g. CH at (0, 0) and AFt at (0, 0.33) in Figure 8.27a. Lines can be drawn between the cloud of C-S-H points and the different phases. The lines are 'tie-lines' and indicate the lines along which measured points are likely to correspond to a binary mixture of C-S-H and the phase in question. In Figure 8.27a, intermixing appears to occur between C-S-H and CH, as well as between C-S-H and AFt. Figure 8.27b also takes sulfate (S/Ca) into account and shows that intermixing of C-S-H with monosulfate occurs as well. Usually, the inner product points are not intermixed as much as those from outer product because the analyses are done in regions around the unreacted clinker grains where we expect to have mainly C-S-H.

C-S-H can include aluminium in place of silicon and also include sulfate, which co-adsorbs with calcium ions. For real systems containing significant sulfate, it is useful to use a corrected value for calcium according to Ca_{corr} = Ca – S, which assumes that every sulfate ion adsorbed on C-S-H has done so with one calcium ion. The plots do not change significantly, but the correction can be useful for comparing different samples or techniques. Examples of such plots can be found elsewhere (Rossen 2014).

Scatter plots are usually sufficient to determine the C-(A)-S-H composition because in systems where the intermixing is not a serious problem, the composition is best estimated using the edge of the cloud of C-(A)-S-H points, i.e. the 'least-intermixed' points. In Figure 8.27a, the edge of the cloud of points is indicated by an arrow and the corresponding atomic ratios for C-(A)-S-H are given. Good agreement is found between this edge and TEM analyses where intermixing can be avoided, as shown in Section 8.7.4.2, Figure 8.53.

To summarise, to properly carry out EDS analyses in polished samples, the following guidelines should be followed:

- In the SEM, the beam current should be about 0.8–1.0 nA to minimise damage and effects on the analyses. Damage is more critical in field emission sources which produce more focussed beams, i.e. higher current density. This can be limited by slightly defocusing the beam. With a tungsten (W) source, the current density is much lower and less problematic in most situations.
- Counting time (or exposure time) is dependent on the combination of SEM and EDS detector. It is a compromise between good statistics per spectrum and not damaging the sample by making holes in it.

The time per point should be set up with respect to the total electron dose (ideally constant between experiments). Because we work at low currents, the counting time should compensate for variations in the number of X rays detected per unit time. This can be done by fixing a total number of X-ray counts as the criteria for stopping an analysis, rather than a time in seconds. The advent of new silicon drift fast detectors allows the counting time to be reduced to around 5–10 s in contrast to the 50–100 s used with old technology detectors.

- 2D scatter plots of atomic ratios are a convenient way to discriminate the intermixed phases present in the sample. In many situations, C-(A)-S-H composition can be estimated using the tie-lines and taking the value at the edge of the cloud of C-(A)-S-H.
- A proper expression for the 'calcium/silicate' ratio is generally (Ca – S)/(Si + Al) because of substitution of silica by alumina and sulfate adsorption occurring in C-(A)-S-H. At the very least, aluminium should be taken into account as it is incorporated in C-S-H.
- Estimating an 'average' C-(A)-S-H atomic composition relies on data acquired by manual choice of the points of interest. Points should be measured in at least 10 different zones across the sample and at sufficiently high magnification (at least ×4000 in our setup). No more than 20 points should be measured in each region. Areas less than about 5 μm wide should not be analyzed as interfaces between phases should be avoided. Automated analyses are not optimal because they cannot distinguish the inner product and outer product regions and are of limited interest because the anhydrous grains, the CH and interfaces are often measured (see Figure 8.28). Figure 8.28 compares

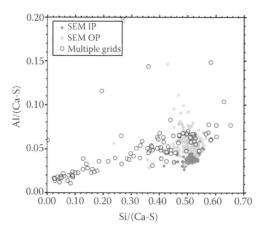

Figure 8.28 Comparison of SEM-EDS data points manually chosen inner product (IP) and outer product (OP) C-S-H regions in a cement paste with the combined data of 25 images (at ×4000 magnification), each containing a rectangular grid of 20 points.

automated analyses to chosen points in the inner and outer products. The automated analyses capture far less of C-S-H, do not distinguish the inner product and outer products and contain far more points not useful for the analyses.

8.6 DIFFERENT APPLICATIONS OF SCANNING ELECTRON MICROSCOPY

8.6.1 Quantification and image analysis for hydration

It was shown by Delesse (1848) that the volume fraction of a given phase is equal to the average surface fraction in a 2D microstructure provided the number of sections analysed is large enough to be statistically representative. This principle is widely used in materials science to quantify elements of microstructures and applies well to cement and concrete observed in SEM by BSE on polished sections.

The use of grey level segmentation based on the histogram of BSE images is possible for many applications. As recorded images are encoded using between 8 and 16 bits, i.e. between $2^8 = 256$ and $2^{16} = 65536$ grey levels, the main challenge is to find useful and reproducible thresholds based on the histograms (two are shown at the bottom of Figure 8.20) to segment the different phases of interest, including the porosity which appears dark in BSE images.

The polishing of the sample is particularly critical for image analysis, as imperfections induced by polishing may cause problems during the segmentation of grains. Special care is therefore required when image analysis is to be carried out.

At the magnification where the features of interest can be easily identified, a single field of view will be well below the representative volume element and the amounts of different phases will vary strongly from one field to another. Consequently, it is very important to analyse enough images to obtain representative results. The number of images needed will depend on the heterogeneity of the sample and mostly on whether aggregates are present or not. Typically for a paste, around 20 images with a field width of a few hundred microns may be sufficient, but 50–100 images would be a better number. For mortars and concrete several hundred images will be required. Acquiring a large number of images can now be done automatically on most instruments, but it is very important that the microscope shows good stability. To estimate the error in the quantitative measurements, the standard error (SE) representative of the whole set of images, rather than the standard deviation (SD) between images should be used:

$$SE = SD/\sqrt{N} \quad (N = \text{number of images}). \tag{8.4}$$

Prior to any batch image acquisition, the user should ensure that the microscope filament is stable enough not to vary during the procedure and select contrast and brightness settings which make good use of the BSE detector's signal range. It should be noted that the optimisation of the grey levels may not simply be to use the full range (between 0 and 255 for 8 bits and between 0 and 65535 for 16 bits) but may need to be adjusted differently from one sample to another in order for the segmentation algorithm to function. Several trials may be necessary for the image analysis to function well on a particular sample.

The user can then proceed to setting up the microscope to record, say, 100 images over the sample without the scale bar. One should use formats such as portable network graphics (PNG), TIFF or bitmap (BMP) and avoid JPEG, which degrades the quality of the image. Text (TXT) files (or any raw format) can also be used. The thresholds can be best defined by plotting a cumulative histogram from all the images.

Only the volume fraction of grains or clusters of material that are easily resolvable can be quantified. Grains of unreacted alite are quite easily segmented and allow the degree of reaction to be determined if the w/c is known. Portlandite, while having a distinct grey level, cannot be reliably quantified as small clusters cause an overestimation or underestimation of the amount present in the microstructure. The absolute amount of porosity is impossible to quantify, but by using arbitrary thresholds, trends may be compared between samples and a good correlation can be established between image analysis and other techniques such as methanol absorption (Scrivener 2004) or ^{1}H-NMR (Muller 2014). Figure 8.29 compares the capillary porosity measured from proton NMR (Chapter 7) to the porosity measured by image analysis on the same samples.

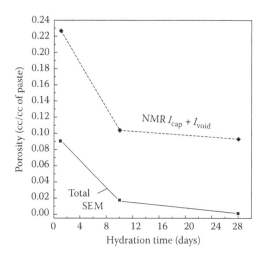

Figure 8.29 Trend in the porosity measured by image analysis compared to measurements of capillary porosity by ^{1}H-NMR (NMR $I_{cap} + I_{void}$).

For estimating the amount of different phases, other methods of bulk analyses may be preferable. One can use XRD with Rietveld analysis for crystalline materials (see Chapter 4) or thermogravimetric analysis (TGA) for the quantification of portlandite (see Chapter 5). Such methods are preferable as they require less time-consuming preparation, are usually faster to analyse and are more precise because errors due to the choice of threshold are avoided. Typically, the relative error on the degree of hydration is about 5%–10% by image analysis compared to only 2%–3% by XRD-Rietveld analysis. Also, XRD can measure alite and belite separately, while the difference in grey levels is usually too insignificant in BSE images. Similarly, TGA is very precise in determining the amount of portlandite and is a very straightforward process compared to SEM sample preparation and analysis.

For SEM it is more useful to focus efforts on quantifying features from the microstructure which may help understand mechanisms or which cannot easily be quantified by bulk methods. Examples include the following:

- The morphology of phases
- Their distribution of phases in the matrix (e.g. are there differences near or far from the ITZ or anhydrous grains?)
- The qualitative size distribution of a given phase (even if the analysis is limited in resolution due to the smallest features not being analysed in BSE)
- The density of C-S-H which is affected, e.g. by temperature (Gallucci et al. 2013) and which can be estimated from the grey level
- The observation of trends for porosity measurements (even though it is not strictly quantitative)
- The characterisation of damage, e.g. in aggregates due to alkali–silica reaction (ASR) (this is discussed later)

8.6.2 Quantification coupling chemistry and backscattered electrons

The reliability and the precision of image analysis are greatly improved by the use of elemental maps coupled with BSE images. It is often possible to find a combination of grey level and element which allows for a very reliable segmentation process. As elemental maps need to be recorded in addition to images, the process takes more time, but it gives access to new information.

With the development of faster silicon drift detectors (SDDs) for EDS, the acquisition of a set of elemental maps can be done within a couple of minutes, in contrast to at least a few hours with older technology. Fully quantifiable spectral maps (a map where the full EDS spectrum is recorded at each pixel) of 1024 × 768 pixels can be recorded in under an hour. Two examples using the combination of BSE and EDS are illustrated in the following.

8.6.2.1 Slag reaction

In cement pastes containing slag, it is of interest to determine the degree of reaction of the slag grains. To this end, both BSE images and Mg maps can be used to discriminate the slag grains and both their reacted rim and unreacted core. This method was developed by Kocaba et al. (2012) for portland cement–slag blends and requires the acquisition of up to 200 BSE images (30 seconds per image) and 200 corresponding Mg maps (90 seconds per map). Only with an SDD detector was it possible to achieve this, as 80,000–100,000 X-ray counts per second was necessary to obtain satisfactory data. J. Bizzozero adapted this methodology to CAC–slag systems (Bizzozero 2014) in which the grey level of the slag also corresponded to the C_2AS anhydrous phase from the cement. A new algorithm was written by Durdziński and Bizzozero using MATLAB (the code is available in the Appendix A-5 in Bizzozero [2014]) and is illustrated in Figure 8.30. The Mg maps are first averaged by using a Hamming window and then segmented to obtain a mask containing slag particles only. In such masks, the pixels have a value of 1 where slag is present and 0 where it is not present. Using the known positions of pixels corresponding to slag by Mg mapping, the grey level thresholds can be determined for the corresponding BSE image. The BSE images are then segmented and finally averaged using a Hamming window to obtain the second mask. The next operation consists of multiplying both masks, resulting in an image containing only slag grains. Then an operation to close the holes is used. This is done for 225 different areas so the result is averaged to obtain a reliable volume fraction according to the Delesse theorem (Delesse 1848) mentioned earlier.

It can be seen from the images in Figure 8.30 that the main error is in the detection of small grains. The consequent error can be estimated from the particle size distribution of the slag if known. Generally the proportion of particles less than the typical threshold of 1–2 μm is a small percentage (<5%). This error will be significant only at early ages. At later ages these small particles have all reacted anyway. Figure 8.31 shows a comparison of the amount of slag reaction measured by image analysis (by grey level alone [BSE] and with the EDS mapping [EDS] and by the partial or no known crystal structure [PONKCS] method with XRD; see Chapter 4). The agreement between the methods is generally good. However, the SEM methods overestimate the degree of reaction at 1 day due to the fact that small grains cannot be well segmented.

8.6.2.2 Fly ash characterisation

Another method was developed by Durdziński et al. (2013) to assess the reactivity of calcareous fly ashes. Here high-resolution (1024 × 768 pixels) spectral maps are recorded and quantified, treated and presented in a

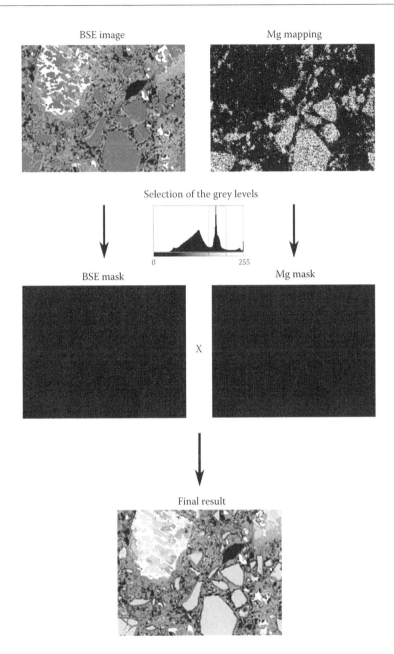

Figure 8.30 Description of the analysis by BSE and Mg mapping for CAC-slag systems. (Courtesy of Julien Bizzozero.)

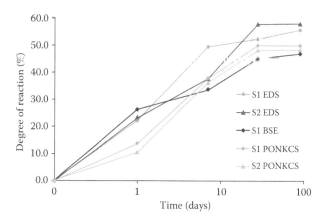

Figure 8.31 Comparison of degrees of reaction of two batches of slag from the same source from polished sections in the SEM (BSE and EDS) and by PONKCS/XRD (Chapter 4).

Ca-Si-Al ternary frequency plot (Figure 8.32c) in order to define relevant subclasses (Figure 8.32b and d).

Once the subclasses are identified, additional criteria can be used (e.g. a BSE mask to remove porosity or the unwanted contribution of distinct phases) to further improve the data set. Finally, each subclass can be monitored over time and the reaction can be shown as a plot (Figure 8.33).

The successful combination of SEM and EDS can allow the user here to get useful information on which type of ash dominates the reaction. This method is highly time consuming as collection of data for each sample takes about 4 hours and quantification of the recorded spectral maps takes about the same time. However, the clear advantage is that millions of points which are plotted allow for a reliable data treatment. This is possible only when using modern SDD detectors.

8.6.3 Degradation reactions

8.6.3.1 Sulfate attack

A good example of combining information from polished sections in the SEM is the study of the mechanism of sulfate attack made by Yu et al. (2013). Here EDS maps of Ca and S were used to determine sulfate and calcium profiles inside mortars immersed in sodium sulfate solutions. These were used to identify regions of interest, which were studied in detail by BSE images and carefully selected EDS analyses plotted in scatter plots.

Using fixed acquisition conditions, maps of Ca and S (Figure 8.34) of 512×384 pixels covering an area of $2500 \times 1870 \ \mu m^2$ were recorded. Using a code for image analysis, they were converted to images with 256 grey levels. A segmented mask excluding the aggregates was made from the Ca

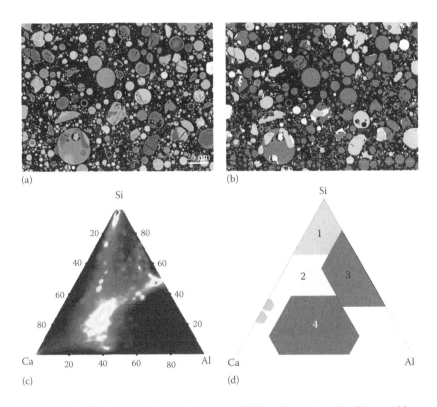

(a)   (b)

(c)   (d)

Figure 8.32 (a) BSE image of fly ash; (b) classes of glass shown in original image; (c) presentation of the full spectral data in a ternary diagram; (d) definition of the different subclasses 1–4. (Courtesy of Paweł Durdziński.)

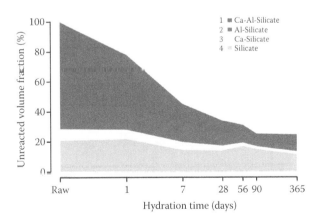

1 ■ Ca-Al-Silicate
2 ■ Al-Silicate
3   Ca-Silicate
4 ■ Silicate

Figure 8.33 Evolution over time of the fraction of all four fly ash subclasses. (Courtesy of Paweł Durdziński.)

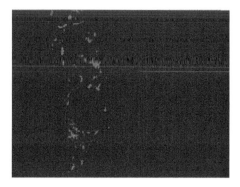

Figure 8.34 S elemental map. The area is 2500 × 1870 μm². (Courtesy of Cheng Yu.)

map by considering that all pixels with a value of 0–4 are aggregates and fixing their intensity to 0, while fixing other pixels to a value of 1. In the S map, the intensities outside the aggregates were summed over each column (parallel to the surface) and divided by the total amount of valid pixels in the column (excluding the aggregates) in order to obtain a profile of sulfur intensity as a function of penetration depth. Then the profiles were calibrated by taking the mean sulfur intensity in the core of the sample as the concentration (in wt.%) measured by EDS in the bulk. By this method profiles of sulfate (expressed as SO_3 in wt.%) as a function of depth could be obtained (Figure 8.35).

At different depths from the surface, the C-S-H was analysed. Figure 8.36 shows the plots at the depths of 2, 3 and 4 mm from the surface for samples

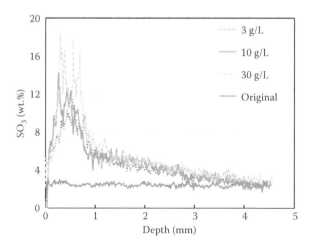

Figure 8.35 Sulfate profiles after 120 days of exposure for cement mortars immersed in different concentrations of Na_2SO_4. (Data from Yu, C. et al., *Cement and Concrete Research*, 43, 105–111, 2013.)

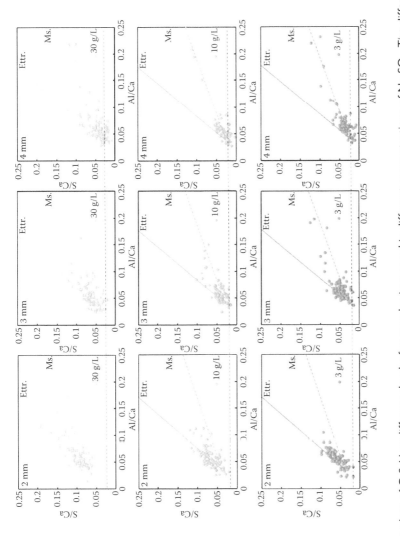

Figure 8.36 Scatter plots of C-S-H at different depths for samples immersed in different concentrations of Na$_2$SO$_4$. The different levels of S/Ca reflect the local concentration of sulfate and are related to the supersaturation of ettringite and its expansive potential. Ettr: ettringite and Ms: monosulfate.

immersed in different concentrations of Na_2SO_4 solutions. The different levels of S/Ca reflect the local concentration of sulfate and are related to the supersaturation of ettringite and its expansive potential.

8.6.3.2 Alkali–silica reaction

ASR is highly problematic in large structures such as dams, particularly because damage is visible only several years after construction. The reactive component of aggregates can react with the alkaline pore solution to form an expansive gel, which causes cracks in the aggregates (see Figure 8.37), and a volume change on the macro scale, which induces strains.

Here the preparation of concrete samples requires additional work. There is a first impregnation under 300 bars of pressure to infiltrate epoxy resin stained with fluorescein, for example. Then it is prepared in the same way as typical polished samples. Concrete usually requires about 2 days of polishing because of the aggregates. The paste is normally not well polished but is not taken into account. What is crucial is getting the aggregates polished well. As discussed in Section 8.4.1.2, it is usually very difficult to have both the paste and the aggregates devoid of polishing damage at the same time.

A set of 12 × 12 BSE images is recorded in the SEM and electronically stitched together (Figure 8.18). Using an image analysis code, e.g. from Dunant (2009), which functions for a wide variety of aggregates, the images are segmented and a morphological selection is applied to extract the aggregates:

- The code first segments the image based on grey levels. The levels can be adjusted for each series.
- Then the code uses the fast blob* detection (Sklansky 1978) to classify features which might be aggregates. As the microstructure can have composite aggregates and similar grey levels both in the cement paste and in the aggregates, the classification is done in two steps.
- Finally, after proper identification of aggregates, the largest single blob which is not an aggregate is considered paste.
- Within the aggregate mask the ASR-damaged areas are then selected by a simple grey level threshold.

It should be noted that the success of the segmentation can be sensitive to the contrast and brightness setup on the SEM. The contrast and brightness should not necessarily be pleasing to look at but must emphasise the contrast between aggregate features and maximise the use of the range of grey levels.

---

* A blob is defined as a set of connected pixels given a rule for deciding if two pixels are connected.

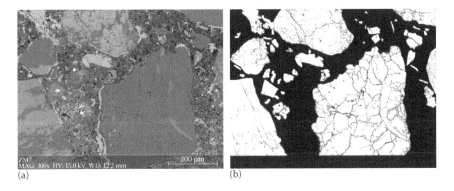

Figure 8.37 (a) concrete sample where cracks have appeared in the aggregates due to ASR; (b) the selection of the ASR-damaged areas at left for analysis. (Courtesy of Lionel Sofia.)

8.7 TRANSMISSION ELECTRON MICROSCOPY

8.7.1 Basics about transmission electron microscopy

TEM allows for very local observation of the sample. Details that cannot be captured in SEM can be studied in TEM, such as the morphology of hydrates and chemical analyses without the problems of intermixing. However, due to the high magnification and specimen preparation, the field of view is limited, and care is needed regarding how representative the observed area is. Also, the high current density used in TEM techniques may easily destroy the sample by excessive heating. Even low doses of electrons may have a quite significant impact on the delicate structure of C-S-H as illustrated by Figure 8.38.

The study of cementitious materials in the TEM is much, much more difficult than the SEM studies discussed previously. Unless you have at least 1 year to devote to learning about sample preparation and observation, we do not recommend even considering the technique. Very few staff in centralised facilities have experience with cementitious materials, which are very different from both classic inorganic materials (metals and ceramics) and from studies of biological materials or polymers.

There are two modes available: classical TEM, where the beam is stationary, and scanning transmission electron microscopy (STEM), where the beam scans the sample as in a SEM.

8.7.1.1 Transmission electron microscopy

In classical TEM, the electron beam is stationary and the image is formed using the transmitted beam (bright-field TEM or TEM-BF) or one of the diffracted beams (dark field or TEM-DF).

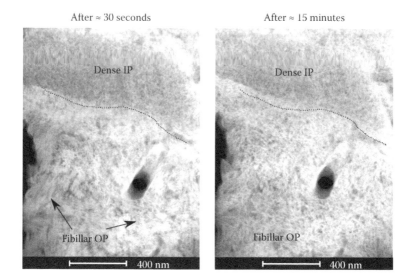

After ≈ 30 seconds   After ≈ 15 minutes

Figure 8.38 Example of moderate beam damage on C-S-H phase from normal examination in the TEM. The fine fibrillar structure visible in the left image is already less distinct in the right image, where the C-S-H appears to have a higher porosity.

8.7.1.2 Scanning transmission electron microscopy

In STEM, the electron probe is focused into a small probe and scanned over the sample. At any point of the specimen the signal generated is detected and reconstructed to form a grey image, as in SEM. In STEM mode, different detectors are used to detect the electrons that form the image. In the examples presented in this chapter, STEM-BF and high-angle annular dark-field (HAADF) detectors were used. The first one picks up the transmitted electrons, while the HAADF detectors detects electrons scattered at high angles, giving images with Z-contrast images (Z being the atomic number).

8.7.1.3 Energy-dispersive spectrometry

In TEM, EDS quantification is done using the standardless Cliff–Lorimer method (Cliff and Lorimer 1975; Lorimer G.W. 1987). This method stipulates that the above-background intensity ratio between two peaks is proportional to their weight fraction as

$$\frac{C_A}{C_B} = k_{AB} \frac{I_A}{I_B}, \tag{8.5}$$

where C_A/C_B is the weight fraction between two elements A and B, k_{AB} is a factor taking into account the microscope parameters and I_A/I_B is the

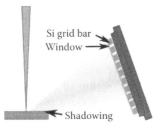

Figure 8.39 Shadowing of the EDS detector by the sample without tilting.

above-background intensities ratio between two peaks of the two elements *A* and *B*. This assumes that the thickness is small and homogeneous through the whole sample. This way, absorption and fluorescence effects (which are present for thick SEM samples and require ZAF correction) can be neglected if all elements in the sample are identified.

For quantitative analysis, the sample must be tilted towards the EDS detector. This ensures that the problem of shadowing of the detector by the sample (as illustrated in Figure 8.39) is avoided but also maximises the collection of X-rays and most importantly limits the generation of spurious X-rays from the thick regions of the sample. If the sample is not tilted, thick regions can generate additional X-rays and greatly change the measured composition.

8.7.2 Sample preparation

The preparation of TEM lamellae of cementitious materials is difficult. There are two main problems. First, the anhydrous grains are extremely hard compared to the hydrates and the embedding resin, which makes homogeneous polishing of the whole compound challenging. Second, the hydrates are beam sensitive; thus, there is a risk of deterioration during preparation and observations (Bensted and Barnes 2002; Richardson and Groves 1993). As for any preparation method, it involves first practising with a trained person and some training before becoming comfortable with any of the following methods. TEM sample preparation is very time consuming, so even if some mishap occurs at any given step, it is always worth finishing the preparation, as the feature of interest may still be observable. In the worst-case scenario, one always learns from mistakes.

Traditionally, most TEM preparation was by first mechanically polishing thin sections down to optical translucency, then cutting discs, which were subsequently thinned to allow electron transmission by ion milling. New equipment, such as the Tripod® polishing method, allows easier preparation before ion beam milling. More recently FIB techniques have opened up new possibilities for more precise sample preparation. The examples here come from the recent studies of Bazzoni (2014) and Rossen (2014).

8.7.2.1 Anhydrous samples

For anhydrous samples shown in this section, the TEM lamellas were produced by two different methods. The first method was the extraction of
a lamella by FIB using a Zeiss NVision 40 dual-beam microscope. In the
FIB microscope two canons are used: electrons are used for imaging and
ions are used for cutting, deposition of material and imaging. The dual-
beam-type FIB allows the user to make a precise selection of the region for
a TEM lamella (5×10 μm^2). The second method also involves the FIB but
avoids a step of transfer within the FIB. Instead, it relies on the milling of
several thin regions in the same embedded sample. This procedure is summarised here for a powder sample of alite and illustrated in the next section
(see Figure 8.41), where it is also used for samples hydrated for short times
(early age).

The preparation of a lamella from a bulk sample of unreacted alite is
shown in Figure 8.40. Here, the selection of the region was done using SEM
and a carbon protective layer was deposited to preserve the surface, as shown
in Figure 8.40a. The sample was then milled to produce a lamella of 5×10 μm^2 in size and 1 μm in thickness. The lamella was then extracted with
a piezo-controlled micromanipulator (Kleindiek) and welded onto a copper
support grid (Figure 8.40b). Finally, the lamella was thinned down to electron transparency (about 100 nm) as required for the TEM (Figure 8.40c).
This method allows the user to make a precise selection of the region of interest. However, its drawback is the limited size of the lamella, especially for the
particles, where only the edges and subsurface regions are of interest.

For powder samples, an example of unreacted alite is shown. The anhydrous powder was mixed with a 'hard polishing' resin and left to harden.
Semicircular discs, 3 mm in diameter, were cut from the blocks of embedded sample. The discs were mechanically polished on the two sides to reach
20–50 μm in thickness. Then some areas ('windows') containing the grain
edges were thinned down by FIB. This method allows a precise selection
of the grain, and several samples can be produced on one semidisc. There
is no need for transferring samples inside the FIB, which reduces the time
for preparation.

Figure 8.40 Production of a TEM lamella of unreacted alite by FIB: (a) selection of
the area where the lamella is extracted; (b) extraction of the lamella and
(c) thinning of the lamella fixed on the copper grid.

8.7.2.2 Early-age pastes

Hydrated samples were prepared according to the second method from the previous section but with some adjustments. Small blocks (about 1 mm in length) of dried samples were first impregnated in a low-viscosity epoxy resin in order to fill the pores to preserve the microstructure. The resin from around the sample was removed and the remaining paste–epoxy block was embedded in a hard resin (Figure 8.41a). Then the same polishing method was used to reduce the thickness to 20–50 µm and to produce thin areas, preferably selecting regions containing anhydrous grains and surrounding hydrates in the FIB microscope (Figure 8.41c). The thin areas must be prepared close to a side of the semidisc, in order to reduce the shadowing effect in EDS analysis due to the remaining walls (Figure 8.41b shows a view from above the thin region). This procedure was chosen because the hard resin was more rigid for the sample manipulation.

8.7.2.3 Later-age pastes

Later-age samples which were stopped by isopropanol solvent exchange are appropriate for mechanical polishing by means of the Tripod method and additional thinning by low-energy ion beam (precision ion polishing system or PIPS™). The brittleness of the samples is usually not a limiting factor after 14–28 days of hydration. Mechanical polishing and ion thinning is far less costly but no less time consuming than the previous methods based on the use of the FIB.

Here, a small piece of dried material is first impregnated using epoxy resin and cut to a thin slice (of dimensions $1.7 \times 1.7 \times 0.7$ mm^3) using a diamond saw with isopropanol as a lubricant. It is thinned down mechanically to a bevel by means of the Tripod method to yield a thickness of about 20–30 µm on the thick side. Diamond lapping film discs of 30, 15, 6, 3, 1 and 0.5 µm are used to progressively reach the final thickness. Isopropanol is used as a lubricant.

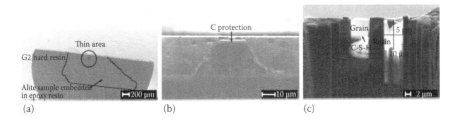

Figure 8.41 Production of a TEM lamella by mechanical polishing and FIB for the production of the thin areas: (a) overall view of the semidisc, polished to 20–50 µm; (b) a view from above the thin area showing the selection of the windows with the FIB; (c) thinning of the lamella.

Figure 8.42 Examples of the thinning of a portland cement (PC)–slag blend: (a) an ion-thinned sample on its copper ring support; (b) thinned region after 6 hours of PIPS.

The sample is then glued to a copper ring (for which a third was cut off to limit redeposition of copper during ion thinning), ion thinned with an argon PIPS operated at a maximum 1.5 keV to achieve electron transparency. As the quality of the mechanical polishing is variable, the time in the PIPS can range from an hour to a whole day. At each step the sample should be checked under an optical microscope for thin regions. A general rule is that the time spent ion thinning should be minimal. Figure 8.42a shows a portland cement–slag sample glued to a copper ring and thinned using the PIPS. Cementitious samples are extremely brittle and easily break. Generally only a small area will be thin. A close-up in Figure 8.42b shows an ideal case of thin region in the same sample. Here thickness fringes are clearly visible in reflected light mode and are located at the very edge. They are suitable for observation.

Finally, the sample is carbon coated with a 5 nm layer. Samples are coated only prior to observation in the TEM, to limit the manipulation of the sample. Samples are stored after observation in the TEM in a high-vacuum desiccator to preserve the samples as long as possible.

8.7.3 Examples of imaging

8.7.3.1 Anhydrous grains (grinding and annealing)

Anhydrous samples were studied by TEM to evaluate the quantity of defects after different stages of the fabrication of pure phases (Bazzoni et al. 2014a). TEM lamellae of C_3S and alite samples were prepared to study the distribution of defects after synthesis, after grinding and after annealing. After the first heating and quenching, there are already many structural defects, as can be seen in Figure 8.43.

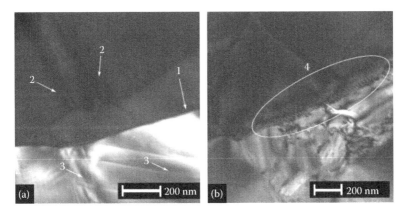

Figure 8.43 TEM-BF images of (a) C_3S and (b) alite after synthesis. (Adapted from Bazzoni et al. 2014a.)

Different types of defects are visible, indicated by numbered arrows. Arrow 1 indicates a grain boundary, identified by the contrasted grey levels on either side which correspond to different crystallographic orientations. Arrows 2 indicate twin boundaries, occurring as pairs of lines with strong fringes. Arrows 3 indicate antiphase boundaries (which is a type of stacking fault): there are always several fringes surrounding them and they can change direction. Arrow 4 indicates phase transformation dislocations which appear when there is a mismatch in the lattice. The identification of the different defects follows Amelinckx and Van Landuyt (1976) and Williams and Carter (2009). The defect types identified above result from displacive transformations, which are typical for samples undergoing polymorphic transitions during cooling (Amelinckx and Van Landuyt 1976). On quenching from 1600°C to room temperature, C_3S undergoes seven polymorphic transformations.

After grinding (Figure 8.44a for C_3S and Figure 8.44b for alite), the same types of defects are observed inside the grains. In addition, at the grain edges,

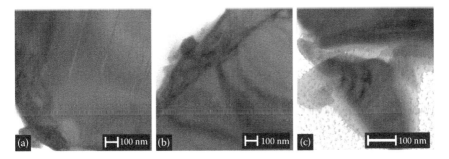

Figure 8.44 TEM-BF images of (a) C_3S and (b) alite after grinding and of (c) debris of alite. (Adapted from Bazzoni et al. 2014a.)

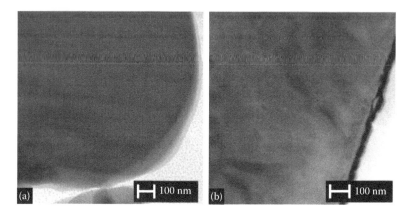

Figure 8.45 TEM-BF images of (a) C₃S and (b) alite after annealing. (Adapted from Bazzoni et al. 2014a.)

there is a region about 50 to 200 nm thick where the many different grey levels indicate a high density of defects, but these could not be characterised further. The deformed zone is not continuous all along the surface and has a variable thickness. This layer forms due to impacts between particles during the grinding procedure. Alite undergoes only one polymorphic transformation, because of the presence of ions; thus, less lattice defects are expected. However, the presence of ions induces lattice strains due to the substitution of calcium by magnesium and of silicon by aluminium. Small grains were also examined, and for both grinding methods, the small grains observed are clearly more damaged than the bigger particles, as shown in Figure 8.44c.

After annealing (Figure 8.45a for C₃S and Figure 8.45b for alite), a reduction in the defect density is observed. The damaged layer at the surface has for the most part disappeared. Defects are still present in the inner part of the grain, but in a lower density. In the case of alite, a recrystallisation is observed (Figure 8.45b).

8.7.3.2 Pastes: Early hydration

The evolution of the microstructure of C₃S paste with a w/c of 0.4, particularly the growth of C-S-H, was followed by STEM images during the first 24 h, as indicated by Figure 8.46. AN indicates the anhydrous C₃S, resin stands for the epoxy resin and the rest is C-S-H.

At the maximum heat flow, C₃S grains are completely covered by outer product, which is characterised by intermixed 'needles'. In terms of morphology, the outer product shows a fibrillar morphology (small needle-shaped particles about 10 nm wide). A gap is observed between the C-S-H and the anhydrous grains. The needles have started to fuse to form a dense layer between the surface of the grains and the outward needles and there is also the start of the formation of the inner product. At 24 h, the inner

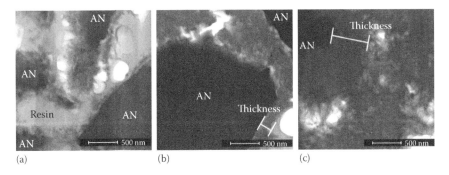

Figure 8.46 STEM-BF images of the evolution of the microstructure of C_3S paste during the first 24 h: (a) 6, (b) 8 and (c) 24 h. w/c = 0.4, cured at 20°C. AN is anhydrous material.

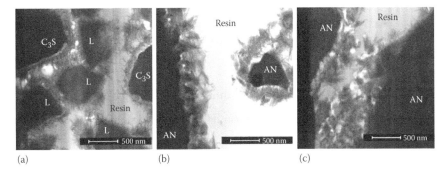

Figure 8.47 STEM-BF images of the evolution of the microstructure of (a) C_3S paste with 40 wt.% limestone water/solid ratio = 0.4, (b) alite paste w/c = 0.4 and (c) clinker paste w/c = 0.4. Curing was done at 20°C. AN is anhydrous material.

product is clearly visible and the gap between the grains and the surrounding C-S-H layer disappears. The inner product of C_3S looks like a dense packing of globules, as described by Richardson (2004). It can be observed that the thickness of the rim of C-S-H around particles increases with time. The surface of the grain is severely etched. A densification of the product is also observed with time. The C-S-H is more compact at 24 h.

As a comparison, at about 8 hours of hydration, Figure 8.47 shows C-S-H formed in a paste of C_3S with 40 wt.% of limestone (water-to-solid ratio = 0.4), C-S-H in alite paste (w/c = 0.4) and C-S-H formed in a clinker paste (w/c = 0.4)

In the presence of limestone, the needles have a fibrillar morphology, but they are thinner and are straight, perpendicular to the surface of the grains. The alite paste is similar to the C_3S paste. In clinker samples, especially on STEM cross sections, it is more difficult to distinguish between C-S-H needles and ettringite needles.

Figure 8.48 STEM-BF images of C-S-H needles embedded in CH crystals for C_3S w/c 0.4 after 28 days.

At 28 days, for a C_3S paste hydrated with a w/c of 0.4 cured at 20°C, the microstructure is filled by hydrates and C-S-H may be embedded in portlandite, as indicated by Figure 8.48.

8.7.3.3 Later microstructure and analysis

This section shows observations on later-age pastes. Alite and cement pastes hydrated for 90 days, at 20°C with w/c = 0.4, are shown in Figure 8.49a and b, respectively. Both inner product (IP) C-S-H regions and fibrillar outer product (OP) C-S-H are visible. As the alite sample (Figure 8.49a) is devoid of aluminate phases, only C-S-H and CH form. In cement systems (Figure 8.49b) relics from AFt or AFm particles removed during preparation are visible close to the border of the original alite grain.

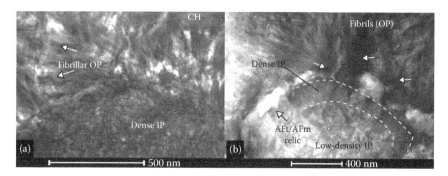

Figure 8.49 STEM-BF images of hydrated samples. (a) Inner (IP) and outer product (OP) regions of a paste of alite hydrated for 90 days at 20°C (w/c = 0.4); (b) inner and outer product region of a CEM I cement hydrated for 90 days at 20°C (w/c = 0.4).

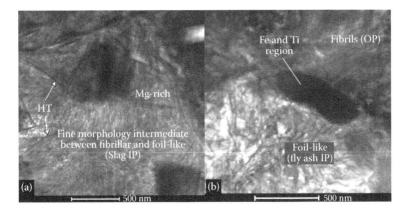

Figure 8.50 STEM-BF images of systems hydrated at 20°C with water/binder = 0.4: (a) portland cement–40% slag blend hydrated for 5 years (HT: hydrotalcite); (b) portland cement–30% fly ash hydrated for 3 years.

Blends of plain cement with SCMs are of particular interest and can reveal the changes in morphology of the C-S-H. Figure 8.50a shows a blend of portland cement with 40% slag where the inner product region in a slag grain has a fine morphology intermediate between fibrillar and foil-like. Figure 8.50b is from a blend of portland cement with 30% of fly ash and shows the more clearly foil-like C-S-H within the fly ash grain. Fibrillar outer product C-S-H is seen on its surface.

8.7.4 Chemical composition

The composition of C-S-H can be determined free from intermixing of other phases in typical paste samples. Unlike in SEM-EDS where the interaction volume is of a few cubic micrometres, STEM-EDS is able to directly show by compositional (spectral) maps which regions are rich in AFt, AFm and CH and other magnesium-rich regions, for example. This allows the user to properly select thin regions, which can then be quantified. Some examples of quantification are given in the following sections.

8.7.4.1 Early-age evolution of C/S in the C-S-H of pure C_3S

The chemical composition of C S H in C_3S paste is characterised by its calcium-to-silicon ratio (C/S). At early age, the C/S varies with the w/c and decreases with time. With the methodology described earlier for early-age paste samples, it was possible to investigate the chemical composition of C-S-H by EDS analysis from as early as a few hours up to 28 days. Figure 8.51 shows the evolution of the C/S in the three C_3S pastes with w/c of 0.4, 0.8 and 1.2.

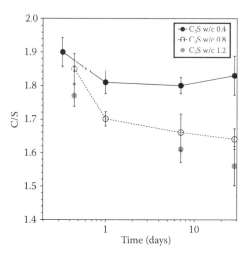

Figure 8.51 Evolution of the C/S in C$_3$S over time for three w/c ratios.

In the case of w/c of 0.4, after 24 h the C/S remains constant with a value around 1.8, which is in good agreement with previous works by Richardson (2000) and Richardson and Groves (1993). For the higher w/c, the C/S ratio seems to decrease with time, which corresponds to previous observations of Lachowski et al. (1980) on ordinary portland cement with a w/c of 0.7 and Taylor et al. (1985) on clinker with a w/c of 0.5.

8.7.4.2 C/S in matured cement pastes

Unlike C$_3$S pastes, there are other elements in the clinker and other minor phases which form. Aluminium for silicon substitution should be taken into account in the composition of C-S-H. Regions containing other very small phases may also be avoided when choosing 'pure' C-S-H regions for quantification. The following example is for a hydrated cement paste matured for 90 days at 20°C, with a w/c of 0.4. Although the BF image (Figure 8.52a) can give indications about the regions which are not C-S-H, only the EDS mapping clearly indicates which areas are truly C-S-H free from sulfur-rich and magnesium-rich phases (Figure 8.52b). It also allows the user to visually observe the detailed distribution of elements. The STEM can also be switched to diffraction mode to verify that no crystalline phases are present, but this is relatively unimportant if detailed compositional maps are used to check all elements of relevance.

With the information from EDS mapping, the user can better select areas for quantification. When this was compared to SEM-EDS data, the STEM-EDS confirms that the values measured by SEM-EDS are biased because of intermixing of C-S-H with phases on the scale smaller than the interaction

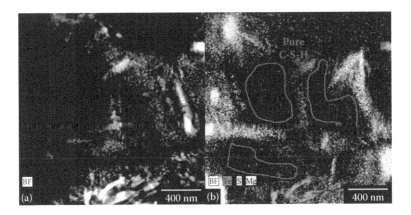

Figure 8.52 (a) Inner product (IP) region of the same cement sample as in Figure 8.49b. There is little contrast between areas of the STEM-BF image. (b) EDS maps reveal the presence of sulfur-rich and magnesium-rich areas or phases in the same inner product region.

volume, as previously discussed, and supports the use of the extreme value of the C-S-H 'cloud' in SEM as the best value for the C-S-H itself. The comparison between the SEM-EDS data with STEM-EDS data is shown in Figure 8.53 for two samples. Figure 8.53a is a white cement and Figure 8.53b is a cement–40% slag blend. Both were hydrated for 5 years at 20°C with water/binder = 0.4. Both indicate that the Si/(Ca-S) (calcium is corrected for sulfate uptake) is best estimated from the edge of the SEM-EDS points because this is the least intermixed region. The Al/(Ca-S) may be slightly overestimated. This is mostly due to the small peaks in the STEM-EDS spectra which are subject to larger errors.

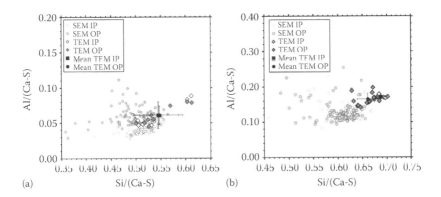

Figure 8.53 (a) EDS analyses from SEM and STEM compared for a white cement sample hydrated for 5 years; (b) same comparison as (a) for a blend with 40% slag, hydrated for 5 years.

8.7.4.3 C/S in matured blended pastes

The problem of intermixing in SEM is especially problematic in blends with SCMs of small particle size, e.g. silica fume and finely ground metakaolin. Estimation of the C-S-H composition from SEM-EDS analyses may be very difficult. STEM-EDS is therefore particularly useful to analyse areas of C-S-H devoid of very small unreacted or hydrated phases whose presence is visible in the STEM-BF image in Figure 8.54a of a cement–metakaolin blend hydrated for 90 days at 20°C. The elemental map (Figure 8.54b) confirms that metakaolin particles (MK) are still present in these images. In Figure 8.54c, the STEM-EDS analyses show where the composition of pure C-S-H is likely to be located compared to the SEM-EDS plots. Pure C-S-H is between analyses intermixed with MK and those intermixed with AFt or AFm.

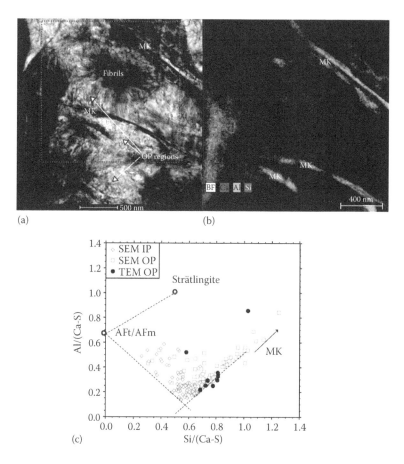

Figure 8.54 (a) Portland cement–20% metakaolin (MK) blend with a heterogeneous microstructure; (b) the corresponding STEM-EDS map of a; (c) comparison of EDS analyses of C-S-H by SEM and STEM.

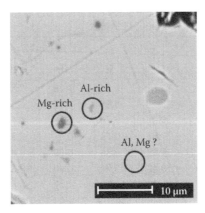

Figure 8.55 SEM-BSE image of an alite grain with Mg-rich and Al-rich clusters.

8.7.4.4 Quantification of minor elements

Low elemental concentrations, below 1 at.%, are difficult to quantify by EDS analysis. The standardless quantification methods used in STEM may mean that low concentrations cannot be as accurately measured as by SEM, as illustrated in the following example. The alite studied contained aluminium and magnesium and Al-rich and Mg-rich clusters and can be identified as indicated by the BSE image of a polished section in Figure 8.55.

XRF gives the total amount without differentiating between the amounts in exsolved regions and in solid solution in alite. The microscopy techniques permit the selection of regions without clusters. With SEM techniques, because of the large interaction volume, subsurface clusters not visible from the surface may be included, but the relatively large distance between exsolved regions makes this unlikely. The STEM technique allows a localised analysis of the composition, since the interaction volume is small because of the small thickness of the sample. Table 8.1 compares the results obtained with the different techniques to the calculated amount of each element from the synthesis recipe.

For the determination of low amounts of an element (less than 1 at.%), the SEM-WDS technique gives the best results, followed by SEM-EDS and STEM-EDS. However, SEM-WDS uses strong beam current. This is not a problem, here, for anhydrous materials, but it is likely to damage hydrated phases. In this case, SEM-EDS technique gives a good compromise between quality of the analysis and beam damage. In this example the STEM-EDS techniques does not really give any advantages, especially when the difficult sample preparation is taken into account. This example emphasises the point that a comparison of several methods is always recommended, particularly for results close to the detection limits.

Table 8.1 Atomic composition of the alites calculated and determined by XRF, SEM-WDS, SEM-EDS and STEM-EDS

	at.%					
	Ca	Si	Mg	Al	O	C/S
Calculated from synthesis	31.89	10.84	1.28	0.50	55.55	2.94
XRF batch 1	29.28	9.85	1.17	0.51	51.46	2.97
XRF batch 2	29.44	9.78	1.19	0.64	51.10	3.01
SEM-WDS batch 1	32.61	10.73	0.83	0.37	55.46	3.04
SEM-EDS batch 1	27.92	9.36	0.98	0.48	60.99	2.98
SEM-EDS batch 2	29.68	9.91	0.90	0.51	58.76	3.00
STEM-EDS batch 1	31.65	11.20	1.12	0.64	55.38	2.83
STEM-EDS batch 2	30.45	10.78	1.03	0.55	57.18	2.82

On the other hand, the STEM-EDS technique has a strong advantage for looking at the presence of minor elements in the hydrates. In order to see if magnesium and zinc enter into solid solution in hydrates, EDS mappings were done on Mg- and Zn-doped alites, hydrated for 24 h. Figure 8.56 shows the selected region, containing an anhydrous grain, C-S-H and portlandite and the resulting elemental mapping for magnesium and for zinc (Bazzoni et al. 2014b). Accurate quantification is simply not possible here, but the distribution of each element is clearly visible, thanks to the very small probe of the STEM mode.

8.7.5 Focussed ion beam microsectioning for three-dimensional reconstruction

FIB nanotomography (FIB-nt) is a recent technique which makes use of the dual beam FIB and a drift-corrected slicing procedure to then reconstruct a 3D data set which can be processed by image analysis on any plane. It has advantages over other methods which reconstruct 3D images. X-ray tomography is only able to provide a voxel (3D pixel) 0.5 to 1 μm in size, thus limiting the smallest resolvable particle size to 5 μm (Garboczi and Bullard 2004). 3D microscopy serial sectioning relies on mechanical polishing, usually not precise enough to obtain a satisfactory reconstruction. Here the interaction volume is also on the order of several cubic micrometres. Single-beam FIB techniques require sample tilting between each milling and imaging step and adds mechanical imprecision, which is overcome with dual-beam techniques provided drift correction is used (Holzer et al. 2004). Compared to TEM tomography, very thick regions can be analysed by sectioning a bulk sample over a large depth, amounting to total volumes of about $50 \times 50 \times 50$ μm³. Although FIB-nt does not reach the same spatial resolution as TEM tomography, particles of a wide range of sizes may be studied on different scales.

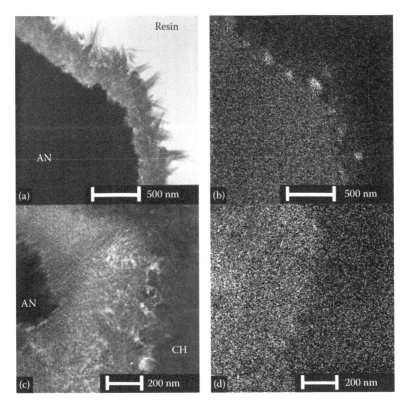

Figure 8.56 (a) STEM-BF image and (b) Mg EDS mapping, after 24 h of hydration, for alite with 1.78 wt.% of Mg; (c) STEM-BF image and (d) Zn EDS mapping, after 24 h of hydration, for alite with 1.16 wt.% Zn. AN: anhydrous grains; CH: portlandite. (Adapted from Bazzoni et al. 2014b.)

The training to use dual-beam FIB techniques has a very steep learning curve as two different beams are used simultaneously for imaging and milling. The same technique was used for sample preparation and is used here to remove slices of material as thin as possible, i.e. with a thickness on the order of the resolution of the SEM column (down to a few nanometres) at the chosen magnification. The magnification depends on the main particle size of interest and so does the thickness of each layer of the stack. This allows the data to be directly reconstructed in 3D stacks with voxel sizes which can go down to ≈10 nm. Thus, all particle sizes may be studied. The technique is illustrated schematically in Figure 8.57.

It is possible to carry out particle shape analysis and topological characterisation of granular textures in microstructures. The technique was used to study unreacted cement (Holzer et al. 2006) impregnated with Wood's metal (Bi 50%, Pb 25%, Sn 12.5%, Cd 12.5%; melting point 70°C) in order to prevent charge effects. Polishing was as standard for preparing

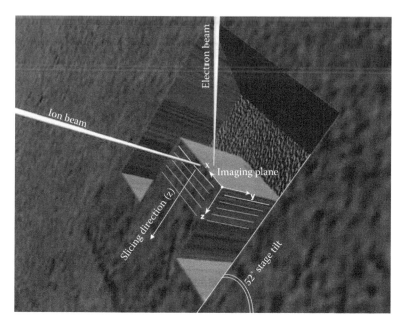

Figure 8.57 A schematic illustration of the FIB-nt technique with dual-beam FIB (52° between both beams). (Reproduced from Holzer, L. et al., *J. Microscopy*, 216(1): 84–95, 2004. With permission.)

SEM samples. The sample was glued onto an aluminium stub and the coating was done with a Pt layer deposited by sputtering.

For hydrated samples, the use of FIB-nt requires that the hydration be stopped. For samples observed during the first stages of hydration (first few hours), this was carried out by high-pressure freezing (HPF) as used in biological studies. Using HPF, water is vitrified to prevent the formation of ice crystals, which can alter the microstructure. For the kinetic and thermodynamic aspects of HPF and vitrification of water, the reader is referred to Bachmann and Mayer (1987) and Moor (1987). Cryo-FIB, as it is called, is combined with observations in cryo-SEM to have both the relatively large area observed by cryo-FIB and the high resolution associated with higher-magnification cryo-SEM techniques.

The procedure to prepare a hydrated paste for cryo-observation requires several additional steps and is beyond the scope of this chapter as the method is extremely rarely used in the field of cement science. The reader is referred to the aforementioned publications (Holzer et al. 2007; Zingg et al. 2008).

For later-age samples prepared using standard methods of stopping hydration and polishing, Trtik et al. (2011) improved upon the FIB-nt technique by making use of an energy-selective BSE detector to take two images or 'views' (in different conditions) per slice and using a 3D diffusion filter to improve the detail in the subsequent reconstruction. In particular, the second view is a prolonged exposure (corresponding to 12 scans in the conditions of the first view), which improves the contrast in part due to beam damage. This helps the segmentation process.

Such techniques based on dual-beam FIB-nt are advantageous to study cement hydration with a high resolution in time and space and resolve complex 3D heterogeneous structures. However, sample preparation is very time consuming, and only small volumes are analysed. The reliability and reproducibility of the case of FIB was discussed by Holzer and Münch (2009). The main problem with all serial sectioning techniques is that the spacing of the sections needs to be smaller than the smallest feature of interest, while the total thickness examined needs to be larger than the largest feature of interest. So while the method allows reproducible microstructural characterisations to be carried out if the size distribution of the microstructural features is not too wide, this is not really the situation which exists in cementitious materials, where the grinding process and the optimisation of rheology and packing favour a wide distribution of particle sizes.

8.8 SUMMARY

The aim of this chapter is to illustrate the use of electron microscopy for the study of cementitious materials. Although imaging of fracture surfaces with SEs is perhaps the most commonly used technique, it is of limited use after the first day or so. The most useful technique is the study of polished sections by BSEs and characteristic X-rays. We have tried to illustrate some of the diverse applications of these methods, including the possibilities to obtain quantitative information by image analysis. However, the essential prerequisite is a well-polished sample, which is not easy to achieve.

TEM is a much more specialised technique recommended only to researchers who have significant time and determination to learn the necessary methods of specimen preparation and observation. Here we have simply tried to illustrate the kind of information which can be obtained.

In Table 8.2 we summarise some of the important experimental parameters. Note that these are not exhaustive and do not necessarily apply to other microscopes than those used by the authors. They should be seen as possible *guidelines*, particularly for the TEM. Nothing replaces a complete training and good knowledge of the techniques. Only then can you conduct proper research. For reporting the main *essential* is to indicate the scale of the image: ideally with a micron marker or at least by an indication of the field width. 'Magnification' depends on the size to which the image is printed so is not generally adequate.

Table 8.2 Examples of experimental parameters for SEM and TEM

Data collection properties and settings		Examples	
		SEM	TEM
Equipment manufacturer	Microscopes	FEI, JEOL, Hitachi, Zeiss, etc.	FEI, JEOL, Hitachi, etc.
	EDS detectors	Bruker, Oxford Instruments, EDAX, SGX Sensortech, Noran, PGT, etc.	Bruker, Oxford Instruments, EDAX, SGX Sensortech, Noran, PGT, etc.
Operating conditions	Electron sources	Tungsten, LaB6, FEG	Tungsten, LaB_6, FEG
	Accelerating voltage	Fracture surfaces: 2–15 kV Polished surfaces: 15 kV	80–120 kV
	Beam current	EDS point analyses: <1 nA EDS mappings: up to 10 nA	EDS point analysis (TEM): < 0.1 nA EDS mappings (STEM): <0.1 nA
	Working distance	Imaging: 2–10 mm EDS: determined by detector	–
Detectors	Electrons	SE, BSE	BF, DF (TEM) BF, HAADF (STEM)
	Characteristic X-rays	EDS, WDS	EDS, WDS
Sample	Dimensions	Slices few millimetres thick up to several centimetres in diameter	3 mm discs polished to electron transparency by ion milling or FIB
Images	Labelling	Scale bar (preferable) or field width	Scale bar (preferable) or field width

Note: These parameters are not exhaustive and do not necessarily apply to microscopes other than those used by the authors. These parameters should be seen as possible *guidelines*, particularly for TEM, where parameters are extremely dependent on the operating conditions.

ACKNOWLEDGEMENTS

The authors of this chapter would like to thank all their colleagues who helped with discussions and images. The people who took images (other than the authors) are named in the figure captions. Amélie Bazzoni would also like to thank Marco Cantoni for his help with the FIB and microscope settings and optimisation for fragile materials such as hydrated cement and Danièle Laub, Colette Vallotton and Fabienne Bobard at the Interdisciplinary Center for Electron Microscopy (CIME), École Polytechnique Fédérale de Lausanne, for their help with the development of the samples preparation for TEM. John E. Rossen also thanks the colleagues at CIME for their support.

REFERENCES

Amelinckx, S., and J. Van Landuyt. 1976. 'Contrast Effects at Planar Interfaces'. In *Electron Microscopy in Mineralogy,* edited by Wenk, H.-R., Berlin, Heidelberg: Springer, 68–112.

Bachmann, L., and E. Mayer. 1987. 'Physics of Water and Ice: Implications for Cryofixation'. In: *Cryotechniques in Biological Electron Microscopy,* edited by Steinbrecht, R. A., and K. Zierold, Berlin, Heidelberg: Springer, 3–34.

Bazzoni, A. 2014. 'Study of Early Hydration Mechanisms of Cement by Means of Electron Microscopy' (École Polytechnique Fédérale de Lausanne, Switzerland, PhD thesis).

Bazzoni, A., M. Cantoni and K. L. Scrivener. 2014a. 'Impact of Annealing on the Early Hydration of Tricalcium Silicate'. *Journal of the American Ceramic Society* 97 (2): 584–591.

Bazzoni, A., S. Ma, Q. Wang, X. Shen, M. Cantoni and K. L. Scrivener. 2014b. 'The Effect of Magnesium and Zinc Ions on the Hydration Kinetics of C_3S'. *Journal of the American Ceramic Society* 97 (11): 3684–3693.

Bensted, J., and P. Barnes, eds. 2002. *Structure and Performance of Cements.* Second Edition. London: Spon Press.

Bizzozero, J. 2014. 'Hydration and Dimensional Stability of Calcium Aluminate Cement Based Systems' (École Polytechnique Fédérale de Lausanne, Switzerland, PhD thesis).

Chatterji, S., and J. W. Jeffery. 1966. 'Three-Dimensional Arrangement of Hydration Products in Set Cement Paste'. *Nature* 209 (5029): 1233–1234.

Cliff, G., and G. W. Lorimer. 1975. 'The quantitative analysis of thin specimens'. *Journal of Microscopy* 103 (2): 203–207.

Delesse, A. 1848. 'Procédé Mécanique Pour Déterminer La Composition Des Roches'. *Annales des Mines* 13 (4): 379–388.

Dunant, C. 2009. 'Experimental and Modelling Study of the Alkali-Silica-Reaction in Concrete' (École Polytechnique Fédérale de Lausanne, Switzerland, PhD thesis).

Durdziński, P., K. Scrivener and M. Ben Haha. 2013. 'Characterisation of Calcareous Fly Ash'. In: Proceedings of the Cement and Concrete Science Conference. University of Portsmouth, UK.

Gallucci, E., X. Zhang and K. L. Scrivener. 2013. 'Effect of Temperature on the Microstructure of Calcium Silicate Hydrate (C-S-H)'. *Cement and Concrete Research* 53: 185–195.

Garboczi, E. J., and J. W. Bullard. 2004. 'Shape Analysis of a Reference Cement'. *Cement and Concrete Research* 34 (10): 1933–1937.

Goldstein, J., D. E. Newbury, D. C. Joy, C. E. Lyman, P. Echlin, E. Lifshin, and J. R. Michael. 2003. *Scanning Electron Microscopy and X-Ray Microanalysis.* Berlin, Heidelberg: Springer.

Gosselin, C. 2009. 'Microstructural Development of Calcium Aluminate Cement Based Systems with and without Supplementary Cementitious Materials' (École Polytechnique Fédérale de Lausanne, Switzerland, PhD thesis).

Harrisson, A. M., N. B. Winter and H. F. W. Taylor. 1987. 'X-Ray-Microanalysis of Microporous Materials'. *Journal of Materials Science Letters* 6 (11): 1339–1340.

Holzer, L., Ph. Gasser, A. Kaech, M. Wegmann, A. Zingg, R. Wepf and B. Muench. 2007. 'Cryo-FIB-Nanotomography for Quantitative Analysis of Particle Structures in Cement Suspensions'. *Journal of Microscopy* **227** (3): 216–228.

Holzer, L., F. Indutnyi, Ph. Gasser, B. Münch and M. Wegmann. 2004. 'Three Dimensional Analysis of Porous BaTiO₃ Ceramics Using FIB Nanotomography'. *Journal of Microscopy* **216** (1): 84–95.

Holzer, L., B. Muench, M. Wegmann, Ph. Gasser and R. J. Flatt. 2006. 'FIB-Nanotomography of Particulate Systems – Part I: Particle Shape and Topology of Interfaces'. *Journal of the American Ceramic Society* **89** (8): 2577–2585.

Holzer, L., and B. Münch. 2009. 'Toward Reproducible Three-Dimensional Microstructure Analysis of Granular Materials and Complex Suspensions'. *Microscopy and Microanalysis* **15** (2): 130–146.

Kjellsen, K. O., and E. Helsing Atlassi. 1998. 'X-Ray Microanalysis of Hydrated Cement: Is the Analysis Total Related to Porosity?' *Cement and Concrete Research* **28** (2): 161–165.

Kocaba, V., E. Gallucci and K. L. Scrivener. 2012. 'Methods for Determination of Degree of Reaction of Slag in Blended Cement Pastes'. *Cement and Concrete Research* **42** (3): 511–525.

Lachowski, E. E., K. Mohan, H. F. W. Taylor and A. E. Moore. 1980. 'Analytical Electron Microscopy of Cement Pastes: II, Pastes of Portland Cements and Clinkers'. *Journal of the American Ceramic Society* **63** (7–8): 447–452.

Lorimer, G. W. 1987. 'Quantitative X-Ray Microanalysis of Thin Specimens in the Transmission Electron Microscope; A Review'. Mineralogical Magazine **51** (359): 49–60.

Moor, H. 1987. 'Theory and Practice of High Pressure Freezing'. In: *Cryotechniques in Biological Electron Microscopy*, edited by Steinbrecht, R. A., and K. Zierold, Berlin, Heidelberg: Springer, 175–911.

Muller, A. 2014. 'Porosity Characterisation across Different Cementitious Binders by a Multi-Technique Approach' (École Polytechnique Fédérale de Lausanne, Switzerland, PhD thesis).

Richardson, I. G. 2000. 'The Nature of the Hydration Products in Hardened Cement Pastes'. *Cement and Concrete Composites* **22** (2): 97–113.

Richardson, I. G. 2004. 'Tobermorite/jennite- and tobermorite/calcium hydroxide-based models for the structure of C-S-H: applicability to hardened pastes of tricalcium silicate, β-dicalcium silicate, Portland cement, and blends of Portland cement with blast-furnace slag, metakaolin, or silica fume'. *Cement and Concrete Research* **34** (9): 1733–1777.

Richardson, I. G., and G. W. Groves. 1993. 'Microstructure and Microanalysis of Hardened Ordinary Portland Cement Pastes'. *Journal of Materials Science* **28** (1): 265–277.

Rossen, J., B. Lothenbach and K. Scrivener. 2013. 'Optimizing the Experimental Conditions for Analyzing Calcium Silicate Hydrates in a White Portland Cement Paste'. In: Proceedings of the 14th Euroseminar on Microscopy applied to Building Materials. Helsingor, Denmark.

Rossen, J. E. 2014. 'Composition and Morphology of C-A-S-H in Pastes of Alite and Cement Blended with Supplementary Cementitious Materials'. (École Polytechnique Fédérale de Lausanne, Switzerland, PhD thesis).

Scrivener, K. L. 1988. 'The Use of Backscattered Electron Microscopy and Image Analysis to Study the Porosity of Cement Paste'. In *Pore Structure and Permeability of Cementitious Materials, Proceedings of Materials Research Society Symposium* **139**: 129–140.

Scrivener, K. L. 2004. 'Backscattered Electron Imaging of Cementitious Microstructures: Understanding and Quantification'. *Cement and Concrete Composites* **26** (8): 935–945.

Scrivener, K. L., A. K. Crumbie and P. Laugesen. 2004. 'The Interfacial Transition Zone (ITZ) Between Cement Paste and Aggregate'. *Interface Science* **12**: 411–421.

Sklansky, J. 1978. 'Image Segmentation and Feature Extraction'. *IEEE Transactions on Systems, Man and Cybernetics* **8** (4): 237–247.

Taylor, H. F. W., K. Mohan and G. K. Moir. 1985. 'Analytical Study of Pure and Extended Portland Cement Pastes: I, Pure Portland Cement Pastes'. *Journal of the American Ceramic Society* **68** (12): 680–685.

Taylor, H. F. W. 1997. *Cement Chemistry*. Second Edition. London: Thomas Telford.

Trtik, P., B. Münch, P. Gasser, A. Leemann, R. Loser, R. Wepf and P. Lura. 2011. 'Focussed Ion Beam Nanotomography Reveals the 3D Morphology of Different Solid Phases in Hardened Cement Pastes'. *Journal of Microscopy* **241** (3): 234–242.

Williams, D. B., and C. B. Carter. 2009. *Transmission Electron Microscopy: A Textbook for Materials Science*. Berlin, Heidelberg: Springer Science+Business Media.

Yu, C., W. Sun and K. Scrivener. 2013. 'Mechanism of Expansion of Mortars Immersed in Sodium Sulfate Solutions'. *Cement and Concrete Research* **43**: 105–111.

Zingg, A., L. Holzer, A. Kaech, F. Winnefeld, J. Pakusch, S. Becker and L. Gauckler 2008. 'The Microstructure of Dispersed and Non-Dispersed Fresh Cement Pastes – New Insight by Cryo-Microscopy'. *Cement and Concrete Research* **38** (4): 522–529.

Chapter 9

Mercury intrusion porosimetry

Elise Berodier, Julien Bizzozero
and Arnaud C. A. Muller

CONTENTS

9.1 INTRODUCTION

Cementitious materials are porous. The size of the pores in hydrated cement materials varies from the micron to the nanoscale. Three main categories of pores are often quoted. The first category is the compaction/air voids. They are the largest pores in a cement matrix with sizes in the range of micrometres to millimetres and result from imperfect placing. The second category is the capillary pores. There is a common consensus for capillary to be the remaining spaces which are not occupied by hydration products or unreacted cement grains. The volume fraction of capillary pores decreases with hydration as cement reacts progressively with capillary water to form hydrates. Thus, the size of capillary pores goes down from initially micrometres to a few nanometres in well-hydrated samples (Muller et al. 2013). The third category of pores is generally denominated as the gel pores. Gel pores are the intrinsic porosity of C-S-H hydrate (or C-S-H gel) and are considered by many authors to be part of it. *Gel pores* is a generic term defining all pores within C-S-H. They are nanometre-sized pores.

Few techniques are able to characterise the complex pore structure of hydrated cementitious materials. One of the most used techniques is mercury intrusion porosimetry (MIP). This technique is based on the intrusion of a nonwetting fluid (mercury) into porous structures under increasing pressure. This simple principle often makes users forget about the underlying assumptions and the limitations of the MIP technique.

This chapter aims to review the MIP technique for cement materials in the light of experimental parameters, sample preparation and theoretical considerations to help users to carry out MIP experiments and to analyse MIP results. Operator safety is an overwhelmingly important issue when using the MIP technique. Section 9.2 is devoted to safety considerations in the context of using and handling mercury. It is essential to read this part before starting MIP experiments. In Section 9.5, the basic components of MIP equipment are detailed and reviewed. The basic theoretical background is introduced. Based on experimental results, this chapter further shows the effect of operational and sample-related parameters on MIP results. Sample preparation routines including methods to stop hydration, sample preconditioning and other important parameters with relevant impacts on the results are discussed. Next, the results obtainable by MIP are discussed and compared to those of other methods which can quantify porosity. Finally, typical MIP curves for characteristic materials, curing conditions, cement type, water-to-cement ratio (w/c) and sample age are shown and described. The key points of MIP technique are summarised at the end of this chapter in Sections 9.11 and 9.12.

9.2 SAFETY CONSIDERATIONS

Mercury is highly toxic and must be handled with care. Exposure to mercury fumes must be strictly limited to the bare minimum. In this section basic safety guidelines for working with mercury are presented. The toxicity and environmental problems related to the disposal of the wastes have moved 147 countries to ban a wide range of mercury-containing products by 2020; this may disrupt mercury supplies for MIP (United Nations Environment Programme 2013).

9.2.1 Background information

Mercury (Hg), also known as quicksilver, is the only metal which is liquid at ambient temperature conditions. Mercury is very volatile (0.056 mg/h cm² at 20°C) and its vapour phase is colourless and odourless. The first rule when using mercury is to avoid fragmentation of the mercury into droplets as the evaporation rate increases with the number of droplets or exposed surface area.

9.2.2 Potential hazards

The effects of exposure to mercury may not be noticeable for months or years. Therefore, the persons working intensively with MIP should be placed under long-term medical surveillance.

Skin contact with mercury is not an important source of exposure. Nevertheless, elemental mercury droplets may be absorbed through eye contact and the use of safety glasses is required. In contrast, mercury vapour (all mercury-based components) is highly hazardous and can affect the central nervous system, lungs and kidneys. Exposure to high concentrations of Hg leads to breathing problems, whereas persons with chronic exposure to lower levels of mercury will develop tremors of the hands, eyelids, sleep disturbance and even emotional instability.

9.2.3 Safety conditions when handling mercury

9.2.3.1 Mercury intrusion porosimetry room

Mercury is not biodegradable and persists in the air. The air containing mercury vapour is highly toxic; thus, it is highly recommended to keep the MIP device in a closed room with an air ventilation system. European Union regulations stipulate that Hg concentration in a room should be lower than 20 µg/m³. In addition to having continuous air ventilation, it is advisable to aerate the room (windows opened) before and after each experiment.

The higher the temperature, the more vapour will be released from liquid metallic mercury. Thus, the temperature of the room should be maintained at 20°C to reduce its evaporation.

9.2.3.2 Protection

Protective rubber or latex laboratory gloves and laboratory coats should be worn at all times when working with equipment, be it the porosimeter or the dilatometer, or during the cleanup process.

Plastic shoe covers should be worn as well when working in the MIP room as mercury vapours are heavier than the air and may linger in higher concentrations close to the floor. As mercury can chemically combine with other metal jewellery (watches, chains, ring), it should be removed before entering in the room.

9.2.3.3 Cleaning and disposal of Hg-contaminated materials

The cleanup process may depend on the MIP device. However, in most of the available devices, the mercury has to be removed from the dilatometer (sample container).

The following points should be carefully considered for a safe cleaning of the contaminated items:

- The cleanup process should be carried out under the fume hood and in a closed box to contain the contaminated area.
- Submersion in water slows down the rate of evaporation.
- All materials and tools used during the MIP experiments and the cleanup procedure should be managed as mercury-contaminated waste. Thus, gloves and shoes protections must be stored in dedicated, sealed trash containers and all contaminated equipment and clothing (lab coats, etc.) must be kept in the closed room.
- Never pour mercury in the sink.

It is imperative that one regards these instructions as the best current practice to handle with MIP. Thus, if the user carefully considers all these safety instructions, working with MIP is quite safe. It is the user's responsibility to check whether the recommendations have been correctly implemented.

9.3 MEASUREMENT PRINCIPLE

9.3.1 Introduction

The MIP technique was first developed in the 1940s to measure pore size distributions over a wide range of diameters (Ritter and Drake 1945) and it was first used with cementitious materials in the 1970s (Diamond and Dolch 1972; Sellevold 1974; Winslow and Diamond 1969). The largest pores that can be detected by MIP are up to about 1 mm. The smallest measurable pores depend on the maximal pressure applied on the mercury. Today, pores with 2 nm radius can be measured, while in the 1970s the smallest pores were about 12 nm (Bager and Sellevold 1975; Diamond and Dolch 1972). The wide popularity of MIP stems from being relatively fast and simple. However, careful sample preparation and data analysis are essential.

One important consideration when doing MIP experiments is that the mercury intrudes only the connected porosity that can be reached by the mercury at given pressures. Thus, MIP results represent pore entry size distributions rather than pore size distributions of porous materials.

9.3.2 Theoretical background

The MIP technique is based on the nonwetting property of mercury. Most liquids easily penetrate pores by capillary pressure. If it is a wetting liquid (i.e. contact angle θ lower than 90°), the surface tension is favourable to the penetration. If it is a nonwetting liquid ($\theta > 90°$), like mercury, the surface tension is opposed to the liquid intrusion, so an external applied pressure P (Equation 9.1) is required to compensate for the pressure difference over the mercury meniscus in the pores. This pressure P is also termed *pore pressure* with

$$P = -\Delta P_c = \gamma\left(\frac{1}{r_m} + \frac{1}{r_{m'}}\right), \tag{9.1}$$

where ΔP_c is the capillary pressure, γ is the surface tension of mercury and r_m and $r_{m'}$ are the meniscus curvature radii. All these variables are undetermined in complex porous media but a simplification can be done following

$$P = \gamma C, \tag{9.2}$$

where C is the curvature of the meniscus. C depends on the contact angle between the mercury and the porous material and the pore geometry. Assuming cylindrical pores (form factor $F = 2$), the curvature is given by

$$C = \frac{2\cos(\theta)}{r}, \tag{9.3}$$

where θ is the contact angle and r is the capillary radius.

The combination of Equations 9.2 and 9.3 gives the Washburn equation (Washburn 1921),

$$P = -\frac{2\gamma\cos(\theta)}{r}, \tag{9.4}$$

for which the relationship between pore radius and pore pressure is illustrated in Figure 9.1.

Figure 9.1 shows the relationship between the pressure applied to the mercury and the corresponding radius of intruded pores. The resulting radii r depend on the parameters and assumptions used in the equation. The surface tension of the mercury (γ) is quite well established and usually researchers use $\gamma = 0.485$ N/m at 25°C (Lide 2003). The contact angle between mercury and cement paste θ is more difficult to determine and has been the topic of many investigations. θ = 140° is generally used as determined by Cook and Hover (1991), although the contact angle might vary with the chemical composition of the sample, the contamination of the surface, etc. (Rigby 2002; Kaufmann et al. 2009).

In practice, the pressure achievable by MIP equipments is limited to a maximal pressure of $P = 400$ MPa, which corresponds to a minimal pore radius of about 2 nm following a cylindrical pore model. As a consequence, pores below this size cannot be measured.

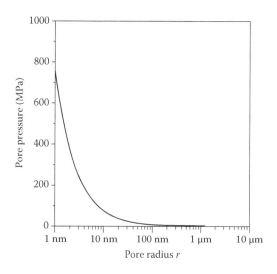

Figure 9.1 Illustration of the Washburn equation relating pore pressure to the radius of the intruded pore.

9.4 RESULT PRESENTATION

The ease of use of the MIP technique and seemingly straightforward interpretation of MIP results often make users forget the assumptions made and the limitations of the MIP technique itself. The large discrepancy of MIP results has sparked fervent discussions on the reliability of the MIP technique.

The first key point to keep in mind when doing MIP experiments is that MIP measures the pore entry sizes and not the real pore sizes of the sample. This effect is more significant than it might be expected as often a high volume of pores becomes accessible through increasingly smaller pore entries (Diamond 2000) (Figure 9.2). This effect is often called the 'ink-bottle effect' but this terminology can be misleading as it reflects the image of only dead end pores rather than large interconnected volumes accessible through smaller channels and cavities.

MIP results are in general plotted as cumulative pore volume versus pore entry radius and normalised per sample volume (as in Figure 9.3) or per gram of material. For more clarity, it is recommended to use the log scale for the x axis where the radii are plotted. Three relevant parameters characterising the pore structure can be determined from classic cumulative MIP curves:

1. Total percolated pore volume: This is the maximum pore volume recorded throughout MIP experiments. This represents the total accessible or percolated pore volume (connected pores) at the maximum

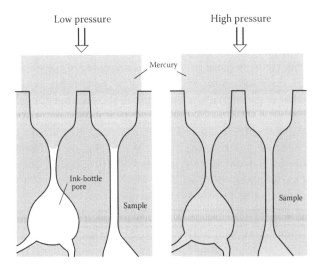

Figure 9.2. Illustration of ink-bottle pores showing large pore volume only accessible through narrow necks.

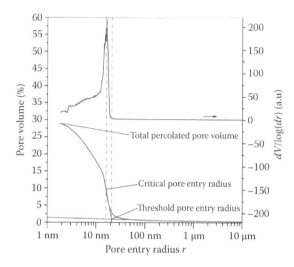

Figure 9.3 Example of typical MIP cumulative and derivative curves obtained when measuring portland cement pastes with a maximum MIP pressure of 400 MPa. The hydration was stopped by isopropanol exchange for 7 days followed by 7 days in a desiccator. The three principal characteristic values of MIP results are identified as the total percolated pore volume, the critical pore entry radius and the threshold pore entry radius.

applied pressure and should not be confused with the total pore volume of the sample (connected pores, nonconnected pores and pores with smaller than 2 nm radius). In Figure 9.3 the total percolated pore volume represents 29% of the sample volume.

2. Threshold pore entry radius: This is defined as the intersection between the two tangents of the cumulative curve, as shown in Figure 9.3. The threshold pore entry radius is interpreted as the minimum radius that is geometrically continuous throughout the whole sample (Winslow and Diamond 1969).

3. Critical pore entry radius: This is the pore size where the steepest slope of the cumulative intrusion curve is recorded. The critical pore entry radius is usually preferred over the threshold pore entry radius since it is defined mathematically by the inflection point of the main intrusion step and can thus be determined accurately.

Figure 9.3 shows a typical MIP curve for a mature portland cement paste where the three aforementioned characteristic parameters are indicated. The derivative curve better shows the critical pore entry radius than it does the cumulative curve.

9.5 EXPERIMENTAL SETUP (DEVICES, INPUT PARAMETERS, ETC.)

9.5.1 Experimental setup

MIP devices are usually composed of a filling station and a low- and a high-pressure unit. The sample is initially placed in the filling station, where it is degassed to remove the air from the pores. Then mercury is introduced into the sample container (also called dilatometer). At this stage, the sample is floating at the surface of the mercury, which remains outside the sample. The mercury is progressively pressurised and the volume intruding the sample is recorded. The pressure is usually applied automatically. The maximal reached pressure depends on the low-/high-pressure units. Most of the modern MIP equipments can reach 400 MPa.

9.5.2 Maintenance and calibrations for dilatometers (sample container)

Maintenance of MIP devices is essential to have reliable results as the accuracy of applied pressures determines the accuracy of calculated pore sizes. It is generally recommended to have a service on the MIP equipment by the manufacturer once a year. Porous glass beads can be used as a standard to check the calibration.

Additionally, a blank correction must be included to compensate for the compressibility of mercury and the deformation of the dilatometer. The blank is specific to each dilatometer and must be carried out after each maintenance and calibration of the machine.

9.5.3 Input parameters

MIP devices are usually supplied with operating software that calculates pore size distribution from the mercury pressure and the intruded volumes. This requires some inputs from the users which depend on the experimental conditions. The main input parameters usually required are as follows.

9.5.3.1 Surface tension

The surface tension of pure mercury is known as $\gamma = 0.485$ N/m at 25°C. This value might change depending on the purity of the mercury and the temperature.

9.5.3.2 Contact angle

The contact angle θ of the mercury on the sample surface is an important parameter of MIP experiments as it plays a direct role in interpreting

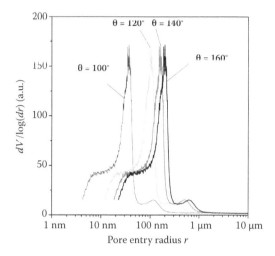

Figure 9.4 MIP first derivative results of a portland cement paste when the contact angle θ is changed to 100°, 120°, 140° and 160°.

mercury pressures into pore sizes (cf. Equation 9.4). The exact value of the contact angle between mercury and a solid surface depends on the surface chemistry and the surface roughness as well as the purity of the mercury. When this solid is as complex as a cementitious paste, the assessment of the contact angle is difficult. Experiments on cement paste provided a range of contact angle from about 120° (Shi and Winslow 1985) to more than 150° (Adolphs et al. 2002; Cook and Hover 1991). The heterogeneous chemistry of the surface, the ageing of the paste, the size of the pores (Cook and Hover 1991) and the drying technique (Hearn and Hooton 1992; Winslow and Liu 1990) are parameters susceptible to change in the contact angle value. Usually these effects are neglected and a contact angle of 140° is used in most intrusion experiments.

Figure 9.4 shows MIP results for a portland cement paste after 7 days of sealed hydration when the contact angle θ is changed to 100°, 120°, 140° and 160°. Increasing the contact angle θ shifts calculated pore sizes towards larger values. This effect is directly explained by Equation 9.4 and highlights the high sensitivity of pore size distributions calculated from MIP experiments on the value of the contact angle.

9.5.3.3 Continuous or incremental mode

The pressure applied on the mercury is automatically controlled in most modern MIP devices. There are two ways of handling increasing pressures: the continuous and the incremental mode (or step mode). The continuous mode applies a constant pressurisation rate throughout time. The incremental mode applies increments of pressure waiting each time until there

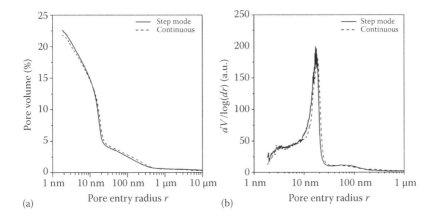

Figure 9.5 Effect of the pressurisation mode on MIP results measuring a sealed portland cement paste mixed at w/c = 0.40 after 7 days of hydration: (a) cumulative pore volumes; (b) first derivatives of the two cumulative curves.

is no further penetration by the mercury. In both cases, the pressurisation rate of the mercury can be adjusted. When using step mode, two options are possible: either volume equilibrium at incremental pressures or pressure equilibrium at incremental (intruded) volumes. The influence of the pressurisation profile is not trivial and depends on the amount of pores at given sizes, the surface-to-volume ratio of the sample, etc. In principle, the mode type and the pressurisation rate can influence both the total pore volume and the pore size distributions of probed samples (Hearn and Hooton 1992; Ma and Li 2013). Most users choose the step method as continuous increase in pressure might underestimate pore volumes due to insufficient time for equilibrium. Figure 9.5 shows MIP results for a portland cement paste after 7 days of sealed hydration for continuous and step mode experiments. In this case study, the differences are negligible between the two pressurisation modes.

9.5.3.4 Pressurisation rate

The pressurisation rate of the mercury can be adjusted to the softness of the sample. Figure 9.6 shows the effect of the pressurisation rate on MIP results measuring a portland cement paste after 7 days of sealed hydration. In this case, changing the pressurisation rate does not change MIP results significantly. This shows that the paste is hard enough to endure high pressurisation rate. Softer pastes, such as very young pastes or young pastes with a high replacement level of SCM, might be more sensitive to this parameter.

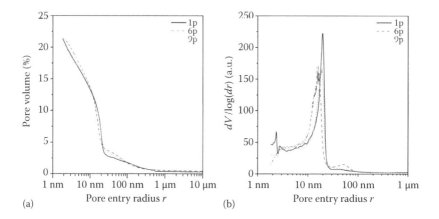

Figure 9.6 Effect of the pressurisation rate on MIP results measuring a sealed portland cement paste mixed at w/c = 0.40 after 7 days of hydration: (a) cumulative pore volumes; (b) first derivatives of the three cumulative curves. The values of pressurisation rates reported do not correspond to physical units. They are relative pressurisation rates, with 9p being the highest rate. These experiments were conducted in the step mode pressurisation method.

9.5.3.5 Sample mass and number of pieces

MIP sample mass and shape (one piece or several) are chosen arbitrarily by most users. However, the morphology of the sample might affect MIP results (Hearn and Hooton 1992; Winslow 1984). Sample mass and shape can be independently changed.

The amount of specimen probed during MIP experiments is restricted by the dimension of the dilatometer itself. Large dilatometers with up to 15 cm^3 volume exist, permitting larger amounts of sample to be measured. Having large specimens is advantageous by avoiding local heterogeneities that can affect MIP results when probing limited sample volumes (Cook and Hover 1999). In addition, large sample volumes in a given dilatometer help reduce mercury compressibility effects (Aligizaki 2005).

For a constant sample mass, increasing the number of pieces of the sample increases its surface-to-volume ratio, which reduces boundary effects. It is generally admitted that too large specimens can corrupt MIP results because of large internal volumes of ink-bottle pores. From a practical point of view, most cement paste samples are cast in cylindrical plastic containers that are cut into slices (disks) and dried prior to MIP experiments. The sample thickness is usually 2–5 mm. Dried slices are then broken into pieces to be placed in the MIP dilatometer. There are different ways of proceeding: (1) the sample may be cut into a few large pieces of irregular shapes and (2) the sample may be cut into small cubes of regular sizes (about the slice thickness). Crushing the sample should be avoided as

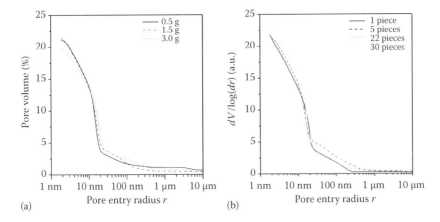

Figure 9.7 (a) Effect of the sample mass on the cumulative pore volume measuring a sealed portland cement paste mixed at w/c = 0.40 after 7 days of hydration; (b) effect of the number of pieces of sample on MIP results at constant sample mass.

it destroys the coarse pore structure and leads to 'external' porosity peaks (Hearn and Hooton 1992).

Figure 9.7a shows the effect of sample mass on MIP results measuring a portland cement paste after 7 days of sealed hydration, keeping the same sample shape. There is no clear tendency for the effect of sample mass on the final pore size distribution of the material. Figure 9.7b shows the effect of the number of pieces on MIP results for the same material, keeping the same sample mass. The same conclusion as for the sample mass can be drawn and the effect of the number of sample pieces on MIP results is relatively small.

Even though there are no apparent sample size effects on the MIP pore size distributions in the above example, it is important to keep the same sample characteristics (amount and number of pieces) in order to improve the overall reproducibility of results. As an example, Figure 9.8 shows cumulative pore volumes and first derivatives of two MIP experiments repeated with the same sample characteristics and the same measuring procedure. This highlights the good repeatability of MIP results when similar experimental conditions are carefully maintained.

9.5.3.6 Temperature

The temperature of the room is important during MIP experiments as it changes the density of the mercury and, therefore, the results for the pore size distribution. It is recommended to have at least one temperature recorder inside the MIP room. Even if the room is temperature controlled, the ambient temperature might vary by several degrees centigrade between midday

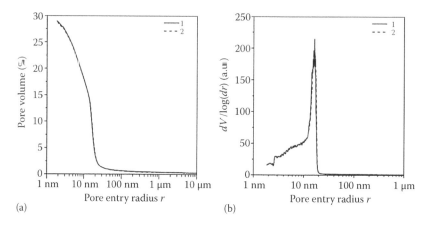

Figure 9.8 (a) Cumulative pore volumes for a portland cement paste cured under water for 3 days; (b) first derivatives of the two cumulative curves.

and evening/night. The input temperature parameter must be adjusted for each experiment.

9.6 PROCEDURE

The procedure adopted for MIP tests highly depends on the equipment. Nevertheless the following steps are independent of the device.

1. Sample preparation: Cement pastes should be mixed, cast and dried according to the recommendations given in Chapter 1 describing the best way to prepare the sample. For MIP, the solvent exchange method is best suited to preserve the pores. The sample should be cut at the required dimension and mass in the bulk of the paste (i.e. the edges of the sample should be discarded).
2. Starting the experiment: The sample is introduced in the sample container and degassed to remove the air from the porosity. Once the sample is under vacuum, the dilatometer (sample container) is progressively filled with liquid mercury.
3. Pressure increase: The experiment and data recording start when the pressure is applied to intrude the mercury in the sample. The increase in pressure sometimes proceeds in two steps depending on the equipment.
4. End of the experiment: The experiment is done when the pressure on the mercury is entirely removed. The sample container is removed from the device. The cleaning step can start (see Section 9.1).

Table 9.1 List of the parameters and data needed to report MIP results in scientific publications

Data collection properties and setting		Examples
Equipment	Manufacturer	Porotec, Micromeritics
	model	Pascal 140, Pascal 440
Input parameter	Contact angle	140°
	mode	Continuous or step
Sample	Sample mass and size	Two pieces, 1 g
	Sample type	Mortar or cement paste
	Drying method	Isopropanol exchange
	Curing	Underwater, sealed

9.7 GUIDELINES FOR REPORTING

Table 9.1 summarises the MIP information to be reported when writing scientific publications.

9.8 EFFECT OF SAMPLE PREPARATION

The sample preparation of cement-based materials is a critical step in many characterisation methods and it becomes particularly important when it comes to probing pore structures. MIP measurements require the removal of pore water beforehand and this step is known to cause damage to the pore structures of interest. Different drying methods, described in Chapter 1, are used as part of the sample preparation routine for MIP tests, among which there are oven drying, vacuum (or desiccator) drying, freeze-drying and solvent exchange drying methods. While the first two involve direct water removal, the freeze-drying and the solvent exchange methods were developed to reduce capillary stresses upon drying.

Different drying methods are known to lead to different porosity results (Gallé 2001). The reasons are manifold and researchers have pointed out microcracking, decomposition of solid phases (Mindess et al. 2003) and the amount of residual water within pores (Moukwa and Aïtcin 1988).

As illustration, Figure 9.9a shows MIP cumulative pore volumes for a white cement paste mixed at w/c = 0.40 after 28 days of sealed hydration using different sample drying methods. Figure 9.9b shows the first derivatives of these results. Taking the solvent exchange method as an example, there are two main intrusion steps which correspond to two main peaks on the derivative. The peak centred at 25 nm pore radius is part of the capillary pore volume. The degree to which the peak located at small pore radii

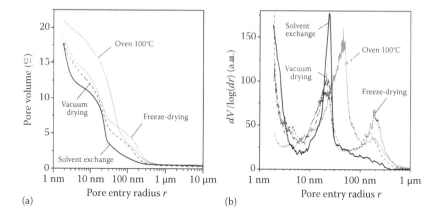

Figure 9.9 (a) Cumulative MIP pore volumes for different drying methods obtained measuring a white cement paste mixed at w/c = 0.40 after 28 days of sealed hydration; (b) first derivatives of the results presented in (a). The different drying methods used were oven drying at 100°C, freeze-drying, vacuum drying (in a desiccator with silica gel at about 10 Pa) and solvent exchange method, for which immersion in isopropanol for 7 days was carried out followed by 7 days in a desiccator.

(~2–5 nm) is associated with C-S-H hydrates is still a matter of debate. This topic will not be further discussed here.

The oven drying method at 100°C significantly increases the total accessible MIP pore volume and coarsens most of the measured porosity compared to the other methods. Freeze-drying and vacuum drying methods give comparable total accessible pore volume as the solvent exchange method but wider overall pore peak distributions. Compared to the solvent exchange, all other tested drying procedures gave rise to a peak at about 200 nm pore radius. This shows that the sample preparation has a high impact on MIP porosity results. What first appears is that oven drying methods need to be avoided for meaningful analysis of pore structures. The solvent exchange method seems most adapted to this purpose, even though it might not entirely be representative of the porosity that exists in never-dried materials. In any case and for a comparison study between different cementitious materials, the same drying method should be used.

9.9 LIMITS OF MERCURY INTRUSION POROSIMETRY, CONTROVERSY AND AGREEMENT WITH OTHERS TECHNIQUES

The translation of MIP results to material bulk properties, such as compressive strength and permeability, has been fervently discussed. The main sources of disagreements between researchers about MIP testing are the potentially untrue nature of pore volumes and pore sizes measured by MIP. Such questions arose from many physical limitations inherent to MIP measurements and from the different assumptions in MIP data interpretation:

- *Pore model*: The first main assumption is that in the Washburn equation cylindrical pore shape is assumed. This is unlikely to be the case in cement pastes (Abell et al. 1999; Diamond 2000). For instance, other characterisation techniques such as scanning electron microscopy clearly reveal that the capillary pore spaces are not cylindrical (see Figure 9.10) (Diamond et al. 1996; Wang and Diamond 1994).
- *Isolated pores*: The mercury can intrude only the porosity that is interconnected and accessible from the outside. Thus, MIP potentially measures the accessible pore volume and not the entire porosity of the sample.
- *Ink-bottle effect*: The main criticism of the MIP technique is the fact that MIP does not measure pore size but pore entry size. This phenomenon, known as the ink-bottle effect, leads to an overestimation of small pores and an underestimation of big pores (Diamond 2000; Moro and Böhni 2002). More simply said, if the only path towards a big pore is a smaller one, the volume of the big cavity will be evaluated at the small pore intrusion pressure. The main evidence of this phenomenon is that not all the mercury intruded in the sample can exit when the pressure is subsequently decreased and at the end of MIP experiments, some mercury remains entrapped in the sample (Van Brakel et al. 1981). Figure 9.11 shows an example of intrusion and extrusion curves when measuring mature cement samples. In the

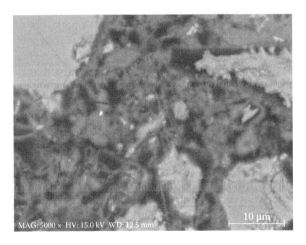

Figure 9.10 BSE micrograph of a polished section of cement paste at 22 hours of hydration.

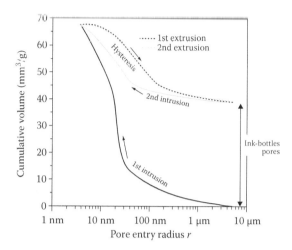

Figure 9.11 Intrusion and extrusion MIP experiments. Extrusion curves illustrate ink-bottle pores where the mercury remains trapped in the sample. In this specific case, half of the intruded pore network can be considered as ink-bottle pores. The second pressurisation/depressurisation cycle shows a hysteresis attributed to a change in the contact angle between the mercury and the sample surface. (Adapted from Kaufmann, J. et al., *Journal of Colloid and Interface Science*, 336, 730–737, 2009.)

presented data, half of the intruded mercury does not exit the sample upon depressurisation. This volume is considered by many researchers as the volume of ink-bottle pores.

Even when a second MIP cycle is carried out, the results show a hysteresis between the second intrusion and second extrusion, although all the ink-bottle pores remain filled after the first cycle. The persistence of a hysteresis upon multiple MIP pressurisation/depressurisation cycles has been attributed to a change in contact angle when the mercury is intruding compared to when the mercury is receding (Kaufmann et al. 2009; Liu and Winslow 1995; Lowell and Shields 1984). Several authors estimated an extrusion contact angle at $\theta = 104°$ instead of the assumed $\theta = 140°$ for the intrusion step (Kaufmann et al. 2009; Moro and Böhni 2002; Salmas and Androutsopoulos 2001). The reason why there is a continuous change in the contact angle between pressurisation and depressurisation experiments is not clear. Wardlaw and Taylor (1976) proposed a change in surface area per unit volume of mercury between intrusion and extrusion, and the fact that surface impurities might remain covered by mercury during depressurisation. In addition, Giesche (2006) stated that not only a change in contact angle is sufficient to entirely explain the MIP hysteresis. This behaviour still remains to be investigated. In any case, extrusion MIP curves are not representative of the pore size distribution of cementitious materials. However, this type of results might give information on the amount of ink-bottle pores.

9.9.1 Comparison with other methods

Comparison of MIP results with other characterisation methods has been the subject of many studies and MIP has been compared several times to nitrogen sorption (Hansen and Almudaiheem 1986; Midgley and Illston 1983; Valckenborg et al. 2001), water sorption (Hansen and Almudaiheem 1986), ethanol resaturation porosity (Day and Marsh 1988), scanning electron microscopy (SEM) image analysis (Diamond 2000; Lange et al. 1994) and ^{1}H-NMR (Valckenborg et al. 2001). Comparing MIP data to results from other characterisation techniques is not an easy task. The reason is that the different aforementioned techniques have different measurement principles, assumptions and operating pore size ranges. While SEM measures only large pores (> 200 nm for conventional equipment), nitrogen sorption experiments determine only accurately pore volumes of small sizes (< 100 nm in diameter). This leads to difficulties when comparing results from different methods as the overlapping zones might be small. Some studies have shown a match between MIP and SEM image analysis for large pore sizes (Lange et al. 1994) but another study reported a significant mismatch between the two (Diamond 2000).

A recent study done by ^{1}H-NMR (Muller et al. 2013) reported the full pore size distribution of a white cement paste cured underwater for 28 days. Figure 9.12 shows this ^{1}H-NMR result compared to a MIP pore size

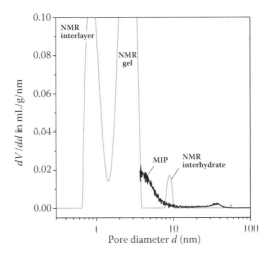

Figure 9.12 Comparison between ^{1}H-NMR and MIP results. MIP pore sizes were calculated with the Washburn equation assuming $\theta = 140°$. ^{1}H-NMR pore sizes were calculated based on the fast-exchange model of relaxation in pores. Measurements were done on white cement pastes cured underwater for 28 days.

distribution of the same material. There is a large discrepancy between pore sizes estimated by the two methods. The MIP result is shifted towards bigger sizes compared to the ^{1}H-NMR result. The reasons for that might be numerous, including, first, the choice of a correct value for the MIP contact angle parameter. It needs also to be specified that the NMR pore size distribution was obtained on a never-dried system, while the MIP sample was formerly dried. In the case of MIP, the drying step might have changed the pore sizes.

9.10 EXAMPLES OF MERCURY INTRUSION POROSIMETRY RESULTS

The porosity of hydrated cementitious materials varies with many factors, such as cement fineness, w/c, mixing procedure and curing conditions. The present section aims to show typical MIP results for different cement-based materials. Contrary to the previous sections, measurement parameters are kept constant (contact angle $\theta = 140°$, step mode of pressurisation, constant pressurisation rate, 1 g sample mass) and only the MIP response to different materials is discussed/shown. This section aims to show general trends of MIP results across different sample systems.

9.10.1 Effect of cement type

Figure 9.13 shows typical MIP results for pastes composed of white cement, grey cement and a blend of calcium aluminate cement with calcium sulfate,

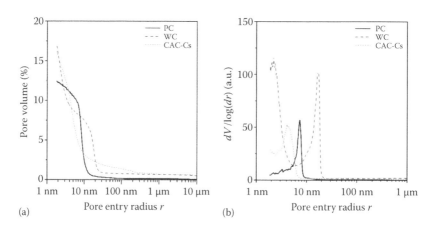

Figure 9.13 (a) Cumulative pore volumes for pastes made of portland cement (PC), white cement (WC) and a mixture of calcium aluminate cement and calcium sulfate (CAC-Cs), all mixed at w/c = 0.4 and cured underwater for 28 days; (b) first derivatives of the three cumulative curves presented in (a).

all cured underwater for 28 days. Derivative results are presented in Figure 9.13b. The white cement paste shows two well-defined peaks centred at 2 and 18 nm pore radii, whereas the grey cement paste has only one peak at about 7 nm pore radius. In comparison, the blend of calcium aluminate cement shows a peak at 4–5 nm. The three mixes under study have largely different threshold pore diameter characteristics. These experiments highlight the potential big difference observed by MIP between pore microstructures resulting from different binders.

9.10.2 Effect of curing conditions

The influence of the curing condition during cement paste hydration is important. Specimens are usually cured underwater (or in water vapour–saturated air) or sealed. Figure 9.14 shows MIP results of white cement pastes cured sealed and underwater for 28 days. While both curing conditions led to similar accessible total pore volumes, there is a refinement of the porosity for samples cured underwater with no large pores in comparison to sealed cured samples. Despite the fact that the cement reaction is usually lower than in underwater cured samples, the presence of large pores for sealed systems is mainly due to self-desiccation. Hydration of cement in sealed systems results in empty spaces devoid of water that remain as porosity. Those voids form within the capillary spaces and according to Kelvin–Laplace laws, the largest pores will empty first. The comparison of sealed and underwater cured pastes with the MIP (Figure 9.14) shows evidence of this phenomenon by the quantification of large pores for the sealed system.

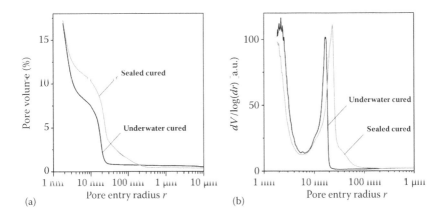

Figure 9.14 (a) Cumulative pore volumes for white cement pastes sealed and underwater cured for 28 days; (b) first derivatives of the two cumulative curves presented in (a).

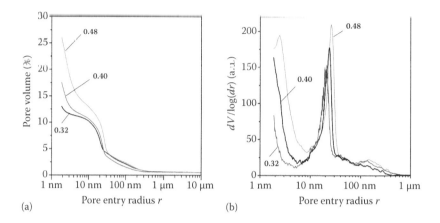

Figure 9.15 (a) Cumulative pore volumes for white cement pastes mixed at w/c = 0.32, 0.40 and 0.48 after 28 days of sealed hydration; (b) first derivatives of the three cumulative curves presented in (a).

9.10.3 Effect of water-to-cement ratio

The w/c is probably the primary factor influencing the porosity of cement-based materials. It defines the spaces between cement grains at mixing and determines the initial capillary porosity of the sample. Figure 9.15a shows examples of MIP results for white cement pastes mixed at w/c = 0.32, 0.40 and 0.48 after 28 days of sealed hydration. The total MIP accessible porosity goes down with decreasing w/c from approximately 25% (w/c = 0.48) to 17.5% for w/c = 0.40 and to 12.5% for w/c = 0.32. The derivative curves presented in Figure 9.15b highlight that the main capillary peak diminishes in volume and in size while decreasing the w/c.

9.10.4 Effect of sample age

Throughout hydration time, products are forming which progressively fill capillary pores. As a result, both total pore volume and critical pore radius of the microstructure decrease. Figure 9.16 shows the pore structure development of a cement paste from 1 to 28 days of hydration. MIP results show the rapid decrease in the total pore volume over the first 14 days of hydration, after which it decreases much more slowly. Both threshold and critical pore radii shift towards smaller sizes but stabilise beyond 14 days of hydration. The stabilisation of the critical pore entry size at about 7 nm might be explained by the increase in the saturation index for growth in the pore (Berodier and Scrivener 2015; Bizzozero et al. 2014; Steiger 2005).

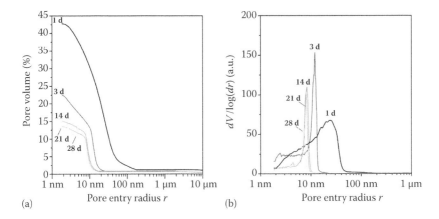

Figure 9.16 (a) Cumulative pore volumes for portland cement pastes mixed at w/c = 0.40 and cured underwater for 1, 3, 14, 21 and 28 days; (b) first derivatives of the five cumulative curves presented in (a).

9.11 CONCLUSIONS

The popularity of MIP experiments for characterising the pore structure of porous media relies on its fast and simple use. Nevertheless, handling mercury requires *strict safety working conditions* and careful mercury storage.

Modern MIP equipment is able to analyse a wide range of pore sizes from millimetres down to about 2 nm pore radius. However, the users must keep in mind that MIP results suffer from the following considerations/limitations:

- The mercury is intruded from the surface of the sample and it accesses the pore structure through continuous paths. Thus, the MIP technique characterises only the connected porosity and determines pore entry sizes rather than real pore sizes.
- A dried sample is required to carry out MIP measurements. Drying methods are known to modify the porosity as it exists in nondried samples. Thus, the users should keep in mind that MIP porosity might not exactly reflect the true porosity of cementitious materials.
- The contact angle θ between the mercury and the sample surface has a great impact on the pore size distribution calculated from MIP experiments (Figure 9.4). Most MIP users adopt θ = 140° but the Hg contact angle may vary with sample properties (e.g. chemical composition and intrusion or extrusion step).
- The Washburn equation assumes cylindrical pores that might not be relevant for cement paste.

Consequently, the aforementioned limitations and assumptions imply careful interpretation of the MIP curves. Nevertheless, MIP is a simple and reliable technique for quantitatively and qualitatively comparing pore structures between different cementitious materials.

9.12 OUTLOOK

The concerns raised by the safe use of mercury in scientific laboratories might lead to the ban of the MIP technique in the future. Alternative techniques or less hazardous nonwetting liquids are currently considered to replace MIP (Moreira 2011; Rouquerol et al. 2011). In the meantime an improved understanding of MIP results can be reached by simulations of porous structures and analyses of path lengths, fractal geometries and different pore shapes and interconnections (Do et al. 2013; Münch and Holzer 2008).

REFERENCES

Abell A., K. Willis and D. A. Lange (1999). 'Mercury intrusion porosimetry and image analysis of cement-based materials'. *Journal of Colloid and Interface Science* **211**: 39–44.

Adolphs J., M. Setzer and P. Heine (2002). 'Changes in pore structure and mercury contact angle of hardened cement paste depending on relative humidity'. *Materials and Structures* **35**: 477–486.

Aligizaki K. K. (2005). *Pore structure of cement-based materials: Testing, interpretation and requirements*, Boca Raton: CRC Press.

Bager D. H. and E. J. Sellevold (1975). 'Mercury porosimetry of hardened cement paste: The influence of particle size'. *Cement and Concrete Research* **5**: 171–177.

Berodier E. S. and K. L. Scrivener (2015). 'Evolution of the pore structure in blended systems'. *Cement and Concrete Research* **73**: 25–35.

Bizzozero J., C. Gosselin and K. L. Scrivener (2014). 'Expansion mechanisms in calcium aluminate and sulfoaluminate systems with calcium sulfate'. *Cement and Concrete Research* **56**: 190–202.

Cook R. A. and K. C. Hover (1991). 'Experiments on the contact angle between mercury and hardened cement paste'. *Cement and Concrete Research* **21**: 1165–1175.

Cook R. A. and K. C. Hover (1999). 'Mercury porosimetry of hardened cement pastes'. *Cement and Concrete Research* **29**: 933–943.

Day R. L. and B. K. Marsh (1988). 'Measurement of porosity in blended cement pastes'. *Cement and Concrete Research* **18**: 63–73.

Diamond S. (2000). 'Mercury porosimetry: An inappropriate method for the measurement of pore size distributions in cement-based materials'. *Cement and Concrete Research* **30**: 1517–1525.

Diamond S. and W. L. Dolch (1972). 'Generalised log-normal distribution of pore sizes in hydrated cement paste'. *Journal of Colloid and Interface Science* **38**: 234–244.

Diamond S., S. Mindess, F. P. Glasser, L. W. Roberts, P. Skalny and L. D. Wakeley (1996). *Microstructure of cement based systems and bonding and interfaces in cementitious materials. Materials Research Society Symposium Proceedings* Vol. 370, pp. 135–136, Warrendale: Materials Research Society.

Do Q., S. Bishnoi and K. L. Scrivener (2013). 'Numerical simulation of porosity in cements'. *Transport in Porous Media* 99: 101–117.

Gallé C. (2001). 'Effect of drying on cement-based materials pore structure as identified by mercury intrusion porosimetry: A comparative study between oven-, vacuum-, and freeze-drying'. *Cement and Concrete Research* 31: 1467–1477.

Giesche H. (2006). 'Mercury porosimetry: A general (practical) overview'. *Particle & Particle Systems Characterisation* 23: 9–19.

Hansen W. and J. Almudaiheem (1986). 'Pore structure of hydrated Portland cement measured by nitrogen sorption and mercury intrusion porosimetry'. In *MRS Proceedings* Vol. 85, p. 105, Cambridge University Press.

Hearn N. and R. Hooton (1992). 'Sample mass and dimension effects on mercury intrusion porosimetry results'. *Cement and Concrete Research* 22: 970–980.

Kaufmann J., R. Loser and A. Leemann (2009). 'Analysis of cement-bonded materials by multi-cycle mercury intrusion and nitrogen sorption'. *Journal of Colloid and Interface Science* 336: 730–737.

Lange D. A., H. M. Jennings and S. P. Shah (1994). 'Image analysis techniques for characterisation of pore structure of cement-based materials'. *Cement and Concrete Research* 24: 841–853.

Lide D. R. (2003). *Handbook of chemistry and physics*, 84th edition, Boca Raton: CRC Press.

Liu Z. and D. Winslow (1995). 'Sub-distributions of pore size: A new approach to correlate pore structure with permeability'. *Cement and Concrete Research* 25: 769–778.

Lowell S. and J. E. Shields (1984). 'Theory of mercury porosimetry hysteresis'. *Powder Technology* 38: 121–124.

Ma H. and Z. Li (2013). 'Realistic pore structure of Portland cement paste: Experimental study and numerical simulation'. *Computers and Concrete* 11: 317–336.

Midgley H. and J. Illston (1983). 'Some comments on the microstructure of hardened cement pastes'. *Cement and Concrete Research* 13: 197–206.

Mindess S., J. F. Young and D. Darwin (2003). *Concrete*, Second edition, Upper Saddle River: Prentice Hall.

Moreira M. (2011). 'Pore structure in blended cement pastes'. PhD thesis, Technical University of Denmark, Kongens Lyngby, Denmark.

Moro F. and H. Böhni (2002). 'Ink-bottle effect in mercury intrusion porosimetry of cement-based materials'. *Journal of Colloid and Interface Science* 246: 135–149.

Moukwa M. and P.-C. Aïtcin (1988). 'The effect of drying on cement pastes pore structure as determined by mercury porosimetry'. *Cement and Concrete Research* 18: 745–752.

Muller A. C. A., K. L. Scrivener, A. M. Gajewicz and P. J. McDonald (2013). 'Use of bench-top NMR to measure the density, composition and desorption isotherm of C–S–H in cement paste'. *Microporous and Mesoporous Materials* 178: 99–103.

Münch B. and L. Holzer (2008). 'Contradicting geometrical concepts in pore size analysis attained with electron microscopy and mercury intrusion'. *Journal of the American Ceramic Society* **91**: 4059–4067.

Rigby S. P. (2002). 'New methodologies in mercury porosimetry'. *Studies in Surface Science and Catalysis* **144**: 185–192.

Ritter H. L. and L. C. Drake (1945). 'Pressure porosimeter and determination of complete macropore-size distributions'. *Industrial & Engineering Chemistry* **17**: 782–786.

Rouquerol J., G. Baron, R. Denoyel, H. Giesche, J. Groen, P. Klobes, P. Levitz, A. V. Neimark, S. Rigby, R. Skudas, K. Sing, M. Thommes and K. Unger (2011). 'Liquid intrusion and alternative methods for the characterisation of macroporous materials (IUPAC Technical Report)'. *Pure and Applied Chemistry* **84**: 107–136.

Salmas C. and G. Androutsopoulos (2001). 'Mercury porosimetry: Contact angle hysteresis of materials with controlled pore structure'. *Journal of Colloid and Interface Science* **239**: 178–189.

Sellevold E. J. (1974). 'Mercury porosimetry of hardened cement paste cured or stored at 97 C'. *Cement and Concrete Research* **4**: 399–404.

Shi D. and D. N. Winslow (1985). 'Contact angle and damage during mercury intrusion into cement paste'. *Cement and Concrete Research* **15**: 645–654.

Steiger M. (2005). 'Crystal growth in porous materials – I: The crystallisation pressure of large crystals'. *Journal of Crystal Growth* **282**: 455–469.

United Nations Environment Programme (2013). Global Mercury Assessment 2013: Sources, Emissions, Releases and Environmental Transport. http://www.unep.org/PDF/PressReleases/GlobalMercuryAssessment2013.pdf.

Valckenborg R. M. E., L. Pel, K. Hazrati, K. Kopinga and J. Marchand (2001). 'Pore water distribution in mortar during drying as determined by NMR'. *Materials and Structures* **34**: 599–604.

Van Brakel J., S. Modrý and M. Svata (1981). 'Mercury porosimetry: State of the art'. *Powder Technology* **29**: 1–12.

Wang Y. and S. Diamond (1994). 'An approach to quantitative image analysis for cement pastes'. In: *MRS Proceedings* **370**: 23.

Wardlaw N. C. and R. Taylor (1976). 'Mercury capillary pressure curves and the interpretation of pore structure and capillary behaviour in reservoir rocks'. *Bulletin of Canadian Petroleum Geology* **24**: 225–262.

Washburn E. W. (1921). 'The dynamics of capillary flow'. *Physical Review* **17**: 273.

Winslow D. and S. Diamond (1969). A mercury porosimetry study of the evolution of porosity in Portland cement: Technical publication.

Winslow D. N. (1984). 'Advances in experimental techniques for mercury intrusion porosimetry'. In: *Surface and Colloid Science*, Berlin, Heidelberg: Springer: 259–282.

Winslow D. and D. Liu (1990). 'The pore structure of paste in concrete'. *Cement and Concrete Research* **20**: 227–235.

Chapter 10

Laser diffraction and gas adsorption techniques

*Marta Palacios, Hadi Kazemi-Kamyab,
Sara Mantellato and Paul Bowen*

CONTENTS

10.1 OVERVIEW

The particle size distribution (PSD) and specific surface area (SSA) are complementary parameters that give information about the fineness of a powder. They can be combined to get information about the state of agglomeration and morphology of a powder. The state of agglomeration of a powder has an influence on its dispersibility, its packing and, hence, the rheology of a suspension; this is of critical importance in the fluidity of cementitious materials.

The particle size and SSA of a cement powder have a direct influence on hydration kinetics. This is illustrated in Figure 10.1, as the particle size of alite (main phase of cement) decreases, and consequently SSA increases, a significant acceleration of the hydration is observed. In addition, these parameters have a direct impact on the rheological and mechanical properties of concrete and on the interaction between the cement and chemical admixtures such as superplasticisers. Both parameters are routinely measured during cement production for quality control purposes.

In concrete, porosity together with SSA values give information about its microstructure and permeability as well as durability. Mindess and Young (1981) established a relationship between the pore sizes and the main properties of concrete they affect, such as strength, permeability or shrinkage.

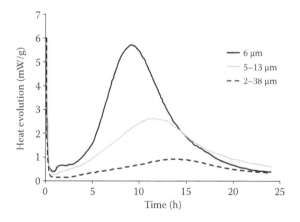

Figure 10.1 Effect of the PSD of alite on the heat evolution rate in hydrated pastes (w/c = 0.4). (Adapted from Costoya, M., Effect of particle size on the hydration kinetics and microstructural development of tricalcium silicate, PhD thesis, École Polytechnique Fédérale de Lausanne, Switzerland, 2008.)

By measuring the SSA and pore size distribution of concrete over time or when it is exposed to different curing conditions, it is possible to get information about the evolution of these properties. However, a reliable measurement of both parameters is not an easy task and will be one of the main features of this chapter.

The International Union of Pure and Applied Chemistry (IUPAC) (Sing et al. 1985) classifies pores according to their internal pore width (or diameter in the case of cylindrical pores) as (1) *micropores*: pores of internal width less than 2 nm; (2) *mesopores*: pores of internal width between 2 and 50 nm and (3) *macropores*: pores of internal width bigger than 50 nm. In this chapter, this is the classification of pores we refer to.

10.2 PARTICLE SIZE DISTRIBUTION

10.2.1 General introduction

Particle size, after chemical composition, is the main characterising parameter of a powder. Size and size distribution influence the handling, storage and domain of use of a powder. As illustrated in Figure 10.1, not only the size of a cement powder particle can influence the hydration kinetics of the cement but also the state of agglomeration has a strong influence on the rheology of a cement paste. Agglomerated particles need more liquid to get a flowable paste as liquid that penetrates into pores is no longer available to lubricate and facilitate movement between particles under the influence of an external force (e.g. gravity or a mixer). The PSD is not enough on its own to characterise a powder and it must be coupled to other techniques (such as – Chapter 4; electron microscopy [SEM and TEM] – Chapter 8; TGA – Chapter 5 and SSA) to correctly interpret and use the measured PSD.

10.2.1.1 Particle diameters

For nonspherical particles there is no single diameter that can be defined as in the case of a perfect sphere. The many methods used to measure PSDs give a characteristic diameter specific to the method used and often referred to as the equivalent spherical diameter (ESD) (Allen 1997). If the particle is not a perfect sphere, then each of the methods will give a different ESD (Jennings and Parslow 1988); e.g. a sieve diameter is the minimum square aperture through which the particle can pass, whereas in sedimentation methods it is the Stokes diameter, which is the diameter of a free falling sphere which would fall at the same rate as the particle in a given fluid. Almost all cementitious powders are not spherical (e.g. ground blast furnace slag in Figure 10.2a); although silica fume (Figure 10.2b) and fly ash (Figure 10.3) have primary particles that are quite spherical, they also have irregular particles or agglomerates. Thus, it is important to note that

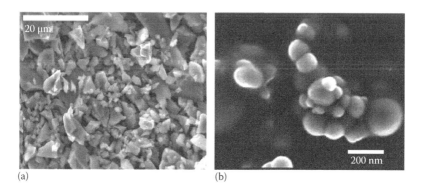

(a)   (b)

Figure 10.2 SEM micrographs illustrating typical morphologies found for cementitious materials: (a) ground blast furnace slag – irregular and angular typical of milled minerals; (b) silica fume – showing almost spherical primary particles and an aggregate.

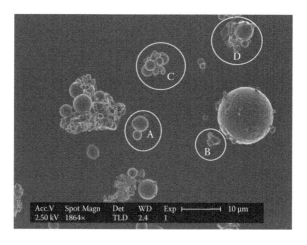

Figure 10.3 SEM micrograph of commercial fly ash showing (A) spherical primary particles, (B) irregular primary particles, (C) agglomerates and (D) aggregates.

the particle diameter is method dependent; i.e. when quoting a diameter, the method used has to be stated (Bowen 2002). The shape of the particle should be assessed using microscopy, and if particles are significantly non-spherical, such as rods or plates, then the ESD given by an instrument can be very misleading and should be treated with caution (Bowen 2002).

Most powders are made up of a series of particles with different diameters and we measure a PSD. When calculating the statistical average or mean, the distribution can be normalised on the total number of particles (Equation 10.1) or on the total volume of the particles (Equation 10.2):

$$D_n = \frac{\sum\limits_{i=1}^{n} D_i N_i}{\sum\limits_{i=1}^{n} N_i}, \qquad\qquad (10.1)$$

$$D_v = \frac{\sum\limits_{i=1}^{n} D_i^4 N_i}{\sum\limits_{i=1}^{n} D_i^3 N_i}, \qquad\qquad (10.2)$$

where D_n is the number length mean diameter, D_v is the volume moment mean diameter, n is the number of divisions or size intervals in the size distribution, D_i is the average diameter of that interval and N_i is the number of particles in that size interval.

For the same distribution, these give significantly different distributions and mean diameters as illustrated in Figure 10.4a. Thus, the second thing to note for PSD measurement is that the base to which the distribution is normalised has to be stated. As well as mean diameters calculated from the whole PSD, certain percentiles of the cumulative distribution are often used to describe the PSD, e.g. D_{10}, D_{50} and D_{90}, where 10%, 50% and 90% of the particles are below these diameters in the cumulative distribution illustrated in Figure 10.4b. The median diameter D_{v50} or D_{n50} (for the volume and number distributions, respectively) are the characteristic diameters where 50% of the particles are above and below the median. When comparing size distributions, care must be taken with the format

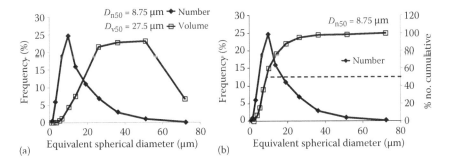

(a)

(b)

Figure 10.4 (a) Frequency distribution for the same size distribution normalised on the total number of particles or the total volume of particles; (b) the number-based distribution presented as a frequency plot and the cumulative distribution showing the % particles less than a given size. The median diameter, D_{n50}, can be read directly from the cumulative distribution plot.

used and presented by the different instruments. When comparing frequency distributions, the y axis should be the density distribution as a function of the division size and not the simple % volume as often given by manufacturers (Sommer 2001) and used in Figure 10.4. If the y axis is not correctly presented and a logarithmic scale is used in the x axis, very misleading distribution shapes can result (see Sommer [2001] for excellent examples). The cumulative distribution does not suffer from such subtleties and is always correct – so it is strongly recommended to use cumulative distributions when comparing PSDs, as will be done throughout this chapter. Frequency distributions (even with the % axis) can be very useful to illustrate how many modes a PSD has (see Section 10.3.4 for illustrations). To summarise:

- The instrument used to make a PSD measurement must be identified.
- The base on which the PSD is normalised must be stated (e.g. number or volume).
- Cumulative distributions should be used for comparing PSDs.

10.2.1.2 Aggregation or agglomeration: Agglomeration factor

Many powders have aggregates (chemical bonds between particles) or agglomerates (physical bonds, e.g. van der Waals forces between particles) from either the processing route or after drying and redispersion. Aggregated particles are more difficult to separate than agglomerates and often need a milling step, whereas sometimes agglomerates can be broken by a simple ultrasonic treatment. To avoid reagglomeration during handling of the dispersion, additives are normally used to modify the surface charge to give electrostatic repulsion or by the adsorption of a polymer to give a steric repulsion (Palacios et al. 2012). When measuring a PSD, it is not possible to distinguish between well-dispersed primary particles, agglomerates or aggregates. An image of the powder is essential for a better interpretation of the PSD, as illustrated in Figure 10.3 for a typical fly ash. The PSD instrument assumes that the particles are dense spheres. One way to estimate the state of agglomeration is to calculate an agglomeration factor F_{AG}, which is the ratio of the D_{v50} measured by the PSD instrument and the D_{BET} – a characteristic diameter that can be calculated from the SSA measured by nitrogen adsorption, assuming monodispersed spherical primary particles:

$$F_{AG} = \frac{D_{v50}}{D_{BET}}, \qquad D_{BET} = \frac{6}{SSA \cdot \rho}. \tag{10.3}$$

D_{v50} is the median volume diameter (in micrometres), D_{BET} is an average diameter (in micrometres) calculated from the specific surface area, SSA (in square metres/gram) measured by nitrogen adsorption (Brunauer–Emmett–Teller [BET] model) and ρ the powder density (in grams/cubic centimetre).

The use of the agglomeration factor allows us to give a number to a particular treatment that aims at reducing agglomeration or aggregation (see the silica fume example in Section 10.2.4.6).

10.2.1.3 Choice of method (laser diffraction versus other methods)

Traditionally the fineness of a cement has been evaluated using techniques such as the Blaine permeability method (ASTM C204), the Wagner sedimentation/turbidity method (ASTM C115) and the sieve residue (ASTM C430) (Ferraris et al. 2005). Although the sedimentation and sieving methods can be used to measure a PSD, they are limited on the fine end, lower than 10 μm, and thus do not give reliable full PSD measurements. There are many different types of PSD instruments for measuring PSDs and even within the same type many different manufacturers (Allen 1997). In 2004 an evaluation of different methods for cement was reported by Ferraris et al. (2004), including laser diffraction (LD), electrical zone sensing, X-ray gravitational sedimentation and scanning electron microscopy (SEM), from two round-robin tests from 21 and 41 laboratories (Ferraris et al. 2002a,b). Their conclusions were that the most popular technique being used was LD and concentrated on giving recommendations for sample preparation and use of LD, which will also be the focus of this section of the chapter. For further information an overview on the different sieving standards used for supplementary cementitious materials (SCMs) has been given in a paper from the RILEM Technical Committee 238-SCM (Arvaniti et al. 2014a).

10.2.2 Importance of a good sampling

When making a particle size measurement, it is important to ensure that the analytical sample placed into the characterisation instrument is representative of the PSD. This is particularly important for wet LD as the sample size is often 1–20 mg, depending on the size and optical properties of the material. In Allen's book (Allen 1997), there are four chapters discussing powder handling and sampling; here is only a very brief presentation of an important aspect of powder technology.

Allen (1997) recommends two golden rules for taking a sample:

- The sample should always be taken from a powder in motion.
- Take several small samples at different intervals of the motion rather than one larger sample in one interval.

Typical sampling devices are rotary samplers or spinning rifflers (Allen 1997) which apply these two golden rules. Allen quotes an example of how many repetitions are needed so that a 95% confidence level can be statistically validated for a powder with a median size of 3.1 ± 0.1 μm. When taken randomly from an unmixed powder, 293 samples are needed; when taken from a mixed sample and a rotary sampler, only 3 samples are needed.

Even if you have taken a representative sample correctly, there is still a minimum weight needed to ensure that you have particles representing sizes from the whole distribution. This is related to the PSD and in particular the largest fraction that will be low in number but high in weight. The minimum sample weight can be calculated using the following equation:

$$W_m = 0.5 \left(\frac{\rho_p}{\sigma_i^2} \right) \left(\frac{1}{w_1} - 2 \right) \left(\frac{D_1^3 - D_2^3}{2} \right) \times 10^3, \tag{10.4}$$

where W_m is the minimum weight (in grams), σ_i^2 is the variance of the tolerated sample error, ρ_p is the powder density (in grams/cubic centimetre), w_1 is the mass fraction of the largest size class sampled, D_1 is the maximum diameter of the largest size class sampled (in centimetres) and D_2 is the minimum diameter of the largest-size class sampled (in centimetres). For a submicron commercial alumina (Alcoa A-16 SG), 0.3 mg is sufficient, but for glass spheres with a broad size distribution from 10 to 1000 μm, 200 g is needed for a 5% tolerated error. For a typical cement this is close to 1 g for a 5% tolerated error and about 300 mg for a 10% tolerated error.

10.2.3 Laser diffraction

10.2.3.1 Theory

LD measures the light scattered, diffracted and adsorbed by a powder dispersed in either a liquid (wet LD) or a gas (dry LD). A laser light passes through a dilute suspension or aerosol and is collected at low angles with respect to the incident light and was thus first termed *low-angle laser light scattering*. The variation in the light intensity I with angle from the forward direction θ for light scattered by diffraction for a powder is given by Azzopardi (1992)

$$I(\theta) = I_0 \int_0^\infty f(R) \left(\frac{RJ1\alpha\theta}{\theta} \right)^2 dR. \tag{10.5}$$

I_0 is the incident light intensity, $\alpha = 2\pi\lambda/R$, λ is the wavelength of the light, R is the particle radius and $J1$ is a Bessel function. A review of the various approaches used to derive particle size from this formulation is well summarised by Azzopardi (1992) and reviewed in an International

Organisation for Standardisation (ISO) document on particle size measurement (ISO 13320-1). In short, two main approaches are used, either Mie theory or Fraunhofer theory. The Mie theory takes into account all the scattering phenomena by spheres of different sizes and needs knowledge of the complex refractive index ($m = n - ik$, where n is the real component and k is the imaginary component [absorption coefficient]) of the material and fluid, whereas the Fraunhofer theory takes into account diffraction only and does not need the refractive index. The Fraunhofer approach can be used for transparent particles with moderate refractive index for sizes > 25 μm and for particles > 4 μm with a large refractive index contrast between fluid and particle and moderate absorption and, e.g. $n = 1.7$, $k = 0.1$. With a good knowledge of the refractive index, the Mie theory can give reliable results down to 100 nm (Allen 1997; Bowen 2002).

The measured diffraction pattern is compared in fact to a theoretical diffraction pattern calculated using the chosen theory and the difference between the measured and theoretical pattern is minimised. Most pieces of software give an estimate on the residual or difference between the two, which gives a guide to how well the theoretical pattern represents the experimental data. A volume distribution is then calculated as the fundamental result and all other information (e.g. SSA and number distribution) are calculated assuming a spherical particle shape. In general the results are very reproducible and accurate to better than 5% over the whole distribution (Khalili et al. 2002), as long as the sampling and dispersion of the powder are well carried out.

10.2.3.2 Assumptions and limitations

The main assumptions in the theoretical approaches are the spherical particle shape and the refractive index for the Mie theory. For pure powder components with a well-known refractive index (e.g. ground quartz), this is not a great limitation, but for cementitious materials (e.g. cement), where composition and purity can vary significantly, care must be taken to make a reasonable assessment. This is discussed in detail in the following.

10.2.3.2.1 Shape

When the particles are spherical, then very good correspondence with image analysis can be found all the way down to 1 μm. For example, silica spheres with a nominal diameter of 1 μm gave a median diameter of 1.04 μm from image analysis and 1.032 μm from wet LD (standard deviation 0.0083 from five repetitions) (Bowen 2002). For irregularly shaped particles, such as cement and ground minerals such as quartz and limestone, it has been shown that LD tends to broaden the distribution (Xu and Di Guida 2002). This is illustrated for a European particle size standard reference material, ground quartz, BCR 66 in Figure 10.5, where the large fraction is particularly overestimated, but the overall PSD is reasonably represented.

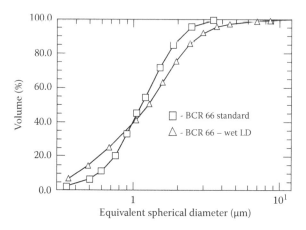

Figure 10.5 LD measurement of European standard quartz powder BCR 66 compared to the standard size distribution.

If particles have significant anisotropic shapes, such as rods or plate-lets, then LD significantly overestimates the breadth and can even give bimodal distributions for monodispersed cylinders (Gabas et al. 1994). This is because the diffraction pattern depends on the geometric shadow of the particle with respect to its orientation with the laser beam. If the anisotropic particles are randomly oriented as they go through the analysis zone, then data from all orientations are produced and only if there is a constant form factor (constant thickness for plates or constant diameter for rods) can some information be carefully extracted from LD measurements (Bowen 2002).

10.2.3.2.2 Refractive index

When using the Mie theory, a knowledge of the complex refractive index ($m = n - ik$, where n is the real component and k is the imaginary component [absorption coefficient]) of the material and fluid are needed. The real part of the refractive index, n, can be found for most of the single phases and standard theoretical values for the typical LD wavelength (620 nm) can be used. Recommendations for cement are discussed in the following and in Section 10.3.4. The absorption coefficient k for the imaginary part of the refractive index is more difficult to assess and has a significant effect on the fine particles in the distribution and if underestimated can give fine fractions lower than those in reality. This sensitivity has been well demon-strated in the studies by Ferraris et al. (2002a,b, 2004) where particle frac-tions under 2 μm tend to be underestimated if $k < 0.1$ is used particularly for real refractive index components of <1.6. Typical values for cementitious materials are given in Table 10.1 along with those for water ($n = 1.33$)

Table 10.1 Parameters and conditions used for wet LD PSD measurement and recommended parameters for measurement and reporting of cementitious materials' PSDs

	CEM I 52.5N	Fly ash	Quartz 1 & 2	Calcined clays		Limestone 2	Silica fume
				Before	After		
Instrument	LD^a	LD^a	LD^a	LD^a	LD^a	LD^a	LD^a
Base	Volume	Volume	Volume	Volume	Volume	Volume	Volume
D_{v50} (μm)	9.31	8.2	2.68/4.56	5.02	7.62	15.0	0.27
Span ($D_{v10}-D_{v90}/D_{v50}$)	2.0	3.9	2.5/3.0	3.6	2.9	2.5	4.2
SSA_{PSD-LD} (m²/g)	0.654	1.13	3.48/2.13	2.36	0.41	0.22	13.04
SSA_{BET} (m²/g)	1.09	1.2	6.21/3.8	19.24	15.8	0.73	21.36
D_{BET} (μm)	1.75	1.87	0.36/0.59	0.12	0.15	3.29	0.13
F_{AG}	5.33	4.39	7.38/7.68	40.2	50.1		2.11
Dispersing liquid (volume [mL])	IPA (100)	IPA (100)	Water (100)	Water/PAA^c (100)	Water/PAA^c (100)	Water (100)	Water/PCE^d (100)
Ultrasonic treatment^b (min)	15	15	15	15	15	15	15

(Continued)

Table 10.1 (Continued) Parameters and conditions used for wet LD PSD measurement and recommended parameters for measurement and reporting of cementitious materials' PSDs

	CEM I 52.5N	Fly ash	Quartz 1 & 2	Calcined clays		Limestone 2	Silica fume
				Before	After		
Powder weight (mg)	500	500	500	500	500	500	200
Stirring rate (LD)	1400	1400	1400	1400	1400	1400	1400
Data acquisition time (s)	4	4	4	4	4	4	4
Obscuration	13.3	12.4	13.8	14.5	12.9	13.2	13.5
Optical model: n, k, liquid	1.7, 0.1, 1.39	1.61, 0.1, 1.40	1.5295, 0.001, 1.33	1.529, 0.01, 1.33	1.529, 0.01, 1.33	1.596, 0.001, 1.33	1.456, 0.001, 1.33

Note: IPA: isopropanol; PAA: polyacrylic acid; PCE: polycarboxylate ether.

[a] Malvern Mastersizer S.
[b] 100 W, 20 kHz.
[c] 0.01 wt.% PAA (pH 10).
[d] PCE-type superplasticiser (3 wt.% from 25 wt.% solution).

and isopropanol ($n = 1.39$), the two most typical liquids used to disperse cementitious materials (both are transparent; thus, k can be approximated to zero). The assessment of the absorption coefficient of the solid is not straightforward and has to be evaluated on a case-by-case basis. General recommendations are for grey powders, such as cement, for which $k = 0.1$ seems to be a good compromise. For white powders $k = 0.001$ can be used. If the user is uncertain, then testing the sensitivity of the changes in the PSD as a function of k from, say, 1 to 0.0001, and comparison of the $SSA_{LD\text{-}PSD}$ with the SSA_{BET} can help for unknown cases. Once chosen for a specific system then the absorption coefficient can be kept constant.

The take-home message from this section, thus, is not to be tempted to overinterpret the tails of the PSD (D_{v10} and D_{v90}) measured by LD if the particles are nonspherical and there is some doubt on the value of the complex refractive index being used.

10.2.3.3 Dry versus wet laser diffraction measurement

In the second round-robin tests discussed earlier (Ferraris et al. 2002b), over 30% of the participants used LD with dry powders (dry LD) and over 60% used wet LD. Overall results after statistical analysis were very comparable over the whole distribution for the cement SRM 114p (Figure 10.6). The main challenge for both wet and dry is a correct dispersion of the powder either in the liquid to form a suspension or in gas to form an aerosol. In liquids, the use of ultrasonic treatment and dispersants can enhance the dispersion and improve reproducibility. In dry LD the compressed air or vacuum pressure used to disperse the particles can be varied. As can be seen from Figure 10.6, when dispersion is optimised, dry and wet

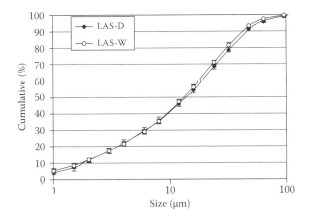

Figure 10.6 Cumulative volume size distributions for the SRM 114p cement powder for wet (LAS-W) and dry LD (LAS-D) methods. (Reprinted from Ferraris, C. F. et al., *Cement, Concrete and Aggregates*, 26(2), 1–11, 2004. With permission.)

methods at least for particle sizes > 1 μm give very similar results (wet in alcohols and both dry and wet used on average $n = 1.7$ and $k = 0.1$). One of the advantages of dry over wet LD is that much more sample is used (grams rather than milligrams for wet), which reduces the errors that may occur from sampling.

10.2.3.4 Guidelines for analysis: Measurement of reliable particle size distributions of cementitious samples

In the round-robin tests (Ferraris et al. 2002a,b, 2004) and the recent RILEM committee paper (Arvaniti et al. 2014a), several practical parameters were discussed with respect to making some sort of general recommendation or standard method for the LD method. They both concentrated on wet LD but most of the things discussed can easily be transposed to the dry method.

10.2.3.4.1 Sample concentration

Most LD instruments have a recommended obscuration factor, i.e. the percentage of laser light that is scattered by the particles. This recommended level will assure an optimum signal-to-noise ratio (S/N) and may vary from instrument to instrument. The NIST (Ferraris et al. 2004) assessment suggested that for wet LD a sample of standard concentration could be prepared outside the instrument and then diluted in the instrument to give the desired obscuration factor. This allows a good control over sample preparation but care must be taken that the aliquots taken to place in the LD instrument sample the whole distribution; i.e. they should be stirred to avoid sedimentation of larger particles (sampling golden rule number 1) and several small samples taken to achieve the desired obscuration (sampling golden rule number 2).

10.2.3.4.2 Dispersion: Ultrasonic treatment (wet)/air pressure (dry)

For wet LD sample dispersion is essential for reproducible measurements – if the produced suspension is unstable and agglomerates form over time, then measurements are difficult to reproduce. Most measurements are made with an ultrasonic treatment to ensure good wetting and to disperse agglomerates. The solid concentration, the volume, the duration, the ultrasonic power and the frequency can all influence which agglomerates are broken or not. Even if this is difficult to standardise, each user should use constant parameters for such treatments and this should assure reproducible measurements. From the NIST (Ferraris et al. 2004) and RILEM (Arvaniti et al. 2014a) papers, treatments longer than 2 minutes showed no modification in the measured PSD, so 2–5 minutes should be sufficient for cements. Other SCMs such as silica fume may be more difficult to disperse and may need more time. To avoid reagglomeration after dispersion, sometimes additives (dispersants) are necessary. The additives will vary with the type of solid and liquid and will be discussed in detail in the individual examples in Section 10.2.4.

Some general recommendations for a first dispersion of a powder include carrying out three to four repetitions in order to verify the colloidal stability of the suspension against time. Once a stable dispersion has been achieved, prepare three dispersions and perform three repetitions with each dispersion if it is reasonable with respect to the measurement time. If you characterise well-known single samples, prepare two dispersions and repeat three times the measurement for each. If you characterise a series of similar samples, prepare one dispersion per sample and perform one or two repetitions.

For dry LD the air or vacuum pressure used to disperse the powder into the laser beam can be varied. This can be increased by steps until no further deagglomeration is observed (e.g. by monitoring D_{v10}). Care must be taken not to use excessive pressures that may induce shear forces greater than those of typical mixers and start fracturing particles. Typical pressures sufficient for deagglomeration for Malvern instruments are 2 to 2.5 bars.

10.2.3.4.3 Stirring speed (wet) and powder flow rate (dry)

Most wet LD instruments circulate or stir the suspension while the instrument is set up for making measurements. Typically these circulation or stirring rates should be sufficient to avoid any sedimentation of the larger particles but not excessively high, whereby air bubbles and other artefacts may be created. Typical values for Malvern instruments are 1500 to 1800 rpm. Flow rates for dry LD again will vary from instrument to instrument and powder to powder – normally these are adjusted to get the correct obscuration and do not have a great influence on the measured PSD.

10.2.3.4.4 Measuring time

This has been discussed in the previous literature (Ferraris et al. 2002b, 2004; Arvaniti et al. 2014a) but it was not clear if this meant data acquisition time defined in the software or the total time for the particle size measurement from introduction of the sample to acquisition of the result. The measuring time – introduction of sample to acquisition of results – is normally very quick in LD measurements and below 5 minutes and often below 2 minutes, making it a very quick analysis method. For data acquisition time – this may be instrument dependent and vary with obscuration or circulation time with each instrument. For Malvern instruments results do not vary significantly for many samples for times between 4 and 8 seconds suggesting data acquisition times of shorter than 10 s are sufficient.

10.2.3.4.5 Refractive index

Cementitious materials have particles in the several micron and submicron ranges and the use of the Mie theory is therefore often used. As discussed previously, a good knowledge of the real and imaginary component is needed. Generally for composite powders such as cement, a volume-weighted average of the individual components is taken if their refractive indices are known. For the more pure materials, such as quartz and limestone, this is more straightforward and reasonably accurate real components are readily available. The absorption component is much more difficult to assess as this may be related to sample purity and modifications in chemical composition. In general ISO 13320 has recommended the use of $k = 0.1$ for irregular shaped particles, even for transparent materials. This may be correct for certain aspects of surface roughness, shape and other parameters not accounted for in the theoretical approach. For standard grey cement, such as SRM 114p (Ferraris et al. 2005), the consensus opinion is the use of $n = 1.7$ and $k = 0.1$. In any case users should always quote the refractive index with an LD PSD measurement and keep it constant for the same materials.

10.2.4 Examples: Wet laser diffraction

Typical protocols for general inorganic powders can be found on the Powder Technology Laboratory website (http://ltp.epfl.ch/page-35598-en.html). Here we will outline typical or recommended protocols used for a series of cementitious materials which help in producing reproducible and reliable PSDs by wet LD.

10.2.4.1 Cement

Grey portland cement is probably the most studied of the cementitious materials with respect to PSD by LD and the thorough investigations by Ferraris et al. (2002a,b, 2004) give us very clear guidelines of the consensus conditions for the measurement, as presented in Table 10.1. Figure 10.7a shows four samples taken from the same dispersion by using isopropanol and external ultrasonic treatment over a 30-minute period. The PSD show very similar results, illustrating that the colloidal stability is good for this cement in isopropanol and the general good reproducibility of the LD method for a stable suspension. Two different commercial CEM I 52.5N portland cements (cements A and B) are shown in Figure 10.7b, again using isopropanol and the standard cement optical model. They give very similar results, with only a minor modification in the tail of the fine particles. For an accurate assessment of the fine tail, comparison between the SSA from nitrogen adsorption is recommended rather than trying to interpret the tail of the PSD. This was carried out and was found to be 0.94 m²/g and for

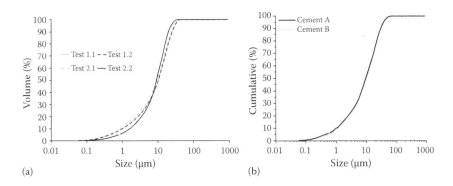

(a)

(b)

Figure 10.7 (a) Cumulative volume size distributions for an portland ordinary cement (CEM I 52.5N) showing very good reproducibility over a 30-minute period; (b) cumulative size distributions for two different commercial CEM I 52.5 cements – showing very similar PSDs.

1.3 m²/g for cements A and B, respectively. This is the opposite of the SSA estimated from the LD (Table 10.1), illustrating the need for complementary methods when assessing the fineness of cements.

10.2.4.2 Fly ash

Figure 10.8 shows the PSD of a fly ash (for morphology see Figure 10.3), dispersed in isopropanol, with two proposed optical models as the refractive index is a priori unknown (model A: $n = 1.53$, $k = 0.1$; model B: $n = 1.61$, $k = 0.1$). X-ray fluorescence (XRF) data show that the fly ash is around 70 wt.% SiO_2 and 30 wt.% other oxides (mainly Al_2O_3). From the real

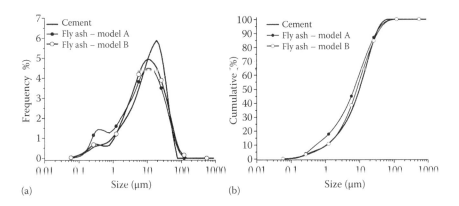

(a)

(b)

Figure 10.8 PSDs of a fly ash with two bounding optical models (see text for discussion) and comparison with a CEM I 52.5 cement: (a) frequency distributions; (b) cumulative distributions.

refractive indices of SiO_2 and Al_2O_3 (1.46 and 1.77, respectively), from using a mass average of the refractive indices for the individual components, the refractive index of the fly ash would be approximately 1.55. In addition, the colour of the fly ash investigated is grey, similar to cement; thus, the imaginary components of its refractive index should be close to 0.1. It can be seen that in both cases there is a bimodal distribution associated with the fly ash particles (Figure 10.8a), one mode below 1 μm and one mode above 1 μm. The PSD with refractive index of 1.53 (model A) estimates the fly ash to have a higher fineness compared to the other model, with 1.61 refractive index (model B). The SSA_{BET} value of 1.20 m²/g indicates, however, that the PSD of this fly ash should be similar to that of CEM I 52.5 cement (SSA_{BET} = 0.94), perhaps a bit finer. Thus, when the refractive index is not known, one cannot extract fine details from the PSD. Nevertheless, the two optical models chosen should give an estimate of the fly ash PSD envelope.

10.2.4.3 Ground quartz

Two quartz powders with different fineness, quartz 1 (SSA_{BET} of 3.80 m²/g) and quartz 2 (SSA_{BET} of 6.23 m²/g), were dispersed in water (Table 10.1) and their PSDs are shown in Figure 10.9. We see that as expected from SSA, quartz 2 has a finer PSD compared to quartz 1, with D_{v50} of 2.68 and 4.56 μm, respectively. The frequency plot (Figure 10.9a) shows clearly a bimodal distribution for both powders, with a higher percentage of particles < 1 mm, ~25% for quartz 2 compared to ~12% for quartz 1. The SEM image in the inset (Figure 10.9a) also shows a large number of fine submicron particles, sometimes agglomerated with the larger particles, confirming the presence of this fine tail of submicron particles seen in the wet LD PSDs.

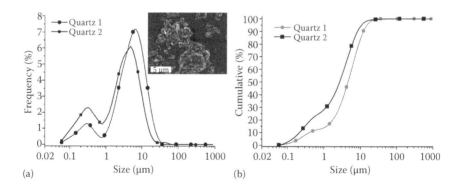

Figure 10.9 PSDs of two quartz powders with different finenesses: (a) frequency distributions showing two modes – the inset shows the irregular angular shape and many fine particles supporting the bimodal analysis by LD; (b) cumulative distributions.

10.2.4.4 Calcined clay

Figure 10.10 shows the PSDs of a natural clay with 79% kaolinite content before and after thermal activation at 700°C. Two optical models (model A, $n = 1.55$, $k = 0.01$, and model B, $n = 1.529$, $k = 0.01$) were used after dispersion in 0.01 wt.% polyacrylic acid aqueous solutions (pH around 10, adjusted with NH_4OH). Further information on the powder chemical composition and thermal treatment can be found in Souri et al. (2015). It can be seen that the two models give comparable results, albeit with the slightly lower refractive index of 1.529; a finer tail is observed before thermal activation (D_{v10} [$n = 1.55$] = 1.00 μm and D_{v10} [$n = 1.529$] = 0.33 μm), again illustrating how difficult it is to precisely measure the tails of a PSD by LD when the refractive index is not known accurately. Thus, care must be taken to not overinterpret. One model (model B, $n = 1.529$, $k = 0.01$), is used for comparison before and after thermal activation. It shows that the D_{v50} of the calcined clay is larger than that of the clay at its natural state (7.62 and 5.02 μm, respectively). It can be also seen that the distribution of the natural clay was bimodal, while after thermal treatment it is monomodal.

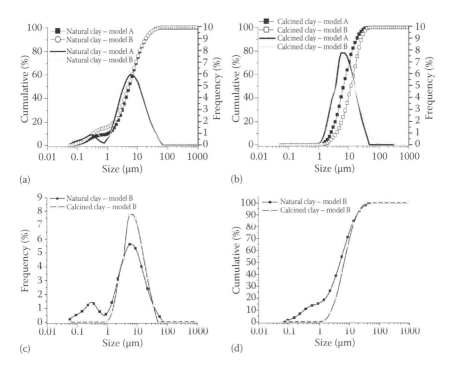

Figure 10.10 Frequency and cumulative distributions of a natural kaolin clay: (a) Untreated. (b) After thermal activation using various optical models. (c) Comparison of frequency distributions. (d) Comparison of cumulative distributions.

The SSA_{BET} value also decreases from 19.2 to 15.8 m²/g for the natural clay to the calcined clay, respectively. The reason for the decrease both in the SSA_{BET} value and shift to a single-mode distribution of particles in the PSD curve of the calcined clay might be due to the partial sintering and particle agglomeration during the thermal activation (700°C) (Souri et al. 2015). This is also reflected in the SSA_{LD-PSD} (Table 10.1; 0.41 and 2.46 m²/g for the calcined and natural clays, respectively). Also, the very low SSA for LD compared to that for nitrogen adsorption illustrates the heavy degree of agglomeration, giving agglomeration factors F_{AG} greater than 40 for both powders, much higher than those for all the other materials in Table 10.1, which have $F_{AG} < 10$.

10.2.4.5 Limestone

The optical model 3PDD ($n = 1.596$, $k = 0.001$, water $n = 1.33$) was used to evaluate the PSD of various limestone powders with different finenesses (Figure 10.11). 0.01% sodium metaphosphate (NaP) solution was used in all cases to disperse the limestone particles. It can be seen (Figure 10.11) that the evaluated PSDs follow the same fineness trend as the SSA_{BET} values (limestones 1, 2, 3 and 4 having SSA_{BET} of 1.75, 0.73, 5.19 and 2.05 m²/g, respectively), with limestone 2 having the highest D_{v50} (15.0 µm) and smallest SSA_{BET} (0.73 m²/g) and limestone 3 with the smallest D_{v50} (2.41 µm) and highest SSA_{BET} (5.19 m²/g).

10.2.4.6 Silica fume

Measuring the PSD of silica fume can be challenging due to its extreme fineness and high SSA (BET = 21.36 m²/g). Such fine powders can agglomerate and strong forces are needed to break them apart. Figure 10.12 shows the PSD evaluation of silica fume using different dispersants, ultrasonic

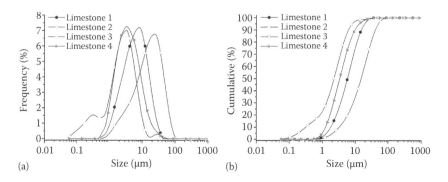

Figure 10.11 PSDs of various limestones of different finesses: (a) frequency distributions; (b) cumulative distributions.

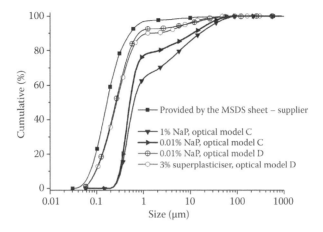

Figure 10.12 Effects of different dispersion conditions on the measured PSD by wet LD for a commercial silica fume. MSDS, Material Safety Data Sheet.

treatment time and high shear stirring before analysing the powders by LD. The use of NaP as a dispersing agent seems to be more efficient at 0.01% concentration rather than at 1 wt.%. After an ultrasonic treatment (100 W) of 30 min, the PSD starts approaching that given by the supplier, reducing the amount of 10 μm agglomerates. The use of a superplasticiser (polycarboxylate type [polycarboxylate ether]) has the same effect after only 15 min of ultrasonic treatment. Here we see how the measured PSD – which is closest to that found in typical cement mixers – significantly depends on the dispersion method and is difficult to evaluate, but from microstructural analysis of cements, when using this silica fume, very little agglomeration was seen, suggesting very high shear rates and something approaching the manufacturer's PSD may be expected in use (Kazemi-Kamyab 2015).

The two models considered, model D, n = 1.4564, k = 0.001, and model C, n = 1.4564, k = 0.1, show significant differences; the very light grey colour of the silica fume used suggests a lower absorption coefficient than that of cement (k = 0.1) and model D (k = 0.001) seems to be the most representative for this particular silica fume.

10.3 GAS ADSORPTION TECHNIQUE

10.3.1 Overview

There are several methods to determine the SSA of cement powder. The method most widely used by the cement industry is the Blaine air permeability as it is a simple and quick method. The SSA determined by the Blaine method is referred to as Blaine fineness and it is expressed as total surface area in square centimetres per gram of cement. In this method, the time needed for a fixed volume of air to pass through a well-packed bed of powder of known dimensions and porosity is determined (Blaine 1943). Surface area is determined considering the Kozeny-Carman theory that assumes a mono-sized and spherical shape of particles. The equipment used, an air permeability apparatus, is empirically calibrated as described in ASTM C204-11 (ASTM International 2011).

The Blaine method suffers from various weaknesses and limitations that can compromise the reliability of the results. The empirical calibration and its poor reproducibility (the value of SSA obtained is clearly influenced by the person who runs the test) are examples of the weakness of this test. In addition, cement particles are not mono-sized and spherical as the calculation of the Blaine fineness assumes.

If Blaine fineness is compared with the SSA determined by other techniques such as nitrogen adsorption (see Figure 10.13), the latter gives higher values as nitrogen can access cracks, crevices and pores that the Blaine test is not able to distinguish (Mantellato et al. 2015b). The Blaine method has also been shown to be unreliable at surface areas higher than 5000 cm²/g (Potgieter and Strydom 1996) and it has great limitations when applied to SCMs. In particular, the air permeability test does not give reliable SSAs of fly ash powders, as it is not able to detect the internal surface of the unburned carbon particles (Kiattikomol et al. 2001; Arvaniti et al. 2014b).

Despite all these limitations, the cement industry extensively relies on the air permeability test to determine the SSAs of cement powders, arguing that the results are more consistent than those of nitrogen adsorption measurements. Recently, Mantellato et al. (2015b) have shown that the main

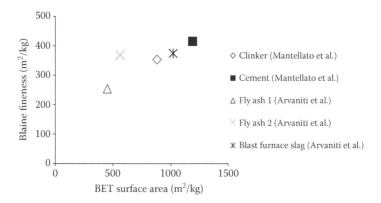

Figure 10.13 Comparison of SSAs determined by the Blaine test and the nitrogen adsorption technique. (Data from Arvaniti, E. C. et al., *Materials and Structures*, 1–15, 2014a; Mantellato, S. et al., Impact of sample preparation on the specific surface area of synthetic ettringite, In preparation, 2015b.)

source of low reproducibility of nitrogen adsorption technique is related to sample preparation and this can be significantly improved by a careful selection of degassing conditions.

LD has been also used to determine the SSA of powder cement. If a spherical shape of cement particles is assumed, the SSA (in square metres/gram) can be determined according to

$$
\mathrm{SSA_{LD}} = \frac{6}{\rho} \cdot \left(\frac{\displaystyle\sum_{i=1}^{n} N_i D_i^2}{\displaystyle\sum_{i=1}^{n} N_i D_i^3} \right),
\tag{10.6}
$$

where n is the number of divisions or size intervals in the size distribution, D_i is the average diameter of that interval, N_i is the number of particles in that size interval and ρ is the powder density (in grams/cubic centimetre).

Unlike the Blaine method, the LD method does not require calibration, but it involves the assumption of the spherical particle geometry. As in the case of the Blaine method, SSAs calculated by LD are significantly lower than the true value as surface roughness and porosity of the particles are not considered. In addition, as explained above, the refractive index of the powder in the dispersive media must be known. The value of refractive index of cement is agreed upon but this value has not been determined accurately for certain supplementary cementitious materials. In some cases, the SSA from LD can give a reasonable assessment (e.g. fly ash in Table 10.1), in other cases it can be very low such as for heavily agglomerated powder (e.g. clays in Table 10.1).

In contrast with the air permeability test and LD techniques, nitrogen adsorption does not postulate the particle shape and size or include semi-empirical equations. Nitrogen adsorption also considers the irregularities and porosity of the particles and that is why it is considered a more reliable technique for the scientific community. In addition, the Blaine method or LD techniques are not suitable to determine the SSA of hydrated cement as they are not able to measure correctly fine agglomerated particles as found in most hydrates (Tamada 2011).

The SSA can also be estimated from the pore size distribution measured by nitrogen adsorption desorption (NAD), mercury intrusion porosimetry (MIP) and proton nuclear magnetic resonance (^{1}H-NMR) relaxometry. The surface area calculation from NAD and MIP needs a pore shape assumption; usually, cylindrical pores are assumed. The approach using ^{1}H-NMR and MIP are described and discussed in detail in Chapters 7 and 9, respectively.

10.3.2 Gas adsorption theory

In the gas adsorption technique, gas molecules (adsorbate) are physically adsorbed onto the surface of a solid (adsorbent). Physical adsorption is due to van der Waals forces between adsorbate gas molecules and the solid surface. Physical adsorption of gases on solids is promoted by decreasing the temperature. For this reason, adsorption measurements are carried out at cryogenic temperatures, typically of liquid nitrogen ($-196°C$).

Automated equipment generally determines the adsorption isotherm by volumetric methods. This consists of adding a known amount of adsorbate gas to the sample cell. The amount of gas adsorbed at the equilibrium pressure is the difference between the amount of gas incorporated and the gas needed to fill the space around the solid, defined as the 'dead space' of the cell. Once the equilibrium is reached, more adsorbate gas is added into the cell and the isotherm is constructed by incorporating increasing amounts of the adsorbate.

Before the test, the solid has to be degassed to remove physisorbed molecules from the surface in order to reduce interferences during the measurement. The degassing is normally done by treating the sample at a high temperature under vacuum or an inert gas flow. As explained in Section 10.3.4.2, the degassing of cementitious samples is not trivial and it is one of the main parameters that can introduce artefacts in the results obtained by this technique.

The adsorption isotherm is obtained by plotting the molar amount of adsorbed gas at a constant temperature as a function of the equilibrium partial gas pressure (P/P_o – where P_o is the saturation pressure of the adsorbing gas at the temperature of the measurement) (Sing et al. 1985). Adsorption and desorption isotherms can be interesting for cement samples to study their porosity, where condensation of the adsorbate in pores occurs as a function of the pore size. According to IUPAC (Sing et al. 1985), adsorption isotherms are classified into six different types (see Figure 10.9). The type of isotherm depends on the nature of the adsorbate gas and the solid to be characterised. Isotherms of type IV are typical for mesoporous materials such as cement samples. The hysteresis loop is associated with the pore condensation and a different meniscus seen on adsorption and desorption, which can also be linked to the pore shape and tortuosity. At high relative pressures a plateau in the isotherm is reached, which indicates all pores are filled. The initial range of type IV isotherms corresponds to the monolayer adsorption used to determine the SSA (point B in Figure 10.14).

10.3.3 Analysis of the adsorption isotherm

10.3.3.1 Specific surface area

The BET theory is the most accepted for determining the SSAs of materials and it is based on the multilayer adsorption of gas molecules onto the adsorbent. This theory is an extension of Langmuir's theory and it involves the following assumptions (Brunauer et al. 1938; Fagerlund 1973):

- The adsorption energy is constant during the formation of the first layer (homogeneous surface).

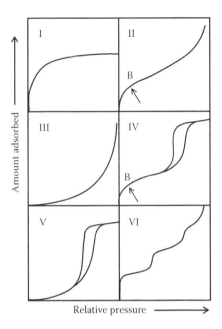

Figure 10.14 Adsorption isotherms classification. (Reprinted from Sing, K. S. W. et al., *Pure and Applied Chemistry*, 57, 603–619, 1985. With permission.)

- At a certain vapour pressure, the adsorption of gas molecules onto previously adsorbed gas molecules occurs. Prior to complete coverage of the surface, a second and higher adsorbed layer of gas will occur.
- At saturation, the number of multilayers is infinite.
- A molecule covered by another molecule cannot evaporate.
- There is no horizontal interaction between adsorbate molecules.
- At dynamic equilibrium, the number of molecules evaporated in the upper layer is equal to the number of molecules condensing in the layer below.

The BET equation can be expressed as the following equation and its derivation is explained in the literature (Brunauer et al. 1938; Lowell et al. 2004):

$$\frac{P}{V_a(P_o - P)} = \frac{1}{V_m C} + \frac{C-1}{V_m C}\left(\frac{P}{P_o}\right),\tag{10.7}$$

where P_o is the saturation pressure of the gas, P is the pressure of the gas in equilibrium with the sample, V_a is the amount of gas adsorbed at pressure P, C is a constant and V_m is the amount of gas needed to cover the surface with a monolayer.

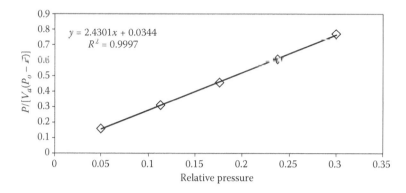

Figure 10.15 Plot used to determine the SSA according to Equation 10.7.

A plot of $P/[V_a(P_o - P)]$ versus P/P_o in the range of partial pressures between 0.05 and 0.35 yields a straight line (see Figure 10.15). It is generally recommended to measure at least three, but better five or more points, in this range of partial pressures to measure reliable SSA values. Considering the slope and intercept of that line, it is possible to determine the volume of a monolayer V_m. The SSA can be determined by

$$S_{BET} = \frac{N_A V_m A_N}{V_o} ,$$
(10.8)

where N_A is the Avogadro constant, A_N is the area of surface occupied by a single adsorbed gas molecule and V_o is the molar volume of gas. Nitrogen is considered the most suitable adsorbate gas and a surface area A_N of 16.2×10^{-20} m^2 is normally assumed.

10.3.3.2 Analysis of the mesoporosity

Mesoporous materials (with pore widths between 2 and 50 nm) experience pore condensation at pressures below the corresponding saturation pressure of the bulk liquid. The volume pore size distribution is generally determined according to the Barret–Joyner–Halenda model. This model considers that condensation occurs in pores when a critical relative pressure is reached according to the modified Kelvin equation

$$\ln\left(\frac{P}{P_o}\right) = \frac{-2 \Upsilon V_o \cos\theta}{RT(r_p - t_c)} ,$$
(10.9)

where Υ is the surface tension at the liquid–vapour interface, V_o is the molar volume of the liquid, θ is the contact angle between liquid and pore wall (taken to be zero for N_2), R is the gas constant, r_p is the pore radius and t_c is the thickness of the adsorbed layer during adsorption. This equation takes into account that a preadsorbed film on the pore wall has already occurred before condensation and it also assumes that the pores have a cylindrical shape.

Pore size distributions can be calculated from the adsorption or desorption curves. Desorption curves have more often been used but adsorption curves are perhaps less influenced by tortuosity and pore shape (Sing 1995). In either case, a condition must be satisfied that all pores have to be considered filled. A simple mean pore diameter (d) can be calculated considering a cylindrical shape of the pores by

$$d = \frac{4V}{A},$$
(10.10)

where A is the BET surface area (SSA_{BET}) and V is the pore volume determined from the adsorption isotherm.

The porosity measured in the mesopore range by NAD gives very similar results to those measured by MIP when both methods use the cylindrical pore assumption (Gregg and Sing 1982). More recently, attempts to analyse pores of different shapes (e.g. slit pores) by using the density functional theory (DFT) to fit the isotherm have shown some promise to move away from the simplistic cylindrical model but the variations in pore size and pore size distributions for hydrated cements did not show significant changes (Costoya 2008). The pore shape (or solid–gas interface curvature) is in fact a big limitation with the transformation of the adsorption or desorption isotherm into a pore size distribution (Scherer 1998). Since the pore shape in cement is really poorly characterised, the pore size distribution from NAD can only at best be indicative and comparative within a series of samples. It is very unlikely to give a pore size similar to those by other techniques, e.g. ¹H-NMR or electron microscopy (using image analysis) except for MIP in the mesopore range when both the NAD and MIP are using the cylindrical pore model. The largest pores that can be analysed by NAD are around 100 to 200 nm. The limitation comes from the very small change in condensation pressure as the pores increase in size. For example, the partial pressure P/P_o changes from 0.9904 to 0.9952 for 100 to 200 nm cylindrical pores, respectively. Microporosity can be evaluated from NAD, but with the limits of the Kelvin equation being reached at 1.78 nm for cylindrical and 2.43 nm for slit pores, the method is really at its limit. The main experimental difficulty for pore size measurement of cementitious materials, as for SSA measurement, is the sample preparation.

10.3.4 Experimental conditions for analysis of cementitious materials

10.3.4.1 Adsorbate selection

Nitrogen, with a purity of at least 99.9%, is currently widely accepted as a standard adsorbate (Sing et al. 1985). This is because of its unique properties: (1) no chemical reactivity with the surface; (2) low and well-defined cross-sectional area at its boiling point (–196°C); (3) no polarity and (4) wide availability. Current volumetric adsorption equipment is able to measure absolute surface areas as low as 1 m^2 using nitrogen as an adsorbate.

Water vapour has also been used as adsorbate to determine the SSA of cement by gravimetric methods (Winslow and Diamond 1974) obtaining significantly higher values than those obtained by using nitrogen (Table 10.2). Some studies postulate this is because water vapour molecules have a lower cross-sectional area compared to nitrogen molecules – 0.162 nm^2 for N_2 and 0.114 nm^2 for water vapour – and can access the ink-bottle pores. However, the difference between the cross-sectional areas of the two molecules does not explain alone the different SSAs measured. In addition to this, water vapour interacts with the C-S-H structure, being able to go in and out of the interlayer space, as has been shown by small-angle scattering measurements (Winslow and Diamond 1974; Pearson and Allen 1985). Also, one should not forget that anhydrous cement particles still present in the sample can hydrate in contact with the water vapour. For all these reasons, water vapour isotherms are not considered suitable for determining the SSA of cementitious systems, N_2 being a more reliable adsorbate.

For absolute surface areas similar or even lower than 1 m^2, results might be too uncertain. To improve the measurements accuracy, one could either increase the quantity of powder in the sample holder (if possible) or change the adsorbate, preferring Kr or Ar with a lower saturation pressure than N_2 at –196°C.

10.3.4.2 Sample preparation

The main limitations of the gas adsorption technique are linked to sample preparation. In the particular case of cementitious samples, these

Table 10.2 SSAs of a mature cement paste with w/c = 0.4

Adsorbate	Surface area (m^2/g dried paste)
H_2O	202.6
N_2	79.4

Source: Mikhail, R. S., and S. A. Selim, Adsorption of organic vapors in relation to the pore structure of hardened Portland cement pastes, *Highway Research Board Special Report*, 90, 1966; Thomas, J. et al., *Concrete Science and Engineering*, 1, 45–64, 1999.

limitations are related to (1) the method used to stop hydration and (2) the degassing conditions.

10.3.4.2.1 Stopping hydration

Gas adsorption is often used to follow the evolution of surface area and porosity of hydrated cement with time and this is correlated afterwards with the degree of hydration or other engineering properties. To measure both parameters at the time of interest, it is required to stop the hydration while preserving the microstructure.

Drying techniques and solvent replacement are the most used methods to stop hydration by removing the capillary water from hydrating cement samples. As summarised in Table 10.3, most of these methods have been shown to modify in some degree the microstructure of hydrated cement

Table 10.3 Summary of impact of different methods of stopping cement hydration on microstructure

Method for stopping hydration		Impact on microstructure	Reference
Drying	Oven drying at temperatures between 60 and 105°C	Removal of structural water from CSH and ettringite; damage on pore structure and microcracking	Korpa and Trettin (2006); Zhang and Scherer (2011)
	Freeze-drying	Degradation of ettringite and AFm	Zhang and Scherer (2011)
	D-drying	Removal of some bound water; preservation of the microstructure of hardened samples; slow method for arresting hydration	Korpa and Trettin (2006); Zhang and Glasser (2000); Zhang and Scherer (2011)
	P-drying	Not total removal of pore water	Zhang and Scherer (2011)
	Vacuum drying	Degradation of ettringite and AFm; damage of the pore structure; slow method for arresting hydration	Zhang and Glasser (2000); Zhang and Scherer (2011)
Solvent replacement	Acetone	Modification of pore structure and occurrence of aldol reaction	Zhang and Scherer (2011)
	Isopropanol	Preservation of the pore structure; long contact times with isopropanol degrade ettringite; short exchange times in cool isopropanol (5°C) does not degrade ettringite	Kocaba (2009); Mantellato et al. (2012)

samples, by modifying the pore structure or some of the hydration products, mainly calcium silicate hydrate (C-S-H), ettringite and monosulfate.

Figure 10.16 highlights the impact of the method to stop hydration on the value of SSA. Solvent exchange with isopropanol gives about two times higher values of SSA of C-S-H samples compared to the D-drying method. The reason that different SSAs are measured is because the surface tension of isopropanol is lower than that of water, which prevents the collapse of the pores when isopropanol is evaporated. Also, during the D-drying not all the water is removed from the pores with a size smaller than 2 nm, and consequently, nitrogen molecules will not be able to access to these pores obtaining smaller SSAs (Korpa and Trettin 2006; Costoya 2008).

Although a definitive method for stopping hydration is still debated, solvent replacement seems to be the most suitable method to be applied before nitrogen adsorption measurements, preserving better the microstructure. Isopropanol is widely used because of its low chemical interaction and adsorption on cement pastes. However, the use of isopropanol has a main drawback as it dehydrates ettringite when long exchange times are applied (Kocaba 2009). The degradation of ettringite may have relevant implications in the measurement of the surface area of young pastes or when calcium sulfoaluminate cements are studied as the main contribution to the surface area in these cases comes from ettringite. Mantellato et al. (2012) showed that it is possible to preserve ettringite crystalline structure when short contact with cooled isopropanol at 5°C is used. As explained in Chapter 5, adsorbed isopropanol can be removed by using a less polar solvent, such as diethyl ether (Deschner et al. 2012).

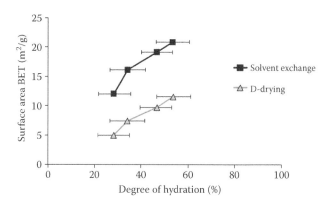

Figure 10.16 Influence of drying method on the BET surface area of C-S-H of hydrated alite. (Reproduced from Costoya, M., Effect of particle size on the hydration kinetics and microstructural development of tricalcium silicate, PhD thesis, École Polytechnique Fédérale de Lausanne, Switzerland, 2008. With permission.)

10.3.4.2.2 Degassing conditions

Before gas adsorption measurement, samples have to be degassed to remove gas physically adsorbed on the surface. This is normally done under nitrogen flow or vacuum and high temperatures are used to accelerate the process. In the case of cementitious samples, the choice of the degassing conditions, involving temperature, pressure and time, is essential to obtain reliable SSA and porosity values of both hydrated and nonhydrated samples. However, they are rarely reported in the literature.

Figure 10.17a shows the impact of the degassing conditions on the SSA of anhydrous portland cement. When the degassing is conducted under vacuum and/or at temperatures higher than 40°C, a partial or total dehydration of gypsum occurs involving an increase in the SSA (see Figure 10.17b). Mantellato et al. (2015b) proposed degassing anhydrous cement samples at 40°C under N_2 flow for 16 hours as an optimum degassing method, since cement composition is preserved. When using this protocol, reproducible results with a scatter of less than 5% can be obtained.

In the case of early hydrated cementitious systems, a degassing process that preserves ettringite should be applied. Yamada (2011) concluded that the SSA of the initial hydrates of cement pastes increases by a factor of 4 when they are degassed at temperatures of 80°C because of dehydration or decomposition of ettringite (see Figure 10.18). In synthetic ettringite samples, Dalas (2014) observed the almost total conversion of ettringite to metaettringite after 14 h of degassing under vacuum at 40°C. Such conversion involved a significant increase in the SSA. In contrast, recent measurements have shown that this conversion occurs in lower degree without

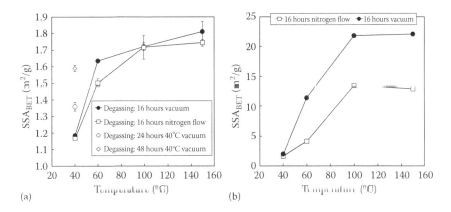

(a)

(b)

Figure 10.17 SSAs of (a) anhydrous cement and (b) gypsum at different degassing conditions. (Reprinted from Mantellato, S. et al., *Cement and Concrete Research*, 67, 286–291, 2015a. With permission.)

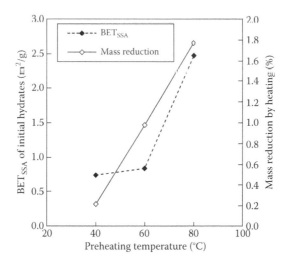

Figure 10.18 Effect of preheating temperature for BET_{SSA} and mass reduction of initial hydrates formed by mixing with water without superplasticiser (paste, w/c = 0.35). (Reprinted from Yamada, K., *Cement and Concrete Research*, 41(7), 793–798, 2011. With permission.)

significant impact on the SSA when degassing is done at 40°C under N_2 flow (Mantellato et al. 2015b).

To measure the porosity of cement pastes, stronger degassing conditions have to be applied to remove the water from the smallest pores and avoid artefacts during the measurement. In this case, degassing under vacuum at

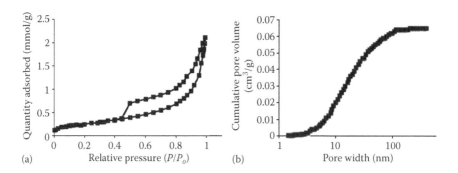

Figure 10.19 (a) Nitrogen adsorption/desorption isotherms for C_3S paste; (b) cumulative pore volume versus pore width for C_3S pastes hydrated for 28 days using the DFT model for slit-shaped pores. (Adapted from Costoya, M., Effect of particle size on the hydration kinetics and microstructural development of tricalcium silicate, PhD thesis, École Polytechnique Fédérale de Lausanne, Switzerland, 2008.)

40°C for 24 hours has been used to obtain reproducible porosity values of C_3S pastes (Costoya 2008). Figure 10.19a shows the adsorption/desorption isotherms of a C_3S paste hydrated for 28 days. Hydration was stopped by solvent exchange with isopropanol and the degassing conditions described previously were applied. Data were transformed using the DFT model for slit-shape pores; the cumulative pore volume of the sample was also determined (Figure 10.19b) (Costoya 2008).

10.3.5 Guidelines for analysis

Sample preparation is the crucial step during surface area and porosity measurement by nitrogen adsorption technique. As explained in Section 10.3.4.2, the reliability of the obtained values will depend mainly on the method employed to stop cement hydration and the degassing procedure (temperature, pressure and time). For each sample, measurements should be repeated at least three times.

In this section, a protocol for analysis of cementitious materials by nitrogen adsorption is proposed and degassing conditions for preserving the microstructure of anhydrous and hydrated cement samples are described. When using these recommended sample preparation conditions, the obtained values have shown to be reproducible and reliable.

The following are the specific guidelines for analysis:

1. Adsorbate gas
   a. Nitrogen with a purity of 99.9% must be used as adsorbate gas.
2. Sample weight
   a. Sample weight is always controlled before and after degassing as well as after the adsorption measurements.
   b. Sample weight should provide at least absolute surface areas equal or higher than 1 m^2 (when nitrogen is used as adsorptive gas).
3. Sample tubes
   a. Tube cells should be immersed to a depth of at least 5 cm below the liquid nitrogen level.
   b. Rod should be inserted into the sample tube to minimise nitrogen consumption.
   c. Isothermal jacket should be used to maintain a constant temperature in the sample tube immersed in the Dewar while the liquid nitrogen evaporates.
4. Stopping of hydration and degassing conditions
      In the case of anhydrous cementitious materials:
      a. For cementitious materials containing gypsum, mild degassing conditions have to be applied to preserve gypsum microstructure and composition, specifically degassing at 40°C under N_2 flow for 16 h. Higher temperatures and/or vacuum treatment

would prompt gypsum dehydration involving significant structural changes.

b. Clinker or supplementary cementitious materials such as slag can be degassed at relatively high temperatures, 200°C, for 1 hour without modifying its microstructure.

In the case of hydrated cementitious materials:

a. Solvent exchange with isopropanol has been proven to be the most suitable way to stop hydration without inducing significant changes on the pore structure. As mentioned above, for fresh cement pastes, cooled (5°C) isopropanol is recommended to preserve ettringite microstructure (Mantellato et al. 2012).

b. For the analysis of the SSA of cementitious systems, degassing conditions should preserve the ettringite microstructure. Degassing at 40°C under N_2 flow partially dehydrates ettringite without a significant impact on the SSA (Mantellato et al. 2015b). In the case of cement pastes cured for a long time (with C-S-H as the main product contributing to the SSA), degassing at 40°C under vacuum has also given reliable results (Costoya 2008).

c. For analysis of porosity, degassing under vacuum is needed in order to remove the water from the smallest pores. Degassing under vacuum for 24 hours at 40°C has given reproducible porosity values of C_3S and alite pastes (Costoya 2008).

Calibration of the temperature of the commercial degassing stations might be needed as a deviation between the set temperature and the effective one has been previously reported (Mantellato et al. 2015a).

When *reporting* SSA and porosity values, the following data should be included:

• Equipment used
• Method for stopping hydration used in the case of hydrated samples
• Degassing conditions (temperature, pressure – if vacuum or N_2 flow was used – and time)
• SSA and, for porosity, the mean pore size and the cumulative pore size distribution including standard deviation from repeated measurements

Table 10.4 shows some examples of the protocols and SSA and porosity values determined for different cementitious systems. In these specific cases, equipment from Micromeritics was always used but several other models and brands are commercially available.

Table 10.4 Summary of degassing conditions prior to nitrogen adsorption measurements

Analysis conditions	Materials					Porosity measurement
	Surface area measurement					C₃S hydrated 28 days
	Anhydrous cement	Gypsum	C₃S	Synthetic ettringite	Anhydrous blast furnace slag	
Weight of sample (g)	1.5–2	1.5–2	1.5–2	0.3–0.5	1.5–2	1.5–3
Equipment	Micromeritics TriStar II 3020	Micromeritics TriStar II 3020	Micromeritics TriStar II 3020	Micromeritics TriStar II 3020	Micromeritics TriStar II 3020	Micromeritics Gemini 2375 V4
Sample tube	Flat bottom, 3.77 cm³	Flat bottom, 3.77 cm³	Flat bottom, 3.77 cm³	Flat bottom, 3.77 cm³	Flat bottom, 3.77 cm³	
Degassing conditions	16 h at 40°C under N₂ flow	16 h at 40°C under N₂ flow	1 h at 200°C under N₂ flow	16 h at 40°C under N₂ flow	1 h at 200°C under N₂ flow	24 h at 40°C under vacuum
Method for stopping hydration	–	–	–	–	–	Isopropanol at room temperature
SSA (m²/g)	1.17 ± 0.01	1.66 ± 0.04	1.33 ± 0.08	8.11 ± 0.23	0.82 ± 0.02	–
Reference	Mantellato et al. (2015a)	Mantellato et al. (2015a)	ETHZ, unpublished results	Mantellato et al. (2015b)	ETHZ, unpublished results	Costoya (2008)

Note: ETHZ: Eidgenössische Technische Hochschule Zürich (Swiss Federal Institute of Technology in Zürich).

10.4 CONCLUSIONS AND OUTLOOK

The PSD, the SSA and the porosity of cementitious materials are frequently measured as they have a direct effect on the kinetics of cement hydration and/or on other engineering properties of concrete. In this chapter the crucial parameters in analysing these parameters by LD and nitrogen adsorption techniques are described.

The main messages to take home for the use of wet LD for particle size measurement are that (1) a reasonable choice of refractive and absorption indices must be made and (2) to not overinterpret the tails of the size distributions if the particles are not spherical. Good, reproducible results can be obtained for cementitious materials, although dispersion conditions for the powder sometimes have to be optimised, particularly for fine supplementary cementitious materials such as silica fume. General recommendations for optical models, dispersion liquids and dispersants have been given.

SSA measurements using nitrogen adsorption have been shown to be accurate and reproducible if care is taken in the drying of the sample before analysis. Because of the limitations of the pore shape models in transforming the NAD isotherms into pore sizes and the limited size range offered, the pore size distributions measured by NAD are of limited use.

By following the proposed measurement guidelines, reliable and reproducible PSD and SSA measurements using nitrogen adsorption can be achieved. Both methods will become cornerstones of the assessment of cementitious material fineness, in both the industry and the academia in the near future.

REFERENCES

Allen, T. (1997). *Particle Size Measurement*. Fifth edition. New York: Chapman and Hall.

Arvaniti, E. C., M. C. G. Juenger, S. A. Bernal, J. Duchesne, L. Courard, S. Leroy, J. L. Provis, A. Klemm and N. De Belie (2014a, October). 'Determination of Particle Size, Surface Area, and Shape of Supplementary Cementitious Materials by Different Techniques'. *Materials and Structures*, 1–15.

Arvaniti, E. C., M. C. G. Juenger, S. A. Bernal, J. Duchesne, L. Courard, S. Leroy, J. L. Provis, A. Klemm and N. De Belie (2014b, October). 'Physical Characterisation Methods for Supplementary Cementitious Materials'. *Materials and Structures*, 1–12.

ASTM International (2011). 'ASTM C204, Test Methods for Fineness of Hydraulic Cement by Air-Permeability Apparatus'.

Azzopardi, B. J. (1992). *Particle Size Analysis*. Cambridge: Royal Society of Chemistry.

Blaine, R. L. (1943). 'A Simplified Air-Permeability Fineness Apparatus'. *ASTM Bulletin*, 51–55.

Bowen, P. (2002). 'Particle Size Distribution Measurement from Millimeters to Nanometers and from Rods to Platelets'. *Journal of Dispersion Science and Technology* 23 (5): 631–662.

Brunauer, S., P. H. Emmet and E. Teller (1938). 'Adsorption of Gases in Multimolecular Layers'. *Journal of American Ceramic Society* **60**: 309–319.

Costoya, M. (2008). 'Effect of Particle Size on the Hydration Kinetics and Microstructural Development of Tricalcium Silicate'. PhD thesis, École Polytechnique Fédérale de Lausanne, Switzerland.

Dalas, F. (2014). 'Influence Des Paramètres Structuraux de Superplastifiants Sur L'hydratation, La Création de Surfaces Initiales et La Fluidité de Systèmes Cimentaires Modèles'. PhD thesis, Université de Bourgogne, Dijon, France.

Deschner, F., F. Winnefeld, B. Lothenbach, S. Seufert, P. Schwesig, S. Dittrich, F. Goetz-Neunhoeffer and J. Neubauer (2012). 'Hydration of Portland Cement with High Replacement by Siliceous Fly Ash'. *Cement and Concrete Research* **42** (10): 1389–1400.

Fagerlund, G. (1973). 'Determination of Specific Surface by the BET Method'. *Matériaux et Construction* **6** (3): 239–245.

Ferraris, C. F., W. Guthrie, A. I. Aviles, R. Haupt and B. S. MacDonald (2005). *Certification of SRM 114q: Part I*. NIST Special Publication, Gaithersburg, Maryland.

Ferraris, C. F., V. A. Hackley, A. I. Aviles and C. E. Buchanan (2002a). 'Analysis of the ASTM Round Robin Test on Particle Size Distribution of Portland Cement: Phase I'. NIST Interagency or Internal Report 6883.

Ferraris, C. F., V. A. Hackley, A. I. Aviles and C. E. Buchanan (2002b). 'Analysis of the ASTM Round Robin Test on Particle Size Distribution of Portland Cement: Phase II'. NIST Interagency or Internal Report 6931.

Ferraris, C. F., V. A. Hackley and A. I. Avilés (2004). 'Measurement of Particle Size Distribution in Portland Cement Powder: Analysis of ASTM Round Robin Studies'. *Cement, Concrete, and Aggregates* **26** (2): 1–11.

Gabas, N., N. Hiquily and C. Laguérie (1994). 'Response of Laser Diffraction Particle Sizer to Anisometric Particles'. *Particle & Particle Systems Characterisation* **11** (2): 121–126.

Gregg, S. J., and K. S. W. Sing (1982). *Adsorption, Surface Area, and Porosity*. Second edition. Waltham: Academic Press.

Jennings, B. R., and K. Parslow (1988). 'Particle Size Measurement: The Equivalent Spherical Diameter'. *Proceedings of the Royal Society of London A: Mathematical, Physical and Engineering Sciences* **419** (1856): 137–149.

Kazemi-Kamyab, H. (2015). Personal Communication.

Khalili, M., W. L. Roricht and S. Y. L. Lee (2002). 'An Investigation to Determine the Precision for Measuring Particle Size Distribution by Laser Diffraction'. In: World Congress on Powder Technology, Paper 111. Sydney.

Kiattikomol, K., C. Jaturapitakkul, S. Songpiriyakij and S. Chutubtim (2001). 'A Study of Ground Coarse Fly Ashes with Different Finenesses from Various Sources as Pozzolanic Materials'. *Cement and Concrete Composites* **23** (4–5): 335–343.

Kocaba, V. (2009). 'Development and Evaluation of Methods to Follow Microstructural Development of Cementitious Systems Including Slags'. PhD thesis, École Polytechnique Fédérale de Lausanne, Switzerland.

Korpa, A., and R. Trettin (2006). 'The Influence of Different Drying Methods on Cement Paste Microstructures as Reflected by Gas Adsorption: Comparison between Freeze-Drying (F-Drying), D-Drying, P-Drying and Oven-Drying Methods. *Cement and Concrete Research* **36** (4): 634–649.

Lowell, S., J. E. Shields, M. A. Thomas and M. Thommes (2004). 'Characterisation of Porous Solids and Powders: Surface Area, Pore Size and Density'. In: *Characterisation of Porous Solids and Powders: Surface Area, Pore Size and Density*, 1–4. Particle Technology Series 16. Dordrecht: Springer Netherlands.

Mantellato, S., M. Palacios and R. J. Flatt (2012). 'Reliable Specific Surface Measurement of Fresh Cement Pastes'. In: 14. GDCh-Tagung Bauchemie. Dübendorf, Switzerland.

Mantellato, S., M. Palacios and R. J. Flatt (2015a). 'Reliable Specific Surface Area Measurements on Anhydrous Cements'. *Cement and Concrete Research* **67**: 286–291.

Mantellato, S., M. Palacios and R. J. Flatt (2015b). 'Impact of Sample Preparation on the Specific Surface Area of Synthetic Ettringite'. In: Preparation.

Mikhail, R. S., and S. A. Selim (1966). 'Adsorption of Organic Vapors in Relation to the Pore Structure of Hardened Portland Cement Pastes'. *Highway Research Board Special Report*, 90.

Mindess, S., and J. F. Young (1981). *Concrete*. Englewood Cliffs: Prentice Hall.

Palacios, M., P. Bowen, M. Kappl, M. Stuer, C. Pecharroman, U. Aschauer and F. Puertas (2012). 'Repulsion Forces of Superplasticizers on Ground Granulated Blast Furnace Slag in Alkaline Media, from AFM Measurements to Rheological Properties'. *Materiales de Construccion* **62** (308): 489–513.

Pearson, D., and A. J. Allen (1985). 'A Study of Ultrafine Porosity in Hydrated Cements Using Small Angle Neutron Scattering'. *Journal of Materials Science* **20** (1): 303–315.

Potgieter, J. H., and C. A. Strydom (1996). 'An Investigation into the Correlation between Different Surface Area Determination Techniques Applied to Various Limestone-Related Compounds'. *Cement and Concrete Research* **26** (11): 1613–1617.

Scherer, G. W. (1998). 'Adsorption in Sparse Networks: I. Cylinder Model'. *Journal of Colloid and Interface Science* **202** (2): 399–410.

Sing, K. S. W. (1995). 'Physisorption of Nitrogen by Porous Materials'. *Journal of Porous Materials* **2** (1): 5–8.

Sing, K. S. W., D. H. Everett, R. A. W. Haul, L. Moscou, R. A. Pierotti, J. Rouquerol and T. Siemieniewska (1985). 'Reporting Physisorption Data for Gas/Solid Systems with Special Reference to the Determination of Surface Area and Porosity (Recommendations 1984)'. *Pure and Applied Chemistry* **57**: 603–619.

Sommer, K. (2001). '40 Years of Presentation Particle Size Distributions – Yet Still Incorrect?' *Particle & Particle Systems Characterisation* **18** (1): 22–25.

Souri, A., H. Kazemi-Kamyab, R. Snellings, R. Naghizadeh, F. Golestani-Fard and K. Scrivener (2015). 'Pozzolanic Activity of Mechanochemically and Thermally Activated Kaolins in Cement'. Submitted.

Thomas, J., H. Jennings and A. Allen (1999). 'The Surface Area of Hardened Cement Paste as Measured by Various Techniques'. *Concrete Science and Engineering* **1**: 45–64.

Winslow, D. N., and S. Diamond (1974). 'Specific Surface of Hardened Portland Cement Paste as Determined by Small-Angle X-Ray Scattering'. *Journal of the American Ceramic Society* **57** (5): 193–197.

Xu, R., and O. A. Di Guida (2002). 'Particle Size and Shape Analysis Using Light Scattering, Coulter Principle, and Image Analysis'. In: World Congress on Powder Technology. Sydney.

Yamada, K. (2011). 'Basics of Analytical Methods Used for the Investigation of Interaction Mechanism between Cements and Superplasticizers'. *Cement and Concrete Research* **41** (7): 793–798.

Zhang, J., and G. W. Scherer (2011). 'Comparison of Methods for Arresting Hydration of Cement'. *Cement and Concrete Research* **41** (10): 1024–1036.

Zhang, L., and F. P. Glasser (2000). 'Critical Examination of Drying Damage to Cement Pastes'. *Advances in Cement Research – Advances in Cement Research* **12** (2): 79–88.

Chapter 11

Ternary phase diagrams applied to hydrated cement

Duncan Herfort and Barbara Lothenbach

CONTENTS

11.1 INTRODUCTION

The phase assemblages that form from the hydration of portland cement over time, or from changes in composition, can be predicted by thermodynamic modelling (Lothenbach 2010), or simply by mass balance calculations with some knowledge of the relative stability of the hydrate phases. Thermodynamic modelling can deal with the complexity of cementitious systems and is well suited to predicting the effect of changes in single or several components at once as, e.g. discussed by Damidot et al. (2011). To successfully apply thermodynamic modelling, some knowledge of geochemistry (e.g. which phases can reasonably precipitate or dissolve under the conditions of equilibrium) as well as practical knowledge of the modelling software being used is needed. For a general overview of the effect of different chemical compositions, ternary phase diagrams provide an effective means of graphical representation of the thermodynamic predictions. The present chapter gives an introduction to how these ternary diagrams are constructed, illustrates their usefulness in predicting the phase assemblage of hydrated cements and offers access to an interactive excel file for performing this in practice.

Ternary diagrams follow the same laws as thermodynamic modelling, including, of course, the phase rule $P + F = C + 2$, where P is the number of phases, F is the number of degrees of freedom and C is the number of components. By fixing temperature and pressure, it can be condensed to $P + F = C$. This chapter illustrates how ternary phase diagrams are constructed and how they can be used and compares them with thermodynamic modelling and experimental results.

Phase diagrams are widely used in all major branches of inorganic chemistry, including geochemistry, ceramics, metallurgy and cement chemistry. For a detailed review of how they are applied in geochemistry, the reader is referred to Ehlers (1972), Krauskopf and Bird (1995) and Winkler (1974). The three most common types of phase diagrams used are the following:

1. The *high-temperature liquidus–solidus* binary and ternary diagrams used in all disciplines (see the example in Figure 11.1)
2. *Activity–activity diagrams* used in the geochemistry of low-grade metamorphic rocks and cement hydration (see the example in Figure 11.2)
3. *Subternary compatibility phase diagrams* widely used in the geochemistry of metamorphic rocks and in cement hydration (see the example in Figure 11.3)

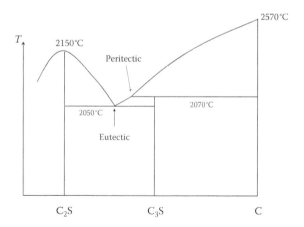

Figure 11.1 Binary phase diagram for the region of the CaO–SiO$_2$ system relevant to portland cement clinker. (Adapted from Welch, J. H., and W. Gutt, *Journal of the American Ceramic Society*, 42, 11–25, 1959.)

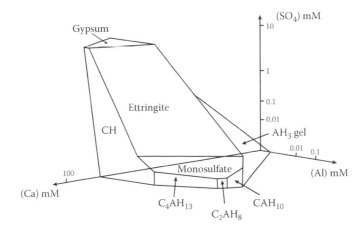

Figure 11.2 Activity–activity diagram for the CaO-CaSO$_4$–Al$_2$O$_3$–H$_2$O system. A general description of activity–activity phase diagrams in metamorphic geology can be found in Krauskopf and Bird (1995). (Reproduced from Damidot, D. et al., *Cement and Concrete Research*, 41(7), 679–695, 2011. With permission.)

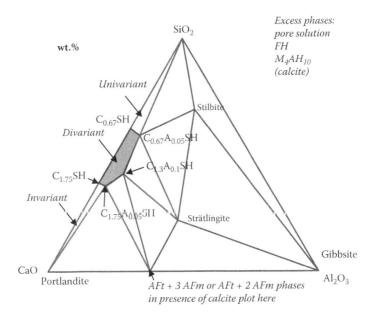

Figure 11.3 The CaO–Al$_2$O$_3$–SiO$_2$ ternary diagram in wt.% in the presence of excess phase pore solution, ferrihydrite and possibly calcite. Hydrogarnet (C$_3$AH$_6$), ettringite and the AFm phases (with the exception of strätlingite) plot on the same point on the CaO–Al$_2$O$_3$ tie-line (C$_3$A). A: Al$_2$O$_3$; C: CaO; S: SiO$_2$. Dark and light gray areas correspond to divariant and univariant regions; white regions are invariant.

Surprisingly, ternary phase diagrams are rarely used in hydrated cement chemistry, despite the similarity to low-grade metamorphic geology. Here it is important to distinguish between true ternary phase diagrams, which are the subject of this chapter, and the ternary diagrams that are commonly used to illustrate general relationships without following the clear rules that apply to phase diagrams as described in the following.

True ternary phase diagrams used in metamorphic geology are a powerful technique for visualising and predicting the phase assemblages and relative phase contents from the three intensive variables needed to define the system studied, usually temperature, pressure and composition. They can and should be applied just as effectively to hydrated portland cement systems. A typical metamorphic rock, or hydrated cement, may contain up to 10 chemical components in high enough amounts to affect the mineralogy. The art of constructing suitable ternary diagrams is to identify the phase relationships of interest and the three chemical components required to describe these relationships. Strictly speaking, because the ternary diagrams occur within multicomponent systems, they are more correctly termed *pseudoternary* or *subternary diagrams*. Well-defined rules are used in their construction, as described in the following sections, where they are applied to the two most useful ternary diagrams for hydrated portland cement, i.e. the $CaO–Al_2O_3–SiO_2$ and $C_3A–CaSO_4–CaCO_3$ diagrams.

II.2 TERNARY DIAGRAMS IN THE CaO–Al₂O₃–(Fe₂O₃–MgO–)SiO₂–SO₃–CO₂–H₂O SYSTEM

These eight major components are needed to define a typical hydrated portland cement. Major components are defined here as the components which are required to form a separate phase under invariant conditions at constant temperature and pressure in which the maximum number of phases are formed at 0 degrees of freedom ($F = 0$). Applying the condensed phase rule $P + F = C$ (constant temperature and pressure),[*] this eight-component system forms a maximum of eight phases at $F = 0$. In hydrated portland cement these phases are C-A-S-H, portlandite, ettringite, monocarbonate, calcite, ferrihydrite, hydrotalcite and pore solution, for a normal portland cement composition. Minor components such as K_2O and Na_2O are not normally present in high enough concentrations to form additional phases, but these add degrees of freedom as they are incorporated in solid solution in existing phases. That is not to say they do not have a major impact on performance; alkalis, for

---

[*] Alternatively, from using the standard phase rule $P + F = C + 2$, the maximum number of phases forms under divariant conditions with two degrees of freedom for temperature and pressure. Of course, reactions can occur at a constant temperature with a higher number of phases than allowed by the condensed phase rule, but since we are concerned only with equilibrium conditions once reactions have gone to completion, this is not considered further in this chapter.

example, normally have the highest concentration in the pore solution, but the pore solution will still be present without the alkalis, albeit with a lower pH. The ternary diagrams shown are constructed from the main oxides, i.e. $CaO–Al_2O_3–SiO_2–SO_3–CO_2–H_2O$, or when appropriate CO_2 and SO_3 are combined with CaO to give $CaCO_3$ and $CaSO_4$ as chemical components and CaO and Al_2O_3 are combined to give C_3A as described in the following sections. Fe_2O_3 is assumed to form ferrihydrite; MgO, a hydrotalcite-like phase $(Mg_4Al_2O_7 \cdot 10H_2O)$ or brucite when there is a large surplus of magnesium.

Various methods can be used to predict the phase assemblage for a given chemical composition, in this case the eight-phase assemblage for a standard portland cement, and to calculate the relative contents of phases. Thermodynamic programs commonly used by cement researchers include PHREEQC, based on the law of mass action, or Gibbs Energy Minimisation Software (GEMS), which minimises the free energy. In any case, thermodynamic equilibrium is assumed, which ensures nonviolation of the phase rule. Metastable equilibrium solutions are also possible by assuming persistent metastability of one phase over another.

11.3 THE $CaO–Al_2O_3–SiO_2$ AND $C_3A–CaSO_4–CaCO_3$ SUBTERNARY PHASE DIAGRAMS

When used together, the diagrams shown in Figures 11.3 and 11.4 represent essentially all the phase assemblages that will form in hydrated systems containing portland cement, pozzolans and limestone. They can also be used to predict the mineralogy resulting from carbonation and sulfate attack. The rules for constructing and using these diagrams or other subternary phase diagrams are as follows:

1. In each diagram the components must be defined as the largest stoichiometric unit needed to describe the phases present, e.g. every phase present in Figure 11.3 can be denoted as a combination of CaO, Al_2O_3 and SiO_2.
2. When the bulk composition of a hydrated cement is plotted onto the ternary diagram, the excess phases that do not plot onto the diagram are removed from the bulk composition; i.e. the oxides associated with these phases are subtracted from the total bulk chemical composition. We can then simply plot the relative amounts of the remaining chemical components onto the diagram to give the relative contents of the phases in the subsystem represented by the ternary diagram and within each Alkemade triangle. The easiest way of performing this is to begin by calculating the total phase assemblage, e.g. from GEMS, and then remove the excess phases. Unreacted anhydrous constituents should also be removed from the bulk composition before performing GEMS calculation.

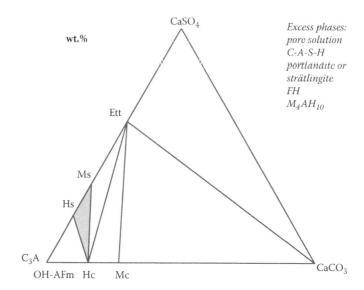

Figure 11.4 The C_3A–$CaCO_3$–$CaSO_4$ ternary diagram in wt.% in the presence of excess phases (pore solution, ferrihydrite and hydrotalcite). The gray area corresponds to a univariant region; white regions are invariant. Ett: ettringite; Hc: hemicarbonate; Hs: hemisulfate; Mc: monocarbonate; Ms: monosulfate. See Table 11.1 for phase composition.

11.3.1 The CaO–Al$_2$O$_3$–SiO$_2$ subternary phase diagram

This diagram shown in Figure 11.3 gives the phase relationships between the C-A-S-H phase, portlandite, strätlingite in systems undersaturated in $Ca(OH)_2$ and the AFm and AFt phases. For systems even more undersaturated in $Ca(OH)_2$, framework silicates will form, approximated in this diagram as stilbite.

The diagram is divided into invariant, univariant and divariant regions. The invariant regions have the maximum number of phases of three in a true ternary system, four in a quaternary system, etc. The invariant regions in the subternary diagram will also contain the maximum number of three phases, since the excess phases can be ignored. The univariant and divariant regions contain one and two phases fewer than the maximum number, respectively. These degrees of freedom are a consequence of the solid solution of the C-A-S-H phase.* A hydrated portland cement with a high

---

* The invariant C-A-S-H compositions are taken from available data for portland cements (Ca/Si = 1.7–2.0; Al/Si = 0.05–0.1), blended cements (Ca/Si = 1–1.5; Al/Si = 0.03–0.2) (Deschner et al. 2013; Girão et al. 2010; Richardson and Groves 1993; Rossen et al. 2015) and alkali-activated slags (Ca/Si = 0.7–1; Al/Si = 0.1–0.3) (Ben Haha et al. 2012; Bernal et al. 2014). See Table 11.1.

silica content and an unusually low alumina content, for example, which plots within the compositional range of the C-A-S-H phase, will have two degrees of freedom, i.e. corresponding to divariant conditions. Normally, however, a hydrated portland cement will contain excess alumina so that with increasing contents of pozzolan, for example, the system will move from invariant to univariant conditions at the point where portlandite is consumed. This degree of freedom will then allow the composition of C-A-S-H to change as the content of pozzolan increases. This will then also allow the composition of the pore solution to change as well, resulting among other things in a reduction in the pH.

For a normal portland cement, since the subternary diagram is constructed within the $CaO-Al_2O_3-Fe_2O_3-MgO-SiO_2-CaSO_4-CaCO_3-H_2O$ (eight-component) system, any bulk composition plotting within the invariant region saturated with respect to $Ca(OH)_2$ will contain the following eight phases: portlandite, C-A-S-H with the fixed composition shown, ferrihydrite, hydrotalcite, pore solution and, for a normally sulfated portland cement with only minor carbonates, monosulfate, ettringite and hemicarbonate. Note that $CaSO_4$ and $CaCO_3$ are chosen as components because SO_3 and CO_2 can be combined with Ca in all phases and therefore represent the largest stoichiometric units for these components. This means that $C_3A\cdot3Cs\cdot32H$, $C_3A\cdot Cs\cdot12H$ and $C_3A\cdot Cc\cdot12H$ all plot onto the same position on this ternary diagram with a C–to–A ratio of 3:1. Strictly speaking, $C_3A\cdot CH\cdot13H$, $C_3A\cdot0.5CH\cdot0.5Cc\cdot12H$ and $C_3A\cdot0.5CH\cdot0.5Cs\cdot12H$ should plot on different positions on the C-A tie-line, but for convenience $Ca(OH)_2$ is chosen as an additional separate component so that these phases also plot onto the same position. In practice this means that the $Ca(OH)_2$ that combines to form the AFm phases must be subtracted alongside $CaCO_3$ and $CaSO_4$ from the bulk composition before the relative content of remaining phases is calculated and/or plotted onto the C-A-S subternary diagram. Hydrogrossular, C_3AH_6, where this is considered thermodynamically stable also plots onto the same position with a C-to-A ratio of 3:1, where H_2O is chosen as the extra component. Of course, since all the AFm and AFt phases plot on the same position on this diagram, no information is given on their relative contents. For this purpose the $C_3A-CaSO_4-CaCO_3$ diagram is used as described in the following section.

11.3.2 The $C_3A-CaSO_4-CaCO_3$ subternary phase diagram

The diagram shown in Figure 11.4 gives the phase relationships for the carbonate- and sulfate-bearing phases. The components C_3A, $CaSO_4$ and $CaCO_3$ are chosen because they represent the largest stoichiometric units occurring in all phases. Apart from one univariant region due to partial solid solution between monosulfate and the hydroxy–AFm phase (Matschei et al. 2007a; Pöllmann 2006), all regions are invariant due to insignificant solid solution between the other phases. A typical portland cement, for example, with only minor carbonate (e.g. from carbonation during grinding in the cement mill) will contain a mixture of monosulfate, ettringite and hemicarbonate, in addition to the C-A-S-H phase and portlandite, which plot onto the $CaO-Al_2O_3-SiO_2$ subternary phase diagram, and the additional excess phases, ferrihydrite, hydrotalcite and the pore solution.

11.4 EXAMPLES OF THE APPLICATION OF THE $CaO-Al_2O_3-SiO_2$ AND $C_3A-CaSO_4-CaCO_3$ DIAGRAMS USED IN CONJUNCTION

11.4.1 Portland–pozzolan–limestone cements

The examples shown in Figures 11.5 through 11.7 are taken from a study on the synergetic effect of limestone and calcined clay in composite portland cements (Steenberg et al. 2011). Laboratory cements were produced by blending portland cement with fine limestone powder and metakaolin and tested after 28 and 90 days. The bulk compositions of the blends are plotted onto the two subternary diagrams in Figures 11.5 and 11.6. Blend 1 is CEM I portland cement, blend 2 contains 30% metakaolin addition, blend 3 contains 30% limestone addition, blend 4 contains 20% metakaolin and

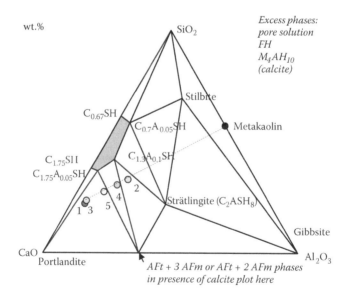

Figure 11.5 Bulk composition from Steenberg et al. (2011) plotted onto the CaO–Al_2O_3–SiO_2 ternary diagram. (From Steenberg, M. et al., Composite cement based on portland cement clinker, limestone and calcined clay. Proceedings of the 13th International Congress on the Chemistry of Cement, Madrid, Spain, 97–104, 2011.)

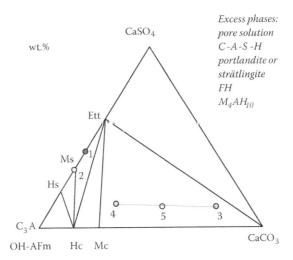

Figure 11.6 Bulk composition from Steenberg et al. (2011) plotted onto the C_3A–$CaCO_3$–$CaSO_4$ ternary diagram. (From Steenberg, M. et al., Composite cement based on portland cement clinker, limestone and calcined clay. Proceedings of the 13th International Congress on the Chemistry of Cement, Madrid, Spain, 97–104, 2011.)

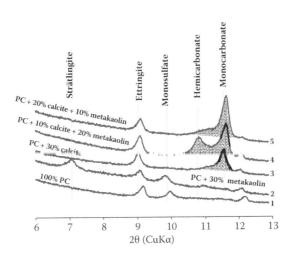

Figure 11.7 Effect of aluminate carbonate reaction with hydrate phase assemblages consistent with predictions from phase diagrams in Figures 11.5 and 11.6. PC: portland cement. (Reproduced from Steenberg, M. et al., Composite cement based on portland cement clinker, limestone and calcined clay. Proceedings of the 13th International Congress on the Chemistry of Cement, Madrid, Spain, 97–104, 2011. With permission.)

10% limestone and blend 5 contains 10% metakaolin and 20% limestone. Assuming complete hydration, the relative contents of phases are shown by the position where the bulk composition plots in each Alkemade triangle.

Figure 11.7 shows the X-ray diffraction (XRD) results after 90 days of hydration for each blend. In addition to C-A-S-H, portlandite and calcite which were also identified, the following assemblages are observed:

1. 100% portland cement: ettringite + monosulfate + C-A-S-H + portlandite
2. 70% portland cement + 30% metakaolin: strätlingite + ettringite + monosulfate + C-A-S-H
3. 70% portland cement + 30% calcite: monocarbonate + calcite + ettringite + C-A-S-H + portlandite
4. 70% portland cement + 20% metakaolin + 10% calcite: monocarbonate + hemicarbonate + ettringite + C-A-S-H
5. 70% portland cement + 10% metakaolin + 20% calcite: monocarbonate + ettringite + C-A-S-H + portlandite + calcite

Apart from some hemicarbonate observed in blend 4, all observed assemblages are consistent with those plotted onto the subternary diagrams.

11.4.2 Comparison with thermodynamic modelling

Another form of graphical representation of the thermodynamic modelling is shown in Figure 11.8. Here the relative phase composition is shown as the bulk composition changes from blend 2 to blend 5 in Figures 11.5 and 11.6. The results of the thermodynamic calculations using GEMS and an updated version of the CEMDATA07 database (Kulik 2011; Lothenbach et al. 2008; Matschei et al. 2007b) are shown in Figure 11.8; detailed descriptions on thermodynamic modelling and examples relevant for cements can be found in Damidot et al. (2011), Lothenbach (2010) and Lothenbach et al. (2010). The ternary diagrams in Figures 11.5 and 11.6 and the complete assemblages calculated from the thermodynamic modelling in Figure 11.8 are representations of the same data as they are based on essentially the same thermodynamic data. Some minor differences are seen due to the slightly different amounts of aluminium in C-A-S-H in the case of the ternary diagrams (see Figure 11.6), whilst for the thermodynamic model used to construct Figure 11.8 a fixed A/S ratio of 0.09 was assumed because of the lack of data to describe aluminium uptake in C-S-H. The direct comparison of the two types of diagrams shows that both methods give equivalent information; the ternary diagrams give a very good overview, allowing predictions to be made for any change in one or more of the major components, whilst the type of diagram shown in Figure 11.8 provides detail on the complete assemblage and how this changes as a result of variations in the bulk composition in a single direction and which can also be used to directly plot the volume or mass changes.

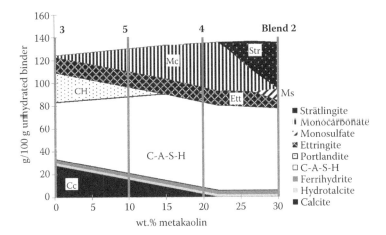

Figure 11.8 Calculated changes using thermodynamic modelling as a function of metakaolin content. 70 wt.% portland cement +30 wt.% (metakaolin + limestone)
based on the data from Steenberg et al. (2011) assuming complete reaction of
metakaolin and A/S = 0.09 in C-S-H. (From Steenberg, M. et al., Composite
cement based on portland cement clinker, limestone and calcined clay.
Proceedings of the 13th International Congress on the Chemistry of Cement,
Madrid, Spain, 97–104, 2011.)

11.4.3 Thaumasite formation

Following the rules described previously for constructing subternary diagrams,
thaumasite (CS·Cs·Cc·12H) plots onto the C_3A–$CaSO_4$–$CaCO_3$ diagram, as
shown in Figure 11.9. This was tested in a study reported by Juel et al. (2003),
in which blends of portland cement, fine limestone powder and gypsum were
hydrated and stored in water at 5°C. Figure 11.10 shows the bulk composition
of the blends in question and a good agreement with the actual assemblages
determined by XRD and Si-NMR. Similar results were also obtained for portland cement limestone blends after storing in a $MgSO_4$ solution at 5°C.

11.4.4 Chloride-containing systems

The introduction of chlorides as either alkali or calcium chlorides results
in binding by the aluminate phases and some adsorption by the C-A-S-H
phase. Ternary phase diagrams were used by Nielsen et al. (2003) to study
the reactions taking place. Figure 11.11 shows the subternary diagram
that we have constructed for the C_3A–$CaSO_4$–$CaCl_2$ subsystem, which is
comparable to the C_3A–$CaSO_4$–$CaCO_3$ diagram already discussed. This
diagram is equally applicable to the introduction of alkali chlorides since
the Cl combines with the aluminate AFm phases as $CaCl_2$. From this it
can be predicted that Friedel's salt will form at the expense of monosulfate as the chloride concentration increases, which also leads to additional

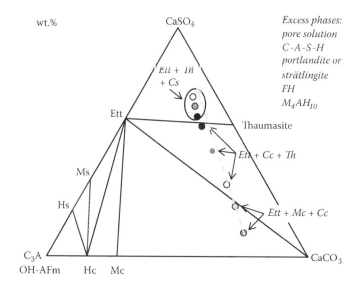

Figure 11.9 Bulk composition from Juel et al. (2003) plotted onto the C₃A–CaCO₃–CaSO₄ ternary diagram at 5°C. The plots are consistent with the compositions designated A through H in Figure 11.10, A with the lowest CaSO₄ content through to H with the highest CaSO₄ content. The addition of CaSO₄ in the calculations leads to hydrate composition, which plots slightly off the direct trend towards the CaSO₄ component, as the amount of $C_{1.75}SA_{0.05}H_4$ decreases and, thus, relatively more C₃A is available to form ettringite and monocarbonate. (From Juel, I. et al., *Cement and Concrete Composites*, 25, 867–872, 2003.)

ettringite formation as SO₃ is released from the monosulfate. Solid solution between the Friedel's salt and monosulfate in this system was found to be insignificant. Nielsen et al. (2003) found complete solid solution between the monocarbonate phase and Friedel's salt phase, as shown in Figure 11.12. These results were later confirmed by Balonis et al. (2010).

11.5 FURTHER COMPONENTS

The addition of minor components can affect the thermodynamic stability of the cement hydrates and change the subternary phase diagrams. Components forming relatively insoluble phases, e.g. Fe₂O₃, MgO and P₂O₅, which form ferrihydrite (or hydroandradite), hydrotalcite (or brucite) and hydroxyapatite, respectively, have a negligible effect on the composition of major phases. Since they do not form significant solid solution with the existing hydrates, they do not affect the application of the subternary phase diagrams. The effect of the presence of Fe₂O₃ and MgO in the diagrams is taken into account by assuming the formation of ferrihydrite and a

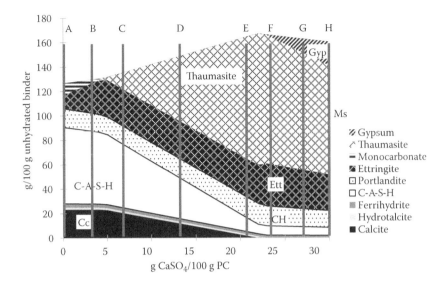

Figure 11.10 Calculated changes using thermodynamic modelling as a function of the $CaSO_4$ content from the trend in bulk compositions shown in Figure 11.9. 30 wt.% portland cement replacement of portland cement by ($CaSO_4$ + $CaCO_3$).

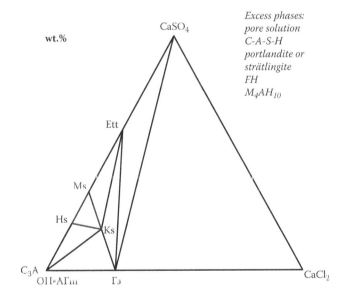

Figure 11.11 Subternary C_3A–$CaSO_4$–$CaCl_2$ ternary diagram from Nielsen et al. (2003). Ett: ettringite; Fs: Friedel's salt; Hs: hemisulfate; Ks: Kuzel's salt; Ms: monosulfate. See Table 11.1 for phase composition. (From Nielsen, E. P. et al., Effect of solid solutions of AFm phases on chloride binding. Proceedings of the 11th International Congress on the Chemistry of Cement, Durban, South Africa, 2003.)

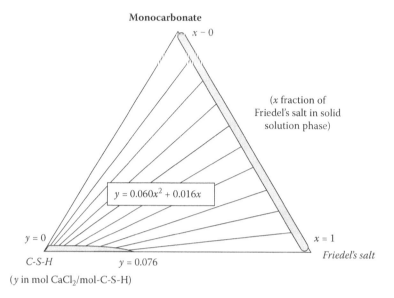

Monocarbonate

$x - 0$

(x fraction of Friedel's salt in solid solution phase)

$y = 0.060x^2 + 0.016x$

$y = 0$

$x = 1$

Friedel's salt

C-S-H $y = 0.076$

(y in mol $CaCl_2$/mol-C-S-H)

Figure 11.12 Subternary CSH–$C_3A \cdot CaCl_2 \cdot 12H$–$C_3A \cdot Cc \cdot 12H$ ternary diagram from Nielsen et al. (2003). (From Nielsen, E. P. et al., Effect of solid solutions of AFm phases on chloride binding. Proceedings of the 11th International Congress on the Chemistry of Cement, Durban, South Africa, 2003.)

hydrotalcite-like phase ($Mg_4Al_2O_7$ $10H_2O$).* Soluble components, in particular the alkalis, on the other hand, can have a major impact on the thermodynamic stability and the phase diagrams. For example, the addition of an alkali adds a degree of freedom to the CaO–Al_2O_3–SiO_2 phase diagram so that the 'invariant' compositions of the C-A-S-H phase becomes 'univariant' allowing the composition of the C-A-S-H phase to change as a function of the alkali content. In addition, very high concentrations of alkali hydroxides (>1 M) can destabilise ettringite (Perkins and Palmer 1999).

11.6 EXCEL FILES FOR PLOTTING SUBTERNARY DIAGRAMS OF HYDRATED PORTLAND– POZZOLAN–LIMESTONE CEMENTS

An Excel file for performing the phase assemblage calculation and automatically plotting the composition onto the two subternary diagrams described

---

* For the sake of simplicity, it was assumed that iron is present as ferrihydrite. However, recent investigations have shown that ferrihydrite precipitates only at early reaction times, while at later ages rather hydroandradite ($C_3FS_{0.8-1.3}H_{4.3-3.3}$) is present (Dilnesa et al. 2014), which results in slightly less C-S-H and portlandite. Hydrotalcite-like phases can have variable compositions, generally in the range $Mg_{4-6}Al_2O_{7-9}$ $10H_2O$, and can contain carbonates instead of hydroxides (Ben Haha et al. 2011; Richardson 2013; Rozov et al. 2011).

previously is provided as supplementary material (available at http://www .empa.ch/cemdata). Typical compositions for portland cement, limestone, fly ash, silica fume and slag are used, but the user is free to enter any composition of interest within the eight-component system and replace the fly ash with other aluminosilicate pozzolans. Gypsum can also be used to test the effect of increased sulfates. There are two different versions of the Excel sheet available, one corresponding to Figure 11.3 and another to Figure 11.4, corresponding to ambient conditions. To include the formation of thaumasite, which is especially important at lower temperatures, a second set of ternary plots was constructed (corresponding to Figure 11.9).

The compositions of the hydrate phases used to construct the diagrams have been taken from the literature and are compiled in Table 11.1. For

Table 11.1 Chemical compositions, molecular weights, molar volumes and densities of the solids used in the ternary plots

Solid	Formula[a]	Mol. weight (g/mol)	Volume[b] (cm³/mol)	Density (g/cm³)
Gypsum	CsH_2	172	75	2.31
Portlandite	CH	74	33	2.24
Calcite	Cc	100	37	2.71
High-Ca C-S-H	$C_{1.75}SH_4$	230	110	2.1
High-Ca C-A-S-H	$C_{1.75}SA_{0.05}H_4$	235	112	2.1
C-A-S-H	$C_{1.3}SA_{0.1}H_3$	197		
Low-Ca C-A-S-H	$C_{0.67}SA_{0.05}H_2$	139		
Low-Ca C-S-H	$C_{0.67}SH_2$	133		
Ettringite	$C_6As_3H_{32}$	1255	707	1.77
Thaumasite	C_3SscH_{15}	623	330	1.89
Monosulfate	C_4AsH_{12}	623	309	2.01
Hemisulfate	$C_4As_{0.5}H_{12.5}$	591		
Monocarbonate	C_4AcH_{11}	568	262	2.17
Hemicarbonate	$C_4Ac_{0.5}H_{12}$	564	285	1.98
OH-AFm	C_4AH_{13}	560	274	2.05
Friedel's salt	$C_4ACl_2H_{10}$	561	272	2.00
Kuzel's salt	$C_4As_{0.5}Cl_2H_{12}$	610	289	2.11
Strätlingite	C_2ASH_8	418	216	1.94
Hydrotalcite	M_4AH_{10}	443	220	2.01
Katoite	C_3AH_6	378	150	2.53
Ca-stilbite	$C_{0.17}SA_{0.17}H_{1.04}$	105	49	2.15
Amorphous SiO_2	S	60	29	2.07
Aluminium hydroxide	AH_3	156	64	2.44

[a] Cement short hand notation is used – A: Al_2O_3; c: CO_2; C: CaO; H: H_2O; M: MgO; s: SO_3; S: SiO_2.

[b] Molar volume from the CEMDATA07 database (Balonis et al. 2010; Lothenbach et al. 2008; Matschei et al. 2007b); densities calculated from molecular weight and molar volumes. The roughly estimated molar volumes and density of C-S-H and stilbite are plotted in italic.

AFt, AFm and hydrotalcite, the compositions suggested by Balonis et al. (2010), Blanc et al. (2010), Lothenbach et al. (2008) and Matschei et al. (2007b) have been used. The boundary compositions for C-A-S-H ($C_{1.75}SH_4$, $C_{1.75}A_{0.05}SH_4$, $C_{1.3}A_{0.1}SH_3$, $C_{0.67}A_{0.05}SH_2$, $C_{0.67}SH_2$) have been chosen based on the range of data determined for portland cements (Ca/Si = 1.7–2.0; Al/Si = 0.05–0.1), blended cements (Ca/Si = 1–1.5; Al/Si = 0.03–0.2) (Deschner et al. 2013; Girão et al. 2010; Richardson and Groves 1993; Rossen et al. 2015) and alkali activated slags (Ca/Si = 0.7–1; Al/Si = 0.1–0.3) (Ben Haha et al. 2012; Bernal et al. 2014) and the water content of C-S-H (H/S including gel water ≈4) in hydrating portland cements (Muller et al. 2013). The invariant C-A-S-H compositions can be modified as new data becomes available, and, of course, these diagrams provide a framework to design experiments for this purpose.

The effect of different alkali contents, which introduce an additional degree of freedom, can also be included in future versions.

Blast furnace slags need to be introduced with some caution, because the model at this stage does not accommodate variable Mg/Al ratios in hydrotalcite. For this a new diagram including hydrotalcite solid solution series needs to be developed. A ternary phase diagram including chlorides is in development, but more work is needed on this before it can be released.

11.7 CONCLUSIONS

As described in this chapter, the two ternary diagrams, $CaO–Al_2O_3–SiO_2$ and $C_3A–CaSO_4–CaCO_3$, can be used to describe and predict the phase assemblages of forming from hydration of portland cement, portland cement containing pozzolans and limestone and portland cement systems subjected to sulfate attack. The method is consistent with current methods of thermodynamic modelling and provides an effective means of representing the results of thermodynamic modelling graphically. A third ternary diagram, $C_3A–CaSO_4–CaCl_2$, is also included to describe the mineralogy resulting from chloride ingress.

REFERENCES

Balonis, M., B. Lothenbach, G. Le Saout and F. P. Glasser (2010). 'Impact of chloride on the mineralogy of hydrated Portland cement systems'. *Cement and Concrete Research* 40(7): 1009–1022.

Ben Haha, M., G. Le Saout, F. Winnefeld and B. Lothenbach (2011). 'Influence of slag chemistry on the hydration of alkali activated blast-furnace slag – Part I: Effect of MgO'. *Cement and Concrete Research* 41(9): 955–963.

Ben Haha, M., G. Le Saout, F. Winnefeld and B. Lothenbach (2012). 'Influence of slag chemistry on the hydration of alkali activated blast-furnace slag – Part II: Effect of Al_2O_3'. *Cement and Concrete Research* 42(1): 74–83.

Bernal, S. A., R. San Nicols, R. J. Myers, R. Mejia de Gutierrez, F. Puertas, J. S. J. Van Deventer and J. L. Provis (2014). 'MgO content of slag controls phase evolution and structural changes induced by accelerated carbonation in alkali-activated binders'. *Cement and Concrete Research* **57**: 33–43.

Blanc, P., X. Bourbon, A. Lassin and E. C. Gaucher (2010). 'Chemical model for cement-based materials: Thermodynamic data assessment for phases other than C–S–H'. *Cement and Concrete Research* **40**(9): 1360–1374.

Damidot, D., B. Lothenbach, D. Herfort and F. P. Glasser (2011). 'Thermodynamics and cement science'. *Cement and Concrete Research* **41**(7): 679–695.

Deschner, F., B. Lothenbach, F. Winnefeld and J. Neubauer (2013). 'Effect of temperature on the hydration Portland cement blended with siliceous fly ash'. *Cement and Concrete Research* **52**: 169–181.

Dilnesa, B. Z., E. Wieland, B. Lothenbach, R. Dähn and K. Scrivener (2014). 'Fe-containing phases in hydrated cements'. *Cement and Concrete Research* **58**: 45–55.

Ehlers, E. G. (1972). *The Interpretation of Geological Phase Diagrams*, New York: W. H. Freeman and Company.

Girão, A. V., I. G. Richardson, R. Taylor and R. M. D. Brydson (2010). 'Composition, morphology and nanostructure of C–S–H in 70% white Portland cement–30% fly ash blends hydrated at 55°C'. *Cement and Concrete Research* **40**(9): 1350–1359.

Juel, I., D. Herfort, R. Gollop, J. Konnerup-Madsen, H. J. Jakobsen and J. Skibsted (2003). 'A thermodynamic model for predicting the stability of thaumasite'. *Cement & Concrete Composites* **25**: 867–872.

Krauskopf, K. B., and D. K. Bird (1995). *Introduction to Geochemistry*, New York: McGraw-Hill.

Kulik, D. A. (2011). 'Improving the structural consistency of C-S-H solid solution thermodynamic models'. *Cement and Concrete Research* **41**(5): 477–495.

Lothenbach, B. (2010). 'Thermodynamic equilibrium calculations in cementitious systems'. *Materials and Structures* **43**: 1413–1433.

Lothenbach, B., D. Damidot, T. Matschei and J. Marchand (2010). 'Thermodynamic modelling: State of knowledge and challenges'. *Advances in Cement Research* **22**(4): 211–223.

Lothenbach, B., T. Matschei, G. Möschner and F. P. Glasser (2008). 'Thermodynamic modelling of the effect of temperature on the hydration and porosity of Portland cement'. *Cement and Concrete Research* **38**(1): 1–18.

Matschei, T., B. Lothenbach and F. P. Glasser (2007a). 'The AFm phase in Portland cement'. *Cement and Concrete Research* **37**(2): 118–130.

Matschei, T., B. Lothenbach and F. P. Glasser (2007b). 'Thermodynamic properties of Portland cement hydrates in the system $CaO-Al_2O_3-SiO_2-CaSO_4-CaCO_3-H_2O$'. *Cement and Concrete Research* **37**(10): 1379–1410.

Muller, A. C., K. L. Scrivener, A. M. Gajewicz and P. J. McDonald (2013). 'Densification of C–S–H measured by ^1H NMR relaxometry'. *The Journal of Physical Chemistry C* **117**(1): 403–412.

Nielsen, E. P., D. Herfort, M. R. Geiker and D. Hooton (2003). Effect of solid solutions of AFm phases on chloride binding. Proceedings of the 11th International Congress on the Chemistry of Cement, Durban, South Africa.

Perkins, R. B., and C. D. Palmer (1999). 'Solubility of ettringite $(Ca_6[Al(OH)_6]_2(SO_4)_3 \cdot 26H_2O)$ at 5–75°C'. *Geochimica et Cosmochimica Acta* **63**(13/14): 1969–1980.

Pöllmann, H. (2006). 'Syntheses, properties and solid solution of ternary lamellar calcium aluminate hydroxy salts (AFm-phases) containing SO_4^{2-}, CO_3^{2-} and OH^-'. *Neues Jahrbuch für Mineralogie (Abhandlungen)* **182**(2): 173–181.

Richardson, I. (2013). 'Clarification of possible ordered distributions of trivalent cations in layered double hydroxides and an explanation for the observed variation in the lower solid-solution limit'. *Acta Crystallographica Section B* **69**(6): 629–633.

Richardson, I. G., and G. W. Groves (1993). 'Microstructure and microanalysis of hardened ordinary Portland cement pastes'. *Journal of Materials Science* **28**(1): 265–277.

Rossen, J., B. Lothenbach and K. Scrivener (2015). 'Composition of C-S-H in pastes with increasing levels of silica fume addition'. *Cement and Concrete Research* **75**: 14–22.

Rozov, K. B., U. Berner, D. A. Kulik and L. W. Diamond (2011). 'Solubility and thermodynamic properties of carbonate-bearing hydrotalcite-pyroaurite solid solutions with a 3:1 Mg/(Al+Fe) mole ratio'. *Clays and Clay Minerals* **59**(3): 215–232.

Steenberg, M., J. Skibsted, S. L. Poulsen, J. S. Damtoft and D. Herfort (2011). Composite cement based on Portland cement clinker, limestone and calcined clay. Proceedings of the 13th International Congress on the Chemistry of Cement, Madrid, Spain. 97–104.

Welch, J. H., and W. Gutt (1959). 'Tricalcium silicate and its stability within the system CaO-SiO_2'. *Journal of the American Ceramic Society* **42**: 11–25.

Winkler, H. G. F. (1974). *Petrogenesis of Metamorphic Rocks*, Berlin, Heidelberg: Springer-Verlag.

Glossary

Acceleration period (cement hydration): Accelerating part of the main cement hydration stage following a period of slow reaction or induction period. Both onset and acceleration kinetics are generally agreed to be related to the nucleation and growth of calcium–silicate–hydrate.

(Apparent) Activation energy: The energy which must be added to a system to allow a chemical reaction to take place. Activation energy can be rigorously determined only for individual chemical reactions/processes. In cement chemistry the term *apparent activation energy* is often used to describe the temperature dependence of the rate of cement hydration as a whole.

Admixtures: Both inorganic and organic materials other than hydraulic cements, water or aggregates that are added immediately before or during mixing; often used to modify the early-age properties of cements. Admixtures are typically classified by their effect on cement reactions or properties as accelerators, retarders, superplasticisers, viscosity modifiers, shrinkage reducing agents or air-entraining admixtures.

Amorphous phase: Noncrystalline solid lacking the long-range order characteristic of a crystal.

Autogenous shrinkage: The volume reduction of the bulk paste due to self-desiccation and subsequent relative humidity change. Depending on the amount of chemical shrinkage, the pore structure and the pore liquid composition, self-desiccation will cause a relative humidity change and autogenous shrinkage.

Blast furnace slag: A by-product of the iron and steelmaking industry produced by rapidly quenching molten iron slag to obtain a largely amorphous, glassy material. Ground granulated blast-furnace slag is commonly used as supplementary cementitious material.

Bleeding: Settling of solid particles in freshly mixed cement pastes producing a gradation of decreasing water-to-cement ratio from top to bottom of the sample. Bleeding tends to increase with water-to-cement ratio and decreases with fineness or early-age reactivity.

Blended cements: Hydraulic cements consisting essentially of an intimate and uniform blend of a number of different constituent materials; commonly portland cement, limestone, fly ash or blast furnace slag.

Blending: Through mixing of separately ground fine materials, e.g. ground granulated blast furnace slag and portland cement.

Bogue calculation: Estimation of the phase composition of portland cement based on the oxide chemical composition.

(Bulk) Deformation: Overall volume or dimensional change after initial hardening of the cement and the total of (drying, autogenous) shrinkage, creep and expansion mechanisms.

Carbonation: Reaction with CO_2 to form carbonate-bearing phases, generally calcium carbonates. Anhydrous and hydrated cements in particular are susceptible by reaction with atmospheric CO_2 during storage. Carbonation should be avoided to preserve (and study) the sample properties.

Casting: Process in which a liquid material (cement paste, mortar, concrete) is poured into a solid mould of the desired shape. Often the material is compacted or vibrated after casting to remove entrained air bubbles.

Cement: A material that sets and hardens and can be used to bind materials together. A distinction is made between hydraulic and nonhydraulic binders. Hydraulic binders are able to set and harden under an excess of water (e.g. portland cement), while nonhydraulic binders require the removal of excess water to harden (e.g. lime).

Cement hydration: Reaction of the anhydrous cement phases with water to form cement hydrates. The reaction is a dissolution–precipitation process involving both the participation and the mediation of an aqueous phase.

Chemical shrinkage: The absolute volume reduction in a cement paste during hydration caused by the difference in volume of products and reactants (including water).

Clinker: Nodular material produced by high-temperature sintering (1450°C) of clays and limestone in a (rotary) kiln. The clinker is mainly composed of alite, belite, aluminate, ferrite and minor phases, such as alkali sulfates, free lime and periclase.

Concrete: Composite material of (portland) cement and fine and coarse aggregates that is fluid and workable when first mixed with water and that subsequently sets and hardens to a solid, rocklike material.

Curing (sealed curing, underwater curing): process in which the cementitious material is protected from water loss and kept within a controlled temperature range to achieve optimal performance and durability. In practise, curing is carried out by covering exposed surfaces with sheets (sheet curing), by sprinkling water (moist curing) or by applying film-forming curing compounds. For laboratory and testing purposes often sealed curing, underwater curing or curing in a 'fog' room (relative humidity of 95% or higher) is used.

D-drying (hydration stoppage): Also called *dry ice drying*; removes water from cement pastes by applying a vacuum over dry ice. At regular conditions this process is very slow and cannot be used for early age samples.

Dead time (nuclear magnetic resonance): Time after pulse when no signal is recorded.

Deceleration period (cement hydration): Period during which the cement hydration rate gradually slows down due to a combination of (1) consumption of reactants, (2) lack of space for hydrates to grow and/or (3) lack of water.

Degree of hydration: Extent of hydration of the cement (binder) compared to full hydration.

Dry mix mortar: Prefabricated dry mix of portland cement, graded sand (fine aggregate) and admixtures sold to end users; ready for application after addition of water.

Enthalpy: Thermodynamic potential. For processes at constant pressure the enthalpy change is equal to the heat adsorbed or released.

Filler effect: Accelerating effect of fine, inert materials on the hydration kinetics of the clinker phases in blended cements because of the presence of additional nucleation sites and the presence of a higher effective water-to-cement ratio; often wrongly assigned to (early) supplementary cementitious material reactivity.

Fly ash: Coal combustion fly ash; a by-product of coal-fired power generation and comprises the fine ash fraction that is carried along with the combustion flue gases. Fly ash is a highly heterogeneous, partially amorphous material with a variable composition that depends on the coal fuel properties. Depending on its quality, it is used as supplementary cementitious material.

Freeze-drying (hydration stoppage): Also called *F-drying*; process in which the hydration reactions are stopped by immersion in liquid nitrogen (–196°C), after freezing, followed by sublimation of the ice in a freeze-dryer at low temperature (–78°C) and pressure. Freeze-drying does not preserve the microstructure of hydrated cements and can degrade ettringite and AFm phases.

Heat capacity: The amount of heat required to result in a unit temperature change of a material. Specific heat capacity is normalised by unit of mass; molar heat capacity, by mole.

Heat of hydration: Heat generated by the hydration reactions of cements, generally exothermal. The rate of heat release is used to follow the kinetics of the hydration reactions. The integrated heat of hydration is used to assess the overall degree of reaction of the cement.

Hydration stoppage: Process of stopping the hydration reaction by removal of water. Typical methods are solvent exchange, freeze-drying, vacuum-drying, D-drying and oven drying.

Induction period (cement hydration): Also called *period of slow reaction*; the minimum occurring between the initial reaction period and the acceleration period.

Initial reaction period (cement hydration): Occurs upon contact of the cement with water and is observed as an early exothermal signal in the calorimetry plot of cement hydration. Cement powder surface wetting and initial rapid dissolution of the clinker phases, CaO and alkali sulfate are considered to be the main contributors to the heat release.

Intergrinding: Cogrinding of two or more components resulting in a uniform mix. Differential grinding often occurs because of differences in hardness.

Kinetics (chemical): Also called *reaction kinetics*; the study of rates of chemical processes.

Limestone: Sedimentary rock largely composed of $CaCO_3$ as calcite and aragonite minerals.

Long-range order: The repetition of a characteristic pattern (e.g. the disposition of atoms in the unit cell) over 'infinitely' great distances compared to interatomic distances (e.g. lattice periodicity in crystals).

Low-heat cement, very low-heat cement: Cement with slow hydration rate and related low rate of heat release. Mostly used to avoid the buildup of internal stress and cracking by thermal expansion in large, voluminous structures such as dams. Low-heat release cements can be achieved by coarse grinding of portland cement and/or by using high levels of replacement of clinker by supplementary cementitious materials.

(Masonry) Mortar: Material used to bind together building blocks. Mortar is workable when freshly mixed with water and hardens over time. Modern mortars are usually a mix of portland cement, sand and water. Mortar bars of specified composition and dimensions are often used as testing material in standardised testing procedures.

Maturity method: Calculation technique to account for the combined effect of time and temperature on cement hydration kinetics and related properties such as strength development.

Metakaolin: A structurally disordered aluminosilicate material produced by dehydroxylation (heating) of kaolin (kaolinite-bearing clay) by calcination; used as supplementary cementitious material.

Oven drying (hydration stoppage): Removal of water from hydrated samples by circulating a warm, dry air flow through the oven. Oven temperatures are between 60 and 105°C. Drying is considered complete when the sample reaches constant mass.

P-drying (hydration stoppage): Removal of water from hydrated samples by drying the sample over magnesium perchlorates (dihydrate and tetrahydrate) ($Mg(ClO_4)_2 \cdot 2H_2O - Mg(ClO_4)_2 \cdot 4H_2O$) in a sealed desiccator (H_2O partial pressure of 1.1 Pa at 25°C).

Prehydration (cement): Slow reaction of cement with atmospheric humidity, usually occurring during storage and resulting in reduced strength development.

Pulse length (nuclear magnetic resonance): Short time during which the magnetic field is applied.

Rheology: Study of the flow of particles (e.g. mortar or concrete).

Secondary aluminate reaction (cement hydration): Rate increase of the C_3A/aluminate hydration reaction related to the depletion of solid calcium sulfate and a drop in sulfate concentrations in the system. To ensure correct setting and strength development, this reaction should occur well after the main hydration peak of alite.

Segregation: Tendency of particulate solids to separate by size, density or shape.

Self-desiccation: Internal drying due to the consumption of water by chemical reactions.

Self-levelling compound, self-levelling mortar: Highly flowable concrete/mortar made by adding admixtures such as superplasticisers and accelerators to purposely optimised mortar or concrete formulations; typically used to make flat and smooth surfaces such as underlayment or toppings of floors.

Set regulator: Calcium sulfate (gypsum, bassanite, anhydrite) added to portland clinker to control cement setting times.

Setting (cement hydration): Change of the cement from a fluid to a consolidated state caused by the first coalescing (interlocking) of cement hydrates upon initial hydration.

Shelf life: Time a product can be stored before becoming unfit (performing below specifications) for use or application.

Short-range order: Order over distances comparable to interatomic distances; a consequence of the chemical bonding between atoms.

Solid solution: Solid in which components are compatible and form a unique phase of mixed composition; often used to denote the compatibility of two similar components in one crystal structure. Solid solutions occur commonly in cement hydrates (and clinkers), e.g. in C-S-H (calcium–silicate–hydrate), AFm and AFt phases.

Solvent exchange (hydration stoppage): Removal of water from hydrated cement samples by exchange of water by an organic solvent. Upon immersion, the solvent replaces the (free) water in the sample. Upon complete exchange, the solvent is removed by evaporation. Isopropanol is most frequently used as solvent. Solvent exchange is usually preferred for microstructural studies.

Sulfate attack: External sulfate attack is caused by the ingress of sulfate-bearing solutions that react with the cement hydrates to form ettringite and, later, gypsum. This leads to strength loss, expansion and cracking of the mortar/concrete.

Sulfate optimisation: Addition of calcium sulfate (set regulator) to portland clinker to achieve optimal early strength development.

Superplasticisers: Also called *water-reducing agents*; chemical admixtures used to disperse particles homogeneously. In concrete or mortar superplasticisers are used to reduce water-to-cement ratios and produce high-strength products or to make highly flowable concrete with self-consolidating and/or self-levelling properties.

Supplementary cementitious materials (SCMs): Materials added to hydraulic binders that chemically react and in this way generate additional hydration (reaction) products. Typical examples are blast furnace slags and fly ashes.

Vacuum drying (hydration stoppage): Removal of water by evaporation in a vacuum chamber (pressure < 0.1 Pa) and low water vapour pressure. Vacuum drying until constant weight does not preserve the paste microstructure and collapses the crystal structures of ettringite and AFm phases by removal of bound water.

Workability: General term that indicates the ease with which a concrete/mortar can be mixed, transported, placed and compacted to give a uniform material.

Index

Page numbers followed by f and t indicate figures and tables, respectively.

A

Accelerating period, 40, 43f
Accuracy
  isothermal calorimetry, 42
  NMR measurements, issues, 343
  for semiadiabatic calorimetry, 46
Acetone
  affecting TGA, 184–185
  in hydration stoppage, 21, 26,
    184–185
Acquisition, of basic MAS NMR
    spectra, 268–269, 270t
Activation energy, defined, 54
Activity–activity diagram, 487f
Additives, on CS development, 94,
    94f
Admixtures
  blending of dry binders and mortars,
    10, 12–13
  cement–admixture interactions, 61,
    62, 63–64, 64f
Adsorbate selection, 472, 472t
Adsorption curves, pore size
    distributions from, 471
Adsorption isotherm, analysis of,
    468–471
  mesoporosity, 470–471
  SSAs, 468–470, 469f, 470f
AFm phases
  hydration stoppage, 22, 23, 24f,
    25–26
  TGA
    measurement, 193–195, 194f
    signals, 185

XRD measurement
  hydration stoppage, 128–130
  sample mounting, 126
AFt phases, in TGA measurement,
    192–193, 192f, 193t
Agglomerates, of microsilica, 18
Agglomeration factor, PSD, 450–451
Air pressure (dry), LD instruments,
    458–459
Air scattering, 141
Alite phases
  in commercial portland cements,
    272–273
  conventional binders *vs.* pure C_3S,
    313
  different morphologies of C-S-H in,
    364, 365f
  phosphorus in, 223
  polymorphs, 113–114, 114f
Alkali phases, 168t
Alkali–silica reaction (ASR), 394, 395f
^{27}Al MAS NMR spectra, 235–237,
    237f, 238f
Aluminosilicates, 166t
American Mineralogist Crystal
    Structure Database
    (AMCSD), 109
Amorphous phases
  artificial detection of, 124
  calcium carbonate, 190–191, 198
  C-S-H, 136, 215
  NMR spectroscopy and, 214–215,
    278
  quantification, 120, 136

H